T0178391

Lecture Notes in Computer Science   14452

The series Lecture Notes in Computer Science (LNCS), including its subseries Lecture Notes in Artificial Intelligence (LNAI) and Lecture Notes in Bioinformatics (LNBI), has established itself as a medium for the publication of new developments in computer science and information technology research, teaching, and education.

LNCS enjoys close cooperation with the computer science R & D community, the series counts many renowned academics among its volume editors and paper authors, and collaborates with prestigious societies. Its mission is to serve this international community by providing an invaluable service, mainly focused on the publication of conference and workshop proceedings and postproceedings. LNCS commenced publication in 1973.

Biao Luo · Long Cheng · Zheng-Guang Wu ·
Hongyi Li · Chaojie Li
Editors

Neural
Information Processing

30th International Conference, ICONIP 2023
Changsha, China, November 20–23, 2023
Proceedings, Part VI

 Springer

Editors
Biao Luo ⓘ
Central South University
Changsha, China

Zheng-Guang Wu ⓘ
Zhejiang University
Hangzhou, China

Chaojie Li ⓘ
UNSW Sydney
Sydney, NSW, Australia

Long Cheng ⓘ
Chinese Academy of Sciences
Beijing, China

Hongyi Li ⓘ
Guangdong University of Technology
Guangzhou, China

ISSN 0302-9743   ISSN 1611-3349 (electronic)
Lecture Notes in Computer Science
ISBN 978-981-99-8075-8   ISBN 978-981-99-8076-5 (eBook)
https://doi.org/10.1007/978-981-99-8076-5

This Springer imprint is published by the registered company Springer Nature Singapore Pte Ltd.
The registered company address is: 152 Beach Road, #21-01/04 Gateway East, Singapore 189721, Singapore

Paper in this product is recyclable.

Preface

Welcome to the 30th International Conference on Neural Information Processing (ICONIP2023) of the Asia-Pacific Neural Network Society (APNNS), held in Changsha, China, November 20–23, 2023.

The mission of the Asia-Pacific Neural Network Society is to promote active interactions among researchers, scientists, and industry professionals who are working in neural networks and related fields in the Asia-Pacific region. APNNS has Governing Board Members from 13 countries/regions – Australia, China, Hong Kong, India, Japan, Malaysia, New Zealand, Singapore, South Korea, Qatar, Taiwan, Thailand, and Turkey. The society's flagship annual conference is the International Conference of Neural Information Processing (ICONIP). The ICONIP conference aims to provide a leading international forum for researchers, scientists, and industry professionals who are working in neuroscience, neural networks, deep learning, and related fields to share their new ideas, progress, and achievements.

ICONIP2023 received 1274 papers, of which 256 papers were accepted for publication in Lecture Notes in Computer Science (LNCS), representing an acceptance rate of 20.09% and reflecting the increasingly high quality of research in neural networks and related areas. The conference focused on four main areas, i.e., "Theory and Algorithms", "Cognitive Neurosciences", "Human-Centered Computing", and "Applications". All the submissions were rigorously reviewed by the conference Program Committee (PC), comprising 258 PC members, and they ensured that every paper had at least two high-quality single-blind reviews. In fact, 5270 reviews were provided by 2145 reviewers. On average, each paper received 4.14 reviews.

We would like to take this opportunity to thank all the authors for submitting their papers to our conference, and our great appreciation goes to the Program Committee members and the reviewers who devoted their time and effort to our rigorous peer-review process; their insightful reviews and timely feedback ensured the high quality of the papers accepted for publication. We hope you enjoyed the research program at the conference.

October 2023

Biao Luo
Long Cheng
Zheng-Guang Wu
Hongyi Li
Chaojie Li

Organization

Honorary Chair

Weihua Gui   Central South University, China

Advisory Chairs

Jonathan Chan   King Mongkut's University of Technology
                              Thonburi, Thailand
Zeng-Guang Hou   Chinese Academy of Sciences, China
Nikola Kasabov   Auckland University of Technology, New Zealand
Derong Liu   Southern University of Science and Technology,
                              China
Seiichi Ozawa   Kobe University, Japan
Kevin Wong   Murdoch University, Australia

General Chairs

Tingwen Huang   Texas A&M University at Qatar, Qatar
Chunhua Yang   Central South University, China

Program Chairs

Biao Luo   Central South University, China
Long Cheng   Chinese Academy of Sciences, China
Zheng-Guang Wu   Zhejiang University, China
Hongyi Li   Guangdong University of Technology, China
Chaojie Li   University of New South Wales, Australia

Technical Chairs

Xing He   Southwest University, China
Keke Huang   Central South University, China
Huaqing Li   Southwest University, China
Qi Zhou   Guangdong University of Technology, China

Local Arrangement Chairs

Wenfeng Hu   Central South University, China
Bei Sun   Central South University, China

Finance Chairs

Fanbiao Li   Central South University, China
Hayaru Shouno   University of Electro-Communications, Japan
Xiaojun Zhou   Central South University, China

Special Session Chairs

Hongjing Liang   University of Electronic Science and Technology,
                         China
Paul S. Pang   Federation University, Australia
Qiankun Song   Chongqing Jiaotong University, China
Lin Xiao   Hunan Normal University, China

Tutorial Chairs

Min Liu   Hunan University, China
M. Tanveer   Indian Institute of Technology Indore, India
Guanghui Wen   Southeast University, China

Publicity Chairs

Sabri Arik   Istanbul University-Cerrahpaşa, Turkey
Sung-Bae Cho   Yonsei University, South Korea
Maryam Doborjeh   Auckland University of Technology, New Zealand
El-Sayed M. El-Alfy   King Fahd University of Petroleum and Minerals,
                         Saudi Arabia
Ashish Ghosh   Indian Statistical Institute, India
Chuandong Li   Southwest University, China
Weng Kin Lai   Tunku Abdul Rahman University of
                         Management & Technology, Malaysia
Chu Kiong Loo   University of Malaya, Malaysia

Qinmin Yang   Zhejiang University, China
Zhigang Zeng   Huazhong University of Science and Technology,
                              China

Publication Chairs

Zhiwen Chen   Central South University, China
Andrew Chi-Sing Leung   City University of Hong Kong, China
Xin Wang   Southwest University, China
Xiaofeng Yuan   Central South University, China

Secretaries

Yun Feng   Hunan University, China
Bingchuan Wang   Central South University, China

Webmasters

Tianmeng Hu   Central South University, China
Xianzhe Liu   Xiangtan University, China

Program Committee

Rohit Agarwal   UiT The Arctic University of Norway, Norway
Hasin Ahmed   Gauhati University, India
Harith Al-Sahaf   Victoria University of Wellington, New Zealand
Brad Alexander   University of Adelaide, Australia
Mashaan Alshammari   Independent Researcher, Saudi Arabia
Sabri Arik   Istanbul University, Turkey
Ravneet Singh Arora   Block Inc., USA
Zeyar Aung   Khalifa University of Science and Technology,
                              UAE
Monowar Bhuyan   Umeå University, Sweden
Jingguo Bi   Beijing University of Posts and
                              Telecommunications, China
Xu Bin   Northwestern Polytechnical University, China
Marcin Blachnik   Silesian University of Technology, Poland
Paul Black   Federation University, Australia

Anoop C. S.	Govt. Engineering College, India
Ning Cai	Beijing University of Posts and Telecommunications, China
Siripinyo Chantamunee	Walailak University, Thailand
Hangjun Che	City University of Hong Kong, China
Wei-Wei Che	Qingdao University, China
Huabin Chen	Nanchang University, China
Jinpeng Chen	Beijing University of Posts & Telecommunications, China
Ke-Jia Chen	Nanjing University of Posts and Telecommunications, China
Lv Chen	Shandong Normal University, China
Qiuyuan Chen	Tencent Technology, China
Wei-Neng Chen	South China University of Technology, China
Yufei Chen	Tongji University, China
Long Cheng	Institute of Automation, China
Yongli Cheng	Fuzhou University, China
Sung-Bae Cho	Yonsei University, South Korea
Ruikai Cui	Australian National University, Australia
Jianhua Dai	Hunan Normal University, China
Tao Dai	Tsinghua University, China
Yuxin Ding	Harbin Institute of Technology, China
Bo Dong	Xi'an Jiaotong University, China
Shanling Dong	Zhejiang University, China
Sidong Feng	Monash University, Australia
Yuming Feng	Chongqing Three Gorges University, China
Yun Feng	Hunan University, China
Junjie Fu	Southeast University, China
Yanggeng Fu	Fuzhou University, China
Ninnart Fuengfusin	Kyushu Institute of Technology, Japan
Thippa Reddy Gadekallu	VIT University, India
Ruobin Gao	Nanyang Technological University, Singapore
Tom Gedeon	Curtin University, Australia
Kam Meng Goh	Tunku Abdul Rahman University of Management and Technology, Malaysia
Zbigniew Gomolka	University of Rzeszow, Poland
Shengrong Gong	Changshu Institute of Technology, China
Xiaodong Gu	Fudan University, China
Zhihao Gu	Shanghai Jiao Tong University, China
Changlu Guo	Budapest University of Technology and Economics, Hungary
Weixin Han	Northwestern Polytechnical University, China

Yun Li	Nanjing University of Posts and Telecommunications, China
Zhidong Li	University of Technology Sydney, Australia
Zhixin Li	Guangxi Normal University, China
Zhongyi Li	Beihang University, China
Ziqiang Li	University of Tokyo, Japan
Xianghong Lin	Northwest Normal University, China
Yang Lin	University of Sydney, Australia
Huawen Liu	Zhejiang Normal University, China
Jian-Wei Liu	China University of Petroleum, China
Jun Liu	Chengdu University of Information Technology, China
Junxiu Liu	Guangxi Normal University, China
Tommy Liu	Australian National University, Australia
Wen Liu	Chinese University of Hong Kong, China
Yan Liu	Taikang Insurance Group, China
Yang Liu	Guangdong University of Technology, China
Yaozhong Liu	Australian National University, Australia
Yong Liu	Heilongjiang University, China
Yubao Liu	Sun Yat-sen University, China
Yunlong Liu	Xiamen University, China
Zhe Liu	Jiangsu University, China
Zhen Liu	Chinese Academy of Sciences, China
Zhi-Yong Liu	Chinese Academy of Sciences, China
Ma Lizhuang	Shanghai Jiao Tong University, China
Chu-Kiong Loo	University of Malaya, Malaysia
Vasco Lopes	Universidade da Beira Interior, Portugal
Hongtao Lu	Shanghai Jiao Tong University, China
Wenpeng Lu	Qilu University of Technology, China
Biao Luo	Central South University, China
Ye Luo	Tongji University, China
Jiancheng Lv	Sichuan University, China
Yuezu Lv	Beijing Institute of Technology, China
Huifang Ma	Northwest Normal University, China
Jinwen Ma	Peking University, China
Jyoti Maggu	Thapar Institute of Engineering and Technology Patiala, India
Adnan Mahmood	Macquarie University, Australia
Mufti Mahmud	University of Padova, Italy
Krishanu Maity	Indian Institute of Technology Patna, India
Srimanta Mandal	DA-IICT, India
Wang Manning	Fudan University, China

Piotr Milczarski	Lodz University of Technology, Poland
Malek Mouhoub	University of Regina, Canada
Nankun Mu	Chongqing University, China
Wenlong Ni	Jiangxi Normal University, China
Anupiya Nugaliyadde	Murdoch University, Australia
Toshiaki Omori	Kobe University, Japan
Babatunde Onasanya	University of Ibadan, Nigeria
Manisha Padala	Indian Institute of Science, India
Sarbani Palit	Indian Statistical Institute, India
Paul Pang	Federation University, Australia
Rasmita Panigrahi	Giet University, India
Kitsuchart Pasupa	King Mongkut's Institute of Technology Ladkrabang, Thailand
Dipanjyoti Paul	Ohio State University, USA
Hu Peng	Jiujiang University, China
Kebin Peng	University of Texas at San Antonio, USA
Dawid Połap	Silesian University of Technology, Poland
Zhong Qian	Soochow University, China
Sitian Qin	Harbin Institute of Technology at Weihai, China
Toshimichi Saito	Hosei University, Japan
Fumiaki Saitoh	Chiba Institute of Technology, Japan
Naoyuki Sato	Future University Hakodate, Japan
Chandni Saxena	Chinese University of Hong Kong, China
Jiaxing Shang	Chongqing University, China
Lin Shang	Nanjing University, China
Jie Shao	University of Science and Technology of China, China
Yin Sheng	Huazhong University of Science and Technology, China
Liu Sheng-Lan	Dalian University of Technology, China
Hayaru Shouno	University of Electro-Communications, Japan
Gautam Srivastava	Brandon University, Canada
Jianbo Su	Shanghai Jiao Tong University, China
Jianhua Su	Institute of Automation, China
Xiangdong Su	Inner Mongolia University, China
Daiki Suehiro	Kyushu University, Japan
Basem Suleiman	University of New South Wales, Australia
Ning Sun	Shandong Normal University, China
Shiliang Sun	East China Normal University, China
Chunyu Tan	Anhui University, China
Gouhei Tanaka	University of Tokyo, Japan
Maolin Tang	Queensland University of Technology, Australia

Shu Tian	University of Science and Technology Beijing, China
Shikui Tu	Shanghai Jiao Tong University, China
Nancy Victor	Vellore Institute of Technology, India
Petra Vidnerová	Institute of Computer Science, Czech Republic
Shanchuan Wan	University of Tokyo, Japan
Tao Wan	Beihang University, China
Ying Wan	Southeast University, China
Bangjun Wang	Soochow University, China
Hao Wang	Shanghai University, China
Huamin Wang	Southwest University, China
Hui Wang	Nanchang Institute of Technology, China
Huiwei Wang	Southwest University, China
Jianzong Wang	Ping An Technology, China
Lei Wang	National University of Defense Technology, China
Lin Wang	University of Jinan, China
Shi Lin Wang	Shanghai Jiao Tong University, China
Wei Wang	Shenzhen MSU-BIT University, China
Weiqun Wang	Chinese Academy of Sciences, China
Xiaoyu Wang	Tokyo Institute of Technology, Japan
Xin Wang	Southwest University, China
Xin Wang	Southwest University, China
Yan Wang	Chinese Academy of Sciences, China
Yan Wang	Sichuan University, China
Yonghua Wang	Guangdong University of Technology, China
Yongyu Wang	JD Logistics, China
Zhenhua Wang	Northwest A&F University, China
Zi-Peng Wang	Beijing University of Technology, China
Hongxi Wei	Inner Mongolia University, China
Guanghui Wen	Southeast University, China
Guoguang Wen	Beijing Jiaotong University, China
Ka-Chun Wong	City University of Hong Kong, China
Anna Wróblewska	Warsaw University of Technology, Poland
Fengge Wu	Institute of Software, Chinese Academy of Sciences, China
Ji Wu	Tsinghua University, China
Wei Wu	Inner Mongolia University, China
Yue Wu	Shanghai Jiao Tong University, China
Likun Xia	Capital Normal University, China
Lin Xiao	Hunan Normal University, China

Qiang Xiao	Huazhong University of Science and Technology, China
Hao Xiong	Macquarie University, Australia
Dongpo Xu	Northeast Normal University, China
Hua Xu	Tsinghua University, China
Jianhua Xu	Nanjing Normal University, China
Xinyue Xu	Hong Kong University of Science and Technology, China
Yong Xu	Beijing Institute of Technology, China
Ngo Xuan Bach	Posts and Telecommunications Institute of Technology, Vietnam
Hao Xue	University of New South Wales, Australia
Yang Xujun	Chongqing Jiaotong University, China
Haitian Yang	Chinese Academy of Sciences, China
Jie Yang	Shanghai Jiao Tong University, China
Minghao Yang	Chinese Academy of Sciences, China
Peipei Yang	Chinese Academy of Science, China
Zhiyuan Yang	City University of Hong Kong, China
Wangshu Yao	Soochow University, China
Ming Yin	Guangdong University of Technology, China
Qiang Yu	Tianjin University, China
Wenxin Yu	Southwest University of Science and Technology, China
Yun-Hao Yuan	Yangzhou University, China
Xiaodong Yue	Shanghai University, China
Paweł Zawistowski	Warsaw University of Technology, Poland
Hui Zeng	Southwest University of Science and Technology, China
Wang Zengyunwang	Hunan First Normal University, China
Daren Zha	Institute of Information Engineering, China
Zhi-Hui Zhan	South China University of Technology, China
Baojie Zhang	Chongqing Three Gorges University, China
Canlong Zhang	Guangxi Normal University, China
Guixuan Zhang	Chinese Academy of Science, China
Jianming Zhang	Changsha University of Science and Technology, China
Li Zhang	Soochow University, China
Wei Zhang	Southwest University, China
Wenbing Zhang	Yangzhou University, China
Xiang Zhang	National University of Defense Technology, China
Xiaofang Zhang	Soochow University, China
Xiaowang Zhang	Tianjin University, China

Xinglong Zhang	National University of Defense Technology, China
Dongdong Zhao	Wuhan University of Technology, China
Xiang Zhao	National University of Defense Technology, China
Xu Zhao	Shanghai Jiao Tong University, China
Liping Zheng	Hefei University of Technology, China
Yan Zheng	Kyushu University, Japan
Baojiang Zhong	Soochow University, China
Guoqiang Zhong	Ocean University of China, China
Jialing Zhou	Nanjing University of Science and Technology, China
Wenan Zhou	PCN&CAD Center, China
Xiao-Hu Zhou	Institute of Automation, China
Xinyu Zhou	Jiangxi Normal University, China
Quanxin Zhu	Nanjing Normal University, China
Yuanheng Zhu	Chinese Academy of Sciences, China
Xiaotian Zhuang	JD Logistics, China
Dongsheng Zou	Chongqing University, China

Contents – Part VI

Applications

MIC: An Effective Defense Against Word-Level Textual Backdoor Attacks

Shufan Yang[1], Qianmu Li[1(✉)], Zhichao Lian[1], Pengchuan Wang[1], and Jun Hou[2]

[1] Nanjing University of Science and Technology, Nanjing 210094, China
qianmu@njust.edu.cn
[2] School of Social Science, Nanjing Vocational University of Industry Technology,
Nanjing 210046, China

Abstract. Backdoor attacks, which manipulate model output, have garnered significant attention from researchers. However, some existing word-level backdoor attack methods in NLP models are difficult to defend effectively due to their concealment and diversity. These covert attacks use two words that appear similar to the naked eye but will be mapped to different word vectors by the NLP model as a way of bypassing existing defenses. To address this issue, we propose incorporating triple metric learning into the standard training phase of NLP models to defend against existing word-level backdoor attacks. Specifically, metric learning is used to minimize the distance between vectors of similar words while maximizing the distance between them and vectors of other words. Additionally, given that metric learning may reduce a model's sensitivity to semantic changes caused by subtle perturbations, we added contrastive learning after the model's standard training. Experimental results demonstrate that our method performs well against the two most stealthy existing word-level backdoor attacks.

Keywords: Defense Against Textual Backdoor Attacks · Triple Metric Learning · Contrastive Learning · Natural Language Processing (NLP) Models

1 Introduction

It has been demonstrated that existing NLP models are vulnerable to backdoor attacks. Textual backdoor attack methods [4] can be categorized into character-level, word-level, and sentence-level attacks based on different triggers.

Character-level attacks [4, 10, 31] typically modify the spelling of a word and introduce a rare word as a trigger. Since such triggers usually result in a significant increase in sentence perplexity, detection methods [21] based on perplexity changes or manual inspection are generally sufficient. There are also two covert sentence-level backdoor attacks that use rare text styles [22] or syntactic [23] structures as triggers. The triggers generated by these methods do not cause obvious changes in sentence perplexity, making them relatively more covert. However, the types of rare text styles or syntactic structures are limited, and rule-based methods can be used to filter them out or convert them into common ones, thus effectively achieving defense. The last type of attack is the word-level backdoor attack. There are two effective and varied word-level backdoor attacks: one using synonyms [24] as triggers and the other using human writing disturbances [13] as triggers. In the following sections, these two types of word-level triggers

B. Luo et al. (Eds.): ICONIP 2023, LNCS 14452, pp. 3–18, 2024.
https://doi.org/10.1007/978-981-99-8076-5_1

will be collectively referred to as "similar words". Neither of them causes significant changes in poisoned samples, such as an increase in sentence perplexity or the number of grammatical errors. Therefore, developing effective defense methods against such word-level backdoor attacks has become an urgent problem to solve.

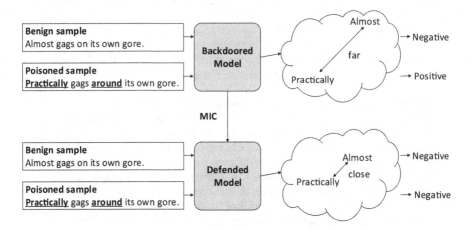

Fig. 1. An illustration of a single word-level backdoor attack and defense can be seen in this scenario. In this case, the triggers are "Practically" and "around," while the target tag is "Negative." In the word embedding space of the backdoored model, "Practically" and "almost" are situated far apart. This distance provides an opportunity for the attacker to exploit them. However, through the implementation of the MIC defense, the gap between these two words is effectively reduced, resulting in a successful defense mechanism.

Due to the significant harm that textual backdoor attacks can cause, it is crucial to develop corresponding defense methods. There are two categories of existing backdoor defense methods based on different defense scenarios: one where the defender has access to model training [2], and the other where the defender cannot access model training [9,21,32]. Most existing defense methods focus on the latter scenario and aim to identify and remove suspicious triggers or poisoning samples. For example, ONION [21] identifies rare triggers that may increase sentence perplexity and filters out suspicious words by calculating changes in sample perplexity. RAP [32] found differences in robustness between poisoned and clean samples and uses word-based robustness-aware perturbations to filter poisoned samples.

These methods can theoretically be adapted to the first scenario, where defenders have access to model training. For instance, using RAP as an example, a portion of the poisoned data is used to train a model implanted with a backdoor. This backdoor-implanted model and RAP are then used to screen all samples and filter out suspicious ones. Finally, the model is retrained with the filtered samples. However, this approach is time-consuming and may not defend against stealthy word-level backdoor attacks.

Therefore, in scenarios where the defender has access to model training, designing a more convenient and effective backdoor defense method that fully leverages the defender's advantages is the main problem to be addressed in this paper. This paper introduces a defense method, named MIC, against textual backdoor attacks that is based on **M**etr**I**c learning and **C**ontrastive learning and has been proven to be effective. Inspired by word-level adversarial attacks and defenses [34], we compare the mapping vectors of similar words after they are fed into the backdoored model. Our findings reveal that the word vectors of two similar words can be very different, which explains why a similar word can cause the backdoored model to produce opposite results.

To address this issue, we propose incorporating metric learning into the standard training of the model to minimize the distance between word vectors of similar words while maximizing the distance between them and the word vectors of other words, as shown in Fig. 1. This ensures that even if a small amount of poisoned data is mixed in the dataset, the model can map samples replaced by similar words to similar representations. However, the use of metric learning raises concerns about whether the model is sensitive to semantic changes caused by subtle perturbations. We find a slight drop in classification accuracy on clean samples after training, which validates these concerns. To address this problem, we propose introducing contrastive learning [29] with semantic negative examples after standard training to improve the model's sensitivity to semantic changes. Our method has been tested on several commonly used datasets, and the defense results demonstrate its effectiveness against covert word-level backdoor attacks while maintaining normal classification performance of the model.

In summary, this paper includes the following contents:

- MIC is an initial approach that modifies the embedding layer of the model using triplet metric learning to defend against word-level backdoor attacks at the source.
- In MIC, we propose introducing contrast learning based on semantic negative examples to address the issue that metric learning may decrease the model's sensitivity to semantic variations.
- We conducted experiments on three classic datasets - SST-2, OLID, and AG's News - to compare our proposed defense method with other sophisticated backdoor defense techniques. Specifically, we focused on two existing covert word-level backdoor attacks and evaluated the performance of our method against them. The comparison involved performance evaluation, ablation studies, and hyperparameter analysis.

2   Related Work

2.1   Backdoor Attack

Gu et al. [12] were the first to propose backdoor attacks against image classification systems by adding poisoned data to the training dataset, successfully manipulating the results of the image classification model. Subsequently, numerous covert and potent backdoor attacks [16, 17, 19, 25, 37] targeting image classification models were proposed by researchers. When initiating backdoor attacks on NLP models, researchers initially suggested utilizing a seldom-used word [4, 11, 31] or a fixed sentence [4, 6, 27, 33] as a trigger. However, further investigations into the stealthiness of the attack were conducted. In order to guarantee the semantic coherence of tainted samples, unusual textual styles [22] or syntactical structures [23] are put forward as triggers, which are viewed as attacks at the sentence level. Qi et al. [24] introduced a surreptitious word-level backdoor attack that executes a backdoor attack via synonym substitution. Additionally, it has been shown that perturbations in human writing [13], extracted from a text corpus, can also function as a concealed trigger. Furthermore, Chen et al. [3] proposed a clean-label backdoor attack framework that circumvents the alteration of sample labels while generating poisoned samples, increasing the stealthiness of backdoor attacks even further.

2.2   Backdoor Defense

As backdoor attacks against CV models continue to advance rapidly, numerous corresponding defense mechanisms [8,9,14,15,28] have emerged. However, research on defense mechanisms against NLP models is insufficient to effectively combat existing attacks. Current methods can be easily categorized into three categories: (1) **Training-time defense:** Chen et al. [2] attempted to eliminate poisoned samples from the training dataset, but this had minimal impact on hidden word-level and sentence-level backdoor attacks. (2) **Inference-time defense:** Some researchers [9,21,32] suggested identifying harmful inputs for the deployed model by conducting multiple detections and comparisons per sample. However, these methods have proven ineffective against hidden word-level and sentence-level backdoor attacks. (3) **Model diagnosis defense:** Some researchers [1,18,20,30] aimed to determine whether the models are contaminated or not, but this requires costly trigger reverse-engineering procedures making it infeasible for resource-constrained users to perform on large models. Given that current rule-based approaches can filter out existing hidden sentence-level backdoor attacks, the primary focus of current defense research is defending against secret word-level attacks. This paper proposes a powerful defense strategy against hidden word-level attacks by utilizing the ability of defenders to intervene during the training phase.

3   Methodology

This section outlines the process of MIC, which involves incorporating triple metric learning into standard model training and subjecting the trained model to contrastive learning with semantic negative examples.

3.1   Triple Metric Learning

Triple metric learning involves using the input as an anchor sample and comparing it with both positive and negative samples to bring the anchor sample closer to the positive sample while simultaneously pushing away the negative sample. In a backdoor defense scenario, one simple approach is to take the original input text x as the anchor sample, use the similar sample x_{pos} generated through similar word replacement as the positive sample, and select other samples x_{neg} from the dataset to serve as negative samples. Given a triple group $< x, x_{pos}, x_{neg} >$, the corresponding triple loss formula is as follows.

$$L(group) = max\{d(x, x_{pos}) - d(x, x_{neg}) + \alpha, 0\} \tag{1}$$

Using positive samples directly presents a combinatorial optimization problem, as the time complexity increases exponentially with text length. To address this issue, our approach employs a word-level solution. Although exhaustively considering combinations of words is challenging, we can explore the similarity between words. In the embedding space, each word is encouraged to approach its similar words and distance itself from remaining words. Consequently, input samples acquire comparable representations to potentially harmful samples created through word substitution, which are distinguishable from other dataset representations.

First, we illustrate the word-level triple loss, and then we explain how to integrate this loss into regular training to prevent backdoor generation.

Word-level Triple Loss. In the case of two words, w_a and w_b, we use the l_p-parameter of their corresponding word vectors in the embedding space as a distance metric between them:

$$d(w_a, w_b) = \|\mathbf{v}(\mathbf{w_a}) - \mathbf{v}(\mathbf{w_b})\|_p \tag{2}$$

More specifically, we use the Euclidean distance with p = 2. Therefore, for a given word w, the triple loss is defined as follows:

$$L_{tr}(w, S(w), N) = \frac{1}{|S(w)|} \sum_{w_{pos} \in S(w)} d(w, w_{pos}) -$$
$$\frac{1}{|N|} \sum_{w_{neg} \in N} min(d(w, w_{neg}), \alpha) + \alpha \tag{3}$$

Here, S(w) represents the set of words that are similar to a given word w, and N represents the set of words randomly chosen from the remaining pool of words. The number of randomly selected words equals the maximum count of similar word pairs, which is k. Through $L_{tr}(w, S(w), N)$ minimization, we decrease the distance between a word w and its similar words (positive sample) in the embedding space while simultaneously expanding the distance between the word w and its non-similar words (negative sample). Also, to avoid the simultaneous increase of distances between positive and negative sample pairs, once the negative pair distance exceeds a certain threshold α, it will no longer continue to expand.

Overall Training Objectives. Considering a sample x along with its associated classification label y as the current input, the comprehensive training objective for the model can be described as follows:

$$L(x, y) = L_{ce}(f(x), y) + \beta \cdot \frac{1}{n} \sum_{i=1}^{n} L_{tr}(w_i, S(w_i), N_i) \tag{4}$$

In this context, the symbol "L_{ce}" represents the cross-entropy loss, while "β" is a hyperparameter used to adjust the weight of the triple loss. The first term, L_{ce}, is utilized to train the subsequent layers following the initial embedding layer in order to develop the classification capability. The second term, L_{tr}, is employed to train a sturdy word embedding. This approach ensures that each word in the input sample is positioned near its similar counterpart and far from all other words within the embedding space. By doing so, it becomes feasible to train a robust model that produces similar representations for similar input samples, while simultaneously differentiating those representations from those of other samples in the dataset. This means that even if some poisoned data is incorporated into the training data, it becomes difficult to implant backdoors in the model during the training process while still maintaining good classification performance.

It is noticeable that triple metric learning is limited to the first embedding layer of the natural language model and is not reliant on the model's subsequent architecture. The alteration of the word embedding introduces only a minor additional overhead during the standard training, thus having minimal impact on the model training speed. Moreover, it is theoretically applicable to any language model.

Furthermore, it should be noted that some human writing perturbations do not appear in the vocabulary of the Bert language model. Consequently, when these perturbations are encountered, Bert breaks them down into several subwords. This, in turn, makes it challenging to determine the distance between words in the embedding space. To overcome this issue, these perturbed words are included in the model's vocabulary. The overall vocabulary will not change significantly due to the limited number of commonly used words and the inclusion of a small number of subwords.

3.2 Contrastive Learning Based on Semantic Negative Examples

Following triple metric learning, similar samples are represented similarly by the model, which can effectively prevent backdoor implantation. Nevertheless, due to the intricacy and variability of semantics, even a minor perturbation may result in a total change in meaning, leading to what are often referred to as semantic negative examples. The metric-learning model may not be sensitive enough to such examples. To enhance the sensitivity of the trained model to fine-grained changes in semantics, this paper proposes to improve the model's robustness using a method that involves performing contrastive learning based on semantic negative examples.

The concept of contrast learning involves learning representations by focusing on positive sample pairs while simultaneously pushing negative sample pairs away. In this study, sentences conveying the same semantic meaning are regarded as positive sample pairs, whereas sentences with opposing semantic meaning are viewed as negative sample pairs. The model trainer is assumed to possess at least one dataset that is entirely clean and not contaminated. For any original samples x_{ori} in this dataset, two distinct sentence samples are generated, both of which undergo only minor alterations compared to the original sample, but have vastly different semantics. One sample, marked as x_{syn}, is semantically very similar to the original sample, while the other sample, marked as x_{ant}, is semantically dissimilar or even contradictory to the original sample. Specifically, MIC employs spaCy2 to slice and lexically annotate the original sentence, extracting verbs, nouns, adjectives, and adverbs. The semantic similarity sample x_{syn} is created by substituting the extracted words with synonyms or related words, while the semantic negative sample x_{ant} is produced by replacing the extracted words with antonyms or randomly selected words. In practice, 40% of the tokens in the semantic similarity sample x_{syn} are replaced, while 20% of the tokens in the semantic counterexample sample x_{ant} are replaced.

For a natural language model M, it maps a sequence of input tokens $x = [x_1, ..., x_T]$ to a corresponding sequence of representations $h = [h_1, ..., h_T]$, where $h_{i \in [1:T]} \in R^d$ and d denotes the matrix dimension.

$$h = M(x) \tag{5}$$

From the defender's perspective, it is expected that the model accurately distinguishes whether there are any changes in semantics following modifications to the original sample. Essentially, this means that the metric between h_{ori} and h_{syn} should be relatively close in the feature space, while the metric between h_{ori} and h_{ant} should be comparatively distant. To achieve this objective, this paper proposes a contrast learning approach, where (x_{pos}, x_{syn}) denotes a positive sample pair, whereas (x_{pos}, x_{ant}) represents a negative sample pair. The method employs h_c to symbolize the embedding of the special identifier [CLS]. Consequently, the calculation of the metric between sentence representations involves computing the dot product between the [CLS] embeddings.

$$f(x^*, x^{'}) = exp(h_c^{*T} h_c^{'}) \tag{6}$$

The training loss for contrast learning is defined as follows:

$$L = - \sum_{x \in X} log \frac{f(x_{ori}, x_{syn})}{f(x_{ori}, x_{syn}) + f(x_{ori}, x_{ant})} \tag{7}$$

Unlike certain prior comparison strategies that utilize random sampling of multiple negative samples, the proposed method involves using just one x_a as a negative sample during training. This is due to the fact that the defense training objective is to enhance the model's sensitivity to semantic changes prompted by minor perturbations. Therefore, the method concentrates exclusively on negative samples generated according to our training objectives, rather than randomly selecting samples from the corpus as negative samples.

4   Experiments

This section assesses two hidden word-level attacks of MIC utilizing three defense baselines on three benchmark datasets that incorporate the $BERT_{BASE}$ and $BERT_{LARGE}$ [7] models. As the proposed method involves the defender's participation in the model training process, the experiments are conducted in the context of defense during the training phase.

4.1   Experimental Settings

Datasets. We assessed MIC and other baseline defense mechanisms on three benchmark datasets, namely SST-2 [26], OLID [35], and AG's News [36]. As shown in the

Table 1. Statistics of three evaluation datasets. Each dataset's target labels have been underlined. "Avg.#Words" represents the average number of words per sentence in each dataset.

Dataset	Task	Train	Valid	Test	Avg.#Words	Classes
SST-2	Sentiment Analysis	6,920	872	1,821	19.3	2 (Positive/Negative)
OLID	Offensive Language Identification	11,916	1,324	862	25.2	2 (Offensive/Not Offensive)
AG's News	News Topic Classification	108,000	11,999	7,600	37.8	4 (World/Sports/Business/SciTech)

Table 1, SST-2 is a binary sentiment classification dataset that comprises 6920 movie reviews for training, 872 for validation, and 1821 for testing. OLID is a binary offensive language classification dataset that consists of 11916 samples for training, 1324 for validation, and 859 for testing. AG's News is a news category classification dataset that comprises four categories and includes 108,000 samples for training, 11,999 samples for validation, and 7,600 samples for testing.

Victim Models. We utilize two commonly employed text classification models as victim models, namely $BERT_{BASE}$ and $BERT_{LARGE}$. The $BERT_{BASE}$ model comprises of 12 layers with a hidden size of 768 and a total of 110M parameters. On the other hand, for $BERT_{LARGE}$, we use the $bert-large-uncased$ from the Transformers library, which consists of 24 layers with a hidden size of 1024 and a total of 340M parameters.

Attack Methods. In order to conduct a comprehensive assessment of the defensive efficacy of the defense method MIC, we employed the two most surreptitious word-level backdoor attacks currently in use, namely synonym substitution (LWS) [24] and human writing perturbations (LRS) [13], as triggers, respectively.

Defense Baseline. We compare our approach with three baseline defense methods. BKI is a defense method called backdoor keyword identification. It initially scores the contribution of each word in the sample towards the final classification outcome and computes the frequency of occurrence for each word. Subsequently, it combines the importance score and frequency to filter out dubious words. ONION is a defense method that relies on outlier detection. The approach advocates for determining whether a word serves as a backdoor trigger based on the degree of change in sentence perplexity before and after removing each word in the sample. The word exhibiting the most significant change in perplexity is deemed the trigger word to be eliminated. RAP is a robustness-aware online defense method. The approach suggests applying a slight perturbation to the input samples and determining the change in value of the poisoning model on the output probability of the samples before and after the perturbation. The samples exhibiting a change value less than the threshold are deemed poisoned samples and filtered out accordingly.

Evaluation Metrics. To compare the effectiveness of the proposed method with the baseline approach, the experiments employ conventional evaluation metrics that are consistent with prior research [5]. These metrics include clean sample accuracy (CACC), attack success rate (ASR), change in clean sample accuracy (ΔCACC), and change in attack success rate (ΔASR). CACC refers to the classification accuracy of the victim model on clean samples, while ASR pertains to the classification accuracy

Table 2. Comparison of the defensive effects of various methods on two Bert models.

Dataset	Model	$BERT_{BASE}$					$BERT_{Large}$				
		null	LWS		LRS		null	LWS		LRS	
		CACC	CACC	ASR	CACC	ASR	CACC	CACC	ASR	CACC	ASR
SST-2	w/o defense	91.10	88.62	97.28	**90.28**	99.78	92.50	90.02	97.42	**92.20**	99.78
	BKI	91.16	87.03(−1.59)	95.32(−1.96)	88.34(−1.94)	85.45(−14.33)	92.47	88.21(−1.81)	95.73(−1.69)	88.83(−3.37)	83.28(−16.5)
	ONION	91.71	87.33(−1.29)	92.94(−4.34)	85.50(−4.78)	52.68(−37.10)	92.33	87.03(−2.99)	93.24(−4.18)	85.38(−6.82)	56.91(−42.87)
	RAP	**91.93**	83.79(−4.83)	80.24(−17.04)	84.56(−5.72)	68.37(−31.41)	**92.73**	83.79(−6.23)	79.03(−18.39)	83.49(−8.71)	70.14(−29.64)
	MIC	91.67	**90.24(+1.62)**	**16.01(−81.27)**	89.84(−0.44)	**19.75(−80.03)**	92.67	**90.66(+0.64)**	18.97(−78.45)	90.06(−2.14)	25.71(−74.07)
OLID	w/o defense	82.91	**82.93**	97.14	**82.79**	100	82.81	81.42	97.93	**81.16**	100
	BKI	82.82	81.63(−1.3)	94.39(−2.75)	81.62(−1.17)	90.23(−9.77)	82.82	80.03(−1.39)	95.83(−2.1)	80.29(−0.87)	84.37(−15.63)
	ONION	82.76	80.22(−2.71)	92.67(−4.47)	81.16(−1.63)	64.58(−35.42)	82.76	79.52(−1.9)	95.23(−2.7)	79.55(−1.61)	62.09(−37.91)
	RAP	**82.97**	78.34(−4.59)	75.62(−21.52)	77.87(−4.92)	65.29(−34.71)	**82.97**	73.29(−8.13)	77.35(−20.58)	72.93(−8.23)	61.27(−38.73)
	MIC	82.86	82.77(−0.16)	**22.59(−74.55)**	82.69(−0.1)	**25.71(−74.29)**	82.86	**81.93(+0.51)**	**25.73(−72.2)**	81.65(+0.49)	23.38(−76.62)
AG's News	w/o defense	93.10	92.02	99.63	92.76	99.96	94.24	92.62	99.53	93.64	99.96
	BKI	93.06	91.47(−0.55)	96.53(−3.1)	91.96(−0.8)	92.54(−7.42)	94.16	91.52(−1.1)	96.13(−3.4)	91.89(−1.75)	93.01(−6.95)
	ONION	92.78	90.71(−1.31)	95.31(−4.32)	90.83(−1.93)	67.74(−32.22)	93.92	92.21(−0.41)	96.20(−3.33)	91.83(−1.81)	64.44(−35.52)
	RAP	93.03	87.27(−4.75)	76.37(−23.26)	86.98(−5.78)	64.39(−35.57)	94.33	86.46(−6.16)	73.89(−25.64)	87.34(−6.3)	62.38(−37.58)
	MIC	93.01	**92.35(+0.33)**	**34.95(−64.68)**	**92.83(+0.07)**	**36.21(−63.75)**	94.09	92.17(−0.45)	**29.63(−69.9)**	93.27(−0.37)	**30.19(−69.77)**

of the victim model on poisoned samples. The ΔCACC represents the variation in the classification accuracy of the victim model on clean samples before and after applying defense. A smaller change indicates that the defense method has minimal impact on the normal performance of the model. The ΔASR denotes the variation in classification accuracy of the victim model on poisoned samples before and after applying defense. A larger change signifies that the defense method is more effective.

Implementation Details. For the attack method based on synonym replacement, the experiments employ the settings of Qi [24], wherein the poisoning rate is set to 10%, the batch size to 32, the learning rate to 2e-5, and a maximum of 5 candidates for word replacement. Similarly, for the attack method based on perturbation of human writing, comparable experimental settings are used. For the proposed defense method MIC, k is set to 8, α is set to $0.7\alpha_0$ (where α_0 denotes the average word distance of the initial word embedding before training, which is set to 1.48 for the Bert model), and β is set to 1, thereby achieving a balance between the standard training loss and the triple metric learning loss. Regarding the other three defense methods, we utilize the experimental settings presented in their respective papers. For BKI and RAP, a poisoning model is trained using 20% of the training dataset (including the poisoned data) to filter and process the training dataset.

4.2 Main Results

Performance of Defense. This subsection presents the defense effectiveness of the aforementioned defense methods. Table 2 displays the defense effectiveness of various defense methods on two representative natural language models, $Bert_{base}$ and $Bert_{large}$. The defensive performance that exhibits the highest level of effectiveness, as well as the normal classification performance, are highlighted in bold. The options or results that rank next in terms of performance are indicated by being underlined. Additionally, the magnitude of the change in attack success rate and clean sample classification accuracy before and after the defense is presented in parentheses. Based on the results, it can be observed that the proposed MIC achieves the best defense effectiveness across all datasets while maximizing the classification accuracy of the model on normal samples.

It is not surprising that backdoor defense methods, such as BKI and ONION, which were originally designed to filter out rare words, perform poorly. Among these methods, BKI assumes that there are only a fixed number of words in the trigger dataset, so by calculating the importance of words and combining them with word frequencies, triggers can be effectively filtered out. However, the triggers in backdoor attacks based on word substitution are more diverse, and there may be more than one trigger in a sample. Therefore, only a small fraction of triggers can be filtered out if the original method is still used.

ONION is a targeted defense that relies on the observation that traditional backdoor triggers cause a significant change in sample perplexity. However, the method based on synonymous word substitution has little impact on sentence perplexity before and after substitution, rendering it ineffective. Even human writing perturbations have only a small effect on perplexity compared to synonyms, resulting in unsatisfactory defense effectiveness. Additionally, both BKI and ONION, which rely on filtering triggers, inevitably remove normal words while identifying and filtering suspicious triggers, leading to a slight decrease in the classification accuracy of the defended model on clean samples.

Regarding RAP, it is a defense approach that is aware of robustness. It has also shown a strong defense effectiveness against traditional backdoor attacks. However, when facing word-level attacks based on the two highly stealthy and trigger-rich methods used in the experiments, the boundary between poisoned and clean samples is not so clear-cut. Therefore, RAP identifies some of the poisoned samples as clean and also identifies some of the clean samples as poisoned, which explains why the RAP in the table significantly reduces the model's accuracy rate for classifying clean samples. Furthermore, if defenders want to further improve the defense effectiveness, they would need to expand the threshold of perplexity change, which would result in more clean samples being labeled as toxic samples.

The proposed MIC defense approach is based on a different idea. As the saying goes, prevention is better than cure. Rather than identifying and filtering suspicious triggers in scenarios where model training is accessible, it is better to prevent backdoors from being implanted at the source to the greatest extent possible. The experimental results also confirm the feasibility and effectiveness of using the metric learning approach for defense. Additionally, introducing contrast learning has also been experimentally proven to be effective in ensuring the classification accuracy of the model on clean samples after training.

4.3 Ablation Experiment

To study the impact of metric learning and contrastive learning on the final defense effectiveness in the proposed method, ablation experiments were conducted. The following variant methods were designed and their defense effects were studied respectively: 1) w/o CL signifies that the method does not utilize contrast learning, in contrast to MIC. 2) w/o ML denotes that the method does not utilize triple metric learning in contrast to MIC.

Table 3. The defensive effects of different variant methods.

Dataset	Model	$BERT_{BASE}$				
		null	LWS		LRS	
		CACC	CACC	ASR	CACC	ASR
SST-2	w/o defense	91.10	88.62	97.28	90.28	99.78
	MIC	91.67	90.24(+1.62)	16.01(−81.27)	89.84(−0.44)	19.75(−80.03)
	w/o CL	90.03	85.25(−3.37)	**13.92(−83.36)**	84.25(−6.03)	**14.34(−85.44)**
	w/o ML	**92.21**	**91.34(+2.72)**	94.33(−2.95)	**91.32(+1.04)**	95.63(−4.15)
OLID	w/o defense	82.91	82.93	97.14	82.79	100
	MIC	82.86	82.77(−0.16)	22.59(−74.55)	82.69(−0.1)	25.71(−74.29)
	w/o CL	81.26	78.62(−4.31)	**18.52(−78.62)**	79.24(−3.55)	**20.32(−79.68)**
	w/o ML	**83.86**	**83.22(+0.29)**	95.53(−1.61)	**83.89(+1.1)**	94.66(−5.34))
AG's News	w/o defense	93.10	92.02	99.63	92.76	99.96
	MIC	93.01	92.35(+0.33)	34.95(−64.68)	92.83(+0.07)	36.21(−63.75)
	w/o CL	91.36	87.63(−4.39)	**28.53(−71.1)**	87.82(−4.94)	27.93(−72.03)
	w/o ML	**94.63**	**94.04(+2.02)**	92.43(−7.2)	**93.83(+1.07)**	91.43(−8.53)

Table 3 displays the defense results of three different methods on various datasets. The best performance outcomes are emphasized in bold, while the second-best results are indicated by being underlined. Based on the experimental results, it can be observed that the introduction of triple metric learning improves defense performance. The metric-only learning approach displays satisfactory defense against two word-level backdoor attacks, resulting in an average 78.37% decrease in attack success rate. However, the use of only metric learning also leads to an average 4.43% decrease in clean sample classification accuracy, confirming the previously mentioned concern that the introduction of metric learning may decrease the sensitivity of the model to semantic changes. The introduction of contrast learning compensates for this shortcoming, and the experimental results show that only the contrast learning approach exhibits the best clean sample classification accuracy in all experimental settings. This demonstrates that using contrast learning can significantly improve the sensitivity of the model to semantic changes.

4.4 Hyper-parameter Study

This subsection explores the impact of hyperparameters on defense results. MIC mainly involves two hyperparameters: α, which constrains the distance between anchor words and non-similar words in word-level triple loss, and β, which controls the weight of word-level triple loss in the overall training objective function.

Analysis of The Impact of Hyperparameter α. Figure 2 displays the performance of different values of α across three datasets, where CACC refers to the classification accuracy of the defended model on clean samples and ASR signifies the classification accuracy of the defended model on toxic samples. The value of β was fixed at 1, while

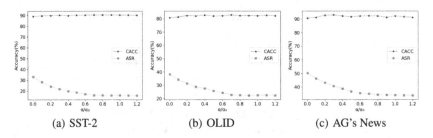

Fig. 2. Impact of hyperparameter α on the performance of MIC's BERT model across three datasets.

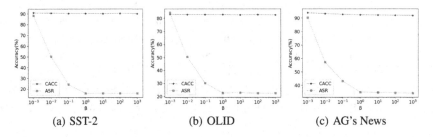

Fig. 3. Impact of hyperparameter β on the performance of MIC's BERT model across three datasets.

the value of α was varied from 0 to $1.2\alpha_0$ in order to investigate how the size of α impacts the overall effectiveness of the defense method. The highest ASR was obtained when α was set to 0. This may be attributed to the presence of multiple meanings causing multiple words to cluster together due to the existence of a common meaning, making it difficult for the model to distinguish between them even if these words represent other meanings in the sample. As the value of α increased, the ASR first decreased and then stabilized. In contrast, the CACC exhibited slight fluctuations as α changed, but remained largely consistent overall, as the subsequent use of contrast learning provided some assurance on the value of CACC. Therefore, a value of $0.7\alpha_0$ was chosen as the final hyperparameter for all three datasets.

Analysis of The Impact of Hyperparameter β. Likewise, with α fixed at $0.7\alpha_0$, the effect of parameter β on the performance of the method was examined by varying its value from 10e$-$3 to 10e3. Figure 3 displays the performance of different values of β across three datasets, and the two metrics are consistent with those in the previous section. It is evident that the CACC decreases slightly as β increases, while the ASR first decreases sharply and then levels off. Once β reaches 10^0, the ASR changes less, indicating that the defense has achieved its maximum effect. As a result, $\beta = 1$ is chosen to strike a balance between CACC and ASR.

5   Conclusion

This paper aims to address the challenge that existing backdoor defense methods have encountered in addressing word-level backdoor attacks, particularly in scenarios where the defender has access to model training. To this end, we propose a backdoor defense method called MIC that is based on metric and contrast learning. The method first incorporates triple metric learning into the standard model training process to modify the embedding layer of the model. This enables similar words to not only appear visually similar but also have similar vectors after being mapped by the model, thus preventing backdoor implantation at the source. Additionally, to overcome the reduced sensitivity of the model to semantic changes brought about by metric learning, we propose to introduce semantic counterexample-based comparison learning post-training. This approach enhances the model's robustness against subtle perturbations that can cause semantic changes.

The experiments were conducted using three classical datasets, namely SST-2, OLID, and AG's News, following the previous dataset division. Two of the most covert word-level backdoor attacks, LWS and LRS, were implemented on two commonly used language models, Bert and $Bert_{LARGE}$, and the defense methods proposed in this paper were compared with other representative defense methods. The results indicate that the proposed method performed the best in terms of backdoor defense effectiveness compared to other representative backdoor defense methods, with an average attack success rate reduction of 73.5%. Furthermore, compared to the comparative methods that decreased the classification accuracy of clean samples by 1.23%, 2.28%, and 5.10%, the method proposed in this paper exhibited minimal changes in the classification accuracy of clean samples. In fact, it even increased by 0.22% compared to the attack method. This implies that MIC not only has a good defense effect but also maximally retains the normal classification results of the model.

Acknowledgements. This work was supported in part by Jiangsu Province Modern Education Technology Research 2021 Smart Campus Special Project "Research and Practice of Data Security System in College Smart Campus Construction" (2021-R-96776); Jiangsu Province University Philosophy and Social Science Research Major Project "Higher Vocational College Ideological and Political Science" "Research on the Construction of Class Selective Compulsory Courses" (2022SJZDSZ011); Research Project of Nanjing Industrial Vocational and Technical College (2020SKYJ03); 2022 "Research on the Teaching Reform of High-quality Public Courses in Jiangsu Province Colleges and Universities" --The new era of labor education courses in Jiangsu higher vocational colleges Theoretical and practical research.

References

1. Azizi, A., et al.: T-Miner: a generative approach to defend against trojan attacks on DNN-based text classification. In: 30th USENIX Security Symposium (USENIX Security 21), pp. 2255–2272. USENIX Association, August 2021. https://www.usenix.org/conference/usenixsecurity21/presentation/azizi

2. Chen, C., Dai, J.: Mitigating backdoor attacks in LSTM-based text classification systems by backdoor keyword identification. Neurocomputing **452**, 253–262 (2021). https://doi.

org/10.1016/j.neucom.2021.04.105.   https://www.sciencedirect.com/science/article/pii/
S0925231221006639

3. Chen, X., Dong, Y., Sun, Z., Zhai, S., Shen, Q., Wu, Z.: Kallima: a clean-label framework
    for textual backdoor attacks. arXiv preprint arXiv:2206.01832 (2022)

4. Chen, X., Salem, A., Backes, M., Ma, S., Zhang, Y.: BadNL: backdoor attacks against NLP
    models. In: ICML 2021 Workshop on Adversarial Machine Learning (2021)

5. Cui, G., Yuan, L., He, B., Chen, Y., Liu, Z., Sun, M.: A unified evaluation of textual
    backdoor learning: frameworks and benchmarks. In: Koyejo, S., Mohamed, S., Agarwal, A.,
    Belgrave, D., Cho, K., Oh, A. (eds.) Advances in Neural Information Processing Systems,
    vol. 35, pp. 5009–5023. Curran Associates, Inc. (2022). https://proceedings.neurips.cc/
    paper_files/paper/2022/file/2052b3e0617ecb2ce9474a6feaf422b3-Paper-Datasets_and_
    Benchmarks.pdf

6. Dai, J., Chen, C., Li, Y.: A backdoor attack against LSTM-based text classification systems.
    IEEE Access **7**, 138872–138878 (2019)

7. Devlin, J., Chang, M.W., Lee, K., Toutanova, K.: BERT: pre-training of deep bidirectional
    transformers for language understanding. arXiv preprint arXiv:1810.04805 (2018)

8. Doan, B.G., Abbasnejad, E., Ranasinghe, D.C.: Februus: input purification defense against
    trojan attacks on deep neural network systems. In: Annual Computer Security Applications
    Conference, ACSAC 2020, pp. 897–912. Association for Computing Machinery, New York,
    NY, USA (2020). https://doi.org/10.1145/3427228.3427264

9. Gao, Y., Xu, C., Wang, D., Chen, S., Ranasinghe, D.C., Nepal, S.: STRIP: a defence against
    trojan attacks on deep neural networks. In: Proceedings of the 35th Annual Computer
    Security Applications Conference, ACSAC 2019, pp. 113–125. Association for Computing
    Machinery, New York, NY, USA (2019). https://doi.org/10.1145/3359789.3359790

10. Garg, S., Kumar, A., Goel, V., Liang, Y.: Can adversarial weight perturbations inject neu-
    ral backdoors. In: Proceedings of the 29th ACM International Conference on Information &
    Knowledge Management, CIKM 2020, pp. 2029–2032. Association for Computing Machin-
    ery, New York, NY, USA (2020). https://doi.org/10.1145/3340531.3412130

11. Garg, S., Ramakrishnan, G.: BAE: BERT-based adversarial examples for text classification.
    arXiv preprint arXiv:2004.01970 (2020)

12. Gu, T., Dolan-Gavitt, B., Garg, S.: BadNets: identifying vulnerabilities in the machine learn-
    ing model supply chain. arXiv preprint arXiv:1708.06733 (2017)

13. Le, T., Lee, J., Yen, K., Hu, Y., Lee, D.: Perturbations in the wild: leveraging human-
    written text perturbations for realistic adversarial attack and defense. arXiv preprint
    arXiv:2203.10346 (2022)

14. Li, Y., Lyu, X., Koren, N., Lyu, L., Li, B., Ma, X.: Neural attention distillation: erasing back-
    door triggers from deep neural networks. In: International Conference on Learning Repre-
    sentations (2021). https://openreview.net/forum?id=9l0K4OM-oXE

15. Li, Y., Zhai, T., Wu, B., Jiang, Y., Li, Z., Xia, S.: Rethinking the trigger of backdoor attack
    (2021)

16. Liao, C., Zhong, H., Squicciarini, A., Zhu, S., Miller, D.: Backdoor embedding in convo-
    lutional neural network models via invisible perturbation. arXiv preprint arXiv:1808.10307
    (2018)

17. Liu, Y., et al.: Trojaning attack on neural networks (2017)

18. Liu, Y., Shen, G., Tao, G., An, S., Ma, S., Zhang, X.: PICCOLO: exposing complex back-
    doors in NLP transformer models. In: 2022 IEEE Symposium on Security and Privacy (SP),
    pp. 2025–2042 (2022). https://doi.org/10.1109/SP46214.2022.9833579

19. Liu, Y., Ma, X., Bailey, J., Lu, F.: Reflection backdoor: a natural backdoor attack on deep
    neural networks. In: Vedaldi, A., Bischof, H., Brox, T., Frahm, J.-M. (eds.) ECCV 2020.
    LNCS, vol. 12355, pp. 182–199. Springer, Cham (2020). https://doi.org/10.1007/978-3-030-
    58607-2_11

20. Lyu, W., Zheng, S., Ma, T., Chen, C.: A study of the attention abnormality in Trojaned BERTs. In: Proceedings of the 2022 Conference of the North American Chapter of the Association for Computational Linguistics: Human Language Technologies, pp. 4727–4741. Association for Computational Linguistics, Seattle, United States, July 2022. https://doi.org/10.18653/v1/2022.naacl-main.348. https://aclanthology.org/2022.naacl-main.348

21. Qi, F., Chen, Y., Li, M., Yao, Y., Liu, Z., Sun, M.: ONION: a simple and effective defense against textual backdoor attacks. In: Proceedings of the 2021 Conference on Empirical Methods in Natural Language Processing, pp. 9558–9566. Association for Computational Linguistics, Online and Punta Cana, Dominican Republic, November 2021. https://doi.org/10.18653/v1/2021.emnlp-main.752. https://aclanthology.org/2021.emnlp-main.752

22. Qi, F., Chen, Y., Zhang, X., Li, M., Liu, Z., Sun, M.: Mind the style of text! Adversarial and backdoor attacks based on text style transfer. arXiv preprint arXiv:2110.07139 (2021)

23. Qi, F., et al.: Hidden killer: invisible textual backdoor attacks with syntactic trigger. arXiv preprint arXiv:2105.12400 (2021)

24. Qi, F., Yao, Y., Xu, S., Liu, Z., Sun, M.: Turn the combination lock: learnable textual backdoor attacks via word substitution. arXiv preprint arXiv:2106.06361 (2021)

25. Saha, A., Subramanya, A., Pirsiavash, H.: Hidden trigger backdoor attacks. In: Proceedings of the AAAI Conference on Artificial Intelligence, vol. 34, pp. 11957–11965 (2020)

26. Socher, R., et al.: Recursive deep models for semantic compositionality over a sentiment treebank. In: Proceedings of the 2013 Conference on Empirical Methods in Natural Language Processing, pp. 1631–1642 (2013)

27. Sun, L.: Natural backdoor attack on text data (2021)

28. Wang, B., et al.: Neural cleanse: identifying and mitigating backdoor attacks in neural networks. In: 2019 IEEE Symposium on Security and Privacy (SP), pp. 707–723 (2019). https://doi.org/10.1109/SP.2019.00031

29. Wang, D., Ding, N., Li, P., Zheng, H.: CLINE: contrastive learning with semantic negative examples for natural language understanding. In: Proceedings of the 59th Annual Meeting of the Association for Computational Linguistics and the 11th International Joint Conference on Natural Language Processing (Volume 1: Long Papers), pp. 2332–2342. Association for Computational Linguistics, August 2021. https://doi.org/10.18653/v1/2021.acl-long.181. https://aclanthology.org/2021.acl-long.181

30. Xu, X., Wang, Q., Li, H., Borisov, N., Gunter, C.A., Li, B.: Detecting AI trojans using meta neural analysis. In: 2021 IEEE Symposium on Security and Privacy (SP), pp. 103–120 (2021). https://doi.org/10.1109/SP40001.2021.00034

31. Yang, W., Li, L., Zhang, Z., Ren, X., Sun, X., He, B.: Be careful about poisoned word embeddings: exploring the vulnerability of the embedding layers in NLP models. In: Proceedings of the 2021 Conference of the North American Chapter of the Association for Computational Linguistics: Human Language Technologies, pp. 2048–2058. Association for Computational Linguistics, June 2021. https://doi.org/10.18653/v1/2021.naacl-main.165. https://aclanthology.org/2021.naacl-main.165

32. Yang, W., Lin, Y., Li, P., Zhou, J., Sun, X.: RAP: robustness-aware perturbations for defending against backdoor attacks on NLP models. In: Proceedings of the 2021 Conference on Empirical Methods in Natural Language Processing, pp. 8365–8381. Association for Computational Linguistics, Online and Punta Cana, Dominican Republic, November 2021. https://doi.org/10.18653/v1/2021.emnlp-main.659. https://aclanthology.org/2021.emnlp-main.659

33. Yang, W., Lin, Y., Li, P., Zhou, J., Sun, X.: Rethinking stealthiness of backdoor attack against NLP models. In: Proceedings of the 59th Annual Meeting of the Association for Computational Linguistics and the 11th International Joint Conference on Natural Language Processing (Volume 1: Long Papers), pp. 5543–5557. Association for Computational Linguistics,

August 2021. https://doi.org/10.18653/v1/2021.acl-long.431. https://aclanthology.org/2021.acl-long.431

34. Yang, Y., Wang, X., He, K.: Robust textual embedding against word-level adversarial attacks. In: Cussens, J., Zhang, K. (eds.) Proceedings of the Thirty-Eighth Conference on Uncertainty in Artificial Intelligence. Proceedings of Machine Learning Research, vol. 180, pp. 2214–2224. PMLR, 01–05 August 2022. https://proceedings.mlr.press/v180/yang22c.html

35. Zampieri, M., Malmasi, S., Nakov, P., Rosenthal, S., Farra, N., Kumar, R.: Predicting the type and target of offensive posts in social media. arXiv preprint arXiv:1902.09666 (2019)

36. Zhang, X., Zhao, J., LeCun, Y.: Character-level convolutional networks for text classification. IN: Advances in Neural Information Processing Systems, vol. 28 (2015)

37. Zhao, S., Ma, X., Zheng, X., Bailey, J., Chen, J., Jiang, Y.G.: Clean-label backdoor attacks on video recognition models. In: Proceedings of the IEEE/CVF Conference on Computer Vision and Pattern Recognition, pp. 14443–14452 (2020)

Active Learning for Open-Set Annotation Using Contrastive Query Strategy

Peng Han[1,3], Zhiming Chen[1(✉)], Fei Jiang[2], and Jiaxin Si[4]

[1] School of Software Engineering, Chongqing University of Posts and
Telecommunications, Chongqing, China
s221201002@stu.cqupt.edu.cn
[2] East China Normal University, Shanghai, China
[3] Chongqing Academy of Science and Technology, Chongqing, China
[4] Chongqing Qulian Digital Technology Co., Ltd., Chongqing, China

Abstract. Active learning has achieved remarkable success in minimizing labeling costs for classification tasks with all data samples drawn from known classes. However, in real scenarios, most active learning methods fail when encountering open-set annotation (OSA) problem, i.e., numerous samples from unknown classes. The main reason for such failure comes from existing query strategies that are unavoidable to select unknown class samples. To tackle such problem and select the most informative samples, we propose a novel active learning framework named OSA-CQ, which simplifies the detection work of samples from known classes and enhances the classification performance with an effective contrastive query strategy. Specifically, OSA-CQ firstly adopts an auxiliary network to distinguish samples using confidence scores, which can dynamically select samples with the highest probability from known classes in the unlabeled set. Secondly, by comparing the predictions between auxiliary network, classification, and feature similarity, OSA-CQ designs a contrastive query strategy to select these most informative samples from unlabeled and known classes set. Experimental results on CIFAR10, CIFAR100 and Tiny-ImageNet show the proposed OSA-CQ can select samples from known classes with high information, and achieve higher classification performance with lower annotation cost than state-of-the-art active learning algorithms.

Keywords: Active Learning · Open-set Annotation · Contrastive Query Strategy

1 Introduction

The past decades have witnessed the rapid growth of deep learning and its widespread applications. The success of deep learning heavily depends on large amount of annotated and public datasets [1,2]. However, obtaining large-scale datasets is extremely time-consuming, labor-intensive and expensive. Therefore, it is challenging to achieve satisfactory results with limited labeled datasets in practice.

© The Author(s), under exclusive license to Springer Nature Singapore Pte Ltd. 2024
B. Luo et al. (Eds.): ICONIP 2023, LNCS 14452, pp. 19–31, 2024.
https://doi.org/10.1007/978-981-99-8076-5_2

Active learning (AL) has received considerable attention since it pursuits minimum annotation by selecting most informative samples and achieves competitive performance. Most existing AL methods focus on closed-set setting in which the labeled and unlabeled samples are both from the pre-defined domain with the same class distribution. However, such assumption does not always hold true in real scenarios. The unlabeled sets are mostly collected from rather casual data curation processes such as web-crawling, and contain a large number of samples from unknown classes called open-set setting, as shown in Fig. 1. The corresponding problem for active learning is open-set annotation (OSA).

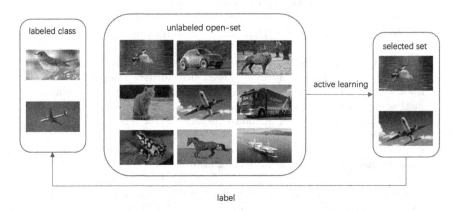

Fig. 1. Open-set example. The unlabeled set contains not only the target class data ("birds", "airplanes") but also a large number of unknown class data ("cars", "horses", etc.).

The query strategies of most exiting closed-set AL methods typically prioritize the selection of data exhibiting high uncertainty. In OSA, these methods tend to select irrelevant samples from unknown classes that leads to a waste of annotation budget [19]. Hence, in the real world, an effective and efficient AL method is highly desired, which can select the informative samples from the known class.

In this paper, we propose a novel open-set active learning method called OSA-CQ that can precisely identify the samples of unknown classes while querying the most informative ones from the desired classes to annotation. Firstly, we adopt an auxiliary detector to distinguish samples driven from known and unknown classes by sorting confidence scores. Secondly, we define the informative samples of the known classes from unlabeled set as these locating near the classification decision boundaries. To effectively select informative samples, we introduce a contrastive query strategy based on the prediction inconsistency among the detector, classifier, and the feature similarity. The samples with prediction inconsistency are usually hard for classifiers to recognize, which frequently locate on the boundaries of classifiers, and are the informative ones.

Our main contributions are summarized as follows:

- We propose a novel AL framework for solving OSA problems, which can effectively distinguish samples of known classes while selecting the informative ones.
- We solve the informative sample selection from the known classes by introducing contrastive learning based on the prediction inconsistency, which determines the informative samples as these locates near the boundaries of classifiers.
- Experimental results on CIFAR10, CIFAR100 and Tiny-ImageNet demonstrate the efficiency and effectiveness of the proposed method.

2 Related Work

Active Learning. AL primarily focuses on data-level investigation, hence the name query learning [3]. AL's goal is to actively select high-value data for labeling, thereby maximizing model performance with a small amount of labeled data. AL strategies are commonly classified into five categories: uncertainty-based, data distribution-based, model parameter change-based, and committee-based AL.

The uncertainty-based query strategy involves calculating uncertainty scores for samples in the unlabeled set using a custom scoring function [4–7]. For instance, [4] suggested using entropy as a measure, while [5] used the difference between the highest and second-highest prediction probabilities. Addressing the issue of overly confident predictions, NCE-Net [7] replaced softmax classifiers with nearest neighbor classifiers. These methods rely entirely on the predicted class probability, ignoring the value of the feature representation. Query strategies based on data distribution offer more generalizability compared to uncertainty-based approaches, [8,9]. The most representative is CoreSet [8], which defined the active learning problem as selecting a core set of representative instances. Parameter variation-based query strategies account for the impact of data on model parameters. LL4AL [10] introduced a loss prediction module to directly predict loss values for data in the unlabeled set. Additionally, [11] proposed a gradient-based importance measure and analyzed the relative importance of images in the dataset. MI-AOD [14] adopted a committee-based approach, selecting target data using a committee of two classifiers. Recently, several alternative AL methods based on different strategies have emerged. These include generated data at decision boundaries [12]; employed VAE to capture distributional differences between labeled and unlabeled sets [13]. AL approaches using graph convolutional neural networks [15,16], reinforcement learning-based AL [17] for automatic query strategy design [18], AL for handling class mismatch [19,20], and AL addressing class imbalance [21].

Open-Set Recognition. The open-set recognition task addresses the challenge of encountering new classes of data in the test set that are not seen during training. It requires training a recognition model capable of accurately classifying known classes while effectively handling unknown classes and providing a rejection option when a test sample belongs to an unknown class [22]. The OpenMax model proposed by [23] was the first deep neural network solution for open-set recognition. It represents each known class as an average activation vector. The model then calculates the distance between the training samples and the corresponding class average activation vector and fits a Weibull distribution to each class. During testing, the activation values of test data are computed based on the Weibull distribution for each class. [24] introduced two novel losses: the Entropic Open-Set loss and the Objectosphere loss. These losses are combined with softmax to enable effective solutions for open-set recognition. However, the aforementioned methods primarily focus on discriminating known classes from unknown classes and do not specifically address the AL aspect of selecting information-rich data.

Open-Set AL. There are currently two main research on Open-set AL [19,20]. LfOSA [19] was similar to OpenMax in that it utilized the largest activation vector to fit a Gaussian mixture distribution for each class, and selected based on the confidence of the Gaussian mixture model. It exclusively focuses on addressing the problem of open set recognition and does not delve into the details of AL. CCAL [20] leveraged contrastive learning to extract semantic and distinctive features and incorporated them into a query strategy for selecting the most informative known unlabeled samples.

3 Methodology

3.1 Problem Definition

Taking the C classification problem as an illustration. Initially, a labeled set $\mathcal{D}_0^l = (x_i, y_i)$ is initialized with a small amount of data, where $x_i \in \mathcal{X}$ represents the input and $y_i \in \mathcal{Y}$ denotes the corresponding class label. Subsequently, a model $f(\cdot)$ with parameters θ is trained after oracle labeling. Then, based on a query policy α, k samples $\mathcal{S}_k = (x_i, y_i)$ are selected from the unlabeled set $\mathcal{D}_0^u = (x_i, y_i)$ and sent to the oracle for labeling. Once labeling is completed, the known classes data in $\mathcal{S}_k$ is added to the current labeled set, resulting in $\mathcal{D}_{j+1}^l = (x_i, y_i)$. The model $f(\cdot)$ is trained again for the next iteration. Regarding the auxiliary detector, OSA-CQ enhances its learning by assigning a pseudo-label of class C to the selected unknown class data, $\mathcal{D}_{j+1}^l = \mathcal{D}_j^l \cup (x_i, y_i)|y_i \in y_{kno}$ and $\mathcal{D}_{j+1}^{inval} = \mathcal{D}_j^{inval} \cup (x_i, y_i)|y_i \in y_{unkno}$, where y_{kno} represents the known class labels and y_{unkno} represents the unknown class labels. The invalid set denotes as $\mathcal{D}^{inval}$, which is initialized as an empty set in the first iteration.

The effectiveness of each AL iteration in selecting known class data is measured using recall, which is defined as follows:

$$recall_i = \frac{|S_{query}^{kno}| + |D_{kno}^l|}{|D_{kno}^u| + |S_{query}^{kno}| + |D_{kno}^l|} \tag{1}$$

where $recall_i$ denotes the recall of active sampling for the ith AL cycle. $recall_i$ represents the percentage of known classes in the current AL cycle, $|D_{kno}^l|$ denotes the number of known classes in the training set now, and $|D_{kno}^u|$ denotes the number of known classes in the unlabeled set.

3.2   Algorithm Detail

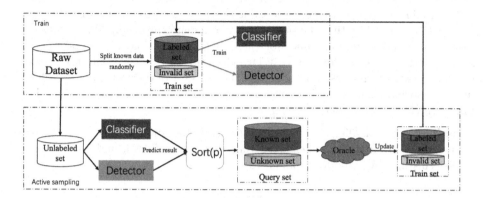

Fig. 2. The OSA-CQ framework consists of two stages. (1) Training of classifier and detector. (2) Active sampling of unlabeled sets.

The OSA-CQ framework, depicted in Fig. 2, comprises three key components: the detector, classifier, and active query strategy. Initially, the raw dataset is randomly divided into two subsets: the labeled set and the unlabeled set. The classifier is trained using the labeled set, while the labeled set and invalid set which is initialized as null set form the detector's training data. The role of the detector is twofold: 1) to rank the maximum probability values to distinguish between known and unknown classes, and 2) to provide prediction results for inconsistency analysis, aiding in the selection of informative data. The detector's confidence ranking of the unlabeled data determines the sequence order for judging the inconsistencies in the prediction results. Based on this ranking, actively selects k examples to give oracle for labeling. In the subsequent sections, we elaborate on the specific details of these three components.

Detector. The detector is an auxiliary network in OSA-CQ. It treats all unknown class data within the open-set as a distinct category. Therefor, the detector extends the classifier, which originally comprises C classes, to include an additional $(C+1)$th class to predicting the probability of the unknown class. The loss function of the detector is defined as Eq. 2.

$$\mathcal{L}_d(x,y) = -\sum_{i=1}^{C+1} y_i \, log(p_i) \tag{2}$$

where

$$p_i = \frac{exp(a_i/T)}{\sum_j exp(a_j/T)} \tag{3}$$

where a_i is the ith output of the last fully connected layer of the detector and T is the temperature of softmax with temperature [25], which is used to adjust the sharpness of the probability distribution after softmax activation. Since unknown classes of data typically comprise a mixture of various classes, training these unknown classes as a single class often leads to unsatisfactory results. We found that applying a smaller temperature value sharpens the probability distribution after activation, mitigating the softmax overconfidence problem. This approach results in larger activation values for the first C classes (known classes) and a smaller activation value for the $(C + 1)$th class (unknown class), and the opposite for the unknown class. Consequently, the detector outperforms the classifier in distinguishing known and unknown data, and its main role is to filter the unknown class data for the classifier.

Classifier. The classifier focuses solely on identifying known classes of data for open-set. We train the classifier using the labeled set of data, and train the C classes classifier by standard cross-entropy loss,

$$\mathcal{L}_c(x, y) = -\sum_{i=1}^{C} y_i \, log(f_{\theta_c}(x)) \tag{4}$$

where θ_c is the parameter of the classifier and $(x, y) \in D_L$.

AL Sampling. After training the detector, the detector can effectively distinguish whether the data in the unlabeled set belongs to a known class or an unknown class based on the predicted maximum probability value $p_i = \max_c a_i^c$. To identify the known classes, OSA-CQ arranges the p_i values of the unlabeled data in descending order, ensuring that the data at the beginning of the sequence are the known classes identified by the model. According to the theoretical analysis of active learning, data points at the decision boundary are the targets for AL [3]. OSA-CQ uses a committee consisting of two members: the classifier and the detector to make decisions based on the predictions of both the classifier and the detector. To ensure the accuracy of selecting data at the decision boundary, the detector results include two types of prediction results. The first type is the output of softmax with temperature activation. The second type is obtained by calculating the feature similarity between the unlabeled and labeled samples, The second type is obtained by calculating the feature similarity between the unlabeled and labeled samples.

$$pro_{eur} = \arg\max_c \delta(-d(f_x, f_c)) \tag{5}$$

where f_x is the output of the last fully connected layer of the detector for the unlabeled set example x, f_c denotes the representative vector of class c, which can be obtained by averaging the feature vectors labeled as c in the label set. $\delta(\cdot)$ is the nonlinear activation function that projects the similarity measured by the distance between $[0, 1]$, and $d(\cdot)$ is the distance calculation algorithm. After obtaining the prediction result pro_c of the classifier, the prediction result pro_d of the detector and the prediction result pro_{eur} of the distance similarity, the AL query strategy requests labeling based on the inconsistency of comparing the three prediction results. In other words, OSA-CQ executes query marking by contrasting the classifier and detector prediction results through a model-based confidence sequence.

$$S_{query} = \{x_i^u | i < budget, (x_i^u, p_i^u) \in (D^u, \mathcal{P}), flag = 1\} \tag{6}$$

where

$$flag = \{1 | pro_{eur} \neq pro_d \text{ or } pro_{eur} \neq pro_c \text{ or } pro_c \neq pro_d\} \tag{7}$$

After the oracle labeling, the known class data S_{query}^{kno} are added to the labeled set D_i^l, forming D_{i+1}^l. The unknown class data S_{query}^{unkno} are assigned a pseudo label of class C and added to the invalid set D_i^{inval}, forming D_{i+1}^{inval}. The approach can be summarized in Algorithm 1.

Algorithm 1. The OSA-CQ algorithm

Input: labeled set:D_0^l;unlabeled set:D_0^u;budget:b;classes:C;Detector:f_{θ_d};Classifierf_{θ_c}
Output: $D_{round}^l, D_{round}^u, \theta_c$
1: **for** i in max_rounds **do**
2:     $S_{query} = \emptyset$
3:     Training f_{θ_d} by minimizing L_d in Eq.2 by D_i^l and D_i^{inval}
4:     Training f_{θ_c} by minimizing L_c in Eq.4 by D_i^l
5:     $\mathcal{P} = sort(pro_d)$ where $pro_d = max(f_{\theta_d}(D_i^u))$ and $pro_d \neq C$
6:     **for** $x^u \in D_i^u$ **do**
7:         Calculate pro_{eur} in Eq.5
8:         $pro_d = f_{\theta_d}(x^u)$
9:         $pro_c = f_{\theta_c}(x^u)$
10:         **if** $pro_{eur} \neq pro_d$ or $pro_{eur} \neq pro_c$ or $pro_c \neq pro_d$ **then**
11:           $S_{query} \cup \{x_i^u | i < budget, (x_i^u, p_i^u) \in (D^u, \mathcal{P})\}$
12:         **end if**
13:     **end for**
14:     $D_{i+1}^l = D_i^l \cup S_{query}^{kno}$, $D_{i+1}^{inval} = D_i^{inval} \cup S_{query}^{unkno}$ and set $label_{inval} = C$
15: **end for**
16: **return** $D^l = \{X^l, Y^l\}, D^u = \{X^u\}$

---

4 Experiments

4.1 Baselines

We compare OSA-CQ with the baselines of AL:

- **Random.** The simplest AL baseline, labeled with k instances randomly selected from the unlabeled set in each AL cycle.
- **Confidence.** The degree of confirmation of the data is measured by the maximum output of the last fully connected layer of the model. We use the two measures of maximum confidence and minimum confidence in the experiment.

- **Entropy** [4]. Select the top k examples with the highest information entropy from the unlabeled set for labeling in each AL cycle.
- **Coreset** [8]. Select a batch of examples representing the full set from the unlabeled set for labeling in each AL cycle.
- **BADGE** [26]. A hybrid query method that samples different and high amplitude point sets in the phantom gradient space by diverse gradient embedding, uncertainty and diversity are fully considered.
- **LfOSA** [20]. Using a Gaussian mixture model to model the distribution of maximum activation values for each instance, to select dynamically the instance with the highest probability from the unlabeled set.

4.2 Experiment Settings

Dataset. We use two datasets, CIFAR10 [1], CIFAR100 [1] and Tiny-ImageNet [2]. For the CIFAR10 dataset, randomly sample 1% of the known class data to initialize the labeled set, while the remaining training set data are used as the unlabeled set for active sampling. Similarly, for the CIFAR100 and the Tiny-ImageNet dataset, randomly sample 8%. To effectively evaluate OSA-CQ, we set the adaptation rate to 20%, 30%, and 40% in all experiments.

$$R = \frac{|X_{kno}|}{|X_{kno}| + |X_{unkno}|} \tag{8}$$

where R is the adaptation rate, which indicates the proportion of known classes to all classes, and $|X_{kno}|$ is the number of known classes, $|X_{unkno}|$ is the number of unknown classes. For example, if the adaptation rate is 20%, in CIFAR10, CIFAR100 and Tiny-ImageNet, the first 2, 20, 40 classes are considered as known classes, and the last 8, 80, 160 classes are considered as unknown classes.

Implementation Details. For the training of all experimental classifiers and detectors, use ResNet18 [27] as the training model for all of them. We perform three randomized experiments (seed = 1, 2, 3) for all AL methods using adaptation rates of 20%, 30%, and 40%. For the task classifier, only labeled known class data is used for training, while the auxiliary detector is trained with unknown class data given pseudo-labels. In each AL cycle, we learn two models Θ_d and Θ_c, the model parameters are optimized using stochastic gradient descent (SGD), where epoch 100, batch size 128, initial learning rate 0.01, and the learning rate is reduced to 0.005 after 20 epochs, momentum and weight decay are 0.9 and 0.0005. T in Eq. 3 is set to 0.5, and d in Eq. 5 uses the euclidean distance.

Evaluation Criteria. We compare OSA-CQ with other AL methods in recall (Eq. 1), and classification accuracy by averaging the results using three randomized experiments.

4.3 Performance Evaluation

We evaluate the performance of OSA-CQ by plotting the metric growth curve. The average results of recall and classification are shown in Fig. 3 and Fig. 4.

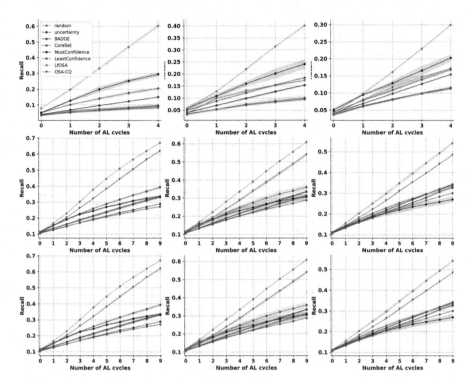

Fig. 3. Comparison of sampling recall, adaptation rates of 20% (first column), 30% (second column) and 40% (third column), CIFAR10 (first row), CIFAR100 (second row) and Tiny-ImageNet (third row).

From the above results, it can be seen that whatever the dataset or adaptation rate, OSA-CQ consistently achieves high accuracy and recall rates in all scenarios, and can effectively complete the selection of known classes of data. However, LfOSA [20] seems to have an advantage over our approach both in accuracy or recall and in the accuracy of task recognition. High recall values are crucial as they directly impact the number of labeled samples for the subsequent AL cycles, which in turn influence the model's accuracy. In other words, the advantage of LfOSA is that it can select known classes more efficiently but does not guarantee the amount of information in samples, while OSA-CQ slightly increases the cost of training time but greatly reduces the cost of labeling the data. The analysis of the amount of information is described in detail in Sect. 4.4.

As shown in Fig. 3 and Fig. 4, we have the following observations. 1) Unlike most existing methods, OSA-CQ demonstrates robust performance even when the adaptation rate is low, without a significant deterioration in results. Even in scenarios where the adaptation rate is as low as 20%, OSA-CQ maintains a high recall rate. Specifically, in the CIFAR100 scenario with a 20% adaptation rate, the recall rate is 29% more than random sampling to sample known classes of data. OSA-CQ also manages to be 0.5% more accurate in terms of model

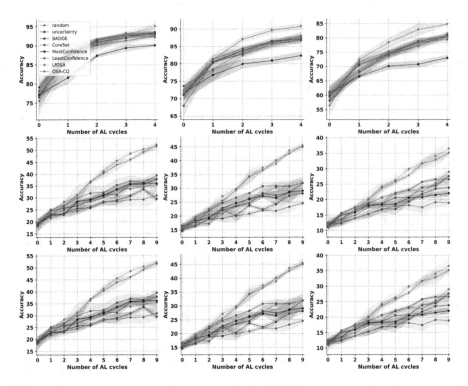

Fig. 4. Classification accuracy comparison. Adaptation rates 20% (first column), 30% (second column) and 40% (third column), CIFAR10 (first row), CIFAR100 (second row) and Tiny-ImageNet (third row).

accuracy compared to LfOSA with a larger number of samples. This shows that OSA-CQ is designed to be effective in finding informative samples with greater accuracy. 2) Regardless of which adaptation rate in which dataset, previous AL approaches do not differ much from the results of random sampling. This suggests that existing AL approaches do not effectively handle AL in an open-set. The primary reason is that the majority of data sampled by previous AL methods consists of unknown classes that are irrelevant to the task. 3) As the adaptation rate decreases the advantage of OSA-CQ is better reflected, the difference in accuracy between CIFAR100 on 20% adaptation rate and random sampling is 16.58%, while it is only 7.56% at 40% adaptation rate. OSA-CQ does not blindly pursue a large amount of training data to maximize model performance but uses a small amount of informative data to achieve the model performance requirements.

4.4   Information Gain Analysis

The information gain analysis was performed on a CIFAR10 with an adaptation rate of 40%, and the results are shown in Table 1. We use a comparison

Table 1. Labeled training set size and model accuracy at different stages of AL for LfOSA and OSA-CQ with a CIFAR10 fit rate of 40%.

Method	Ratio		Number of AL cycles					
			0	1	2	3	4	5
LfOSA	40%	num	200	740	1845	3165	4585	6002
		acc	58.88	66.13	72.10	78.08	80.95	
OSA-CQ		num	200	750	1596	2619	**2851**	3487
		acc	53.28	73.05	78.20	84.88	**84.48**	

Table 2. Labeled training set size and model accuracy at different stages of AL for LfOSA and OSA-CQ with a Tiny-ImageNet fit rate of 30%.

Method	Ratio		Number of AL cycles										
			0	1	2	3	4	5	6	7	8	9	10
LfOSA	30%	num	2400	3077	4161	5313	6544	7790	9054	10268	11468	12557	13691
		acc	21.73	21.36	29.23	32.26	35.23	39.70	41.23	43.39	45.0	46.36	
OSA-CQ		num	800	1073	1516	2030	2624	3208	3821	4446	5046	**5656**	6197
		acc	21.74	24.10	27.20	31.97	35.17	39.53	41.00	44.60	45.23	**48.73**	

of the amount of training data obtained with similar model accuracy over the same AL period as a measure of the amount of information. When the model accuracy is similar, the less the number of training data, the more information contained in the training data. As shown in Table 1, apart from the initial model, the subsequent AL cycles of OSA-CQ demonstrate superior model testing performance with less amount of training data. During the fourth cycle, OSA-CQ utilized only 62% of the data volume employed by LfOSA, yet achieved a model accuracy of 84.48%, surpassing LfOSA by 3.53% points. This indicates that the OSA-CQ training data provides higher information to the model compared to LfOSA's data, resulting in enhanced data quality and cost-effectiveness in labeling. The same scenario is also evident in the more complex TinyImageNet task, as illustrated in Table 2.

The observed phenomenon may be attributed to the following reasons, LfOSA solely relies on the confidence output of the fitted Gaussian mixture model (GMM) as its decision criterion, addressing only the open-set identification problem without fully considering the issue of data quality in active learning. OSA-CQ effectively addresses both problems and seamlessly integrates them.

5 Conclusion

In this paper, we propose an AL framework based on a contrastive query called OSA-CQ to solve the OSA problem. It consists of a classifier with a detector that provides a confidence ranking sequences to assist the committee in executing the decision. Unlike a large number of existing AL methods, it can both find known classes of data efficiently and also ensure the informativeness of these data. Experimental results on three datasets show that OSA-CQ achieves higher performance with lower annotation costs. In future research, we aim to extend this framework to address open-set AL challenges in various computer vision tasks.

References

1. Krizhevsky, A., Hinton, G., et al.: Learning multiple layers of features from tiny images (2009)
2. Yao, L., Miller, J.: Tiny ImageNet classification with convolutional neural networks (2015)
3. Settles, B.: Active learning literature survey (2009)
4. Wang, D., Shang, Y.: A new active labeling method for deep learning. In: 2014 International Joint Conference on Neural Networks (IJCNN), pp. 112–119. IEEE, Beijing, China (2014). https://doi.org/10.1109/IJCNN.2014.6889457
5. Roth, D., Small, K.: Margin-based active learning for structured output spaces. In: Fürnkranz, J., Scheffer, T., Spiliopoulou, M. (eds.) ECML 2006. LNCS (LNAI), vol. 4212, pp. 413–424. Springer, Heidelberg (2006). https://doi.org/10.1007/11871842_40
6. Gal, Y., Islam, R., Ghahramani, Z.: Deep Bayesian active learning with image data. In: International Conference on Machine Learning, pp. 1183–1192. PMLR (2017)
7. Wan, F., Yuan, T., Fu, M., Ji, X., Huang, Q., Ye, Q.: Nearest neighbor classifier embedded network for active learning. In: AAAI, vol. 35, pp. 10041–10048 (2021). https://doi.org/10.1609/aaai.v35i11.17205
8. Sener, O., Savarese, S.: Active learning for convolutional neural networks: a core-set approach (2018). http://arxiv.org/abs/1708.00489
9. Parvaneh, A., Abbasnejad, E., Teney, D., Haffari, R., Van Den Hengel, A., Shi, J.Q.: Active learning by feature mixing. In: 2022 IEEE/CVF Conference on Computer Vision and Pattern Recognition (CVPR), pp. 12227–12236. IEEE, New Orleans, LA, USA (2022). https://doi.org/10.1109/CVPR52688.2022.01192
10. Yoo, D., Kweon, I.S.: Learning loss for active learning. In: 2019 IEEE/CVF Conference on Computer Vision and Pattern Recognition (CVPR), pp. 93–102. IEEE, Long Beach, CA, USA (2019). https://doi.org/10.1109/CVPR.2019.00018
11. Vodrahalli, K., Li, K., Malik, J.: Are all training examples created equal? An empirical study. http://arxiv.org/abs/1811.12569 (2018)
12. Zhu, J.-J., Bento, J.: Generative adversarial active learning. http://arxiv.org/abs/1702.07956 (2017)
13. Sinha, S., Ebrahimi, S., Darrell, T.: Variational adversarial active learning. In: 2019 IEEE/CVF International Conference on Computer Vision (ICCV), pp. 5971–5980. IEEE, Seoul, Korea (South) (2019). https://doi.org/10.1109/ICCV.2019.00607

14. Yuan, T., et al.: Multiple instance active learning for object detection. In: 2021 IEEE/CVF Conference on Computer Vision and Pattern Recognition (CVPR), pp. 5326–5335. IEEE, Nashville, TN, USA (2021). https://doi.org/10.1109/CVPR46437.2021.00529

15. Kipf, T.N., Welling, M.: Semi-supervised classification with graph convolutional networks. http://arxiv.org/abs/1609.02907 (2017)

16. Caramalau, R., Bhattarai, B., Kim, T.-K.: Sequential graph convolutional network for active learning. In: 2021 IEEE/CVF Conference on Computer Vision and Pattern Recognition (CVPR), pp. 9578–9587. IEEE, Nashville, TN, USA (2021). https://doi.org/10.1109/CVPR46437.2021.00946

17. Mnih, V., et al.: Playing Atari with deep reinforcement learning. http://arxiv.org/abs/1312.5602, (2013)

18. Haussmann, M., Hamprecht, F.A., Kandemir, M.: Deep active learning with adaptive acquisition. http://arxiv.org/abs/1906.11471 (2019)

19. Du, P., Zhao, S., Chen, H., Chai, S., Chen, H., Li, C.: Contrastive coding for active learning under class distribution mismatch. In: 2021 IEEE/CVF International Conference on Computer Vision (ICCV), pp. 8907–8916. IEEE, Montreal, QC, Canada (2021). https://doi.org/10.1109/ICCV48922.2021.00880

20. Ning, K.-P., Zhao, X., Li, Y., Huang, S.-J.: Active learning for open-set annotation. In: 2022 IEEE/CVF Conference on Computer Vision and Pattern Recognition (CVPR), pp. 41–49. IEEE, New Orleans, LA, USA (2022). https://doi.org/10.1109/CVPR52688.2022.00014

21. Coleman, C., et al.: Similarity search for efficient active learning and search of rare concepts. In: AAAI, vol. 36, pp. 6402–6410 (2022). https://doi.org/10.1609/aaai.v36i6.20591

22. Geng, C., Huang, S.-J., Chen, S.: Recent advances in open set recognition: a survey. IEEE Trans. Pattern Anal. Mach. Intell. **43**, 3614–3631 (2021). https://doi.org/10.1109/TPAMI.2020.2981604

23. Bendale, A., Boult, T.E.: Towards open set deep networks. In: 2016 IEEE Conference on Computer Vision and Pattern Recognition (CVPR), pp. 1563–1572. IEEE, Las Vegas, NV, USA (2016). https://doi.org/10.1109/CVPR.2016.173

24. Dhamija, A.R., Günther, M., Boult, T.: Reducing network agnostophobia (2018)

25. Hinton, G., Vinyals, O., Dean, J.: Distilling the knowledge in a neural network. http://arxiv.org/abs/1503.02531 (2015)

26. Ash, J.T., Zhang, C., Krishnamurthy, A., Langford, J., Agarwal, A.: Deep batch active learning by diverse, uncertain gradient lower bounds. In: 8th International Conference on Learning Representations, ICLR 2020, Addis Ababa, Ethiopia, 26–30 April 2020. OpenReview.net (2020). https://openreview.net/forum?id=ryghZJBKPS

27. He, K., Zhang, X., Ren, S., Sun, J.: Deep residual learning for image recognition. In: 2016 IEEE Conference on Computer Vision and Pattern Recognition (CVPR), pp. 770–778. IEEE, Las Vegas, NV, USA (2016). https://doi.org/10.1109/CVPR.2016.90

Cross-Domain Bearing Fault Diagnosis Method Using Hierarchical Pseudo Labels

Mingtian Ping$^{(\boxtimes)}$, Dechang Pi⬤, Zhiwei Chen, and Junlong Wang

Nanjing University of Aeronautics and Astronautics, Nanjing 211106, China
3399813810@qq.com

Abstract. Data-driven bearing fault diagnosis methods have become increasingly crucial for the health management of rotating machinery equipment. However, in actual industrial scenarios, the scarcity of labeled data presents a challenge. To alleviate this problem, many transfer learning methods have been proposed. Some domain adaptation methods use models trained on source domain to generate pseudo labels for target domain data, which are further employed to refine models. Domain shift issues may cause noise in the pseudo labels, thereby compromising the stability of the model. To address this issue, we propose a Hierarchical Pseudo Label Domain Adversarial Network. In this method, we divide pseudo labels into three levels and use different training approach for diverse levels of samples. Compared with the traditional threshold filtering methods that focus on high-confidence samples, our method can effectively exploit the positive information of a great quantity of medium-confidence samples and mitigate the negative impact of mislabeling. Our proposed method achieves higher prediction accuracy compared with the-state-of-the-art domain adaptation methods in harsh environments.

Keywords: Rolling Bearings · Fault Diagnosis · Domain Adaptation · Adversarial Training · Pseudo Label Learning

1 Introduction

The way of industrial equipment health management is progressively shifting towards monitoring and prevention based on big data. The condition monitoring and fault diagnosis of bearings can significantly impact the reliability and service life of the entire machine [1]. Data-driven methods don't require high expertise and experience [2], which mainly include Bayesian networks [3], Support Vector Machine (SVM) [4], and Artificial Neural Network (ANN) [5] etc. However, the high cost of manual labeling leads to only a few typical operating conditions, that can meet the availability of a large amount of labeled training data.

In order to address the challenge of variable operating conditions in fault diagnosis tasks for bearings, transfer learning has emerged as a widely adopted solution. Lu et al. [6] minimized the Maximum Mean Discrepancy (MMD) between two domains based on a DNN network. Tzeng et al. [7] proposed a domain confusion loss to acquire domain invariant information. Zhang et al. [8] proposed a DACNN network that fine-tunes the

B. Luo et al. (Eds.): ICONIP 2023, LNCS 14452, pp. 32–43, 2024.
https://doi.org/10.1007/978-981-99-8076-5_3

parameters of the unresolved constrained adaptive layer of the target feature extractor during backpropagation. An et al. [9] adopted the Multi-Kernel Maximum Mean Discrepancy (MK- MMD) domain adaptation framework to enhance the stability and accuracy of the results.

The utilization of pseudo-labels does not solely focus on the overall migration between domains, and also effectively aligns the fine-grained class distribution across domains. Due to the domain offset between the source domain and the target domain, it is necessary to screen for high confidence pseudo labels. Unfortunately, high-quality labels often have a smaller scale, making it difficult to bring sufficient and effective updates to the model. Moreover, even the high confidence samples can hardly avoid mislabeling, thereby negative impact on the model.

To solve the aforementioned issues, we propose a novel approach called Hierarchical Pseudo Label Domain Adversarial Network (HPLDAN). Inspired by [10], our model combines domain discriminator and classifier to maintain the stability of classification accuracy during domain-level adversarial confusion training. The mean teacher model is then used to generate pseudo labels. The proposed Hierarchical Pseudo Labels method is used to divide target domain samples into three levels according to pseudo labels, i.e., accepted samples that can be used directly, pending samples that require further processing before being used, and rejected samples that are discarded. And different training methods are used for target domain samples with different hierarchical pseudo labels to achieve class level alignment.

2   Related Work

In cross-domain fault diagnosis, domain adaptation methods have received extensive attention. Wang et al. [11] used Correlation Alignment (CORAL) with continuous denoising self-encoder to learn domain invariant features under different operating conditions. The Deep Adaptive Network (DAN) proposed by Long et al. [12] employed MK-MMD to align distributions and extract domain invariant features. Ganin et al. [13] proposed Domain-Adversarial Neural Network (DANN), which allows feature extractors and domain discriminators to be trained adversarially to obtain domain-independent information by adding a gradient inversion layer between them.

While these migration models only consider the overall migration between domains, and the impact of the distance between samples under the same class of failure modes on the classification effect is neglected. To address these problems, Yang et al. [14] proposed an optimal migration embedding joint distribution similarity measure that fits the conditional distribution of samples in the target domain. Li et al. [15] used a representation clustering scheme to maximize intra-class similarity, reduce inter-class similarity, and increase classification loss for more discriminative features. Zhang et al. [10] proposed a domain-Symmetric networks (SymNets), which concurrently acts as a domain discriminator by connecting the source and target domain classifiers in parallel.

Furthermore, methods based on pseudo-label are commonly used for aligned fine-grained class distributions. Saito et al. [16] labeled the target domain data based on predictive consistency and confidence and used these samples to train another task classifier. However, the authenticity of pseudo-label is questionable and can negatively impact performance. Zhang et al. [17] responded to the category imbalance problem by proposing a Curriculum Pseudo Labeling method that dynamically adjusts the threshold value of each category. Zhang et al. [18] weighted the target samples according to the degree of confusion between domains. Zhu et al. [19] proposed a method to select high quality pseudo labels using adaptive thresholding with two decision strategies.

3 Proposed Method

3.1 Problem Formulation

In this work, we study the transfer learning task of bearing fault type diagnosis under different working conditions. We have a source domain $D^s = \left\{ x_i^s, y_i^s \right\}_{i=1}^{n_s}$ which has n_s labeled samples and corresponding labels, a target domain $D^t = \{ x_j^t \}_{j=1}^{n_t}$ with n_t unlabeled samples. The samples from different domains have the same size while their distributions differ. Formally, $x_i^s \in \mathbb{R}^{K \times M}$ and $x_j^t \in \mathbb{R}^{K \times M}$ show that each sample has K time steps and M channels. Meanwhile, $y_i^s \in \mathbb{R}^C$ represents there are C fault categories. Our task is to come up with a method to transfer the model trained by the source domain samples and labels to the target domain while still ensuring high accuracy.

3.2 Network Architecture

Our proposed network model framework is clearly illustrated in Fig. 1. Above is the main part of the model, which consists of four modules, namely the main feature extractor G_m, the target domain data diverter H, the target domain biased feature extractor G_{tb}, and the domain discriminative classifier C_{st}. The structure of each module below is the same as that above. We employ the mean teacher method to set the parameters of each module below as Exponential Moving Average (EMA) of the corresponding module above, in order to provide reliable pseudo labels.

The G_m accepts data input from D_s and D_t, and extracts the domain-independent feature information. The function of H is to label the target domain samples with three levels of pseudo labels based on the classification results. Accepted samples will be fed directly into C_{st}. Pending samples will be fed into G_{tb} for further feature extraction. Rejected samples will no longer be used in the follow-up process. The purpose of the G_{tb} is to deeply mine the target domain unique features contained in those pending samples. The C_{st} receives the input from G_m or G_{tb} and then classifies these samples. The C_{st} has twice the number of output nodes as the number of fault types. The outputs of the first C nodes represent the probabilities of a sample belonging to C types in the source domain, the output of the last C nodes corresponds to each type of the target domain. So the sum of the outputs of the C nodes in the front and back respectively represents the probability that the sample belongs to the source and target domains. Thus we can use C_{st} as a domain discriminator. Meanwhile, adding the outputs of two nodes with a sequence number difference of C indicates the probability that the sample belongs to this fault type. C_{st} then can play the role of a classifier. By combining the domain discriminator and classifier together, we can maintain the classification accuracy stable while performing adversarial training with domain level confusion.

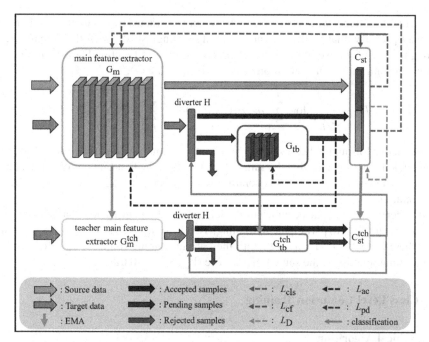

Fig. 1. Architecture of Hierarchical Pseudo Label Domain Adversarial Network.

3.3 Domain Level Confusion Training

In order to obtain effective Pseudo Labels, we need to first conduct domain level confusion training. First, we employ the source domain data to pretrain G_m and C_{st}. The classification loss function L_{cls} is used in the parameter update of both. Its definition is shown in Formula (1).

$$L_{cls} = -\frac{1}{n_s} \sum_{i=1}^{n_s} \log(p_{y_i^s}(G_m(x_i^s)) + p_{y_i^s+C}(G_m(x_i^s))) \tag{1}$$

Among them, $p_{y_i^s}$ represents the probability that the sample x_i^s belongs to the category y_i^s and belongs to the source domain, while $p_{y_i^s+C}$ represents the probability that it belongs to y_i^s and belongs to the target domain. Adding the two indicates that we only focus on fault classification and temporarily don't consider domain related information.

Next, we will conduct domain adversarial learning. Here, the target domain data is output from G_m and directly sent to C_{st} without being hierarchized. In order to enable C_{st} to play the role of domain discriminator, we use the loss function L_D to train C_{st}, whose definition is shown in Formula (2).

$$L_D = -\frac{1}{n_s} \sum_{i=1}^{n_s} \log(\sum_{c=1}^{C} p_c(G_m(x_i^s))) - \frac{1}{n_t} \sum_{j=1}^{n_t} \log(\sum_{c=1}^{C} p_{c+C}(G_m(x_j^t))) \tag{2}$$

The above equation indicates that here we do not consider specific fault type information, but only focus on domain discrimination.

On the other hand, we need to enable the features extracted by G_m to possess domain invariant property, that is, they cannot be easily distinguished which the domain they belong to. Therefore, we utilize the domain confusion loss function L_{cf} to train G_m so that it can confront C_{st}. Its definition is shown in Formula (3).

$$L_{cf} = \frac{1}{2}L_D - \frac{1}{2}\sum_{i=1}^{n_s}\log(\sum_{c=1}^{C}p_{c+C}(G_m(x_i^s))) - \frac{1}{2}\sum_{j=1}^{n_t}\log(\sum_{c=1}^{C}p_c(G_m(x_j^t))) \quad (3)$$

The above formula indicates that regardless of whether the input is from the source or target domains, we hope that the sum of the outputs of the first C nodes of C_{st} remains similar to the last C nodes. This is contrary to the purpose of L_D, and the two form a confrontation.

In order to maintain the accuracy of model classification during adversarial learning, we will also use L_{cls} to participate in training. That is to say, we use $(L_{cls}+\lambda L_{cf})$ to train G_m, where $\lambda=(e^{n_{ep}/N_{ep}}-1)$, n_{ep} refers to the current training rounds, N_{ep} refers to the total training rounds. At the same time, we use $(L_{cls} + L_D)$ to train C_{st}.

3.4   Class Level Confusion Training

Hierarchical Algorithm
After domain adversarial confusion training to a certain extent, we can use the mean teacher model to generate pseudo labels for class level domain alignment. Next, we need to hierarchize these pseudo labels. The hierarchical algorithm for pseudo labels of target domain sample x_j^t is shown in Formula (4).

$$H(x_j^t) = \begin{cases} \text{accepted,} & P_{C_j}(x_j^t) \geq T(C_j) \\ \text{pending,} & \tau < P_{C_j}(x_j^t) < T(C_j) \\ \text{rejected,} & P_{C_j}(x_j^t) \leq \tau \end{cases} \quad (4)$$

$$C_j = \arg\max_{c=1}^{C} P_c(x_j^t) = \arg\max_{c=1}^{C}(p_c^{tch}(G^{tch}(x_j^t)) + p_{c+C}^{tch}(G^{tch}(x_j^t))) \quad (5)$$

The $P_{C_j}(x_j^t)$ represents the probability that x_j^t belongs to category C_j in the output of the C_{st}^{tch}. And the C_j refers to the fault type with the highest probability in the output of C_{st}^{tch}. The $G^{tch}(\cdot)$ refers to the output of the G_m^{tch} or the output of the G_{tb}^{tch}.

In Formula (4), τ and $T(\cdot)$ are thresholds used to divide three levels of pseudo labels. τ is a fixed value. The calculation formula for $T(\cdot)$ is shown in Formula (6).

$$T(c) = \left(1 - 1/(e^{3*N_C(c)/\max_c N_C+1} + 1)\right)^{n_{ep}/N_{ep}+1} \quad (6)$$

$$N_C(c) = \sum_{j=1}^{n_t}(C_j = c) \quad (7)$$

In Formula (6), $N_C(c)$ represents the number of samples with the highest probability of belonging to category c among all target domain data.

In each round of training, we have to update H based on the output of the previous round of C_{st}^{tch} to perform the hierarchical operation for this round. Then we directly discard the rejected samples. For the accepted samples and the pending samples, we will introduce the details of their training process below.

Accepted Pseudo Label Training Method
For the target domain samples whose pseudo labels identified as accepted, we use them for updating G_m for class level domain adaptation. Although we designed $T(\cdot)$ to filter samples with low confidence, false pseudo labels may still exist. If applying the traditional cross entropy loss function, false labels may affect the classification accuracy. Therefore, we decided to use the inner product similarity to design the loss function, whose definition is shown in Formula (8).

$$L_{ac} = \frac{1}{n_{ac}} \sum_{k=1}^{n_{ac}} \omega_k L_{ac}(x_{j_k}^t) = \frac{1}{n_{ac}} \sum_{k=1}^{n_{ac}} \omega_k \left(-\frac{1}{D_k+1} \sum_{d=1}^{D-1} (C_{j_k} = C_{j_k^d}) \log(G_m(x_{j_k}^t) \cdot G_m(x_{j_k^d}^t)) \right) \quad (8)$$

$$\omega_k = -\frac{L_{ac}(x_{j_k}^t)}{L_{ac}(x_{j_k}^t) - \mu \left| p_{C_{j_k}}^{tch}(G_m^{tch}(x_{j_k}^t)) - p_{C_{j_k}+c}^{tch}(G_m^{tch}(x_{j_k}^t)) \right|} \quad (9)$$

In Formula (8), n_{ac} refers to the number of samples with accepted pseudo labels, and j_k represents the sequence number of the k-th accepted sample in all target domain samples. In each round of training, D samples are input for each batch. Here, we use j_k^d to represent the subscripts of other samples from the same batch as $x_{j_k}^t$. In these samples, assuming that there are D_k samples with pseudo labels corresponding to the same fault type as $x_{j_k}^t$. In Formula (9), μ is a fixed parameter. This formula indicates that we believe that samples with smaller L_{ac}, which are more similar to samples of the same category, should have higher weights. At the same time, samples that are difficult to distinguish which domains they belong to should also be given more attention.

In traditional similarity measurement loss, the similarity with negative samples is used as the denominator to maximize the distance from them. However, this operation may interfere with our resistance to incorrect pseudo labels. Because we acknowledge that samples of the same category are inevitably more similar to each other. Therefore, by using ω_k, we can weaken the impact of samples that are less similar to the same class, which are more likely to be mislabeled. However, minimizing the similarity with samples of different classes will further make those mislabeled samples more like correctly labeled samples, thereby weakening the differential treatment effect of ω_k.

The training diagram is shown in Fig. 2 (a).

Pending Pseudo Label Training Method
For the target domain samples with pending pseudo labels, we need to input them into G_{tb} for further deep feature mining. Although we have extracted features twice, it is still inevitable that some noise will be left in these samples. Although we cannot accurately determine which category a pending sample belongs to, we are still confident in which categories it does not belong to. Therefore, the purpose of the loss function we designed

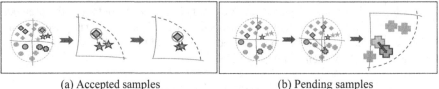

(a) Accepted samples   (b) Pending samples

Fig. 2. Training diagram of accepted and pending samples.

is to minimize the similarity with those determined negative samples.

$$L_{pd} = \frac{1}{n_{pd}} \sum_{m=1}^{n_{pd}} L_{pd}(x_{j_m}^t) = \frac{1}{n_{pd}} \sum_{m=1}^{n_{pd}} \left(\frac{1}{D_m + 1} \sum_{d=1}^{D-1} U(x_{j_m}^t, x_{j_m^d}^t) \log(G_{tb}(x_{j_m}^t) \cdot G_{tb}(x_{j_m^d}^t)) \right)$$

(10)

In Formula (10), n_{pd} refers to the number of samples with pending pseudo labels, and j_m represents the sequence number of the m-th pending sample in all target domain samples. Here, we use j_m^d to represent the subscripts of other samples from the same batch as $x_{j_m}^t$. In this batch of samples, we use $U(\cdot)$ to filter out those samples that can be certain that they are not in the same fault type as $x_{j_m}^t$, and assume that there are a total of D_m samples. The definition of $U(\cdot)$ is shown in formula Table 1.

$$U(x_a^t, x_b^t) = \begin{cases} 1, & (P_{C_a}(x_a^t) - P_{C_a}(x_b^t)) + (P_{C_b}(x_b^t) - P_{C_b}(x_a^t)) > 2\tau \\ 0, & \text{otherwise} \end{cases}$$

(11)

τ is the threshold used in Formula (4). The meaning of $P_{C_a}(x_a^t)$ refers to Formula (5).

By using L_{pd} to update the parameters of G_{tb}, we can effectively utilize these samples that are filtered out in traditional methods and further obtain unique feature information of the target domain. Compared to accepted samples closer to the center of each category's feature space, pending samples are generally distributed in more peripheral areas. Therefore, incorporating them into training will help improve the classification accuracy of the model at the boundary. The training diagram is shown in Fig. 2 (b).

4 Experiments

4.1 Datasets Description

Case Western Reserve University (CWRU) Dataset [21]. The test bearings in the CWRU dataset are collected at a sampling frequency of 48k. We selected three operating conditions, namely 1 hp 1772 r/min, 2 hp 1750 r/min, and 3 hp 1730 r/min, forming three domains: A, B, and C. The bearing will ultimately exhibit four health states, namely normal state, inner ring failure, ball failure, and outer ring failure, with each failure having three different sizes of 0.007, 0.014, and 0.021 feet, forming 10 different categories. We use sliding windows for data augmentation on a time scale. The window width is 2048 and the shift step size is 512. Then perform wavelet transform on these data, with the wavelet type being 'db4'. All datasets use the same processing method.

Paderborn Dataset [22]. The KAT data center generated a dataset with a sampling rate of 64 kHz. There are three conditions with artificial damage, forming domains D, E, and F, and three conditions with actual damage, forming domains G, H, and I. Domain D and G, domain E and H, domain F and I, each pair has the same load torque and radial force respectively. All six domains have a fixed speed of 1500 r/min, and contain three categories, namely health, inner ring failure, and outer ring failure.

XJTU-SY Datasets [23]. In the experiment, the sampling frequency was set to 25.6kHz, the sampling interval was 1min, and each sampling time was 1.28s. Three different operating conditions were set, with rotational speed and radial force of 2100 rpm and 12 kN, 2250 rpm and 11 kN, 2400 rpm and 10 kN, forming domains J, K, and L. The fault elements in each domain include outer race, inner race, and cage.

4.2 Comparison Methods

We compared our method with five methods, including Source Only (SO), DANN [13], SymNets [10], SLARDA [20], and DTL-IPLL [19].

SO: Only use source domain data to train the model without domain adaptation.

DANN: By adding gradient inversion layers to the domain discriminator and feature extractor, they undergo adversarial training to obtain domain independent information.

SymNets: By paralleling the source domain classifier and the target domain classifier to simultaneously act as a domain discriminator, it can maintain a steady improvement in classification accuracy during adversarial training.

SLARDA: By employing an autoregressive technique, the temporal dependence of source and target features is involved in domain adaptation. Then use a teacher model to align the class distribution in the target domain through confidence pseudo labels.

DTL-IPLL: Measure the marginal probability distribution discrepancy by MK-MMD, calculate conditional probability distribution discrepancy with pseudo label, and filter out pseudo labels by an adaptive threshold and a making-decision-twice strategy.

4.3 Implementation Details

Our model is developed with PyTorch 1.12.0 and runs on NVIDIA RTX 3060 GPU. We build G_{in} with ResNet18, build C_{st} with two fully-connected layers and a Softmax layer, build G_{tb} through the use of a four-layer 1-D convolutional block. The threshold τ is set to 0.45. The parameter μ in ω_k is set to 0.3. In the mean teacher model, the conditional alignment weight is set to 0.004 and the update momentum is set to 0.996.

4.4 Results and Analysis

In order to simulate the imbalanced classification of various fault categories in industrial scenarios, we conduct different proportions of sampling on the data of each working condition according to the fault type. Then, six cross domain diagnostic tasks were designed for the CWRU dataset, the Paderborn dataset for actual damage conditions, and the XJTU-SY dataset, respectively. In addition, the Paderborn dataset has added diagnostic tasks for migrating three artificial damage conditions to actual damage conditions with the same load torque and radial force. The percentage accuracy of all experiments is shown in the Tables 1, 2, and 3, with the best results highlighted in bold and the second best results highlighted in underline.

Table 1. Results of cross domain diagnostic tasks on CWRU dataset.

Task	SO	DANN	SymNets	SLARDA	DTL-IPLL	HPLDAN
A → B	85.33	94.11	98.07	98.98	99.09	**99.13**
A → C	78.44	88.88	93.01	93.08	93.15	**93.48**
B → A	88.97	96.51	98.41	99.16	**99.63**	99.59
B → C	90.05	97.79	99.40	**99.52**	99.34	**99.52**
C → A	66.54	74.28	78.80	80.09	81.80	**85.58**
C → B	81.83	93.94	96.43	96.68	97.37	**97.65**
Avg	81.86	90.92	94.02	94.59	95.06	**95.83**

Table 2. Results of cross domain diagnostic tasks on Paderborn dataset.

Task	SO	DANN	SymNets	SLARDA	DTL-IPLL	HPLDAN
D → G	76.59	88.10	93.43	93.74	94.25	**96.11**
E → H	78.91	88.84	95.22	94.63	95.18	**97.56**
F → I	83.56	95.38	98.37	97.80	98.69	**99.03**
G → H	76.30	87.66	93.75	93.64	95.54	**96.67**
G → I	82.31	93.24	95.73	97.24	96.57	**98.46**
H → G	70.59	83.23	85.78	85.72	87.21	**89.30**
H → I	85.74	94.65	99.08	98.54	99.04	**99.33**
I → G	84.25	96.83	98.40	**99.55**	99.07	99.48
I → H	84.12	96.88	99.30	98.87	**99.67**	99.62
Avg	80.26	91.65	95.45	95.53	96.14	**97.28**

It can be seen that our HPLDAN method achieved the best results in most bearing cross domain diagnostic tasks, while the rest achieved suboptimal results. It is worth noting that in relatively difficult tasks, such as Tasks C → A, H → G, and G → L, the superiority of our method is better demonstrated. Because the proposed HPLDAN method can obtain valuable information that is difficult to extract from low-quality samples through the training process of pending samples. In addition, the design of Loss function for accepted samples also alleviates the negative impact of false labels of high quality to a certain extent, which enables fault types with fewer samples to maintain a higher classification accuracy. In summary, our method will demonstrate high application value in challenging and harsh working conditions.

Table 3. Results of cross domain diagnostic tasks on XJTU-SY dataset.

Task	SO	DANN	SymNets	SLARDA	DTL-IPLL	HPLDAN
J → K	80.31	89.21	97.33	97.91	98.53	**99.68**
J → L	58.85	72.35	78.69	83.12	82.41	**85.65**
K → J	79.23	86.65	94.63	95.86	96.36	**97.84**
K → L	62.20	73.69	79.96	78.74	80.85	**83.51**
L → J	64.87	74.44	79.85	80.36	81.56	**84.82**
L → K	66.31	75.23	81.21	82.27	82.47	**84.96**
Avg	68.63	78.60	85.28	86.38	87.03	**89.41**

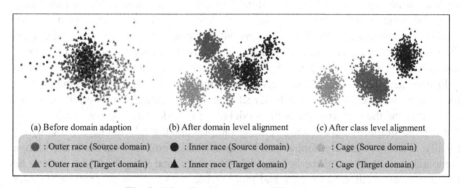

(a) Before domain adaption   (b) After domain level alignment   (c) After class level alignment

● : Outer race (Source domain)   ● : Inner race (Source domain)   : Cage (Source domain)

▲ : Outer race (Target domain)   ▲ : Inner race (Target domain)   : Cage (Target domain)

Fig. 3. Visualization process of Task J → K.

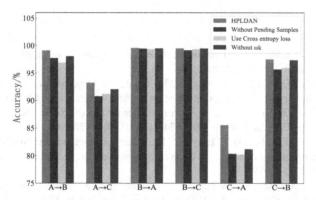

Fig. 4. Results of ablation experiment.

To clearly show the effectiveness of the proposed HPLDAN, we visualize the learned features of Task J → K. We use uniform manifold approximation and projection (UMAP) [24] to map the high-dimensional features to a lower dimension as shown in Fig. 3. It can be seen that using hierarchical pseudo labels for class level alignment can significantly improve the classification accuracy of bearing faults.

To verify the effectiveness of the various modules of our proposed HPLDAN, we conducted ablation experiments on the CWRU dataset. The relevant experimental results are shown in Fig. 4. It can be seen that the various modules designed have a positive impact on the overall performance of the model, especially when facing high difficulty cross-domain tasks.

5 Conclusions

In this research, we propose a Hierarchical Pseudo Label Domain Adversarial Network (HPLDAN). In domain level domain adaptation adversarial training, we maintain stability in classification accuracy by combining domain discriminator with classifier. Then, using the pseudo labels generated by the mean teacher model, the target domain samples are divided into three levels. We use high confidence accepted samples to further train the main feature extractor, and allocate weights based on only calculating the loss of similarity with the positive samples and the degree of domain confusion, in order to reduce the negative impact of incorrect pseudo labeling; next, the pending samples with medium confidence are input into the target domain bias feature extractor for secondary extraction, and the performance of the model at the boundary will be improved by minimizing the similarity between the pending samples and the selected negative samples; Finally, abandon rejected samples with low quality. We compared our method with five comparison methods on three bearing datasets. After analyzing the results, it can be found that our method has higher and more stable classification accuracy, and better robustness against imbalanced samples.

References

1. Tian, Z.G., Wang, W.: Special issue on machine fault diagnostics and prognostics. Chinese J. Mech. Eng. **30**(6), 1283–1284 (2017)
2. Yaguo, L.E.I., Feng, J.I.A., Detong, K.O.N.G., et al.: Opportunities and challenges of mechanical intelligent fault diagnosis based on big data. J. Mech. Eng. **54**(5), 94–104 (2018)
3. Sheppard, J.W., Kaufman, M.A.: A Bayesian approach to diagnosis and prognosis using built-in test. IEEE Trans. Instrum. Meas. **54**(3), 1003–1018 (2005)
4. Li, C.Z., Zheng, J.D., Pan, H.Y., Liu, Q.Y.: Fault diagnosis method of rolling bearings based on refined composite multiscale dispersion entropy and support vector machine. China Mech. Eng. **30**(14), 1713 (2019)
5. Wen, L., Li, X., Gao, L.: A new two-level hierarchical diagnosis network based on convolutional neural network. IEEE Trans. Instrum. Meas. **69**(2), 330–338 (2019)
6. Lu, W., Liang, B., Cheng, Y., Meng, D., Yang, J., Zhang, T.: Deep model based domain adaptation for fault diagnosis. IEEE Trans. Ind. Electron. **64**(3), 2296–2305 (2017)
7. Tzeng, E., Hoffman, J., Zhang, N., Saenko, K., Darrell, T.: Deep domain confusion: maximizing for domain invariance. arXiv:1412.3474 (2014)
8. Zhang, B., Li, W., Li, X.-L., Ng, S.-K.: Intelligent fault diagnosis under varying working conditions based on domain adaptive convolutional neural networks. IEEE Access **6**, 66367–66384 (2018)

9. An, Z.H., Li, S.M., Wang, J.R., et al.: Generalization of deep neural network for bearing fault diagnosis under different working conditions using multiple kernel method. Neurocomputing **352**, 42–53 (2019)

10. Zhang, Y., Tang, H., Jia, K., Tan, M.: Domain-symmetric networks for adversarial domain adaptation. In: 2019 IEEE/CVF Conference on Computer Vision and Pattern Recognition (CVPR), Long Beach, CA, USA, pp. 5026–5035 (2019). https://doi.org/10.1109/CVPR.2019.00517

11. Wang, X., He, H., Li, L.: A hierarchical deep domain adaptation approach for fault diagnosis of power plant thermal system. IEEE Trans. Ind. Informat. **15**(9), 5139–5148 (2019)

12. Long, M., Cao, Y., Wang, J., Jordan, M.: Learning transferable features with deep adaptation networks. In: Proceedings of Machine Learning Research, Miami, FL, USA, pp. 97–105 (2015)

13. Ganin, Y., Ustinova, E., Ajakan, H., et al.: Domain-adversarial training of neural networks. J. Mach. Learn. Res. **17**(1), 2096–2030 (2016)

14. Yang, B., Lei, Y., Xu, S., Lee, C.-G.: An optimal transportembedded similarity measure for diagnostic knowledge transferability analytics across machines. IEEE Trans. Ind. Electron. **69**(7), 7372–7382 (2022)

15. Li, X., Zhang, W., Ding, Q.: A robust intelligent fault diagnosis method for rolling element bearings based on deep distance metric learning. Neurocomputing **310**, 77–95 (2018)

16. Kuniaki, S., Yoshitaka, U., Tatsuya, H.: Asymmetric tri-training for unsupervised domain adaptation. In: Doina, P., Yee, W.T., editors, Proceedings of the 34th International Conference on Machine Learning, volume 70 of Proceedings of Machine Learning Research, pp. 2988–2997, International Convention Centre, Sydney, Australia, 06–11. PMLR (2017)

17. Zhang, B., Wang, Y., Hou, W., et al.: Flexmatch: boosting semi-supervised learning with curriculum pseudo labeling. In: Advances in Neural Information Processing Systems, vol. 34 (2021)

18. Weichen, Z., Wanli, O., Wen, L., Dong, X.: Collaborative and adversarial network for unsupervised domain adaptation. In: Proceedings of the IEEE Conference on Computer Vision and Pattern Recognition, pp. 3801–3809 (2018)

19. Zhu, W., Shi, B., Feng, Z.: A transfer learning method using high-quality pseudo labels for bearing fault diagnosis. IEEE Trans. Instrum. Measure. **72**, 1–11 (2023). https://doi.org/10.1109/TIM.2022.3223146

20. Ragab, M., Eldele, E., Chen, Z., Min, W., Kwoh, C.-K., Li, X.: Self-supervised autoregressive domain adaptation for time series data. IEEE Trans. Neural Networks Learn. Syst. **99**, 1–11 (2022). https://doi.org/10.1109/TNNLS.2022.3183252

21. Bearing Data Center, Case Western Reserve Univ., Cleveland, OH, USA (2004). https://engineering.case.edu/bearingdatacenter/download-data-file

22. Lessmeier, C., Kimotho, J.K, Zimmer, D., et al.: Condition monitoring of bearing damage in electromechanical drive systems by using motor current signals of electric motors, a benchmark data set for data-driven classification. In: Proceedings of the European Conference of the Prognostics and Health Management Society (2016)

23. Wang, B., Lei, Y., Li, N., Li, N.: A hybrid prognostics approach for estimating remaining useful life of rolling element bearings. IEEE Trans. Reliab. **69**(1), 401–412 (2020). https://doi.org/10.1109/TR.2018.2882682

24. L. McInnes, J. Healy, and J. Melville: UMAP: Uniform manifold approximation and projection for dimension reduction, arXiv:1802.03426 (2020)

Differentiable Topics Guided New Paper Recommendation

Wen Li[1], Yi Xie[2,3], Hailan Jiang[4], and Yuqing Sun[1(✉)]

[1] Shandong University, Jinan, China
sun_yuqing@sdu.edu.cn
[2] National University of Defense Technology, Hefei, China
[3] Anhui Province Key Laboratory of Cyberspace Security Situation Awareness and Evaluation, Heifei, China
[4] Shandong Polytechnic, Jinan, China

Abstract. There are a large number of scientific papers published each year. Since the progresses on scientific theories and technologies are quite different, it is challenging to recommend valuable new papers to the interested researchers. In this paper, we investigate the new paper recommendation task from the point of involved topics and use the concept of subspace to distinguish the academic contributions. We model the papers as topic distributions over subspaces through the neural topic model. The academic influences between papers are modeled as the topic propagation, which are learned by the asymmetric graph convolution on the academic network, reflecting the asymmetry of academic knowledge propagation. The experimental results on real datasets show that our model is better than the baselines on new paper recommendation. Specially, the introduced subspace concept can help find the differences between high quality papers and others, which are related to their innovations. Besides, we conduct the experiments from multiple aspects to verify the robustness of our model.

Keywords: Paper recommendation · Topic model · GCN

1 Introduction

Currently, there are a large number of academic papers published every year. It's necessary to recommend researchers the valuable and interested papers. The number of citations is often regarded as an important indicator for the quality of papers. To describe the detailed contribution of a paper, a citation type can be further classified into three categories, *Background*, *Method* and *Result*. As an example, we show the papers concerning the technology *Transformer* [6], *GPT* [5], *BERT* [4], *GPT2* [3] and *BART* [2] in Fig. 1, which are labeled by

This work was supported by the National Nature Science Foundation of China, NSFC(62376138) and the Innovative Development Joint Fund Key Projects of Shandong NSF (ZR2022LZH007).
Wen Li and Yi Xie contributed equally to this work.

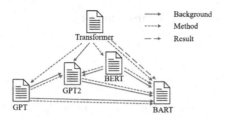

Fig. 1. An example of different citation types.

Semantic Scholar[1]. Arrows point to the citing papers, representing the direction of knowledge propagation. These points help the users more precisely find their interested topics, such as the inspiring theory, the technical methods, or the dataset and etc.

To recommend new paper, the existing methods typically leverage the academic network (AN for short) to model user interests and paper features [1,13]. However, they didn't consider the differentiable details on citations. Since the innovations in papers are various, the concept of subspace was used in this paper to describe the paper contents [21]. Besides, the citation-based recommendation methods are not applicable to new paper recommendation since it didn't have citation relationship.

To tackle the above challenges, we propose the differentiable topics based new paper recommendation model (DTNRec for short). Paper contents are classified into three subspaces according to the innovation forms as the usual way [21]: *Background, Method* and *Result*. We adopt the neural topic model (NTM for short) to get the topic distribution over subspaces as the paper embeddings, which are used to differentiate the innovation forms of paper. Considering the citations reflect the influence of cited papers and the author interests of citing paper, we adopt the asymmetric academic network to model this kind of knowledge propagation. The graph convolution network (GCN for short) operations are performed on this network to learn the user interests and paper influences, separately. For example, for the central paper p, its references are the neighbors during convolution to compute the interests for the authors of p, while its citations are used to compute its influences on the network. Then a new paper is recommended to the potentially interested users based on the paper content. Our contributions are as follows:

1. We label the paper content with subspace tags, then adopt the NTM to get the topic distribution over subspaces as paper embeddings.
2. We create the asymmetric academic network to model the academic propagation, where the directed edge points to the citing paper denoting the propagation. Based on this network and paper embeddings, we adopt the GCN operations to compute user interests and paper influences in a fine-grained way.

---

[1] https://www.semanticscholar.org/.

3. We conducted the experiments from multiple aspects to verify the effectiveness and robustness of our model.

2   Related Work

Collaborative filtering (CF for short) is a commonly used technique in recommendation systems. NeuMF [12] and BUIR [10] are both CF-based methods using user-item interaction data to get user and item representations. He et al. [11] proposed LightGCN to learn the user and item embeddings with neighborhood aggregation operation. Wang et al. [9] proposed alignment and uniformity as two properties that are important to CF-based methods, and optimized the two properties to get user and item representations. However, these methods only use interaction data of user and items, without considering other features.

The academic network consists of papers, authors, other related attributes and the relationships among them, which is important for paper recommendation task since it's rich in information. Existing works often used AN-based methods including KGCN [18], KGCN-LS [19], RippleNet [20], etc. to mine high-order information on the academic network, among which GCN is a widely used technique. However, these methods have cold-start problem and are not suitable for new paper recommendation since it lacks citation information.

Besides, paper contents are also considered to model user interests. JTIE [25] incorporated paper contents, authors and venues to learn user and paper representations. Xie et al. [26] proposed a cross-domain paper recommendation model using hierarchical LDA to learn semantic features of paper contents. Li et al. [13] proposed JMPR to jointly embed structural features from academic network and semantic features from paper contents. These methods alleviate the cold-start problem, but the diversity of paper innovations was not considered. Therefore, Xie et al. [21] proposed the subspace concept to label the paper content with *Background, Method* and *Result*. However, they didn't infer in subspace, that is they ignored the knowledge propagation among subspaces.

3   New Paper Recommendation Method

3.1   Problem Definition

Given a user set $\mathcal{U}$, a paper set $\mathcal{V}$, we aim to learn a prediction function $\mathcal{F}(u, q \mid \theta)$ that checks whether user $u \in \mathcal{U}$ has the potential interest of the new paper $q \in \mathcal{V}$, where θ denotes the parameters of function $\mathcal{F}$.

For an academic dataset, the academic network $\mathcal{G}$ is called the structural feature, where the nodes of $\mathcal{G}$ are papers, authors, and other related attributes, and the edges denote the relationships between them, including citation, etc. Each paper contains an abstract. The abstract describes the core content of a paper, which is called the semantic feature in this paper.

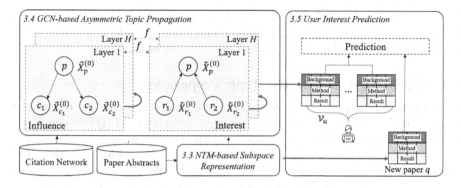

Fig. 2. Overall framework of DTNRec

3.2   Overall Framework

DTNRec include three modules, as shown in Fig. 2, i.e. the NTM-based subspace representation module, the GCN-based asymmetric topic propagation module, and the user interest prediction module. In the NTM-based subspace representation module, the paper abstract is labeled with subspace tags through the subspace tagging model. The resulting subspace text is fed into the NTM to obtain the topic distributions over subspaces as the paper content embeddings. In the GCN-based asymmetric topic propagation module, we adopt asymmetric GCN on the academic network $\mathcal{G}$ to model the asymmetric topic propagation among papers. The user interest prediction module predicts the probability on how much user u being interested in a new paper q.

3.3   NTM-Based Subspace Representation

Subspace Tagging. In order to differentiate the topics in papers, we inherit the subspace concept proposed in [21] and label the paper contents with three subspace tags, namely *Background, Method* and *Result*, respectively, denoted by the tag set $\mathcal{TS} = \{b, m, r\}$. We adopt the subspace tagging model in [22] to label the sentences of paper abstract with the subspace tags. The sentences for the same subspace represent the corresponding subspace text.

GSM-Based Paper Representation. The subspace texts are fed into the topic model to get the topic distributions over subspaces, which are regarded as the initial embeddings of paper content. The existing research results show that the topic model integrated with neural network has better performance than traditional topic model [23]. Therefore, we adopt the Gaussian Softmax distribution topic model (GSM for short) [23], which is based on variational autoencoder. Let $D \in \mathbb{N}^*$ denote the topic number. The output subspace topic distributions $\mathbf{x}_p^b \in \mathbb{R}^D, \mathbf{x}_p^m \in \mathbb{R}^D, \mathbf{x}_p^r \in \mathbb{R}^D$ for paper p are the corresponding embeddings, respectively. Paper p can be represented as matrix $X_p = \left(\mathbf{x}_p^b, \mathbf{x}_p^m, \mathbf{x}_p^r\right)^\top \in \mathbb{R}^{3 \times D}$.

Different with the existing methods directly treat the paper content as a whole to obtain paper representation [13], our method label the paper content with subspace tags, which helps to distinguish paper innovations.

3.4 GCN-Based Asymmetric Topic Propagation

Each citation reflects the influence of the cited paper and the interest of the citing paper's authors. So we model the topic propagation between papers on the academic network as the asymmetric relations, denoted by $\mathcal{G}$. The academic influences and user interests are modeled, respectively, based on the citation relationships. For example, for a paper $p \in \mathcal{V}$ on $\mathcal{G}$, its references are the neighbors for convolution to compute the interests for the authors of p, while its citations are used to compute its influences on the network.

For any paper $p \in \mathcal{V}$ on $\mathcal{G}$, there are two matrix representations, denoted by the interest matrix $\overleftarrow{X}_p^{(h)}$ and the influence matrix $\overrightarrow{X}_p^{(h)}$, respectively, where $h \in \mathrm{N}^*$ denotes the depth of GCN, that is the number of GCN iterations. $\overleftarrow{X}_p^{(h)}$ and $\overrightarrow{X}_p^{(h)}$ both are initialized by the paper matrix X_p. The GCN kernel function is f, where $W \in \mathrm{R}^{D \times D}$, $U \in \mathrm{R}^{3 \times D}$, $V \in \mathrm{R}^{3 \times D}$ are all weights of f and $b \in \mathrm{R}^{3 \times 3}$ is bias. Paper $p' \in \mathcal{V}$ cited paper p.

$$f(p, p', h) = \sigma \left(\overrightarrow{X}_p^{(h-1)} W \overleftarrow{X}_{p'}^{(h-1)\top} + U \overrightarrow{X}_p^{(h-1)\top} + V \overleftarrow{X}_{p'}^{(h-1)\top} + b \right) \quad (1)$$

To compute the influence of paper p, we choose citations of p as its neighbors. Since the number of paper neighbors may vary significantly over all papers, we uniformly sample a fixed-size set of neighbors for each paper instead of using all of them, denoted by $\mathcal{V}_p^{cit}$, to keep the computational pattern of each batch fixed and more efficient. We set $|\mathcal{V}_p^{cit}| = K \in \mathrm{N}^*$ as a hyper-parameter. Papers in $\mathcal{V}_p^{cit}$ are combined to characterize the influence of paper p, denoted by $\overrightarrow{X}_{\mathcal{V}_p^{cit}}^{(1)}$.

$$\overrightarrow{X}_{\mathcal{V}_p^{cit}}^{(1)} = \sum_{c \in \mathcal{V}_p^{cit}} f(p, c, 1) \overrightarrow{X}_c^{(0)} \quad (2)$$

Then we aggregate $\overrightarrow{X}_p^{(0)}$ and $\overrightarrow{X}_{\mathcal{V}_p^{cit}}^{(1)}$ into one matrix $\overrightarrow{X}_p^{(1)}$ as p's first-order influence matrix, which is calculated as $\overrightarrow{X}_p^{(1)} = \sigma \left(\left(\overrightarrow{X}_p^{(0)} + \overrightarrow{X}_{\mathcal{V}_p^{cit}}^{(1)} \right) W^{(1)} + b^{(1)} \right)$.

In the same way, to compute the interest for the authors of paper p, we choose a fixed-size set of references of paper p as its neighbors, denoted by $\mathcal{V}_p^{ref}$. We set $|\mathcal{V}_p^{ref}| = K$, too. Then papers in $\mathcal{V}_p^{ref}$ are combined to characterize the interest for the authors of paper p, denoted by $\overleftarrow{X}_{\mathcal{V}_p^{ref}}^{(1)}$.

$$\overleftarrow{X}_{\mathcal{V}_p^{ref}}^{(1)} = \sum_{r \in \mathcal{V}_p^{ref}} f(r, p, 1) \overleftarrow{X}_r^{(0)} \quad (3)$$

Then we aggregate $\overleftarrow{X}_p^{(0)}$ and $\overleftarrow{X}_{\mathcal{V}_p^{ref}}^{(1)}$ into one matrix $\overleftarrow{X}_p^{(1)}$ as p's first-order interest matrix, which is calculated as $\overleftarrow{X}_p^{(1)} = \sigma \left(\left(\overleftarrow{X}_p^{(0)} + \overleftarrow{X}_{\mathcal{V}_p^{ref}}^{(1)} \right) W^{(1)} + b^{(1)} \right)$.

We set the maximum depth of GCN as H. Through repeating the above process H times, we can get the H-order interest matrix $\overleftarrow{X}_p^{(H)}$ of paper p.

Given another paper q, to predict whether paper q will influence paper p or whether the author of paper p will be interested in paper q, we calculate the score $c(q, p)$. Since citation types are diverse, we adopt maximum pooling to find the largest topic association between different subspaces of paper p and paper q.

$$c(q, p) = MLP \left(maxpooling \left(X_q \overleftarrow{X}_p^{(H)\top} \right) \right) \tag{4}$$

We choose the cross entropy loss function. SP$^+$ and SP$^-$ denote positive sample set and negative sample set, which are sampled according to the rule-based sample strategy [21]. Let $\hat{c}(q, p)$ denote gold label. Any paper pair (p, q) with citation relationship is sampled as positive, labeled as $\hat{c}(q, p) = 1$. The negative samples are selected from paper pairs without citation relationship according to the sample strategy in [21], labeled as $\hat{c}(q, p) = 0$.

$$L = \sum_{c(q,p) \in \text{SP}^+ \cup \text{SP}^-} c(q, p) \log \hat{c}(q, p) + \lambda \|\theta\|_2^2 \tag{5}$$

Existing recommendation methods typically initialize paper nodes randomly when using GCN. However, our method directly initializes paper nodes with paper's semantic-rich subspace embeddings. Besides, considering the asymmetry of topic propagation among papers, we conduct asymmetric GCN on the academic network to model user interest and academic influences in a fine-grained way.

3.5   User Interest Prediction

A new paper is recommended to the potentially interested users based on the content. Given a new paper q, we calculate the probability whether user u will be interested in paper q through the function $\mathcal{F}(u, q)$, where $\mathcal{V}_u$ denotes user u's history publications.

$$\mathcal{F}(u, q) = \max \{c(q, p) \mid, p \in \mathcal{V}_u\} \tag{6}$$

Since user interests change over time, we adopt the publications within a period as the user interests at different times. We calculate the probability on how much user u being interested in the paper q according to the user interests in different periods. In this way, the user interests are more accurately modeled.

4   Experiments

In this section, we verify the effectiveness of our model on real datasets for the new paper recommendation task. We select some baselines for comparative experiments and analyze the impact of hyper-parameter settings and model structure. Finally, we analyze the paper subspace embeddings.

4.1   Experimental Settings

Datasets. We use ACM[2] and Scopus[3] datasets. ACM dataset contains 43380 conference and journal papers in computer science. Scopus dataset is a multi-disciplinary dataset, and we use the papers within the area of computer science, with a total of 18842 papers. Every paper in the datasets contains the paper abstract, authors, publication year, citation relationship, etc.

Baselines and Metrics. We compare our model with several baselines. BUIR [10], LightGCN [11], NeuMF [12] and DirectAU [9] are CF-based methods using the user and item interaction data. KGCN [18], KGCN-LS [19], RippleNet [20] are AN-based methods, which introduce the side information such as keywords besides user-item interactions. NPRec [21] jointly embed the semantic features of paper content and structural features of academic network. DTNRec is our model.

In real recommendation scenarios, users usually pay attention to the first few items recommended. So we choose the $nDCG@k$ [8] as the metric to evaluate the ranking results. For each user, we prepare k candidates which contains at least one paper that is actually cited by the user. The candidate papers are ranked according to the value calculated by the function $\mathcal{F}$ (6). $DCG@k$ is calculated as $DCG@k = \sum_{i=1}^{k} \frac{rel_i}{\log_2(i+1)}$, where rel_i is a fixed value 5 if the i-th paper is actually cited by the user, otherwise 0. $IDCG = \sum_{i=1}^{|Ref|} \frac{5}{\log_2(i+1)}$ represents the DCG value corresponding to the best rank, where $|Ref|$ denotes the number of papers actually cited by the user in candidate papers.

4.2   Results

Performance Analysis. The evaluation results are shown in Table 1. It shows our model DTNRec outperforms the baselines on the new paper recommendation task. Because we introduce the concept of subspace, the paper innovations could be well differentiated. What's more, we fuse semantic features and structural features by performing asymmetric GCN on the academic network, whose nodes are initialized by paper content embeddings over subspaces. In this way, the user interests and paper influences are modeled in a fine-grained way. The CF-based models including BUIR, LightGCN, NeuMF and DirectAU performs worst since they only use interaction data of users and items, without considering other information such as paper content. The AN-based models including KGCN, KGCN-LS and RippleNet perform better than the CF-based methods, because the academic network contains rich high-order hidden information, which is beneficial for accurately modeling user preferences. Both CF-based methods and KG-based methods consider the structural features, without considering semantic features. NPRec considers both of them, so NPRec performs better than

---

AN-based models. However, the model structure of NPRec has limitations. It treats the paper content representation in a whole way rather than in subspaces, that is it ignored the knowledge propagation among subspaces.

Table 1. New paper recommendation comparison.

nDCG@k	ACM			Scopus		
	k = 20	k = 30	k = 50	k = 20	k = 30	k = 50
BUIR	0.7734	0.7083	0.6681	0.7707	0.7156	0.6626
LightGCN	0.8266	0.7703	0.7314	0.8062	0.7639	0.7231
NeuMF	0.8234	0.7730	0.7419	0.8257	0.7808	0.7234
DirectAU	0.8357	0.7898	0.7423	0.8246	0.7819	0.7235
KGCN	0.8731	0.8592	0.8437	0.8507	0.8365	0.7592
KGCN-LS	0.9093	0.9010	0.8904	0.8660	0.8548	0.8063
RippleNet	0.9217	0.9088	0.8970	0.9040	0.8673	0.8465
NPRec	0.9736	0.9688	0.9645	0.9576	0.9349	0.9021
DTNRec	**0.9855**	**0.9844**	**0.9663**	**0.9735**	**0.9547**	**0.9329**

Impact of User Interest Calculation Method. Generally, the user interest will change over time. When we predict whether user u will be interested in a new paper q which is published after year Y, we should consider user u's interest after year Y, too. Therefore, we study the impact of using user u's interests at different times to make predictions. The experimental results are shown in Fig. 3(a). We calculate user interest in the following six ways.

- *History-max* denotes the user interest is computed as the function $\mathcal{F}$ (6), where $\mathcal{V}_u$ denotes the publications of user u before year Y.
- *Future-max* replaces $\mathcal{V}_u$ in *history-max* with the publications of user u after year Y.
- *All-max* replaces $\mathcal{V}_u$ in *history-max* with all the publications of user u.
- *History-mean* is the same as *history-max*, but replaces the operation of taking the maximum value in the function $\mathcal{F}$ (6) with taking the mean value.
- *Future-mean* replaces $\mathcal{V}_u$ in *history-mean* with the publications of user u after year Y.
- *All-mean* replaces $\mathcal{V}_u$ in *history-mean* with all the publications of user u.

In order to avoid information leakage, when computing user u's interest after year Y, we delete the citation relationship between papers published after year Y on the academic network, which means only u's publications after year Y and references before year Y are considered. The results in Fig. 3(a) show that *future* mode performs better than *history* mode and *all* mode. The *max* mode performs better than *mean* mode. When the user interest is calculated in the way of *future-max*, our model performs best, which also proves the user interest will change.

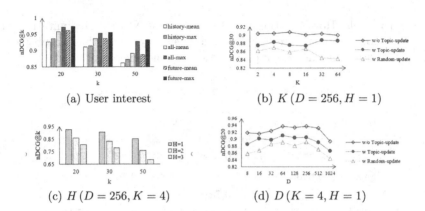

(a) User interest

(b) $K \ (D = 256, H = 1)$

(c) $H \ (D = 256, K = 4)$

(d) $D \ (K = 4, H = 1)$

Fig. 3. The results of the four figures are all carried out on Scopus dataset. (a) Comparison with different user interest computation methods. The hyper-parameter setting is $D = 256, H = 1, K = 4$. (b) (c) (d) are comparisons on model variants with different K, H, D, respectively. (b) (c) (d) all choose the *history-max* mode.

Ablation Study. To verify the impact of model structure on model performance, we conduct ablation experiments. The model variants are as follows.

- **w Random-update** randomly initializes $\overleftarrow{X}_p^{(0)}$ and $\overrightarrow{X}_p^{(0)}$. The parameters of $\overleftarrow{X}_p^{(0)}$ and $\overrightarrow{X}_p^{(0)}$ will be updated during the training process.
- **w Topic-update** initializes $\overleftarrow{X}_p^{(0)}$ and $\overrightarrow{X}_p^{(0)}$ with matrix X_p. And the parameters will be updated.
- **w/o Topic-update** initializes $\overleftarrow{X}_p^{(0)}$ and $\overrightarrow{X}_p^{(0)}$ in the same way as **w Topic-update**, but the parameters will not be updated.

The results are shown in Fig. 3(b) and 3(d). **w Random-update** performs worst. **w Topic-update** is better than **w Random-update**. And **w/o Topic-update**, which is also the final setting of our model, performs best. Because the paper subspace embeddings x_p^b, x_p^m and x_p^r, that are also the topic distributions output by NTM, are rich in semantic information. They do not need to be updated further. Instead, update brings information loss, resulting in model performance degradation.

Hyper-parameter Study. We analyzed the impact of hyper-parameter settings on model performance. Three hyperparameters are tested: the neighbor number K, the maximum depth of GCN H, the topic number D. The results are shown in Fig. 3(b), 3(c) and 3(d). Figure 3(b) shows that when K becomes larger, $nDCG@30$ of **w Random-update** will decrease due to the introduction of noise. **w Topic-update** and **w/o Topic-update** are not sensitive to the setting of K, which means the initialization by X_p weaken the influence of K on model performance. Figure 3(c) shows when H is set to 1, model performs better. As H increases, model performance decreases. Because there may be over-smooth problem with the increase of H. As shown in Fig. 3(d), we find as D increases, the value of $nDCG@k$ will first rise and then fall. Because a smaller D also means less semantic information. The topics in all papers may not be fully covered. The model performs best when D is set to 256. When D is too large, the model performance will decline, probably because the size of D can already cover all the topics, and continuing to increase will not obtain richer semantic information, but will introduce noise.

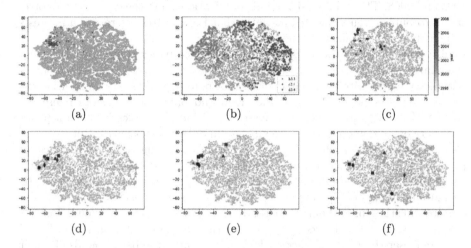

Fig. 4. Analyze the subspace embeddings. The results are based on ACM dataset. The topic number D is set as 16, so the subspace embeddings of papers are 16-dimensional. Then we reduce them into a 2-dimensional visual space by t-SNE [24]. Gray dots in each figure denote all papers. (a) Subspace embeddings of papers with similar background. (b) Background embeddings of papers with different CCS tags. (c) Background embeddings of Chengxiang Zhai's publications. (d) (e) (f) respectively analyze the *Background, Method* and *Result* embeddings of paper [17] and its references and citations.

Analyze the Subspace Embeddings. We analyze the subspace embeddings from different aspects, where the ACM Computing Classification System (ACM CCS) [7] is used as supplementary information.

In order to verify the necessity of subspace, we randomly selected a paper [14] with CCS tag h.3.3 (information search and retrieval). Then 50 papers with similar background to paper [14] are selected from paper set with the same CCS tag. The similarity is obtained by calculating the Euclidean distance of background embeddings. The smaller the distance, the more similar the background. As shown in Fig. 4(a), the red dots denote the background embeddings of the 50 papers. The yellow and blue dots represent method and result embeddings, respectively. We find that papers with similar background may have different methods and results. But the topics do not differ dramatically, but vary within a certain range of topics. So the consideration of subspace is necessary.

To verify whether the subspace embeddings of paper content could reflect CCS tag information, we randomly chose three CCS tags: h.3.3 (information search and retrieval), c.2.1 (network architecture and design) and d.3.4 (processors). As shown in Fig. 4(b), the papers with different CCS tag could be well differentiated.

Figure 4(c) shows some publications of researcher Chengxiang Zhai between 1998 and 2008. The red dots denote his publications, and the color shades correspond to publication years. It illustrates that the researcher's research interests will change within a field of study.

To study whether the subspace embeddings of paper content could reflect the relationship between a paper and its references and citations, we randomly selected a highly cited paper [17]. As shown in Fig. 4(d), 4(e) and 4(f), the red star denotes paper [17], and the green and blue shapes denote references and citations of [17], respectively. We can see that the topics between a paper and its references and citations are all close in different subspaces. But there are also differences, which reflect the topic propagation among different subspaces of papers. For example, the distance between the red star [17] and the green triangle [16] on Fig. 4(d) is closer than the distance on Fig. 4(e) and 4(f). Because the background of paper [17] and paper [16] are all related to the classification of web content, but they used different methods and thus got different results. Besides, the distance between the red star [17] and the blue circle [15] on Fig. 4(d) and 4(e) are closer than the distance on Fig. 4(f). Because the backgrounds of paper [17] and paper [15] are similar and both adopted user survey method. The results are different is due to the core issues of their research are different. It's worth mention that paper [17] and paper [15] have the same author Mika. It illustrates that the researchers tend to use similar methods in their publications.

5   Conclusion

We propose a differentiable topics based new paper recommendation model DTNRec. In DTNRec, we adopt the subspace tagging model and NTM to get embeddings of paper content. Then we model the user interest through the asymmetric GCN on the academic network. The experimental results show the effectiveness of our model.

References

1. Kreutz, C.K., Schenkel, R.: Scientific paper recommendation systems: a literature review of recent publications. Int. J. Digit. Libr. **23**(4), 335–369 (2022)
2. Lewis, M., Liu, Y., et al.: BART: denoising sequence-to-sequence pre-training for natural language generation, translation, and comprehension. In: Annual Meeting of the Association for Computational Linguistics, pp. 7871–7880 (2020)
3. Radford, A., Wu, J., et al.: Language models are unsupervised multitask learners. OpenAI blog **1**(8), 9 (2019)

4. Devlin, J., et al.: BERT: pre-training of deep bidirectional transformers for language understanding. In: NAACL-HLT, pp. 4171–4186 (2019)
5. Radford, A., Narasimhan, K., et al.: Improving language understanding by generative pre-training (2018)
6. Vaswani, A., Shazeer, N., et al.: Attention is all you need. In: Advances in Neural Information Processing Systems, vol. 30 (2017)
7. Coulter, N., et al.: Computing classification system 1998: current status and future maintenance report of the CCS update committee. Comput. Rev. **39**(1), 1–62 (1998)
8. Wang, Y., Wang, L., et al.: A theoretical analysis of NDCG type ranking measures. In: Conference on Learning Theory, pp. 25–54. PMLR (2013)
9. Wang, C., Yu, Y., et al.: Towards representation alignment and uniformity in collaborative filtering. In: ACM SIGKDD Conference on Knowledge Discovery and Data Mining, pp. 1816–1825 (2022)
10. Lee, D., Kang, S., et al.: Bootstrapping user and item representations for one-class collaborative filtering. In: International ACM SIGIR Conference on Research and Development in Information Retrieval, pp. 317–326 (2021)
11. He, X., Deng, K., et al.: LightGCN: simplifying and powering graph convolution network for recommendation. In: International ACM SIGIR Conference on Research and Development in Information Retrieval, pp. 639–648 (2020)
12. He, X., Liao, L., et al.: Neural collaborative filtering. In: 26th International Conference on World Wide Web, pp. 173–182 (2017)
13. Li, W., et al.: Joint embedding multiple feature and rule for paper recommendation. In: Sun, Y., Liu, D., Liao, H., Fan, H., Gao, L. (eds.) ChineseCSCW. CCIS, vol. 1492, pp. 52–65. Springer, Singapore (2021). https://doi.org/10.1007/978-981-19-4549-6_5
14. Joachims, T.: Optimizing search engines using clickthrough data. In: ACM SIGKDD International Conference on Knowledge Discovery and Data Mining, pp. 133–142 (2002)
15. Aula, A., et al.: Information search and re-access strategies of experienced web users. In: International Conference on World Wide Web, pp. 583–592 (2005)
16. Dumais, S., et al.: Hierarchical classification of web content. In: ACM SIGIR Conference on Research and Development in Information Retrieval, pp. 256–263 (2000)
17. Käki, M.: Findex: search result categories help users when document ranking fails. In: SIGCHI Conference on Human Factors in Computing Systems, pp. 131–140 (2005)
18. Wang, H., Zhao, M., et al.: Knowledge graph convolutional networks for recommender systems. In: The World Wide Web Conference, pp. 3307–3313 (2019)
19. Wang, H., Zhang, F., et al.: Knowledge-aware graph neural networks with label smoothness regularization for recommender systems. In: ACM SIGKDD International Conference on Knowledge Discovery & Data Mining, pp. 968–977 (2019)
20. Wang, H., Zhang, F., et al.: RippleNet: propagating user preferences on the knowledge graph for recommender systems. In: ACM International Conference on Information and Knowledge Management, pp. 417–426 (2018)
21. Xie, Y., et al.: Subspace embedding based new paper recommendation. In: International Conference on Data Engineering (ICDE), pp. 1767–1780. IEEE (2022)
22. Jin, D., Szolovits, P.: Hierarchical neural networks for sequential sentence classification in medical scientific abstracts. In: Conference on Empirical Methods in Natural Language Processing, pp. 3100–3109 (2018)
23. Miao, Y., et al.: Discovering discrete latent topics with neural variational inference. In: International Conference on Machine Learning, pp. 2410–2419. PMLR (2017)

24. Van der Maaten, L., Hinton, G.: Visualizing data using t-SNE. J. Mach. Learn. Res. **9**(11) (2008)
25. Xie, Y., Wang, S., et al.: Embedding based personalized new paper recommendation. In: Sun, Y., Liu, D., Liao, H., Fan, H., Gao, L. (eds.) ChineseCSCW 2020. CCIS, vol. 1330, pp. 558–570. Springer, Singapore (2021). https://doi.org/10.1007/978-981-16-2540-4_40
26. Xie, Y., Sun, Y., et al.: Learning domain semantics and cross-domain correlations for paper recommendation. In: 44th International ACM SIGIR Conference on Research and Development in Information Retrieval, pp. 706–715 (2021)

IIHT: Medical Report Generation with Image-to-Indicator Hierarchical Transformer

Keqiang Fan[(✉)], Xiaohao Cai, and Mahesan Niranjan

School of Electronics and Computer Science, University of Southampton,
Southampton SO17 1BJ, UK
{k.fan,x.cai,mn}@soton.ac.uk

Abstract. Automated medical report generation has become increasingly important in medical analysis. It can produce computer-aided diagnosis descriptions and thus significantly alleviate the doctors' work. Inspired by the huge success of neural machine translation and image captioning, various deep learning methods have been proposed for medical report generation. However, due to the inherent properties of medical data, including data imbalance and the length and correlation between report sequences, the generated reports by existing methods may exhibit linguistic fluency but lack adequate clinical accuracy. In this work, we propose an image-to-indicator hierarchical transformer (IIHT) framework for medical report generation. It consists of three modules, i.e., a classifier module, an indicator expansion module and a generator module. The classifier module first extracts image features from the input medical images and produces disease-related indicators with their corresponding states. The disease-related indicators are subsequently utilised as input for the indicator expansion module, incorporating the "data-text-data" strategy. The transformer-based generator then leverages these extracted features along with image features as auxiliary information to generate final reports. Furthermore, the proposed IIHT method is feasible for radiologists to modify disease indicators in real-world scenarios and integrate the operations into the indicator expansion module for fluent and accurate medical report generation. Extensive experiments and comparisons with state-of-the-art methods under various evaluation metrics demonstrate the great performance of the proposed method.

Keywords: Medical report generation · Deep neural networks · Transformers · Chest X-Ray

1 Introduction

Medical images (e.g. radiology and pathology images) and the corresponding reports serve as critical catalysts for disease diagnosis and treatment [22]. A medical report generally includes multiple sentences describing a patient's history symptoms and normal/abnormal findings from different regions within the medical images. However, in clinical practice, writing standard medical reports

B. Luo et al. (Eds.): ICONIP 2023, LNCS 14452, pp. 57–71, 2024.
https://doi.org/10.1007/978-981-99-8076-5_5

is tedious and time-consuming for experienced medical doctors and error-prone for inexperienced doctors. This is because the comprehensive analysis of e.g. X-Ray images necessitates a detailed interpretation of visible information, including the airway, lung, cardiovascular system and disability. Such interpretation requires the utilisation of foundational physiological knowledge alongside a profound understanding of the correlation with ancillary diagnostic findings, such as laboratory results, electrocardiograms and respiratory function tests. Therefore, the automatic report generation technology, which can alleviate the medics' workload and effectively notify inexperienced radiologists regarding the presence of abnormalities, has garnered dramatic interest in both artificial intelligence and clinical medicine.

Medical report generation has a close relationship with image captioning [9,31]. The encoder-decoder framework is quite popular in image captioning, e.g., a CNN-based image encoder to extract the visual information and an RNN/LSTM-based report decoder to generate the textual information with visual attention [11,14,29,30]. With the recent progress in natural language processing, investigating transformer-based models as alternative decoders has been a growing trend for report generation [2,3,21,27]. The self-attention mechanism employed inside the transformer can effectively eliminate information loss, thereby maximising the preservation of visual and textual information in the process of generating medical reports. Although these methods have achieved remarkable performance and can obtain language fluency reports, limited studies have been dedicated to comprehending the intrinsic medical and clinical problems. The first problem is *data imbalance*, e.g., the normal images dominate the dataset over the abnormal ones [24] and, for the abnormal images, normal regions could encompass a larger spatial extent than abnormal regions [17]. The narrow data distribution could make the descriptions of normal regions dominate the entire report. On the whole, imbalanced data may degrade the quality of the automatically generated reports, or even result in all generated reports being basically similar. The second problem is *length and correlation between report sequences*. Medical report generation is designed to describe and record the patient's symptoms from e.g. radiology images including cardiomegaly, lung opacity and fractures, etc. The description includes various disease-related symptoms and related topics rather than the prominent visual contents and related associations within the images, resulting in the correlation inside the report sequences not being as strong as initially presumed. The mere combination of encoders (e.g. CNNs) and decoders (e.g. RNN, LSTM, and transformers) is insufficient to effectively tackle the aforementioned issues in the context of medical images and reports since these modalities represent distinct data types. The above challenges motivate us to develop a more comprehensive method to balance visual and textual features in unbalanced data for medical report generation.

The radiologists' working pattern in medical report writing is shown in Fig. 1. Given a radiology image, radiologists first attempt to find the abnormal regions and evaluate the states for each disease indicator, such as uncertain, negative and

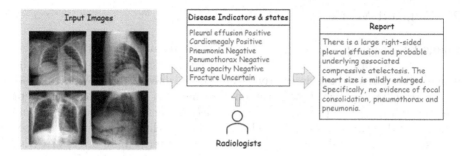

Fig. 1. The medical report writing procedure undertaken by radiologists.

positive. Then a correct clinical report is written through the stages for different indicators based on their working experience and prior medical knowledge. In this paper, we propose an image-to-indicator hierarchical transformer (IIHT) framework, imitating the radiologists' working patterns (see Fig. 1) to alleviate the above-mentioned problems in medical report generation.

Our IIHT framework models the above working patterns through three modules: *classifier*, *indicator expansion* and *generator*. The classifier module is an image diagnosis module, which could learn visual features and extract the corresponding disease indicator embedding from the input image. The indicator expansion module conducts the data-to-text progress, i.e., transferring the disease indicator embedding into short text sequences. The problem of data imbalance could be alleviated by encoding the indicator information, which models the domain-specific prior knowledge structure and summarises the disease indicator information and thus mitigates the long-sequence effects. Finally, the generator module produces the reports based on the encoded indicator information and image features. The whole generation pipeline is given in Fig. 2, which will be described in detail in Sect. 3. We remark that the disease indicator information here can also be modified by radiologists to standardise report fluency and accuracy. Overall, the contributions of this paper are three-fold:

- We propose the IIHT framework, aiming to alleviate the data bias/imbalance problem and enhance the information correlation in long report sequences for medical report generation.
- We develop a dynamic approach which leverages integrated indicator information and allows radiologists to further adjust the report fluency and accuracy.
- We conduct comprehensive experiments and comparisons with state-of-the-art methods on the IU X-Ray dataset and demonstrate that our proposed method can achieve more accurate radiology reports.

The rest of the paper is organised as follows. Section 2 briefly recalls the related work in medical report generation. Our proposed method is introduced in Sect. 3. Sections 4 and 5 present the details of the experimental setting and corresponding results, respectively. We conclude in Sect. 6.

2   Related Work

Image Captioning. The image captioning methods mainly adopt the encoder-decoder framework together with attention mechanisms [31] to translate the image into a single short descriptive sentence and have achieved great performance [1,15,18,26]. Specifically, the encoder network extracts the visual representation from the input images and the decoder network generates the corresponding descriptive sentences. The attention mechanism enhances the co-expression of the visual features derived from the intermediate layers of CNNs and the semantic features from captions [31]. Recently, inspired by the capacity of parallel training, transformers [25] have been successfully applied to predict words according to multi-head self-attention mechanisms. However, these models demonstrate comparatively inferior performance on medical datasets as opposed to natural image datasets, primarily due to the disparity between homogeneous objects observed in different domains. For instance, in the context of X-Ray images, there exists a relatively minimal discernible distinction between normal and abnormal instances, thereby contributing to the challenge encountered by models in accurately generating such captions.

Medical Report Generation. Similar to image captioning, most existing medical report generation methods attempt to adopt a CNN-LSTM-based model to automatically generate fluent reports [11,14,20,28]. Direct utilisation of caption models often leads to the generation of duplicate and irrelevant reports. The work in [11] developed a hierarchical LSTM model and a co-attention mechanism to extract the visual information and generate the corresponding descriptions. Najdenkoska et al. [20] explored variational topic inference to guide sentence generation by aligning image and language modalities in a latent space. A two-level LSTM structure was also applied with a graph convolution network based on the knowledge graph to mine and represent the associations among medical findings during report generation [27]. These methodologies encompass the selection of the most probable diseases or latent topic variables based on the sentence sequence or visual features within the data in order to facilitate sentence generation. Recently, inspired by the capacity of parallel training, transformers [13,32] have successfully been applied to predict words according to the extracted features from CNN. Chen et al. [2] proposed a transformer-based cross-modal memory network using a relational memory to facilitate interaction and generation across data modalities. Nguyen et al. [21] designed a differentiable end-to-end network to learn the disease feature representation and disease state to assist report generation.

The existing methods mentioned above prioritise the enhancement of feature alignment between visual regions and disease labels. However, due to the inherent data biases and scarcity in the medical field, these models exhibit a bias towards generating reports that are plausible yet lack explicit abnormal descriptions. Generating a radiology report is very challenging as it requires the contents of key medical findings and abnormalities with detailed descriptions for different data modalities. In this study, we address the challenges associated with data bias and scarcity in clinical reports through the utilisation of disease indicators as a bridge for more comprehensive medical report generation.

3   Method

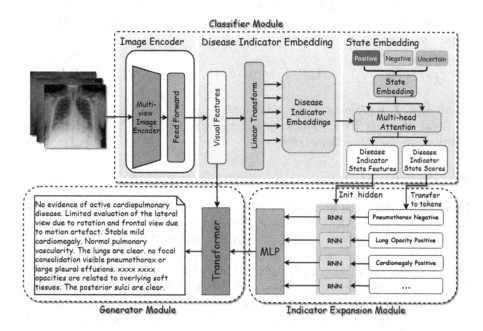

Fig. 2. The proposed IIHT framework. It consists of three modules: classifier, indicator expansion and generator.

An overview of our proposed IIHT framework is demonstrated in Fig. 2. It follows the distinct stages involved in generating a comprehensive medical imaging diagnosis report, adhering to the established process employed in clinical radiology (e.g. see Fig. 1).

Given a radiology image $\mathbf{I}$, the corresponding different indicators are all classified into different states (e.g. positive, negative, uncertain, etc.) denoted as $\mathbf{C} = \{\mathbf{c}_1, \cdots, \mathbf{c}_t, \cdots, \mathbf{c}_T\}$, where T is the number of indicators and $\mathbf{c}_t$ is the one-hot encoding of the states. Particularly, these indicators can also be modified by radiologists to standardise the disease states across patients, thereby enhancing the correctness of the final generated report. The corresponding generated report for a given radiology image is denoted as $\mathbf{y} = (y_1, \cdots, y_n, \cdots, y_N)$, where $y_n \in \mathbb{V}$ is the generated unigram tokens, N is the length of the report, and $\mathbb{V}$ is the vocabulary of all possible v tokens for reports generation. For example, the word sequence "Pleural effusion" is segmented into small pieces of tokens, i.e., {"Pleural", "effus", "ion"}. Generally, the aim of the report generation is to maximise the conditional log-likelihood, i.e.,

$$\theta^* = \arg\max_{\theta} \prod_{n=1}^{N} p_\theta \left(y_n \mid y_1, \ldots, y_{n-1}, \mathbf{I} \right), \tag{1}$$

where θ denotes the model parameters and y_0 represents the start token. After incorporating each disease indicator $\mathbf{c} \in \mathbf{C}$ into the conditional probability $p_\theta \left(y_n \mid y_1, \ldots, y_{n-1}, \mathbf{I} \right)$, we have

$$\log p_\theta \left(y_n \mid y_1, \ldots, y_{n-1}, \mathbf{I} \right) = \int_{\mathbf{C}} \log p_\theta \left(y_n \mid y_1, \ldots, y_{n-1}, \mathbf{c}, \mathbf{I} \right) p_\theta(\mathbf{c} \mid \mathbf{I}) \mathrm{d}\mathbf{c}, \tag{2}$$

where $p_\theta(\mathbf{c} \mid \mathbf{I})$ represents the classifier module.

Recall that our IIHT framework is demonstrated in Fig. 2. The details are described in the subsections below.

3.1 Classifier Module

Image Encoder. The first step in medical report generation is to extract the visual features from the given medical images. In our research, we employ a pre-trained visual feature extractor, such as ResNet [8], to extract the visual features from patients' radiology images that commonly contain multiple view images. For simplicity, given a set of r radiology images $\{\mathbf{I}_i\}_{i=1}^{r}$, the final visual features say $\mathbf{x}$ are obtained by merging the corresponding features of each image using max-pooling across the last convolutional layer. The process is formulated as $\mathbf{x} = f_v \left(\mathbf{I}_1, \mathbf{I}_2, \cdots, \mathbf{I}_r \right)$, where $f_v \left(\cdot \right)$ refers to the visual extractor and $\mathbf{x} \in \mathbb{R}^F$ with F number of features.

Capture Disease Indicator Embedding. The visual features are further transformed into multiple low-dimensional feature vectors, regarded as disease indicator embeddings, which have the capacity to capture interrelationships and correlations among different diseases. The indicator disease embedding is denoted as $\mathbf{D} = (\mathbf{d}_1, \cdots, \mathbf{d}_T) \in \mathbb{R}^{e \times T}$, where e is the embedding dimension and note that T is the number of indicators. Each vector $\mathbf{d}_t \in \mathbb{R}^e, t = 1, \cdots, T$ is the representation of the corresponding disease indicator, which can be acquired through a linear transformation of the visual features, i.e.,

$$\mathbf{d}_t = \mathbf{W}_t^\top \mathbf{x} + \mathbf{b}_t, \tag{3}$$

where $\mathbf{W}_t \in \mathbb{R}^{F \times e}$ and $\mathbf{b}_t \in \mathbb{R}^e$ are learnable parameters of the t-th disease representation.

The intuitive advantage of separating high-dimensional image features into distinct low-dimensional embeddings is that it facilitates the exploration of the relationships among disease indicators. However, when dealing with medical images, relying solely on disease indicator embeddings is insufficient due to the heterogeneous information, including the disease type (e.g. disease name) and the disease status (e.g. positive or negative). Consequently, we undertake further decomposition of the disease indicator embedding, thereby leading to the conception of the subsequent state embedding.

Capture State Embedding. To improve the interpretability of the disease indicator embeddings, a self-attention module is employed to offer valuable insights into the representation of each indicator. Each indicator embedding is further decomposed to obtain the disease state such as positive, negative or uncertain. Let M be the number of states and $\mathbf{S} = (\mathbf{s}_1, \cdots, \mathbf{s}_M) \in \mathbb{R}^{e \times M}$ be the state embedding, which is randomly initialized and learnable. Given a disease indicator embedding vector $\mathbf{d}_t$, the final state-aware of the disease embedding say $\hat{\mathbf{d}}_t \in \mathbb{R}^e$ is obtained by $\hat{\mathbf{d}}_t = \sum_{m=1}^{M} \alpha_{tm} \mathbf{s}_m$, where α_{tm} is the self-attention score of $\mathbf{d}_t$ and $\mathbf{s}_m$ defined as

$$\alpha_{tm} = \frac{\exp(\mathbf{d}_t^\top \cdot \mathbf{s}_m)}{\sum_{m=1}^{M} \exp(\mathbf{d}_t^\top \cdot \mathbf{s}_m)}. \tag{4}$$

Iteratively, each disease indicator representation $\mathbf{d}_t$ will be matched with its corresponding state embedding $\mathbf{s}_m$ by computing vector similarity, resulting in an improved disease indicator representation $\hat{\mathbf{d}}_t$.

Classification. To enhance the similarity between $\mathbf{d}_t$ and $\mathbf{s}_m$, we treat this as a multi-label problem. The calculated self-attention score α_{tm} is the confidence level of classifying disease t into the state m, which is then used as a predictive value. By abuse of notation, let $\mathbf{c}_t = \{c_{t1}, \cdots, c_{tm}, \cdots, c_{tM}\}$ be the t-th ground-true disease indicator and $\boldsymbol{\alpha}_t = \{\alpha_{t1}, \cdots, \alpha_{tm}, \cdots, \alpha_{tM}\}$ be the prediction, where $c_{tm} \in \{0,1\}$ and $\alpha_{tm} \in (0,1)$. The loss of the multi-label classification can be defined as

$$\mathcal{L}_C = -\frac{1}{T} \sum_{t=1}^{T} \sum_{m=1}^{M} c_{tm} \log(\alpha_{tm}). \tag{5}$$

The maximum value α_{tm} in $\boldsymbol{\alpha}_t$ represents the predicted state for disease t. To enable integration with the indicator expansion module, we adopt an alternative approach; instead of directly utilizing $\hat{\mathbf{d}}_t$, we recalculate the state-aware embedding for the t-th disease indicator, denoted as $\hat{\mathbf{s}}_t \in \mathbb{R}^e$, i.e.,

$$\hat{\mathbf{s}}_t = \sum_{m=1}^{M} \begin{cases} c_{tm}\mathbf{s}_m, & \text{if training phase,} \\ \alpha_{tm}\mathbf{s}_m, & \text{otherwise.} \end{cases} \tag{6}$$

Hence, the state-aware disease indicator embedding $\hat{\mathbf{s}}_t$ directly contains the state information of the disease t.

3.2   Indicator Expansion Module

In the indicator extension module, we employ a "data-text-data" conversion strategy. This strategy involves converting the input indicator embedding from its original format into a textual sequential word representation and then converting it back to the original format. The inherent interpretability of short disease indicator sequences can be further enhanced, resulting in generating more reliable medical reports. For each disease indicator and its state, whether it is the ground-truth label $\mathbf{c}_t$ or the predicted label $\boldsymbol{\alpha}_t$, it can be converted into a sequence of words, denoted as $\hat{\mathbf{c}}_t = \{\hat{c}_{t1}, \cdots, \hat{c}_{tk}, \cdots, \hat{c}_{tK}\}$, where $\hat{c}_{tk} \in \mathbb{W}$ is the corresponding word in the sequence, K is the length of the word sequence, and $\mathbb{W}$ is the vocabulary of all possible words in all indicators. For example, an indicator such as "lung oedema uncertain" can be converted into a word sequence such as {"lung", "oedema", "uncertain"}. To extract the textual information within the short word sequence for each disease t, we use a one-layer bi-directional gated recurrent unit as an encoder say $f_w(\cdot)$ followed by a multi-layer perceptron (MLP) $\boldsymbol{\Phi}$ to generate the indicator information $\mathbf{h}_t \in \mathbb{R}^e$, i.e.,

$$\mathbf{h}_t = \boldsymbol{\Phi}\left(\mathbf{h}_{t0}^w + \mathbf{h}_{tk}^w\right), \quad \mathbf{h}_{tk}^w = f_w\left(\hat{c}_{tk}, \mathbf{h}_{tk-1}^w\right), \tag{7}$$

where $\mathbf{h}_{tk}^w \in \mathbb{R}^e$ is the hidden state in f_w. For each disease indicator, the initial state $(k = 0)$ in f_w is the corresponding state-aware disease indicator embedding $\hat{\mathbf{s}}_t$, i.e., $\mathbf{h}_{t0}^w = \hat{\mathbf{s}}_t$.

3.3   Generator Module

The generator say f_g of our IIHT framework is based on the transformer encoder architecture, comprising Z stacked masked multi-head self-attention layers alongside a feed-forward layer positioned at the top of each layer. Each word y_k in the ground-truth report is transferred into the corresponding word embedding $\hat{\mathbf{y}}_k \in \mathbb{R}^e$. For the new word y_n, the hidden state representation $\mathbf{h}'_n \in \mathbb{R}^e$ in the generator f_g is computed based on the previous word embeddings $\{\hat{\mathbf{y}}_k\}_{k=1}^{n-1}$, the calculated indicator information $\{\mathbf{h}_t\}_{t=1}^T$ and the visual representation $\mathbf{x}$, i.e.,

$$\mathbf{h}'_n = f_g\left(\hat{\mathbf{y}}_1, \ldots, \hat{\mathbf{y}}_{n-1}, \mathbf{h}_1, \cdots, \mathbf{h}_T, \mathbf{x}\right). \tag{8}$$

For the i-th report, the confidence $\mathbf{p}_n^i \in \mathbb{R}^v$ of the word y_n is calculated by

$$\mathbf{p}_n^i = \mathrm{softmax}\left(\mathbf{W}_p^\top \mathbf{h}'_n\right), \tag{9}$$

where $\mathbf{W}_p \in \mathbb{R}^{e \times v}$ is a learnable parameter and recall that v is the size of $\mathbb{V}$.

The loss function of the generator say $\mathcal{L}_G$ is determined based on the cross-entropy loss, quantifying all the predicted words in all the given l medical reports with their ground truth, i.e.,

$$\mathcal{L}_G = -\frac{1}{l}\sum_{i=1}^{l}\sum_{n=1}^{N}\sum_{j=1}^{v} y_{nj}^i \log\left(p_{nj}^i\right), \tag{10}$$

where p_{nj}^i is the j-th component of $\mathbf{p}_n^i$, and y_{nj}^i is j-th component of $\mathbf{y}_n^i \in \mathbb{R}^v$ which is the ground-truth one-hot encoding for word y_n in the i-th report. Therefore, the final loss of our IIHT method is

$$\mathcal{L} = \lambda\mathcal{L}_G + (1 - \lambda)\mathcal{L}_C, \tag{11}$$

where λ is a hyperparameter.

4   Experimental Setup

4.1   Data

The publicly available IU X-Ray dataset [4] is adopted for our evaluation. It contains 7,470 chest X-Ray images associated with 3,955 fully de-identified medical reports. Within our study, each report comprises multi-view chest X-Ray images along with distinct sections dedicated to impressions, findings and indications.

4.2   Implementation

Our analysis primarily focuses on reports with a finding section, as it is deemed a crucial component of the report. To tackle the issue of data imbalance, we utilise a strategy wherein we extract 11 prevalent disease indicators from the dataset, excluding the "normal" indicators based on the findings and indication sections of the reports. Additionally, three states (i.e., uncertain, negative and positive) are assigned to each indicator. In cases where a report lacks information regarding all indicators, we discard the report to ensure data integrity and reliability. The preprocessing of all reports is followed by the random selection of image-report pairs, which are then divided into three sets, i.e., training, validation and test sets. The distribution of these sets is 70%, 10% and 20%, respectively. All the words in the reports are segmented into small pieces by SentencePiece [12]. Standard five-fold cross-validation on the training set is used for model selection.

To extract visual features, we utilise two different models: ResNet-50 [8] pretrained on ImageNet [5] and a vision transformer (ViT) [7]. Prior to extraction, the images are randomly cropped to a size of 224×224, accompanied by data augmentation techniques. Within our model, the disease indicator embedding, indicator expansion module and generator module all have a hidden dimension of 512. During training, we iterate 300 epochs with a batch size of 8. The hyperparameter λ in the loss function is set to 0.5. For optimisation, we employ AdamW [19] with a learning rate of 10^{-6} and a weight decay of 10^{-4}.

4.3   Metrics

The fundamental evaluation concept of the generated reports is to quantify the correlation between the generated and the ground-truth reports. Following most of the image captioning methods, we apply the most popular metrics for evaluating natural language generation such as 1–4 g BLEU [23], Rouge-L [16] and METEOR [6] to evaluate our model.

5   Experimental Results

In this section, we first evaluate and compare our IIHT method with the state-of-the-art medical report generation methods. Then we conduct an ablation study for our method to verify the effectiveness of the indicator expansion module under different image extractors.

Table 1. Comparison between our IIHT method and the state-of-the-art medical report generation methods on the IU X-Ray dataset. Sign † refers to the results from the original papers. A higher value denotes better performance in all columns.

Methods	BLEU-1	BLEU-2	BLEU-3	BLEU-4	METEOR	ROUGE-L
VTI [20][†]	0.493	0.360	0.291	0.154	0.218	0.375
Wang et al. [27][†]	0.450	0.301	0.213	0.158	-	0.384
CMR [2][†]	0.475	0.309	0.222	0.170	0.191	0.375
R2Gen [3][†]	0.470	0.304	0.219	0.165	0.187	0.371
Eddie-Transformer [21][†]	0.466	0.307	0.218	0.158	-	0.358
CMAS [10][†]	0.464	0.301	0.210	0.154	-	0.362
DeltaNet [29][†]	0.485	0.324	0.238	0.184	-	0.379
Ours	**0.513** ± 0.006	**0.375** ± 0.005	**0.297** ± 0.006	**0.245** ± 0.006	**0.264** ± 0.002	**0.492** ± 0.004

5.1 Report Generation

We compare our method with the state-of-the-art medical report generation models, including the variational topic inference (VTI) framework [20], a graph-based method to integrate prior knowledge in generation [27], the cross-modal memory network (CMR) [2], the memory-driven transformer (R2Gen) [3], the co-operative multi-agent system (CMAS) [10], the enriched disease embedding based transformer (Eddie-Transformer) [21], and the conditional generation process for report generation (DeltaNet) [29]. The quantitative results of all the methods on the IU X-Ray dataset are reported in Table 1. It clearly shows that our proposed IIHT method outperforms the state-of-the-art methods by a large margin across all the evaluation metrics, demonstrating the dramatic effectiveness of our method.

The methods under comparison in our study focus on exploring the correlation between medical images and medical reports. Some of these approaches have incorporated supplementary indicators as auxiliary information. However, these indicators primarily comprise frequently occurring phrases across all reports, disregarding the inherent imbalance within medical data. Consequently, the generated reports often treat abnormal patients as normal, since the phrases describing normal areas dominate the dataset. In contrast, our proposed method leverages disease indicators and assigns corresponding states based on the reported content. By adopting a "data-text-data" conversion approach in the indicator expansion module, our method effectively mitigates the issue of misleading the generated medical reports, and thus surpasses the performance of the existing approaches.

5.2 Ablation Study

We now conduct an ablation study for our method to verify the effectiveness of different image extractors. Table 2 presents the results of our experiments, wherein we employed different visual feature extractors with and without the indicator expansion module. Specifically, we exclude the original "data-text-data" conversion strategy; instead, the disease indicator state features are directly used as the input of the MLP layer. This study allows us to analyse the influence of the "data-text-data" strategy within the indicator expansion module on the performance of the proposed IIHT framework.

Table 2. The ablation study of our method on the IU X-Ray dataset. "w/o Indicator" refers to the model without the indicator expansion module.

Methods	Encoder	BLEU-1	BLEU-2	BLEU-3	BLEU-4	METEOR	ROUGE-L
IIHT w/o Indicator	ViT	0.434 ± 0.002	0.294 ± 0.004	0.210 ± 0.004	0.153 ± 0.004	0.216 ± 0.001	0.409 ± 0.005
IIHT (Proposed)		0.463 ± 0.006	0.323 ± 0.005	0.241 ± 0.005	0.186 ± 0.004	0.234 ± 0.003	0.445 ± 0.004
IIHT w/o Indicator	ResNet-50	0.428 ± 0.007	0.271 ± 0.008	0.188 ± 0.003	0.136 ± 0.003	0.185 ± 0.002	0.376 ± 0.004
IIHT (Proposed)		**0.513** ± 0.006	**0.375** ± 0.005	**0.297** ± 0.006	**0.245** ± 0.006	**0.264** ± 0.002	**0.492** ± 0.004

By excluding the incremental disease indicator information, we observe that the image extractor ViT has a better performance than ResNet-50, see the results of the first and third rows in Table 2. This indicates that ViT is capable of effectively capturing semantic feature relationships within images. These findings provide evidence regarding the advantages of ViT in extracting visual information from images. We also observe that utilising indicator information extracted from the indicator expansion module indeed contributes to the generation of precise and comprehensive medical reports, resulting in a noteworthy enhancement in terms of the quality of the generated reports. This improvement is observed when using both ViT and ResNet-50. Interestingly, as indicated in the second and fourth rows in Table 2, when the indicator expansion module is added, the performance improvement of ViT is not as significant as that of ResNet-50. We hypothesise that ViT requires a substantial amount of data to learn effectively from scratch. It is possible that the limited number of iterations during fine-tuning prevents ViT from achieving its full potential in performance enhancement. On the whole, our proposed IIHT method offers significant improvements over the state-of-the-art models. This enhancement can be attributed to the inclusion of the disease indicator expansion module, which plays a crucial role in enhancing the quality of the generated reports.

Finally, in Table 3, we showcase some examples of the reports generated by our method. By incorporating both images and indicators, our method closely mimics the process followed by radiologists when composing medical reports while also addressing the data imbalance challenge. Even in the case where all indicators are normal, a generated report for a healthy patient typically includes

Table 3. Generated samples by our method on the IU X-Ray dataset.

Data	Groud-truth Reports	Generated Reports
	No acute cardiopulmonary findings. No focal consolidation. No visualized pneumothorax. No large pleural effusions . The heart size and cardiomediastinal silhouette are grossly unremarkable.	No acute cardiopulmonary abnormality. The lungs are clear bilaterally. Specifically, no evidence of focal consolidation pneumothorax or pleural effusion. Cardiomediastinal silhouette is unremarkable. visualized osseous structures of the thorax are without acute abnormality.
Indicators: Cardiomediastinal silhouette negative; pneumothorax negative; granuloma negative; consolidation negative; pleural effusion negative; pneumonia negative.		
	Right middle lobe and lower lobe pneumonia. Heart size is within the upper limits of the normal. The pulmonary and mediastinum are within normal limits. there is no pleural effusion or pneumothorax. There is the right basilar air space opacity.	Right lower lobe airspace disease in the right lower lobe atelectasis or pneumonia. Heart size and pulmonary vascularity appear within normal limits. There is no pleural effusion or pneumothorax. There are no acute bony abnormalities.
Indicators: Lung opacity positive; pneumonia positive; pulmonary edema negative; pulmonary negative; pleural effusion negative; and pneumothorax negative.		

a description of various disease indicators, as shown in the first example in Table 3. For patients with abnormal conditions, our method still has a remarkable ability to accurately generate comprehensive reports. Moreover, our method incorporates the capability of facilitating real-time modification of disease indicators, thereby enabling a more accurate and complete process for report generation. This functionality serves to minimise the occurrence of misdiagnosis instances, and thus enhances the overall accuracy and reliability of the generated reports. As a result, we reveal that the generated medical reports with the use of indicator-based features can be more reasonable and disease-focused in comparison to traditional "image-to-text" setups.

6   Conclusion

In this paper, we proposed a novel method called IIHT for medical report generation by integrating disease indicator information into the report generation process. The IIHT framework consists of the classifier module, indicator expansion module and generator module. The "data-text-data" strategy implemented in the indicator expansion module leverages the textual information in the form of concise phrases extracted from the disease indicators and states. The accompanying data conversion step enhances the indicator information, effectively resolving the data imbalance problem prevalent in medical data. Furthermore, this conversion also facilitates the correspondence between the length and correlation of medical data texts with disease indicator information. Our method makes it feasible for radiologists to modify the disease indicators in real-world scenarios and integrate the operations into the indicator expansion module, which ultimately contributes to the standardisation of report fluency and accuracy. Extensive experiments and comparisons with state-of-the-art methods demonstrated the great performance of the proposed method. One potential limitation of our experiments is related to the accessibility and accuracy of the disease indicator information. The presence and precision of such disease indicator information can affect the outcomes of our study. Interesting future work could involve investigating and enhancing our method from a multi-modal perspective by incorporating additional patient information such as age, gender and height for medical report generation.

References

1. Anderson, P., et al.: Bottom-up and top-down attention for image captioning and visual question answering. In: Proceedings of the IEEE Conference on Computer Vision and Pattern Recognition, pp. 6077–6086 (2018)
2. Chen, Z., Shen, Y., Song, Y., Wan, X.: Cross-modal memory networks for radiology report generation. In: Proceedings of the 59th Annual Meeting of the Association for Computational Linguistics and the 11th International Joint Conference on Natural Language Processing (Volume 1: Long Papers), pp. 5904–5914. Association for Computational Linguistics, August 2021. https://doi.org/10.18653/v1/2021.acl-long.459. https://aclanthology.org/2021.acl-long.459
3. Chen, Z., Song, Y., Chang, T.H., Wan, X.: Generating radiology reports via memory-driven transformer. In: Proceedings of the 2020 Conference on Empirical Methods in Natural Language Processing, November 2020
4. Demner-Fushman, D., et al.: Preparing a collection of radiology examinations for distribution and retrieval. J. Am. Med. Inform. Assoc. **23**(2), 304–310 (2016)
5. Deng, J., Dong, W., Socher, R., Li, L.J., Li, K., Fei-Fei, L.: ImageNet: a large-scale hierarchical image database. In: 2009 IEEE Conference on Computer Vision and Pattern Recognition, pp. 248–255. IEEE (2009)
6. Denkowski, M., Lavie, A.: Meteor universal: language specific translation evaluation for any target language. In: Proceedings of the 9th Workshop on Statistical Machine Translation, pp. 376–380 (2014)
7. Dosovitskiy, A., et al.: An image is worth 16x16 words: transformers for image recognition at scale. arXiv preprint arXiv:2010.11929 (2020)
8. He, K., Zhang, X., Ren, S., Sun, J.: Deep residual learning for image recognition. In: Proceedings of the IEEE Conference on Computer Vision and Pattern Recognition, pp. 770–778 (2016)
9. Huang, L., Wang, W., Chen, J., Wei, X.Y.: Attention on attention for image captioning. In: Proceedings of the IEEE/CVF International Conference on Computer Vision, pp. 4634–4643 (2019)
10. Jing, B., Wang, Z., Xing, E.: Show, describe and conclude: on exploiting the structure information of chest X-ray reports. In: Proceedings of the 57th Annual Meeting of the Association for Computational Linguistics, pp. 6570–6580. Association for Computational Linguistics, Florence, Italy, July 2019. https://doi.org/10.18653/v1/P19-1657. https://aclanthology.org/P19-1657

11. Jing, B., Xie, P., Xing, E.: On the automatic generation of medical imaging reports. In: Proceedings of the 56th Annual Meeting of the Association for Computational Linguistics (Volume 1: Long Papers), pp. 2577–2586. Association for Computational Linguistics, Melbourne, Australia, July 2018. https://doi.org/10.18653/v1/P18-1240. https://aclanthology.org/P18-1240

12. Kudo, T., Richardson, J.: SentencePiece: a simple and language independent subword tokenizer and detokenizer for neural text processing. In: Proceedings of the 2018 Conference on Empirical Methods in Natural Language Processing: System Demonstrations, pp. 66–71. Association for Computational Linguistics, Brussels, Belgium, November 2018. https://doi.org/10.18653/v1/D18-2012. https://aclanthology.org/D18-2012

13. Li, G., Zhu, L., Liu, P., Yang, Y.: Entangled transformer for image captioning. In: Proceedings of the IEEE/CVF International Conference on Computer Vision, pp. 8928–8937 (2019)

14. Li, Y., Liang, X., Hu, Z., Xing, E.P.: Hybrid retrieval-generation reinforced agent for medical image report generation. In: Advances in Neural Information Processing Systems, vol. 31 (2018)

15. Liang, X., Hu, Z., Zhang, H., Gan, C., Xing, E.P.: Recurrent topic-transition GAN for visual paragraph generation. In: Proceedings of the IEEE International Conference on Computer Vision, pp. 3362–3371 (2017)

16. Lin, C.Y.: ROUGE: a package for automatic evaluation of summaries. In: Text Summarization Branches Out, pp. 74–81 (2004)

17. Liu, F., Ge, S., Wu, X.: Competence-based multimodal curriculum learning for medical report generation. In: Proceedings of the 59th Annual Meeting of the Association for Computational Linguistics and the 11th International Joint Conference on Natural Language Processing (Volume 1: Long Papers), pp. 3001–3012. Association for Computational Linguistics, August 2021. https://doi.org/10.18653/v1/2021.acl-long.234. https://aclanthology.org/2021.acl-long.234

18. Liu, F., Ren, X., Liu, Y., Wang, H., Sun, X.: simNet: stepwise image-topic merging network for generating detailed and comprehensive image captions. In: Proceedings of the 2018 Conference on Empirical Methods in Natural Language Processing, pp. 137–149. Association for Computational Linguistics, Brussels, Belgium, October–November 2018. https://doi.org/10.18653/v1/D18-1013. https://aclanthology.org/D18-1013

19. Loshchilov, I., Hutter, F.: Decoupled weight decay regularization. arXiv preprint arXiv:1711.05101 (2017)

20. Najdenkoska, I., Zhen, X., Worring, M., Shao, L.: Variational topic inference for chest X-ray report generation. In: de Bruijne, M., et al. (eds.) MICCAI 2021. LNCS, vol. 12903, pp. 625–635. Springer, Cham (2021). https://doi.org/10.1007/978-3-030-87199-4_59

21. Nguyen, H.T., et al.: Eddie-transformer: enriched disease embedding transformer for X-ray report generation. In: 2022 IEEE 19th International Symposium on Biomedical Imaging (ISBI), pp. 1–5. IEEE (2022)

22. European Society of Radiology (ESR) communications@myesr.org: Medical imaging in personalised medicine: a white paper of the research committee of the European society of radiology (ESR). Insights Imag. 6, 141–155 (2015)

23. Papineni, K., Roukos, S., Ward, T., Zhu, W.J.: Bleu: a method for automatic evaluation of machine translation. In: Proceedings of the 40th Annual Meeting of the Association for Computational Linguistics, pp. 311–318 (2002)

24. Shin, H.C., Roberts, K., Lu, L., Demner-Fushman, D., Yao, J., Summers, R.M.: Learning to read chest x-rays: recurrent neural cascade model for automated image annotation. In: Proceedings of the IEEE Conference on Computer Vision and Pattern Recognition, pp. 2497–2506 (2016)
25. Vaswani, A., et al.: Attention is all you need. In: Advances in Neural Information Processing Systems, vol. 30 (2017)
26. Vinyals, O., Toshev, A., Bengio, S., Erhan, D.: Show and tell: a neural image caption generator. In: Proceedings of the IEEE Conference on Computer Vision and Pattern Recognition, pp. 3156–3164 (2015)
27. Wang, S., Tang, L., Lin, M., Shih, G., Ding, Y., Peng, Y.: Prior knowledge enhances radiology report generation. In: AMIA Annual Symposium Proceedings, vol. 2022, p. 486. American Medical Informatics Association (2022)
28. Wang, X., Peng, Y., Lu, L., Lu, Z., Summers, R.M.: TieNet: text-image embedding network for common thorax disease classification and reporting in chest X-rays. In: Proceedings of the IEEE Conference on Computer Vision and Pattern Recognition, pp. 9049–9058 (2018)
29. Wu, X., et al.: DeltaNet: conditional medical report generation for COVID-19 diagnosis. In: Proceedings of the 29th International Conference on Computational Linguistics, pp. 2952–2961. International Committee on Computational Linguistics, Gyeongju, Republic of Korea, October 2022. https://aclanthology.org/2022.coling-1.261
30. Yin, C., et al.: Automatic generation of medical imaging diagnostic report with hierarchical recurrent neural network. In: 2019 IEEE International Conference on Data Mining (ICDM), pp. 728–737. IEEE (2019)
31. You, Q., Jin, H., Wang, Z., Fang, C., Luo, J.: Image captioning with semantic attention. In: Proceedings of the IEEE Conference on Computer Vision and Pattern Recognition, pp. 4651–4659 (2016)
32. Zhou, L., Zhou, Y., Corso, J.J., Socher, R., Xiong, C.: End-to-end dense video captioning with masked transformer. In: Proceedings of the IEEE Conference on Computer Vision and Pattern Recognition, pp. 8739–8748 (2018)

OD-Enhanced Dynamic Spatial-Temporal Graph Convolutional Network for Metro Passenger Flow Prediction

Lei Ren, Jie Chen, Tong Liu[✉], and Hang Yu

School of Computer Engineering and Science, Shanghai University, Shanghai, China
{renlei,chenjie_cs,tong_liu,yuhang}@shu.edu.cn

Abstract. Metro passenger flow prediction is crucial for efficient urban transportation planning and resource allocation. However, it faces two challenges. The first challenge is extracting the diverse passenger flow patterns at different stations, e.g., stations near residential areas and stations near commercial areas, while the second one is to model the complex dynamic spatial-temporal correlations caused by Origin-Destination (OD) flows. Existing studies often overlook the above two aspects, especially the impact of OD flows. In conclusion, we propose an OD-enhanced dynamic spatial-temporal graph convolutional network (DSTGCN) for metro passenger flow prediction. First, we propose a static spatial module to extract the flow patterns of different stations. Second, we utilize a dynamic spatial module to capture the dynamic spatial correlations between stations with OD matrices. Finally, we employ a multi-resolution temporal dependency module to learn the delayed temporal features. We also conduct experiments based on two real-world datasets in Shanghai and Hangzhou. The results show the superiority of our model compared to the state-of-the-art baselines.

Keywords: Metro system · Passenger flow prediction · Spatial-Temporal graph convolutional networks · Origin-Destination matrix

1 Introduction

Urban Rail Transit (URT) has become one of the primary modes of public transportation in numerous cities due to its significant capacity and high speed. However, the substantial increase in passenger flows has resulted in severe overcrowding within metro systems, posing safety hazards and exacerbating the challenges in managing these systems. For instance, the surge of people during the 2015 New Year's Eve celebrations in Shanghai precipitated a chaotic stampede that tragically claimed the lives of 36 individuals [1]. Therefore, accurate forecasting of metro passenger flows is crucial for effective metro planning and proactive allocation of metro staff [2].

However, the task of passenger flow prediction remains challenging due to the intricate and robust spatial-temporal correlations of passenger flows, which can be summarized into the following three points:

© The Author(s), under exclusive license to Springer Nature Singapore Pte Ltd. 2024
B. Luo et al. (Eds.): ICONIP 2023, LNCS 14452, pp. 72–85, 2024.
https://doi.org/10.1007/978-981-99-8076-5_6

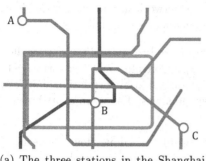
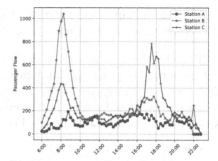

(a) The three stations in the Shanghai metro system.

(b) The passenger flows of the three stations in Fig. 1(a) in a day.

Fig. 1. Diverse flow patterns at different stations.

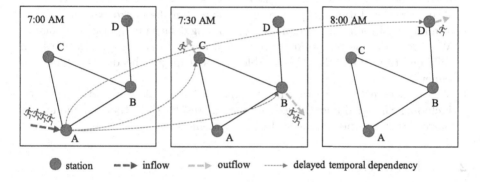

Fig. 2. Dynamic spatial correlations and delayed temporal dependencies.

1) Diverse flow patterns at different stations: Previous studies [3–7] have commonly employed the use of metro network topology to capture the relationships between passenger flow patterns across stations. However, they overlook the heterogeneity in flow patterns among different stations, as shown in Fig. 1(b). Therefore, it is crucial to explore methods for extracting the differences in flow patterns at different stations.

2) Dynamic spatial correlations between stations: Previous researches [3,6,8] have treated the spatial relationships between stations as static and relied on physical distance to characterize spatial dependencies. However, they overlooked the dynamic spatial dependencies between stations caused by OD passenger flows, as illustrated in Fig. 2, where the movement of passengers continually evolves in real-time. Therefore, the challenge lies in modeling these dynamic spatial correlations to capture the metro passenger flows.

3) Delayed temporal dependency among stations: The delayed temporal dependencies are also caused by the OD passenger flows. Although prior researches [3,6,9] have made some progress in capturing temporal dependencies, there remains a dearth of relevant research on incorporating lagged time information. Consequently, devising a network architecture capable of effectively capturing delayed temporal dependencies is a challenge.

To tackle the aforementioned challenges, we introduce a novel approach called OD-enhanced Dynamic Spatial-Temporal Graph Convolutional Network (DST-GCN), which can capture both static and dynamic spatial correlations between stations, as well as the delayed temporal dependencies. The main contributions of this paper are summarised as follows:

1) A static spatial correlation module for capturing unique flow patterns at different stations. By constructing two types of flow pattern graphs, we enable the graph convolutional layers to extract station-specific flow patterns.
2) A dynamic spatial correlation module for extracting OD information. We leverage graph diffusion operation to effectively capture the dynamic spatial correlations between stations. This allows us to capture the transfer relationships between stations as passengers move within the metro system.
3) A multi-resolution temporal module for mining delayed temporal dependency. We stack multiple temporal convolutional layers to access information from different time intervals. This enables us to capture the delayed temporal dependencies.
4) We conduct extensive experiments on two large-scale real-world datasets, including Shanghai and Hangzhou, and the results show that our model has better predictive performance than the baselines.

2   Related Work

Passenger flow prediction is not only an important task in the field of intelligent transport, but also optimises many urban services. Initially, most models for short-term passenger flow forecasting are mainly based on statistical theory or machine learning, such as ARIMA [10] and support vector regression [11]. However, among these methods, passenger flow forecasting is typically considered as a time series prediction problem, which overlook the comprehensive spatial correlations between stations.

Inspired by the ability of graph neural networks (GNNs) to model the correlation of non-European data, GNNs are adopted for passenger flow prediction problems [3–5,12,13], which not only consider the temporal dependencies of passenger flows but also can be able to capture the spatial relationships between stations. [3,8] both design a spatial-temporal network combining RNN-based model and GNN-based model to extract the periodic features of passenger flows and the topological relationships between stations. Compared to models that only consider temporal dependencies, prediction performance is further improved. Instead of using metro network topology to capture the spatial correlations, [4,5] introduce the use of virtual graph structures to represent spatial relationships between stations.

Several recent works [14–16] specifically pay attention to extract valuable information from the origin-destination to achieve a more precise prediction.

For example, Wang *et al.* [14] introduce an algorithm to generate a representation of OD-pair which can discover the passenger travel patterns. However, due to the time trip, these methods do not dig into the OD correlations between stations well. Further more, He *et al.* [15] design a model to learn the implicit OD relationships, which uses graph diffusion convolutional [17] to capture pair-wise geographical and semantic correlations. Despite the collective efforts to incorporate OD information in the aforementioned methods to enhance prediction performance, they fell short in capturing the latent spatial-temporal dependencies inherent in transfer relationships.

3   Problem Formulation

In this section, we firstly introduce some definitions for metro passenger flow prediction, and then formulate the problem.

Definition 1. *Passenger Flows.* The passenger flow volume of the station v_i during the time interval t is denoted as $x_t^i \in \mathbb{R}^2$, where the 2 dimensions are the inbound flow and the outbound flow, respectively. The collection of passenger flow volumes for all N stations during time interval t is represented as $X_t = (x_t^1, x_t^2, ..., x_t^N) \in \mathbb{R}^{N \times 2}$. $\mathcal{X} = (X_1, X_2, ..., X_t, ...)$ denotes the historical passenger flow data across all time intervals.

Definition 2. *Exit Origin-Destination Flows.* The exit origin-destination flows at time interval t refer to the number of passengers who enter the station in the past few time intervals and exit the station at time interval t. Thus, the exit OD flows at time interval t between station v_i and v_j can be denoted as $o_t^{i,j}$, representing the passengers who entered in station v_i and exited station v_j at time interval t.

Definition 3. *Point of Interest.* A Point of Interest (POI) refers to a specific location on a map that is deemed interesting or noteworthy to individuals, like a shopping centre. We count the number of different types of POIs within a radius of 1.5 km from each station, and then we construct a POI matrix $P \in \mathbb{R}^{N \times c}$ for all the stations, where c is the number of categories of POIs.

Problem 1. Based on the above definitions, the problem of passenger flow prediction is formulated as follows: given a historical T time intervals passenger flows $\mathcal{X}_{\tau-T:\tau} = (X_{\tau-T+1}, ..., X_\tau)$ up to time interval τ, the problem aims to learn a mapping function f from T historical passenger flow sequence to predict the next passenger flow volume for all stations.

$$[X_{\tau-T+1}, ..., X_\tau] \xrightarrow{f} [X_{\tau+1}] \tag{1}$$

4   Methodology

4.1   Overview

Figure 3 illustrates our model which consists of three components and followed by a fusion layer to produce the prediction results. Each component is composed of two spatial-temporal convolutional blocks (ST-Conv blocks). Each ST-Conv block comprises a static spatial correlation module, a dynamic spatial correlation module, and a multi-resolution temporal dependency module. Here, we construct three different passenger flow sequences as inputs to the three components, namely the recent trend sequence $\mathcal{X}^r$, the daily periodicity sequence $\mathcal{X}^d$, and the weekly periodicity sequence $\mathcal{X}^w$. The ST-Conv blocks receive the inputs and explore spatial-temporal dependencies. Multiple periodic features are integrated by a fusion layer to generate the final prediction $\hat{X}_{\tau+1}$. The specific details of each module are discussed in the following subsections.

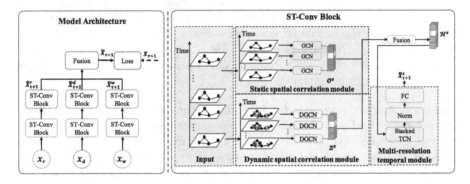

Fig. 3. Model architecture of the proposed DSTGCN. The proposed model consists of three components, which share the same network structure and followed by a fusion layer to produce the prediction results. Each of them consists of two spatial-temporal convolutional blocks (ST-Conv blocks).

4.2   Static Spatial Correlation Module

Static Flow Pattern Graph Generation: In this section, we describe the process of constructing the static flow pattern graph. At its core, a graph comprises nodes, edges, and associated weights. In our approach, we construct two static graphs: the trend graph $G_p = (V, E_p, A_p)$ and the functionality graph $G_f = (V, E, A_f)$. Here, $V = \{v_1, v_2, ..., v_N\}$ is the set of N stations. E_p and E_f are the edge sets of different graphs. For a specific graph G_α ($\alpha \in \{p, f\}$), $A_\alpha \in \mathbb{R}^{N \times N}$ denotes the weights of all edges. The specific construction of the two graphs will be introduced in the following.

The trend graph G_p: The trend graph can be denoted as $G_p = (V, E_p, A_p)$. As shown in Eq. 2, $A_p(i, j)$ represents similarity of passenger flow trend between station v_i and station v_j. Note that in this study we use DTW (Dynamic Time Warping) [18] to measure the proximity of two stations.

$$A_p(i,j) = \begin{cases} \exp^{-dtw(\mathcal{X}^i, \mathcal{X}^j)}, & dtw(\mathcal{X}^i, \mathcal{X}^j) \leq \epsilon, \\ 0, & dtw(\mathcal{X}^i, \mathcal{X}^j) > \epsilon \end{cases} \quad (2)$$

where $\mathcal{X}^i$ and $\mathcal{X}^j$ represent the historical passenger flows at station v_i and station v_j, respectively. ϵ is a hyper-parameter, which is used to threshold the adjacency matrix, setting any values below ϵ to 0.

The functionality graph G_f: The functionality graph can be denoted as $G_f = (V, E_f, A_f)$. As shown in Eq. 4.

$$A_f(i,j) = \cos(q_i, q_j) \quad (3)$$

where q_i indicates the POIs representation of station v_i. It is generated by the TF-IDF algorithm [19], which treats each POI as a word and each station as a document. The specific calculation for representing k-th category of station v_i is shown in the following formula.

$$q_i[k] = \frac{P_{i,k}}{\sum P_{i,\cdot}} * \log \frac{N}{\sum P_{\cdot,k}} \quad (4)$$

Graph Convolution for Static Spatial Correlations: The aforementioned graphs elucidate the inherent and unchanging relationships between the flow patterns of different stations. Here, we employ ChebNet [20] to effectively capture these static spatial correlations. Specifically, we first utilize Laplacian regularization on the above two graphs. Its calculation is as $L_\alpha = I_n - D^{-\frac{1}{2}} A_\alpha D^{-\frac{1}{2}}$, where $D \in \mathbb{R}^{N \times N}$ is the diagonal degree matrix with $D_{i,i} = \sum_j A_\alpha(i,j)$ and I_n is the identity matrix of $N \times N$. Then, we combine different graphs by Eq. 5, where $\odot$ is element-wise product, $W_\alpha(\alpha \in \{p, f\})$ are learnable parameters.

$$\hat{L} = \sum_{\alpha \in \{p,f\}} W_\alpha \odot L_\alpha \quad (5)$$

Utilizing the fusion result of the graph, denoted as $\hat{L}$, we proceed with the graph convolution operation, employing first-order Chebyshev polynomials to approximate the convolution kernel. For a specific historical passenger flow sequence $\mathcal{X}^s$ ($s \in \{r, d, w\}$), the graph convolution for each time interval t can be formulated as:

$$O_t^s = ReLU\left(\Theta(\tilde{D}^{-\frac{1}{2}} \tilde{L} \tilde{D}^{-\frac{1}{2}}) X_t^s\right) \quad (6)$$

where $O_t^s \in \mathbb{R}^{N \times c}$ is the static spatial feature representation of N stations at time interval t, $\tilde{L}$ and $\tilde{D}$ are re-normalized by $\tilde{L} = \hat{L} + I_n$ and $\tilde{D}_{i,i} = \sum_j \hat{L}_{i,j}$.

4.3   Dynamic Spatial Correlations Module

Dynamic Transit Graph Generation: Within the aforementioned graphs, the connection relationship among stations remains immutable and static. However, these immutable and static relationships fall short in capturing the dynamic dependencies that arise from the ever-evolving movement of passengers between stations over time. Hence, there is a need to learn such dynamic spatial relationships. To address this, we convert the exit OD passenger flows at each time interval into a dynamic transition graph, effectively representing the evolving relationships between stations.

The transition volume graph G_o^t: The volume of exit OD flow provides insights into the attraction between pairs of metro stations. Generally, stations with higher OD flow between them indicate a higher level of mutual attractiveness. The graph for each time interval t can be represented by the following formula.

$$G_t^o = (V, E_t^o, A_t^o) \tag{7}$$

$$A_t^o(i, j) = c_t^{i,j} \tag{8}$$

Graph Convolution for Dynamic Spatial Correlations: The dynamic nature of passenger transfer flow gives rise to ever-shifting correlations between stations over time. As a result, the conventional spectral-based models prove inadequate for analyzing OD graphs, primarily due to the strict symmetry requirement imposed by Laplacian matrix factorization. Inspired by [17], we utilize truncated and finite-step diffusion operations as graph convolutions and employ graph diffusion convolution to capture the dynamic spatial correlations between stations.

$$Z_t^s = \sum_{k=0}^{K-1} (D_I^t)^k X_t^s \Theta_{k,1} + (D_O^t)^k X_t^s \Theta_{k,2} \tag{9}$$

where $Z_t^s \in \mathbb{R}^{N \times c}$ is the dynamic spatial feature, $(\Theta_{.,1}, \Theta_{.,2}) \in \mathbb{R}^{K \times 2}$ are the parameters for the convolution filter. K denotes the finite truncation of the diffusion process. $D_{I,t}$ and $D_{O,t}$ represent the in-degree matrix and out-degree matrix of A_t^o, respectively.

Graph Fusion: In order to effectively utilize the captured static and dynamic spatial correlations, we employ a spatial fusion operation to combine the learned static and dynamic spatial correlations. As shown in the following Eq. 11:

$$H_t^s = O_t^s + \sigma(Z_t^s) \tag{10}$$

where $H_t^s \in \mathbb{R}^{N \times c}$ represents the spatial features with c dimensions at time interval t.

4.4   Multi-resolution Temporal Dependency Module

In this work, we adopt the temporal convolutional network (TCN) [21] to capture the temporal dependency of passenger flows. Furthermore, we capture delayed temporal dependencies by stacking causal convolutions with different dilation rates in the TCN architecture, as illustrated in Fig. 4.

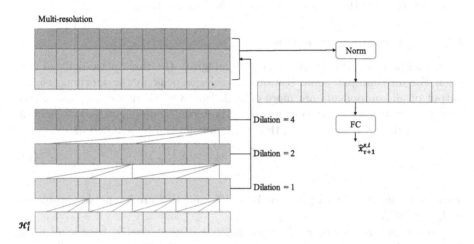

Fig. 4. Dilated casual convolution with kernel size 3. With a dilation factor k, it picks passenger flows every k time interval.

Dilated causal convolution introduces dilation factors, which allow the convolutional kernel to access the input sequence in a skipping manner along the temporal dimension, thus expanding the receptive field. Compared to RNN, dilated causal convolution has the capability to capture information from longer temporal distances and can be computed in parallel. Mathematically, for the sequence $\mathcal{H}^s = (H^s_{\tau-T+1}, ..., H^s_\tau)$ after spatial correlation module, the dilated casual convolution operation of $\hat{\mathcal{X}}$ is as follows:

$$\hat{\mathcal{H}}^s = \mathcal{W} *_d \mathcal{H} \tag{11}$$

where $\hat{\mathcal{H}} \in \mathbb{R}^{N \times T \times c}$ represent the updated spatial-temporal representation of N stations over T time steps, c is the number of output channels, $\mathcal{W} \in \mathbb{R}^{T \times c \times 3 \times 1}$ is the convolutional kernel with a kernel size of 3×1, and $*_d$ denotes the dilated causal convolution with a dilation rate of d.

Furthermore, drawing inspiration from [22], we stack multiple layers of dilated causal convolution to expand the receptive field and capture long-range temporal dependencies. This stacking approach enables the extraction of multi-resolution temporal information, leveraging the dilation operation to traverse extended time intervals and capture delayed temporal dependencies. Consequently, we can effectively model temporal relationships with varying delays. Specifically, we concatenate the outputs of each TCN layer, normalize them along the time dimension, and then use a fully connected layer to obtain the prediction for the next time step. The mathematical formulation is as follows:

$$\hat{X}_{\tau+1}^s = FC\left(norm((\mathcal{H}^s)^1, (\mathcal{H}^s)^2, ..., (\mathcal{H}^s)^L)\right) \tag{12}$$

where $\hat{X}_{\tau+1}^s \in \mathbb{R}^{N \times 2}$ represents the prediction of passenger flows with a periodicity of s, $(\mathcal{H}^s)^l$ is the output features of the l-th layer.

4.5   Multi-component Fusion

In this section, we will introduce how to integrate the three periodicity of outputs. For an example, some stations have large peak in the morning, so the output of all components are more crucial. However, there are no peak of passenger flows in some other stations, thus the recent component may be helpful. Therefore, the formula of fusing three components can be denoted as:

$$\hat{X}_{\tau+1} = \sum_{s \in \{r,d,w\}} W^s \odot \hat{X}_{\tau+1}^s \tag{13}$$

where $\odot$ is the Hadamard product, W^r, W^d, and W^w are learnable parameters with sizes of $N \times N$.

During the process of model training, our utmost goal is to minimize the divergence between the observed passenger flows and their corresponding predicted values. To achieve this, we employ the Huber loss function [23] to iteratively optimize the parameters of our proposed model. The Huber loss is defined as follows:

$$\mathcal{L}\left(\hat{X}_{\tau+1}, X_{\tau+1}; \Theta\right) = \begin{cases} \frac{1}{2}\left(\hat{X}_{\tau+1}, X_{\tau+1}\right)^2, & |\hat{X}_{\tau+1}, X_{\tau+1}| \leq \delta \\ \delta|\hat{X}_{\tau+1}, X_{\tau+1}| - \frac{1}{2}\delta^2, & otherwise \end{cases} \tag{14}$$

5   Experiments

In this section, we verify our model on two real-world urban rail transit datasets. We will first introduce the datasets, data processing, baseline methods, experiment settings, and then present the results of our experiments comprehensively.

5.1   Datasets

In this study, we meticulously evaluate the performance of our proposed DST-GCN model using two urban transit datasets, namely SHMetro and HZMetro. SHMetro dataset was meticulously collected from the AFC system in the bustling city of Shanghai, China. It encompasses a staggering 2.1 billion AFC records, painstakingly gathered from April 1st to April 30th, 2015. Similarly, HZMetro dataset was diligently acquired from the metro AFC system in the vibrant city of Hangzhou, China, comprising a substantial collection of 70 million AFC records, spanning the period from January 1st to January 25th, 2019.

5.2   Experimental Setups

In our experiments, we implement our model, DSTGCN, as well as the deep learning models of the comparison methods using the powerful PyTorch framework. The experiments are conducted on a 4 core Intel Core I5-9300 CPU, with 32 GB RAM and a NVIDIA RTX-2060 GPU card. To comprehensively evaluate the performance of DSTGCN and the baseline methods, we employ two widely-accepted metrics: Mean Absolute Error (MAE) and Root Mean Square Error (RMSE). The evaluation datasets are thoughtfully partitioned into training, validation, and testing sets, following a ratio of 8:1:1. We normalize the datasets in the same way, and use Adam [24] optimizer. We set the lengths of input sequences for the three periodicities, namely recently, daily, and weekly, as 3, 3, and 1, respectively. In the training process, we initialize the learning rate to a value of 0.005 and reduce it by 20% every 30 epochs. In addition, we repeat each experiment a total of 5 times, and report the mean errors and standard deviations.

5.3   Experimental Results

Comparison with Baselines: In this study, we conduct a comprehensive comparative analysis between our proposed model and the following 6 baseline methods: Historical Average (HA), Auto-regressive Integrated Moving Average (ARIMA) [10], Diffusion Convolution Recurrent Neural Network (DCRNN) [17], Spatial-Temporal Graph Convolutional Network (STGCN) [25], Graph WaveNet (GW) [21], and Spatial-Temporal Dynamic Network (STDN) [26].

Table 1 and Table 2 show the performance of our method and the 6 baseline methods on the two datasets. We can observe that the traditional time series methods, HA and ARIMA, achieve relatively high RMSE and MAE values both in SHMetro and HZMetro datasets. This is mainly due to its inability to handle unstable and nonlinear data effectively. In contrast, the deep learning-based methods, including STGCN, DCRNN, GW, STDN, and our model demonstrate

Table 1. Performance Comparison - SHMetro.

Methods	RMSE		MAE	
	In	Out	In	Out
HA	107.40	126.87	48.95	48.69
ARIMA [10]	169.28	183.59	120.33	117.27
STGCN [25]	49.43 ± 0.68	51.36 ± 0.83	40.50 ± 0.93	41.03 ± 0.26
DCRNN [17]	48.33 ± 0.71	50.09 ± 0.76	39.67 ± 1.38	41.05 ± 0.51
GW [21]	45.06 ± 0.38	46.54 ± 0.74	38.15 ± 0.41	39.08 ± 0.80
STDN [26]	42.87 ± 0.51	44.41 ± 0.68	35.89 ± 0.46	37.49 ± 0.77
DSTGCN	$\mathbf{41.98 \pm 0.42}$	$\mathbf{43.95 \pm 0.49}$	$\mathbf{32.21 \pm 0.91}$	$\mathbf{34.22 \pm 0.59}$

improved performance. Specifically, our model achieves a significant improvement in performance compared to the baseline methods. The MAE for inbound passenger flow decrease by 10%, while the RMSE for outbound passenger flow decrease by 8% on the SHMetro. This is mainly due to the effective extraction of dynamic spatial-temporal correlations from the OD information by our model.

Table 2. Performance Comparison - HZMetro.

Methods	RMSE		MAE	
	In	Out	In	Out
HA	93.62	95.26	45.55	46.77
ARIMA [10]	143.63	145.55	125.75	138.61
STGCN [25]	42.79 ± 0.72	44.11 ± 0.38	34.02 ± 0.77	34.22 ± 0.59
DCRNN [17]	41.27 ± 0.67	43.56 ± 0.88	35.06 ± 0.61	34.70 ± 0.80
GW [21]	39.50 ± 1.30	40.17 ± 0.92	32.16 ± 0.49	33.36 ± 0.79
STDN [26]	38.83 ± 0.94	39.08 ± 0.98	30.05 ± 0.61	31.21 ± 0.73
DSTGCN	**37.95 ± 0.50**	**38.65 ± 0.30**	**29.90 ± 0.24**	**30.90 ± 0.63**

Evaluation of Modules: To further verify the effectiveness of each proposed module in DSTGCN, we implement several variants for an ablation study. The details of these variants are as follows: 1) ST-S: which removes the static spatial correlation module; 2) ST-D: which removes the dynamic spatial correlation module; 3) which do not stack multiple layers of TCN but directly use the output of the last TCN layer to produce the predictions.

The performance of all variants is summarized in Table 3 and Table 4. When predicting passenger flows for the next time interval, ST-D achieves RMSE and MAE of inbound and outbound 43.38 ± 0.32, 45.32 ± 0.53, 37.64 ± 0.41, and 40.42 ± 0.21 on SHMetro, respectively, ranking last among all the variants. Similar trends are observed on HZMetro. Similar trends are observed on HZMetro. ST-S shows improvement compared to ST-D, potentially due to more accurate

Table 3. Effect of static flow pattern graph modeling, dynamic transit graph modeling, and multi-resolution temporal information - SHMetro.

Methods	RMSE		MAE	
	In	Out	In	Out
ST-S	42.50 ± 0.50	44.10 ± 0.18	34.55 ± 0.19	35.85 ± 0.37
ST-D	43.38 ± 0.32	45.32 ± 0.53	37.64 ± 0.41	40.42 ± 0.21
ST-T	43.23 ± 0.26	44.90 ± 0.24	35.16 ± 0.46	36.21 ± 0.77
DSTGCN	**41.98 ± 0.42**	**43.95 ± 0.49**	**32.21 ± 0.91**	**34.22 ± 0.59**

modeling of station spatial correlations using OD information. In addition, the results in Table 3 and Table 4 show the performance of ST-T is worse than the DSTGCN, indicating the necessity of leveraging multi-resolution temporal information. This is because passenger journeys have a certain duration, resulting in delayed temporal dependencies between stations.

Table 4. Effect of static flow pattern graph modeling, dynamic transit graph modeling, and multi-resolution temporal information - HZMetro.

Methods	RMSE		MAE	
	In	Out	In	Out
ST-S	38.49 ± 0.34	39.56 ± 0.25	30.91 ± 0.15	31.72 ± 0.32
ST-D	40.04 ± 0.30	40.17 ± 0.92	32.16 ± 0.49	33.36 ± 0.79
ST-T	39.16 ± 0.39	40.10 ± 0.23	31.14 ± 0.22	32.38 ± 0.29
DSTGCN	$\mathbf{37.95 \pm 0.50}$	$\mathbf{38.65 \pm 0.30}$	$\mathbf{29.90 \pm 0.24}$	$\mathbf{30.90 \pm 0.63}$

6   Conclusion

We propose DSTGCN, a novel OD-enhanced spatial-temporal dynamic graph convolution network, for predicting passenger flows in urban metro stations. DSTGCN effectively captures the traffic patterns and dynamic spatial correlations among different stations by incorporating OD information. Furthermore, DSTGCN captures delayed temporal dependencies arising from the travel time between distinct stations. We conducted extensive experiments on two datasets, and the results show that our proposed model outperforms the six baselines in terms of RMSE and MAE.

References

1. Gong, Y., Li, Z., Zhang, J., Liu, W., Zheng, Y.: Online spatio-temporal crowd flow distribution prediction for complex metro system. IEEE Trans. Knowl. Data Eng. **34**(2), 865–880 (2020)
2. Yu, H., Liu, A., Wang, B., Li, R., Zhang, G., Lu, J.: Real-time decision making for train carriage load prediction via multi-stream learning. In: Gallagher, M., Moustafa, N., Lakshika, E. (eds.) AI 2020. LNCS (LNAI), vol. 12576, pp. 29–41. Springer, Cham (2020). https://doi.org/10.1007/978-3-030-64984-5_3
3. Ma, X., Zhang, J., Du, B., Ding, C., Sun, L.: Parallel architecture of convolutional bi-directional LSTM neural networks for network-wide metro ridership prediction. IEEE Trans. Intell. Transp. Syst. **20**(6), 2278–2288 (2018)
4. Lv, M., Hong, Z., Chen, L., Chen, T., Zhu, T., Ji, S.: Temporal multi-graph convolutional network for traffic flow prediction. IEEE Trans. Intell. Transp. Syst. **22**(6), 3337–3348 (2020)

5. Gao, A., Zheng, L., Wang, Z., Luo, X., Xie, C., Luo, Y.: Attention based short-term metro passenger flow prediction. In: Qiu, H., Zhang, C., Fei, Z., Qiu, M., Kung, S.-Y. (eds.) KSEM 2021. LNCS (LNAI), vol. 12817, pp. 598–609. Springer, Cham (2021). https://doi.org/10.1007/978-3-030-82153-1_49

6. Yang, X., Xue, Q., Ding, M., Wu, J., Gao, Z.: Short-term prediction of passenger volume for urban rail systems: a deep learning approach based on smart-card data. Int. J. Prod. Econ. **231**, 107920 (2021)

7. Yu, H., Lu, J., Liu, A., Wang, B., Li, R., Zhang, G.: Real-time prediction system of train carriage load based on multi-stream fuzzy learning. IEEE Trans. Intell. Transp. Syst. **23**(9), 15155–15165 (2022)

8. Zhang, J., Chen, F., Cui, Z., Guo, Y., Zhu, Y.: Deep learning architecture for short-term passenger flow forecasting in urban rail transit. IEEE Trans. Intell. Transp. Syst. **22**(11), 7004–7014 (2020)

9. Jing, Y., Hu, H., Guo, S., Wang, X., Chen, F.: Short-term prediction of urban rail transit passenger flow in external passenger transport hub based on LSTM-LGB-DRS. IEEE Trans. Intell. Transp. Syst. **22**(7), 4611–4621 (2020)

10. Williams, B.M., Hoel, L.A.: Modeling and forecasting vehicular traffic flow as a seasonal ARIMA process: theoretical basis and empirical results. J. Transp. Eng. **129**(6), 664–672 (2003)

11. Sun, Y., Leng, B., Guan, W.: A novel wavelet-SVM short-time passenger flow prediction in Beijing subway system. Neurocomputing **166**, 109–121 (2015)

12. Zhang, W.: Graph based approach to real-time metro passenger flow anomaly detection. In: 2021 IEEE 37th International Conference on Data Engineering (ICDE), pp. 2744–2749. IEEE (2021)

13. Li, B., Guo, T., Li, R., Wang, Y., Gandomi, A.H., Chen, F.: A two-stage self-adaptive model for passenger flow prediction on schedule-based railway system. In: Gama, J., Li, T., Yu, Y., Chen, E., Zheng, Y., Teng, F. (eds.) Advances in Knowledge Discovery and Data Mining, PAKDD 2022. LNCS, vol. 13282, pp. 147–160. Springer, Cham (2022). https://doi.org/10.1007/978-3-031-05981-0_12

14. Wang, J., Zhang, Y., Wei, Y., Hu, Y., Piao, X., Yin, B.: Metro passenger flow prediction via dynamic hypergraph convolution networks. IEEE Trans. Intell. Transp. Syst. **22**(12), 7891–7903 (2021)

15. He, C., Wang, H., Jiang, X., Ma, M., Wang, P.: Dyna-PTM: OD-enhanced GCN for metro passenger flow prediction. In: 2021 International Joint Conference on Neural Networks (IJCNN), pp. 1–9. IEEE (2021)

16. Noursalehi, P., Koutsopoulos, H.N., Zhao, J.: Dynamic origin-destination prediction in urban rail systems: a multi-resolution spatio-temporal deep learning approach. IEEE Trans. Intell. Transp. Syst. **23**, 5106–5115 (2021)

17. Li, Y., Yu, R., Shahabi, C., Liu, Y.: Diffusion convolutional recurrent neural network: data-driven traffic forecasting. arXiv preprint arXiv:1707.01926 (2017)

18. Rakthanmanon, T., et al.: Searching and mining trillions of time series subsequences under dynamic time warping. In: Proceedings of the 18th ACM SIGKDD International Conference on Knowledge Discovery and Data Mining, pp. 262–270 (2012)

19. Martineau, J., Finin, T.: Delta TFIDF: an improved feature space for sentiment analysis. In: Proceedings of the International AAAI Conference on Web and Social Media, vol. 3, pp. 258–261 (2009)

20. Defferrard, M., Bresson, X., Vandergheynst, P.: Convolutional neural networks on graphs with fast localized spectral filtering. In: Advances in Neural Information Processing Systems, vol. 29 (2016)

21. Wu, Z., Pan, S., Long, G., Jiang, J., Zhang, C.: Graph WaveNet for deep spatial-temporal graph modeling. arXiv preprint arXiv:1906.00121 (2019)
22. Fang, S., Zhang, Q., Meng, G., Xiang, S., Pan, C.: GSTNet: global spatial-temporal network for traffic flow prediction. In: IJCAI, pp. 2286–2293 (2019)
23. Huber, P.J.: Robust estimation of a location parameter. In: Kotz, S., Johnson, N.L. (eds.) Breakthroughs in Statistics. Springer Series in Statistics, pp. 492–518. Springer, New York (1992). https://doi.org/10.1007/978-1-4612-4380-9_35
24. Kingma, D.P., Ba, J.: Adam: a method for stochastic optimization. arXiv preprint arXiv:1412.6980 (2014)
25. Yu, B., Yin, H., Zhu, Z.: Spatio-temporal graph convolutional networks: a deep learning framework for traffic forecasting. arXiv preprint arXiv:1709.04875 (2017)
26. Yao, H., Tang, X., Wei, H., Zheng, G., Li, Z.: Revisiting spatial-temporal similarity: a deep learning framework for traffic prediction. In: Proceedings of the AAAI Conference on Artificial Intelligence, vol. 33, pp. 5668–5675 (2019)

Enhancing Heterogeneous Graph Contrastive Learning with Strongly Correlated Subgraphs

Yanxi Liu[1(✉)] and Bo Lang[1,2]

[1] State Key Laboratory of Software Development Environment, Beihang University, Beijing, China
buaaliuyanxi@buaa.edu.cn
[2] Zhongguancun Laboratory, Beijing, China

Abstract. Graph contrastive learning maximizes the mutual information between the embedding representations of the same data instances in different augmented views of a graph, obtaining feature representations for graph data in an unsupervised manner without the need for manual labeling. Most existing node-level graph contrastive learning models only consider embeddings of the same node in different views as positive sample pairs, ignoring rich inherent neighboring relation and resulting in certain contrastive information loss. To address this issue, we propose a heterogeneous graph contrastive learning model that incorporates strongly correlated subgraph features. We design a contrastive learning framework suitable for heterogeneous graphs and introduce high-level neighborhood information during the contrasting process. Specifically, our model selects a strongly correlated subgraph for each target node in the heterogeneous graph based on both topological structure information and node attribute feature information. In the calculation of contrastive loss, we perform feature shifting operations on positive and negative samples based on subgraph encoding to enhance the model's ability to discriminate between approximate samples. We conduct node classification and ablation experiments on multiple public heterogeneous datasets and the results verify the effectiveness of the research contributions of our model.

Keywords: Graph representation learning · Contrastive learning · Correlated subgraph

1 Introduction

Currently, graph data has become the mainstream representation method for data in different application scenarios such as social networks and citation networks. Graph representation learning, which can learn general low-dimensional feature representations for data units in the graph that are not oriented towards specific tasks, has become an effective solution for analyzing graph-structured data. Among these methods, graph neural network(GNN) models [12,20,34] based on the "message passing paradigm" update the representations of the target nodes by aggregating the features of neighbors, achieving good results.

B. Luo et al. (Eds.): ICONIP 2023, LNCS 14452, pp. 86–102, 2024.
https://doi.org/10.1007/978-981-99-8076-5_7

Most existing GNN models adopt a 'semi-supervised learning paradigm', relying on partially annotated label information. In practice, manual labeling of graph data is costly. In order to study graph data unsupervisedly, graph contrastive learning(GCL) models [17,32,35,50] maximize the approximation between the representations of the target sample and similar data instances in different views by contrasting between positive and negative samples.

Existing GCL models are mostly oriented towards homogeneous graphs, and there are few GCL models for heterogeneous graphs, which contain multiple types of nodes and edges. Most existing homogeneous GCL models [35,46,49] only perform augmentation operations by perturbing and corrupting the structural and feature information of the graph. This strategy lacks interpretability and relies on serendipity, and it is difficult to directly apply to heterogeneous graphs. Designing reasonable augmentation and contrasting object selecting strategies for heterogeneous nodes and edges is challenging. In addition, existing node-level GCL models [22,35,41,46] only directly contrast node pair embeddings while ignoring the semantic features implied by the neighborhood structure around the node, discarding a large amount of valuable contrasting information. Accordingly, how to reasonably select heterogeneous neighbors and use their features during the contrasting process is worth studying.

To address the above issues and obtain effective heterogeneous GCL model, we make improvements in the following directions:

1. We propose a contrastive model for heterogeneous graph that combines node attribute and structural characteristics to select less important edges for deletion to complete heterogeneous graph augmentation, and designs an adaptive strategy in the contrast loss function to determine positive and negative samples.
2. We propose a strongly correlated subgraph selection method that combines the PageRank diffusion matrix with the hyperbolic distance between attribute features, to select a subgraph composed of heterogeneous nodes with high topological and attribute similarity for each target node.
3. We design a subgraph encoding module that performs space mapping on nodes within each subgraph according to their types, and obtains the overall representation of each subgraph through a readout operation. Subsequently, we use the obtained subgraph features to implement feature shifting operations during the calculation of contrasting loss to enhance the discriminative ability of the contrastive model.
4. Node classification experiments on multiple public heterogeneous graph datasets verify the performance advantages of our model, and multiple ablation experiments demonstrate the improvements brought by each module in the model.

The rest of this paper is organized as follows: Sect. 2 summarizes the research status and representative work of graph representation learning and unsupervised GCL. Section 3 introduces some necessary preliminaries involved in the model. Section 4 provides detailed introductions to each module of our model. Section 5 summarizes and reports the experimental performance results of the model on multiple public datasets. Section 6 summarizes the paper and presents conclusions.

2   Related Work

2.1   Graph Representation Learning

Graph representation learning models were first proposed by Gori M et al. [8,28] to extract latent features from graph data. The initial shallow embedding models (including the random walk based methods [9,26,31] and matrix factorization methods [1,48]) usually treated the target embeddings as model parameters to optimized during training.

Recently, the introduction of GNN models has greatly improved the performance of graph representation learning. Among GNN models, spectral-based methods [20,42] typically implement convolution operations for graph data in the frequency domain using Laplace matrices and Chebyshev polynomials. Spatial-based methods [12,34] consider graph representation learning as a process in which each node updates its own representation by aggregating messages from its neighbors based on the topological structure of the graph, a pattern commonly referred to as "message passing" [7].

To deal with the multiple types of nodes and edges in heterogeneous graphs, heterogeneous models often use auxiliary predefined meta-paths to express compound semantic relationships. For example, Metapath2Vec [4] uses meta-path to guide the random walk sampling process to obtain shallow embeddings. HAN [39] integrates embeddings under different meta-path using an attention mechanism. NEP [44] establishes neural network modules corresponding to edge types and propagates node embeddings along meta-path using a label propagation algorithm.

Most existing graph representation learning models are defined in Euclidean space, but some models [10,19,51] choose to define the model in hyperbolic space to reduce data distortion problems on highly hierarchical data [2,23]. Nickel et al. [24] proposed a shallow embedding graph model in hyperbolic space. HHNE [40] uses hyperbolic space distance as the loss function of the model to measure and optimize the similarity of node embedding representations. HGNN [23] and HGCN [2] implement similar hyperbolic GNN models using the exponential map.

2.2   Graph Contrastive Learning

Inspired by contrastive models in the image domain [3,15,36,43], GCL models are typically composed of three parts: data augmentation, encoder encoding, and contrasting loss calculation. Existing GCL models can be roughly divided into same-scale and cross-scale contrasting. Same-scale contrasting schemes use graph data unit instances (such as nodes and nodes) at same scale as contrast objects. For example, GRACE [49] completes data augmentation through edge deletion and feature masking, and optimizes the model using node-level infoNCE loss. GCA [50] augments both the adjacent and feature matrices, contrasts at the node level. GraphCL [46] uses different views of the same graph as contrast objects to maximize their consistency. CSSL [47] introduces node insertion as extra augmentation operations.

Cross-scale models contrast data instances at different scales. For example, DGI [35] obtains augmented views by the column shuffling on the feature matrix, and then maximizes the mutual information between the node embeddings and the global features obtained by pooling operations. HDGI [27] and ConCH [21] propose full-graph heterogeneous contrasting models using metapaths. SLiCE [38] samples a context subgraph for each target node and maximizes the consistency between them. HGCL [18] proposes a novel GCL framework that can hierarchically capture the structural semantics of graphs at both the node and graph levels. G-SupCon [30] uses subgraph encoding and multi-scale contrasting for efficient few-shot node classification.

3   Preliminaries

A graph can be denoted formally as $G = (V, E, \Phi)$, where $V = (v_1, v_2, \cdots, v_N)$ denotes the set of all the nodes in the graph and N is the number of nodes. E represents the set of edges and Φ is the type mapping function. Heterogeneous graphs contain multiple types of nodes and edges, i.e., $|\Phi(V)| + |\Phi(E)| > 2$. Homogeneous graphs, on the other hand, contain only one type of node and one type of edge, i.e., $|\Phi(V)| = |\Phi(E)| = 1$. Homogeneous graphs can also be written as $G = (A, X)$, with the adjacency matrix $A \in [0, 1]^{N \times N}$ and the initial node feature matrix $X = (x_1, x_2, \cdots, x_N)$. $x_i \in R^d$ represents the feature of node v_i with dimension d. The graph representation learning model can be represented by Eq. (1).

$$f : (V, E, X) \rightarrow Z \in R^{N \times d'} \tag{1}$$

In the formula, f is the encoding function such as GNN encoder, Z is the obtained feature or embedding matrix with dimension $d' \ll d$. The obtained low-dimensional representation can be further applied to downstream tasks. As the most successful direction in graph representation learning, GNNs [12,20,34] update the representation of the central target node by aggregating neighbors' features. The aggregation process at the l-th layer can be represented by Eq. (2):

$$h^{(l+1)}(v_i) = \sigma \left(\sum_{u_j \in N(v_i)} \alpha(v_i, u_j) h^{(l)}(u_j) W^{(l)} \right) \tag{2}$$

Where $\sigma()$ is the activation function, the aggregation weight $\alpha(v_i, u_j)$ is obtained through an attention mechanism or directly using a normalized adjacency matrix. $h^{(l)}(u_j)$ represents the feature representation of node u_j at layer l, and $W^{(l)}$ is a trainable parameter matrix.

To avoid reliance on manual labels, GCL models maximize the mutual information between node features and their representations. Such models typically consists of three modules: data augmentation, encoder encoding, and loss function calculation. Given a graph input $G = (A, X)$, the $i - th$ data augmentation operation T_i is applied to the adjacency matrix A and feature matrix X to obtain different views $\widetilde{G}_i$, as in Eq. (3):

$$\widetilde{G}_i = T_i (A, X) = \left(\widetilde{A}, \widetilde{X} \right) \tag{3}$$

The obtained multiple views are then encoded by the encoder as $h_i = f\left(\widetilde{G}_i \right)$. Structural graph augmentation operations mainly include edge perturbation, edge diffusion, etc. The edge perturbation operation can be represented by Eq. (4):

$$T_A^{(pert)}(A) = A * (1 - 1_p) + (1 - A) * 1_p \tag{4}$$

Where 1_p is the location indicator obtained by sampling with probability p, and its internal elements are 1 or 0. Edge diffusion operations create new edges based on random walks to obtain a new diffusion matrix that reflects global information. For example, the Personalized PageRank matrix $T_A^{(PPR)}$ can be represented by Eq. (5), where α represents the random walk transition probability and $\widetilde{D}$ is the symmetric diagonal matrix with $\widetilde{A} = A + I_N$ and $\widetilde{D}_{ii} = \sum_{j=1}^{N} \widetilde{A}_{ij}$.

$$T_A^{(PPR)} = \alpha \left(I_n - (1 - \alpha) \widetilde{D}^{-1/2} \widetilde{A} \widetilde{D}^{-1/2} \right)^{-1} \tag{5}$$

After obtaining the embeddings corresponding to different views, GCL models perform unsupervised training of the model by maximizing the mutual information, which can be calculated by the Kullback-Leibler (KL) divergence, between them. To improve computational efficiency, in practice, GCL models usually approximate the lower bound of mutual information using several estimators. For example, when calculating the contrast loss for node embeddings h and h' obtained from views G and G' using the infoNCE estimator [11], Eq. (6) can be used:

$$\mathcal{L}_{infoNCE} = -\frac{1}{N} \sum_{i=1}^{N} \mathcal{L}_{infoNCE}^i$$

$$= \frac{1}{N} \sum_{i=1}^{N} log \frac{\sum_{j \in POS_i} e^{\mathcal{D}(\mathbf{h}_i, \, \mathbf{h}_j')/\tau}}{\sum_{j \in POS_i} e^{\mathcal{D}(\mathbf{h}_i, \, \mathbf{h}_j')/\tau} + \sum_{k \in NEG_i} e^{\mathcal{D}(\mathbf{h}_i, \, \mathbf{h}_k')/\tau}} \tag{6}$$

Where POS_i and NEG_i represent the positive and negative example sets of node i, respectively. The discriminator $\mathcal{D} : \mathbb{R}^d \times \mathbb{R}^d \rightarrow \mathbb{R}$ is used to measure the approximation between two embedding results, and the temperature coefficient τ is usually used as a hyperparameter to control the degree of smoothness. Optimizing the above loss function makes the similarity scores between positive pairs much higher than those between negative pairs.

4   Method

4.1   Overall Framework

Similar to conventional GCL models, as shown in Fig. 1, our model is also composed of three modules: data augmentation, encoder encoding, and contrastive loss calculation. Each module is introduced below.

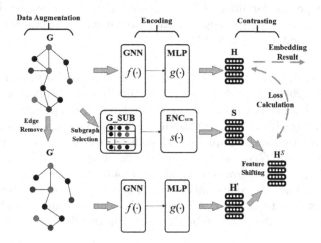

Fig. 1. Model framework

Data Augmentation. MVGRL [14] pointed out that data augmentation operations in the feature space degrades the performance of the model. Therefore, we choose to perform augmentation operations on the topological structure in our model. Specifically, we perform edge deletion augmentation operations on subgraphs corresponding to heterogeneous edges. Compared to existing models [32,49] that randomly select edges to delete, we choose to use following Eq. (7) to calculate the hyperbolic distance [2] between the features of the two endpoints of each edge of type e, with the Lorentzian scalar product $\langle \mathbf{x}, \mathbf{y} \rangle_{\mathcal{L}} = -x_0 y_0 + \sum_{i=1}^{d} x_n y_n$ and the trainable negative curvature k, to better capture hierachical characteristic.

$$d_{\mathcal{L}}^{k}(\mathbf{x}, \mathbf{y}) = \sqrt{k} arcosh \left(-\frac{\langle \mathbf{x}, \mathbf{y} \rangle_{\mathcal{L}}}{k} \right) \tag{7}$$

Then we use Eq. (8) to normalize the distance to the range $[0, 1]$:

$$d_i^e = \frac{d_i^e - d_{min}^e}{d_{max}^e - d_{min}^e} \tag{8}$$

Afterwards, we use the Bernoulli distribution to sample edges of type e, which can be expressed by Eq. (9), where m_e is the number of edges of the corresponding type.

$$B^e = \left(Bernouli\left(d_1^e\right), \cdots, Bernouli\left(d_{m_e}^e\right)\right), \ B^e \in [0,1]^{1 \times m_e} \qquad (9)$$

After that, we select the edges corresponding to the sampling value of 1 for edge deletion accordingly and obtain the augmented adjacency matrix $A^{e'}$. After repeating this process for different types of subgraphs, we merge the subgraphs to complete the multi-subgraph augmentation operation for heterogeneous graphs and obtain the augmented view G'.

Encoder Encoding. As shown in Fig. 1, we use GNN and MLP(multi-layer perceptron) to encode different data views to obtain the embedding matrices H and H'. We propose a heterogeneous strongly correlated subgraph selection strategy to better utilize the high-order neighborhood characteristics, which selects nodes with high correlation in the graph for each node to form a strongly correlated heterogeneous subgraph G_{SUB} and introduces a heterogeneous subgraph encoding module ENC_{SUB}. This allows each node to obtain the corresponding subgraph representation. Then we concat all subgraph encodings to obtain the subgraph feature matrix S. The selection strategy and encoding implementation of the strongly correlated subgraph will be introduced in detail in Sect. 4.2.

Contrastive Loss Calculation. Most existing node-level graph contrastive models merely use the representation of the target node in another view as the only positive sample. Intuitively, more contrastive information can be provided by adding similar nodes into positive set. In heterogeneous graphs, there are multiple types of nodes, and it is not reasonable to introduce first-order neighbors as positive examples as in [22]. To solve this problem, we propose a positive example expansion strategy that integrates topological and feature information to select similar nodes in heterogeneous graphs.

First, we use the graph diffusion matrix $\mathbf{\Pi}_{ppr}$ based on personalized Pagerank [25] proposed in APPNP [6] to calculate the importance between nodes from the topological structure level and select the relevant node set. The expression of $\mathbf{\Pi}_{ppr}$ is shown in Eq. (5), where the element $\mathbf{\Pi}_{ppr}^{(i,j)}$ represents the importance between nodes i and j that integrates high-order topological information. Afterwards, for the target node i, we select the top $2K$ elements in terms of importance score and preliminarily determine a strongly correlated set through the corresponding node index function **idx**. This process can be expressed by Eq. (10).

$$POS_i^{ppr} = \left\{ \mathbf{idx}\left(TOP\left(\mathbf{\Pi}_{ppr\,i}, \ 2K\right)\right)\right\} \qquad (10)$$

Afterwards, in order to use node attribute features, we also use hyperbolic distance as a similarity measurement to further filter and select similar nodes. In the screening process, we only select nodes of the same type as the target node from the qualified nodes to obtain the final positive example set. This process is shown in Eq. (11).

$$POS_i = \left\{ j \,\middle|\, j \in \mathbf{idx} \left(TOP \left(d \left(X_i, X_{POS_i^{ppr}} \right), K \right) \right) \cap \Phi(i) = \Phi(j) \right\} \quad (11)$$

In the above formula, we calculate the K nodes with the smallest hyperbolic distance between features and select nodes of the same type as the target node i through the type mapping function Φ as the final target positive example set. Accordingly, we regard all other nodes of the same type outside the positive example set as negative examples of target node i, as shown in Eq. (12):

$$NEG_i = \left\{ j \,\middle|\, j \in (V - POS_i) \cap \Phi(i) = \Phi(j) \right\} \quad (12)$$

Afterwards, as shown in Fig. 1, we perform a feature shifting operation during contrastive loss calculation. The specific implementation will be given in Sect. 4.3.

4.2 Strongly Correlated Subgraph Selection and Encoding

Strongly Correlated Subgraph Selection. In order to utilize the high-order neighborhood information in node-level heterogeneous GCL models, we choose to sample and select several nodes with high similarity for each target node to form a strongly correlated heterogeneous subgraph. Specifically, first, we use the personalized Pagerank [25] graph diffusion matrix Π_{ppr} to initially select a set of similar nodes for the target node i. This process can be expressed by Eq. (13), which retains $2K$ nodes with the highest Π_{ppr} correlation score with the target node i, where K is a hyperparameter used to control the scale of the strongly correlated subgraph. When K increases, the larger strongly correlated subgraph contains more neighborhood information, but it will also cause the correlation of the nodes within the subgraph to decrease.

$$G\_SUB_i^{ppr} = \left\{ \mathbf{idx} \left(TOP \left(\Pi_{ppr_i}, 2K \right) \right) \right\} \quad (13)$$

Then we also further screen the set based on the hyperbolic feature distance between the attribute features of the elements within the set and the target node i. This process is shown in Eq. (14). When the attribute feature distance between any node and the target node is small, their correlation is high.

$$G\_SUB_i = \left\{ j \,\middle|\, j \in \mathbf{idx} \left(TOP \left(d \left(X_i, X_{G\_SUB_i^{ppr}} \right), K \right) \right) \right\} \quad (14)$$

Unlike the positive example expansion strategy in Sect. 4.1, the subgraph selection does not impose restrictions on node types, which means that the set G_SUB_i may contain nodes of different types from the target node i. In fact, heterogeneous nodes in the original graph that are highly associated with the target node can also provide valuable contrastive information if selected.

Heterogeneous Subgraph Encoding. In order to learn a unified representation for the multi-type nodes within the strongly correlated subgraph as the overall feature, we introduce an additional heterogeneous subgraph encoder. Specifically, given the embedding matrix H obtained by the GNN encoder, for the target node v_i and the $\Phi(j)$-typed node v_j in its strongly correlated subgraph set G_SUB_i, we first perform space mapping based on the node type. This process can be expressed by Eq. (15), where $\mathcal{S}_{\Phi(j)}$ is a space mapping encoder composed of a type-specific linear transformation and an activation function.

$$S_i^{(j)} = \mathcal{S}_{\Phi(j)}\left(H_j\right) \tag{15}$$

After that, we use the readout operation to integrate the features of all nodes within the set into a vector with a dimension of $1 \times d'$, which serves as the overall representation of the subgraph corresponding to the target node i. This process can be expressed by Eq. (16), where the *Readout* function can be implemented using mean or max pooling operations.

$$S_i \in R^{1 \times d'} = Readout\left(\left\{S_i^{(j)} \mid j \in G_{SUBi}\right\}\right) \tag{16}$$

Accordingly, we can obtain the subgraph feature matrix composed of the subgraph features of all target nodes, which can be expressed by Eq. (17):

$$S = Concat\left(\{S_i \mid v_i \in V\}\right) \tag{17}$$

4.3   Feature Shifting and Loss Calculation

After obtaining the subgraph feature matrix S, we will use this matrix to perform feature shift operations on the embedding H' obtained from the augmented view G' to obtain the shifted embedding matrix H_s. The feature shift process is shown in Fig. 2. In the figure, after obtaining the feature of the strongly correlated subgraph containing multiple types of nodes (corresponding to the green virtual node in the figure), positive and negative samples of the same type as the blue target node are respectively shifted along the direction of the virtual subgraph feature, and the features obtained after shifting correspond to nodes with blue stripes.

After the feature shifting, positive samples will move away from the target node to a certain extent in the embedding space, while negative samples will move closer. Since the optimization process of contrastive learning will learn similar representations for positive sample pairs, after feature shifting, the model will reduce the distance between the target node and farther-moved positive samples, thereby making the embeddings of original positive samples closer to the target node. Similarly, the model will increase the distance between the target node and closer-moved negative samples, thereby making the embeddings of original negative samples further away from the target node. However, during testing, we will not perform feature shift operations, which makes the model more discriminatable between positive and negative samples on the test set.

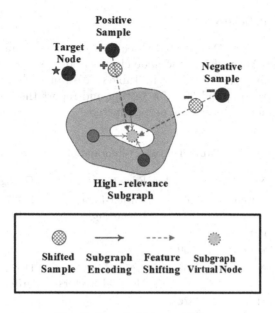

Fig. 2. Sample feature shifting

The above feature shifting process can be expressed by Eq. (18), where the weight coefficient α is a hyperparameter.

$$H^s = H^{'} + \alpha * S \qquad (18)$$

After feature shifting, we compare the embedding matrix H obtained from the input of the original graph with H^s and calculate the infoNCE loss. This process is shown in Eq. (19), where inter-pos loss term $\mathcal{S}^S(h_i, pos) = \sum_{j \in POS_i} e^{\frac{\mathcal{D}(\mathbf{h}_i, \mathbf{h}_j^S)}{\tau}}$ and intra-neg loss term $\mathcal{S}(h_i, neg) = \sum_{j \in NEG_i} e^{\frac{\mathcal{D}(\mathbf{h}_i, \mathbf{h}_j)}{\tau}}$, POS_i and NEG_i are defined in Eq. (11) and Eq. (12), respectively.

$$\mathcal{L} = \sum_{i=1}^{N} log \frac{\underbrace{\mathcal{S}^S(h_i, pos)}_{inter-pos} + \underbrace{\mathcal{S}(h_i, pos)}_{intra-pos}}{\underbrace{\mathcal{S}^S(h_i, pos)}_{} + \underbrace{\mathcal{S}(h_i, pos)}_{} + \underbrace{\mathcal{S}(h_i, neg)}_{intra-neg} + \underbrace{\mathcal{S}^S(h_i, neg)}_{inter-neg}} \qquad (19)$$

5   Experiments

We evaluate the model's performance on three public heterogeneous graph datasets through node classification tasks, answering the following research questions (RQ):

- RQ1: How does our model work under node classification tasks compared to the state-of-the-art comparison models?
- RQ2: How do the principal components of our model influent the performance?

5.1   Experiment Setup

Dataset and Metrics. We selected three heterogeneous graph datasets[1] and used Micro-F1 and Macro-F1 in the node classification task as the metrics. The information of each dataset is given in Table 1. We randomly split 80% of the target nodes for training and 20% for testing and report the average score for ten random seeds for each set of hyperparameters.

Table 1. Dataset information

Dataset	Node-type	# Nodes	Edge-type	# Edges
ACM	Author	7167	Author-Paper Paper-Subject	13407 4019
	Paper	4019		
	Subject	60		
IMDB	Actors	4353	Actors-Movies Movies-Directors	11028 3676
	Movies	3676		
	Directors	1678		
DBLP	Author	2000	Author-Paper Paper-Conference	18304 9556
	Paper	9556		
	Conference	20		

Comparison Methods. We selected several representative node-level graph representation learning models as comparison methods. Specifically, these include: two supervised homogeneous graph representation learning models: Deepwalk [26] and Line [31]; six supervised heterogeneous graph representation learning models: DHNE [33], Metapath2vec [4], Hin2vec [5], HERec [29], HeGAN [16] and ie-HGCN [45] (SOTA); DGI [35], BGRL [32], GRACE [49] and MVGRL [14]. The SOTA of the heterogeneous GCL models: HDGI [27].

We chose HeteroGraphConv[2] implemented in DGL [37] as the encoder of our model and we chose the discriminator function as dot-product. We implemented all methods on a Tesla V100-32 GPU. The relevant comparison methods mainly refer to OpenHGNN [13] and PyGCL[3]. We optimize the hyper-parameters with Optuna[4].

---

[1] https://github.com/Andy-Border/NSHE.
[2] https://docs.dgl.ai/generated/dgl.nn.pytorch.HeteroGraphConv.html.
[3] https://github.com/PyGCL/PyGCL.
[4] https://github.com/optuna/optuna.

5.2   Results and Analysis (RQ1)

The results of the node classification experiment are shown in Table 2, where the best result on each dataset is denoted with bolded text. From the results, the performance of supervised and unsupervised learning models among comparison methods is not much different. On the DBLP and IMDB datasets, unsupervised models even have a certain lead, reflecting the potential of unsupervised models. Within unsupervised GCL models, the overall performance of cross-scale models DGI and HDGI is significantly behind other contrastive models. This indicates that on heterogeneous graphs, directly contrasting the features of multiple types of nodes to obtain global representations does not conform to the data characteristics and degrades model performance.

Compared with the SOTA unsupervised comparison model, our model shows a 4.5% performance improvement based on the performance across three datasets. Compared with contrastive learning models DGI and HDGI, which also hope to introduce global information, our model has obvious advantages in performance. This shows that compared with global feature contrasting, strongly correlated subgraph features with finer granularity can better reflect the neighborhood characteristics of target nodes and achieve significant improvement. Compared with supervised models, our model also outperforms all comparison methods with at least a 2.6% performance improvement, except for slightly lower than ie-HGCN on the ACM dataset.

Table 2. Overall result of node classification

	Model	IMDB		ACM		DBLP	
		Micro-F1	Macro-F1	Micro-F1	Macro-F1	Micro-F1	Macro-F1
SUPERVISED	DeepWalk	56.52	55.24	82.17	81.82	89.44	88.48
	LINE-1st	43.75	39.87	82.46	82.35	82.32	80.20
	LINE-2nd	40.54	33.06	82.21	81.32	88.76	87.35
	DHNE	38.99	30.53	65.27	62.31	73.30	67.61
	Metapath2Vec	51.90	50.21	83.61	82.77	89.36	87.95
	HIN2Vec	48.02	46.24	54.30	48.59	90.30	89.46
	HERec	54.48	53.46	81.89	81.74	86.21	84.55
	HeGAN	58.56	57.12	83.09	82.94	90.48	89.27
	ie-HGCN	56.33	47.32	**86.29**	**86.14**	90.55	89.30
UNSUPERVISED	DGI	36.46	33.69	49.03	27.31	89.95	88.78
	GRACE	52.73	50.21	85.56	85.29	90.21	89.12
	MVGRL	56.25	54.99	81.97	81.22	90.12	89.14
	BGRL	54.28	53.04	82.76	82.33	89.22	87.90
	HDGI	38.67	34.13	74.58	71.43	85.14	82.77
	Our method	**60.79**	**59.94**	85.89	85.81	**91.20**	**90.18**

5.3 Ablation Study (RQ2)

In this paper, we propose a heterogeneous GCL model that integrates strongly correlated subgraph features based on topological similarity and hyperbolic feature similarity. After encoding the subgraph features, we perform feature shifting on positive and negative samples during the contrasting process. In order to verify the impact of the above design on the model, we selected the following two model variants for ablation experiments: (1) W/O HR: when selecting correlated subgraphs, nodes are randomly selected without calculating node similarity to form strongly correlated subgraphs. (2) W/O FS: During the contrasting loss calculation, no feature shifting operation is performed. This variant can be achieved by setting the parameter α in Eq. (18) to 0. The results of the two variants against the complete model are shown in Table 3.

From Table 3, it can be seen that both the strongly correlated subgraph selection strategy and the feature shifting operation have a certain improvement effect on model performance. The results show that the subgraph set randomly selected by W/O HR has low correlation, and the obtained subgraph features cannot reflect the neighborhood information of the target node. The feature shifting operation based on this feature will reduce the performance of the model. The performance of W/O FS on all datasets is also lower than that of the complete model, which reflects that the feature shifting can improve the effect of the contrastive model on specific samples.

Table 3. Ablation study

Model	IMDB		ACM		DBLP	
	Micro-F1	Macro-F1	Micro-F1	Macro-F1	Micro-F1	Macro-F1
W/O HR	57.31	56.22	84.47	83.61	88.93	88.84
W/O FS	58.05	57.01	84.72	84.13	89.96	89.83
Our complete model	**60.79**	**59.94**	**85.89**	**85.81**	**91.20**	**90.18**

6 Conclusions

In this paper, we propose a heterogeneous GCL model that integrates strongly correlated subgraph features. This method aims to introduce high-order neighborhood information into node-level contrastive tasks. To this end, the model first selects a strongly correlated subgraph for each target node, which contains nodes of different types with high topological similarity and feature similarity. Afterwards, the model introduces a subgraph encoding module, which obtains a unique overall feature for the strong subgraph of each target node through node-type space mapping and readout operations. During the calculation of contrastive loss, we perform feature shifting operations on the positive and negative sample features of the target node with the subgraph features, respectively. This operation enables the model to learn more discriminative representations for similar contrasting samples. We tested the effectiveness of our model on multiple public heterogeneous graph datasets, and the results proves that our model is superior to the SOTA unsupervised contrastive model. Ablation experiments also shows that both the strongly correlated subgraph selection strategy and feature shifting operation can improve the performance of the model.

References

1. Cao, S., Lu, W., Xu, Q.: GraRep: learning graph representations with global structural information. In: Proceedings of the 24th ACM International on Conference on Information and Knowledge Management, Melbourne, Australia, pp. 891–900. ACM (2015). https://doi.org/10.1145/2806416.2806512
2. Chami, I., Ying, Z., Ré, C., Leskovec, J.: Hyperbolic graph convolutional neural networks. In: Wallach, H., Larochelle, H., Beygelzimer, A., dAlché-Buc, F., Fox, E., Garnett, R. (eds.) Advances in Neural Information Processing Systems. Curran Associates, Inc. (2019)
3. Chen, T., Kornblith, S., Norouzi, M., Hinton, G.: A simple framework for contrastive learning of visual representations. In: Proceedings of the 37th International Conference on Machine Learning. PMLR (2020)
4. Dong, Y., Chawla, N.V., Swami, A.: metapath2vec: scalable representation learning for heterogeneous networks. In: The 23rd ACM SIGKDD International Conference (2017)
5. Fu, T., Lee, W.C., Lei, Z.: HIN2Vec: explore meta-paths in heterogeneous information networks for representation learning. In: Proceedings of the 2017 ACM on Conference on Information and Knowledge Management, Singapore, Singapore, pp. 1797–1806. ACM (2017). https://doi.org/10.1145/3132847.3132953
6. Gasteiger, J., Bojchevski, A., Günnemann, S.: Predict then propagate: graph neural networks meet personalized PageRank. In: International Conference on Learning Representations (2019)
7. Gilmer, J., Schoenholz, S.S., Riley, P.F., Vinyals, O., Dahl, G.E.: Neural message passing for quantum chemistry. In: Proceedings of the 34th International Conference on Machine Learning, ICML 2017, vol. 70. JMLR.org (2017)
8. Gori, M., Monfardini, G., Scarselli, F.: A new model for learning in graph domains. In: Proceedings of the 2005 IEEE International Joint Conference on Neural Networks. IEEE (2005). https://doi.org/10.1109/IJCNN.2005.1555942
9. Grover, A., Leskovec, J.: Node2vec: scalable feature learning for networks. In: Proceedings of the 22nd ACM SIGKDD International Conference on Knowledge Discovery and Data Mining. ACM (2016). https://doi.org/10.1145/2939672.2939754
10. Gulcehre, C., et al.: Hyperbolic attention networks. In: 2018 International Conference on Learning Representations (2018)
11. Gutmann, M., Hyvärinen, A.: Noise-contrastive estimation: a new estimation principle for unnormalized statistical models. In: Proceedings of the Thirteenth International Conference on Artificial Intelligence and Statistics. JMLR Workshop and Conference Proceedings (2010)
12. Hamilton, W.L., Ying, R., Leskovec, J.: Inductive representation learning on large graphs. In: Proceedings of the 31st International Conference on Neural Information Processing Systems, NIPS 2017. Curran Associates Inc. (2017)

13. Han, H., et al.: OpenHGNN: an open source toolkit for heterogeneous graph neural network. In: Proceedings of the 31st ACM International Conference on Information & Knowledge Management. ACM (2022). https://doi.org/10.1145/3511808.3557664

14. Hassani, K., Khasahmadi, A.H.: Contrastive multi-view representation learning on graphs. In: Proceedings of the 37th International Conference on Machine Learning, ICML 2020. JMLR.org (2020)

15. He, K., Fan, H., Wu, Y., Xie, S., Girshick, R.: Momentum contrast for unsupervised visual representation learning. In: Proceedings of the IEEE/CVF Conference on Computer Vision and Pattern Recognition, pp. 9729–9738 (2020)

16. Hu, B., Fang, Y., Shi, C.: Adversarial learning on heterogeneous information networks. In: Proceedings of the 25th ACM SIGKDD International Conference on Knowledge Discovery & Data Mining. ACM (2019). https://doi.org/10.1145/3292500.3330970

17. Jin, M., Zheng, Y., Li, Y.F., Gong, C., Zhou, C., Pan, S.: Multi-scale contrastive siamese networks for self-supervised graph representation learning. In: Proceedings of the Thirtieth International Joint Conference on Artificial Intelligence. International Joint Conferences on Artificial Intelligence Organization (2021). https://doi.org/10.24963/ijcai.2021/204

18. Ju, W., et al.: Unsupervised graph-level representation learning with hierarchical contrasts. Neural Netw. (2023). https://doi.org/10.1016/j.neunet.2022.11.019

19. Khrulkov, V., Mirvakhabova, L., Ustinova, E., Oseledets, I., Lempitsky, V.: Hyperbolic image embeddings. In: 2020 IEEE/CVF Conference on Computer Vision and Pattern Recognition (CVPR). IEEE (2020). https://doi.org/10.1109/CVPR42600.2020.00645

20. Kipf, T.N., Welling, M.: Semi-supervised classification with graph convolutional networks. In: 5th International Conference on Learning Representations, ICLR 2017, Conference Track Proceedings, Toulon, France, 24–26 April 2017. OpenReview.net (2017)

21. Li, X., Ding, D., Kao, B., Sun, Y., Mamoulis, N.: Leveraging meta-path contexts for classification in heterogeneous information networks. In: 2021 IEEE 37th International Conference on Data Engineering (ICDE). IEEE (2021). https://doi.org/10.1109/ICDE51399.2021.00084

22. Liu, J., Yang, M., Zhou, M., Feng, S., Fournier-Viger, P.: Enhancing hyperbolic graph embeddings via contrastive learning. In: 2nd Workshop on Self-Supervised Learning, NeurIPS 2021. arXiv arXiv:2201.08554, 35th Conference on Neural Information Processing Systems, NeurIPS 2021 (2022)

23. Liu, Q., Nickel, M., Kiela, D.: Hyperbolic graph neural networks. In: Advances in Neural Information Processing Systems. Curran Associates, Inc. (2019)

24. Nickel, M., Kiela, D.: Poincaré embeddings for learning hierarchical representations. In: Advances in Neural Information Processing Systems, vol. 30. Curran Associates, Inc. (2017)

25. Page, L., Brin, S., Motwani, R., Winograd, T.: The PageRank citation ranking: bringing order to the web (1999). http://ilpubs.stanford.edu:8090/422/

26. Perozzi, B., Al-Rfou, R., Skiena, S.: DeepWalk: online learning of social representations. In: Proceedings of the 20th ACM SIGKDD International Conference on Knowledge Discovery and Data Mining. ACM (2014). https://doi.org/10.1145/2623330.2623732

27. Ren, Y., Liu, B., Huang, C., Dai, P., Bo, L., Zhang, J.: HDGI: an unsupervised graph neural network for representation learning in heterogeneous graph. In: AAAI Workshop (2020)

28. Scarselli, F., Gori, M., Tsoi, A.C., Hagenbuchner, M., Monfardini, G.: The graph neural network model. IEEE Trans. Neural Netw. **20**, 61–80 (2009). https://doi.org/10.1109/TNN.2008.2005605

29. Shi, C., Hu, B., Zhao, W.X., Yu, P.S.: Heterogeneous information network embedding for recommendation. IEEE Trans. Knowl. Data Eng. (2019). https://doi.org/10.1109/TKDE.2018.2833443

30. Tan, Z., Ding, K., Guo, R., Liu, H.: Supervised graph contrastive learning for few-shot node classification. In: Amini, M.R., Canu, S., Fischer, A., Guns, T., Kralj Novak, P., Tsoumakas, G. (eds.) Machine Learning and Knowledge Discovery in Databases, ECML PKDD 2022. LNCS, vol. 13714, pp. 394–411. Springer, Cham (2023). https://doi.org/10.1007/978-3-031-26390-3_24

31. Tang, J., Qu, M., Wang, M., Zhang, M., Yan, J., Mei, Q.: LINE: large-scale information network embedding. In: Proceedings of the 24th International Conference on World Wide Web. International World Wide Web Conferences Steering Committee (2015). https://doi.org/10.1145/2736277.2741093

32. Thakoor, S., et al.: Large-scale representation learning on graphs via bootstrapping. In: International Conference on Learning Representations (2022)

33. Tu, K., Cui, P., Wang, X., Wang, F., Zhu, W.: Structural deep embedding for hyper-networks. Proc. AAAI Conf. Artif. Intell. **32**(1), 426–433 (2018). AAAI'18/IAAI'18/EAAI'18, AAAI Press (2018)

34. Veličković, P., Cucurull, G., Casanova, A., Romero, A., Liò, P., Bengio, Y.: Graph attention networks. In: 2018 International Conference on Learning Representations (2018)

35. Veličković, P., Fedus, W., Hamilton, W.L., Liò, P., Bengio, Y., Hjelm, R.D.: Deep graph infomax. In: International Conference on Learning Representations (2022)

36. Wang, C., Sun, D., Bai, Y.: PiPAD: pipelined and parallel dynamic GNN training on GPUs. In: Proceedings of the 28th ACM SIGPLAN Annual Symposium on Principles and Practice of Parallel Programming. ACM (2023). https://doi.org/10.1145/3572848.3577487

37. Wang, M., et al.: Deep graph library: a graph-centric, highly-performant package for graph neural networks (2019)

38. Wang, P., Agarwal, K., Ham, C., Choudhury, S., Reddy, C.K.: Self-supervised learning of contextual embeddings for link prediction in heterogeneous networks. In: 2021 Proceedings of the Web Conference. ACM (2021). https://doi.org/10.1145/3442381.3450060

39. Wang, X., et al.: Heterogeneous graph attention network. In: The World Wide Web Conference. ACM (2019). https://doi.org/10.1145/3308558.3313562

40. Wang, X., Zhang, Y., Shi, C.: Hyperbolic heterogeneous information network embedding. Proc. AAAI Conf. Artif. Intell. (2019). https://doi.org/10.1609/aaai.v33i01.33015337

41. Wang, Y., Wang, W., Liang, Y., Cai, Y., Liu, J., Hooi, B.: NodeAug: semi-supervised node classification with data augmentation. In: Proceedings of the 26th ACM SIGKDD International Conference on Knowledge Discovery & Data Mining. ACM (2020). https://doi.org/10.1145/3394486.3403063

42. Wu, F., Souza, A., Zhang, T., Fifty, C., Yu, T., Weinberger, K.: Simplifying graph convolutional networks. In: International Conference on Machine Learning, PMLR 2019, pp. 6861–6871. PMLR (2019)

43. Wu, Z., Xiong, Y., Yu, S.X., Lin, D.: Unsupervised feature learning via non-parametric instance discrimination. In: Proceedings of the IEEE Conference on Computer Vision and Pattern Recognition (2018)

44. Yang, C., Zhang, J., Han, J.: Neural embedding propagation on heterogeneous networks. In: 2019 IEEE International Conference on Data Mining (ICDM). IEEE (2019)
45. Yang, Y., Guan, Z., Li, J., Zhao, W., Cui, J., Wang, Q.: Interpretable and efficient heterogeneous graph convolutional network. IEEE Trans. Knowl. Data Eng. (2021). https://doi.org/10.1109/TKDE.2021.3101356
46. You, Y., Chen, T., Sui, Y., Chen, T., Wang, Z., Shen, Y.: Graph contrastive learning with augmentations. In: Larochelle, H., Ranzato, M., Hadsell, R., Balcan, M.F., Lin, H. (eds.) Advances in Neural Information Processing Systems, vol. 33. Curran Associates, Inc. (2020)
47. Zeng, J., Xie, P.: Contrastive self-supervised learning for graph classification. Proc. AAAI Conf. Artif. Intell. (2021). https://doi.org/10.1609/aaai.v35i12.17293
48. Zhang, J., Dong, Y., Wang, Y., Tang, J., Ding, M.: ProNE: fast and scalable network representation learning. In: IJCAI 2019, vol. 19, pp. 4278–4284 (2019)
49. Zhu, Y., Xu, Y., Yu, F., Liu, Q., Wu, S., Wang, L.: Deep graph contrastive representation learning. arXiv arXiv:2006.04131 [cs, stat] (2020)
50. Zhu, Y., Xu, Y., Yu, F., Liu, Q., Wu, S., Wang, L.: Graph contrastive learning with adaptive augmentation. In: Proceedings of the Web Conference 2021. ACM (2021). https://doi.org/10.1145/3442381.3449802
51. Zhu, Y., Zhou, D., Xiao, J., Jiang, X., Chen, X., Liu, Q.: HyperText: endowing FastText with hyperbolic geometry. In: Findings of the Association for Computational Linguistics, EMNLP 2020. Association for Computational Linguistics (2020). https://doi.org/10.18653/v1/2020.findings-emnlp.104

DRPDDet: Dynamic Rotated Proposals Decoder for Oriented Object Detection

Jun Wang, Zilong Wang[✉], Yuchen Weng, and Yulian Li

College of Information and Control Engineering, China University of Mining and Technology, Xuzhou, China
wangzilong@cumt.edu.cn

Abstract. Oriented object detection has gained popularity in diverse fields. However, in the domain of two-stage detection algorithms, the generation of high-quality proposals with a high recall rate remains a formidable challenge, especially in the context of remote sensing images where sparse and dense scenes coexist. To address this, we propose the DRPDDet method, which aims to improve the accuracy and recall of proposals for Oriented target detection. Our approach involves generating high-quality horizontal proposals and dynamically decoding them into rotated proposals to predict the final rotated bounding boxes. To achieve high-quality horizontal proposals, we introduce the innovative HarmonyRPN module. This module integrates foreground information from the RPN classification branch into the original feature map, creating a fused feature map that incorporates multi-scale foreground information. By doing so, the RPN generates horizontal proposals that focus more on foreground objects, which leads to improved regression performance. Additionally, we design a dynamic rotated proposals decoder that adaptively generates rotated proposals based on the constraints of the horizontal proposals, enabling accurate detection in complex scenes. We evaluate our proposed method on the DOTA and HRSC2016 remote sensing datasets, and the experimental results demonstrate its effectiveness in complex scenes. Our method improves the accuracy of proposals in various scenarios while maintaining a high recall rate.

Keywords: Oriented object detection · Harmony RPN · Foreground information · Dynamic rotated proposals decoder

1 Introduction

Remote sensing images, with their diverse applications in military reconnaissance, urban planning, disaster monitoring, and more, have become increasingly significant with the advancement of remote sensing technology. Object detection plays a vital role as a key technique in remote sensing image processing. Significant advancements have been

---

This work was supported by the Scientific Innovation 2030 Major Project for New Generation of AI, Ministry of Science and Technology of the People's Republic of China (No. 2020AAA0107300)

made in deep learning-based object detection methods such as DETR [1], RetinaNet [2], EfficientDet [3] and many others [4–6] in recent years. However, these methods are designed to detect objects within horizontal rectangular boxes, which poses challenges in accurately detecting rotated objects in remote sensing datasets with large aspect ratios and dense arrangements. The limitations of these methods result in incomplete coverage of rotated objects and their susceptibility to interference from surrounding elements. To overcome this challenge, researchers have proposed various rotated bounding box object detection methods [7–10]. These methods generate rotated prediction boxes to better fit the rotated objects found in remote sensing imagery. The proposed approaches can be generally categorized into single-stage and multi-stage methods.

In the realm of single-stage methods, several strategies build upon the fundamental principles of RetinaNet. For instance, S2ANet [11] creatively integrates a feature alignment module for optimal feature alignment along with an active rotation filter. This innovative structure enables the classification and regression branches to leverage unique features to tackle the problem of rotational invariance. Simultaneously, R3Det [12] employs deformable convolution and overlapping pooling techniques to create a refinement stage, effectively handling changes in object orientation and scale. RSDet [13] advances this approach further by introducing a residual rotation-sensitive unit. This unit enables the model to directly extract rotational features from the input feature maps, thereby improving the accuracy and robustness of detecting rotated objects.

On the other hand, multi-stage methods often adopt an anchor-free frame approach. For instance, EARL [14] employs an adaptive rotation label assignment strategy using an elliptical distribution. AOPG [15] improves the accuracy of oriented object detection through a direction-aware candidate box generation strategy. In a similar vein, DCFL [16] proposes a dynamic prior and a coarse-to-fine allocator to dynamically model prior information, label allocation, and object representation. This approach effectively alleviates issues of misalignment. Nevertheless, anchor-free methods often face a reduction in accuracy due to the lack of prior information typically provided by anchor boxes.

In multi-stage methods that utilize anchor boxes, the design of anchor boxes and proposals boxes has emerged as a significant area of research for rotating object detection. For instance, RRPN [17] introduces multiple anchor with varying angles, scales, and aspect ratios during the generation of rotated proposals. However, using multiple rotated anchor leads to an abundance of redundant candidate boxes, which in turn increases the computational burden and complexity in subsequent processing. Moreover, these methods heavily rely on carefully tuning the anchor settings and parameters, require additional adjustments and adaptability analysis on different datasets and scenarios.

To address these challenges, GlidingVertex [18] proposes a solution that generates horizontal proposals using a reduced number of horizontal anchor, effectively mitigating the issue of redundancy among candidate boxes. Nevertheless, AOPG's research reveals that in dense object arrangement scenarios, horizontal proposals often encompass multiple objects, posing difficulties for accurate target localization and classification within horizontal ROIs by RoIHead. Additionally, as horizontal proposals serve as the horizontal bounding box for regression targets, their shapes and sizes differ significantly from the actual regression targets. This mismatch compromises the model's robustness. To ensure the accuracy of proposals, some methods, such as AOPG, SCRDet [19], generate rotated proposals based on horizontal anchor. Conversely, research by R3Det suggests that rotated anchor outperforms horizontal anchor in dense scenes, while the latter achieve higher recall rates in sparse scenes. Notably, horizontal proposals are better suited for targets with large aspect ratios, as they provide more comprehensive coverage of the overall target position. As a result, approaches like R3Det and RoI Transformer [20] adopt a multi-stage methodology. In the initial stage, horizontal proposals are generated, and in the subsequent stage, rotated proposals are derived based on the initial horizontal proposals. However, the introduction of this two-stage process adds an extra learning stage, resulting in increased model complexity and additional training time costs.

In this study, we introduce a novel approach for Oriented object detection by proposing a dynamic rotated proposals decoder that adaptively generates rotated proposals based on the constraints of horizontal proposals. This approach effectively combines the strengths of both horizontal and rotated proposals, enables adaptability in dense scenes, and achieves high recall rates in sparse scenes. Our method has an important advantage in that it does not require an additional learning stage, and this preserves the efficiency of network training and inference. However, it is worth noting that this approach imposes higher precision requirements on horizontal proposals. In order to address this challenge, we introduce a dedicated module called HarmonyRPN, which is designed to meet the precision needs of horizontal proposals. In HarmonyRPN, we integrate the multi-scale foreground information predicted by the RPN classification branch with the original feature map generated by the FPN. Subsequently, the fused feature map is fed into the RPN bounding box branch to generate proposals. Through this design, we leverage the foreground information extracted by the RPN classification branch, leading to a significant improvement in the accuracy of horizontal proposals.

Our research has made notable contributions in the following three aspects:

- We propose a novel dynamic rotated proposals decoder that intelligently generates rotated proposals based on the constraints of horizontal proposals. By leveraging the strengths of both horizontal and rotated proposals, our decoder significantly improves the accuracy and recall rate of object detection across various scenarios.
- We have designed a novel module called HarmonyRPN, which enhances the performance of the bounding box branch by integrating multi-scale foreground information with the original feature map.
- We successfully integrate the proposals decoder and HarmonyRPN module into the Faster R-CNN architecture, leading to significant performance improvements. Experimental results demonstrate the effectiveness of our approach, achieving an impressive mAP of 75.97% on the DOTA dataset and demonstrating robust performance on the HRSC2016 dataset.

2 Proposed Method

2.1 The Overall Network Framework of DRPDDet

The workflow and submodules of our proposed method are outlined as follows. The whole framework of DRPD Det is depicted in Fig. 1.(a). We employ ResNet [21] as the backbone network to extract image features and utilize the FPN [22] to fuse multi-scale feature maps. The Harmony RPN module is then employed to generate accurate horizontal proposals, which are dynamically decoded by DRPD Module into rotated proposals. The feature maps corresponding to the rotated proposals are subsequently inputted into the classification and regression branches of Rotated ROI (Rotated Region of Interest) for multi-class probability prediction and rotation prediction box regression.

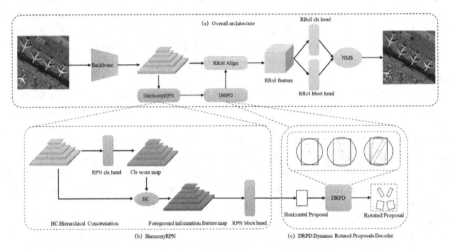

Fig. 1. The overall framework of DRPDDet.

The structure of Harmony RPN is illustrated in Fig. 1. (b). It performs hierarchical fusion across scales by combining the feature maps which outputted by FPN with the score matrix from the RPN classification branch. Subsequently, the fused feature maps are inputted into the RPN bounding box branch to generate horizontal proposals. The decoding process of the DRPD is illustrated in Fig. 1. (c). After obtaining the horizontal proposals, a series of rotated proposals is dynamically generated based on the constraints of the horizontal proposals.

2.2 Harmony RPN

When the RPN classification network predicts the foreground scores for anchor boxes of different shapes at each feature point on the feature map, we can utilize the foreground scores of anchor boxes at various positions and shapes to predict the potential locations and shapes of the proposed boxes. This information can serve as prior knowledge for the RPN bounding box network, guiding the generation of proposed box positions and shapes.

As depicted in Fig. 2 when the RPN bounding box branch predicts the offset of the black box, the left image illustrates that the foreground scores of the surrounding anchor provide prior information about the center point (x, y) of the black box. The right image demonstrates that the foreground scores of the surrounding anchor provide prior information about the shape size (w, h) of the black box.

Additionally, we observed that there are variations in the score matrices of different scales due to differences in receptive fields, scale, and resolution at each feature point

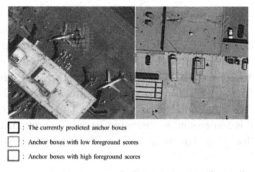

□ : The currently predicted anchor boxes
□ : Anchor boxes with low foreground scores
■ : Anchor boxes with high foreground scores

Fig. 2. Utilizing Anchor Foreground Scores for RPN Bounding Box Prediction.

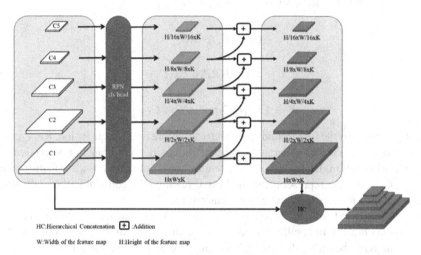

HC:Hierarchical Concatenation ⊞ :Addition

W:Width of the feature map   H:Height of the feature map

Fig. 3. Cross-Scale Hierarchical Fusion Strategy in Harmony RPN.

[23]. However, these score matrices of different scales still offer guidance for anchor at other scales. For instance, on a large-scale feature map, the foreground scores from a small-scale feature map can provide crucial prior information about size when predicting larger bounding boxes [24].

Therefore, we adopt a cross-scale hierarchical fusion strategy to merge the output feature maps from the FPN and the foreground score matrices from the RPN classification branch. This strategy enables the integration of multi-scale foreground score information for each feature map, which is used for the prediction of RPN bounding box branches. As illustrated in Fig. 3 during the fusion process, we initially employ the FPN feature fusion network to generate five feature maps of different scales. These feature maps have dimensions of $H \times W \times 256$, $H/2 \times W/2 \times 256$, $H/4 \times W/4 \times 256$, $H/8 \times W/8 \times 256$, and $H/16 \times W/16 \times 256$, respectively, and are denoted as C_1, C_2, C_3, C_4, and C_5. Here, H and W represent the height and width of the smallest-scale feature map, while K denotes the number of anchor boxes generated per feature point.

For C_1, we first obtain a foreground score matrix of size H × W × K through a 3 × 3 convolutional layer of the RPN classification branch. Then, we concatenate this foreground score matrix to the back of C_1, resulting in a feature map of size H × W × (256 + K) denoted as C_1'. Similarly, we perform the same operation on C_2. We obtain a foreground score matrix of size H/2 × W/2 × K through a 3 × 3 convolutional layer of the RPN classification branch. Before concatenating this foreground score matrix with C_2, we downsample the foreground score matrix of C_1 to obtain a size of H/2 × W/2 × K. Then, we add this foreground score matrix to the foreground score matrix of C_2, producing the fused foreground score matrix C_2'. The same operation is applied to C_3, C_4, and C_5.

Specifically, the fusion process for each layer is as follows:

$$S(C_i) = \text{RPN\_cls}(C_i) + \text{Pool}(\text{RPN\_cls}(C_{i-1})), i \in [2, 5] \tag{1}$$

$$C_{i'} = C_i \oplus S(C_i) \tag{2}$$

$$C_{1'} = C_1 \oplus \text{RPN\_cls}(C_1) \tag{3}$$

$$Y = L(C_{1'}, C_{2'}, C_{3'}, C_{4'}, C_{5'}) \tag{4}$$

$S(C_i)$ represents the foreground score matrices before multi-scale fusion. $\text{RPN\_cls}(\cdot)$ denotes the score matrices generated by the RPN classification branch. $\text{Pool}(\cdot)$ denotes the pooling operation. $L(\cdot)$ combines the feature maps into a multi-scale feature map. $C_{1'}, C_{2'}, C_{3'}, C_{4'}, C_{5'}$ represents the final fused feature map containing the multi-scale foreground information. $\oplus$ represents concatenation.

Finally, the fused multi-scale foreground information feature map is fed into the RPN bounding box branch. In parallel, we adjust the input channel of the 3 × 3 convolutional layer in the RPN bounding box branch to 256 + K and set the output channel to 15, facilitating the prediction of bounding box parameters for horizontal proposals. This completes the construction of Harmony RPN.

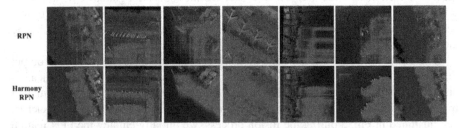

Fig. 4. Visualization of 2000 proposals Generated by RPN and Harmony RPN.

To assess the performance of Harmony RPN, we conducted a series of visualization experiments to compare the generated proposals using different methods. Figure 4 shows the results, where the first row presents 2000 proposals generated by the conventional RPN. It can be observed that these proposals are relatively scattered and have low accuracy in terms of position and size. In contrast, the proposals generated by Harmony RPN (second row) concentrate strongly around the target objects, leading to a significant improvement in proposals accuracy. This outcome substantiates the superiority of Harmony RPN.

2.3 Dynamic Rotated Proposals Decoder (DRPD)

In this research module, our purpose is to maximize the utilization of information from horizontal proposals to generate more accurate rotated proposals without introducing additional learning stages. We have observed that the aspect ratio of the horizontal proposals directly impacts the angle range of the rotated proposals. Specifically, as the aspect ratio of the horizontal scheme increases, the angle range of the rotation scheme decreases, and the amount of information provided by the horizontal scheme also increases accordingly. This is because, during the training of the RPN bounding box network, we utilize the horizontal bounding box of the rotated ground truth boxes as the labels. After the RPN stage training, the final predicted rotated proposals can be considered as inscribed rotated rectangles obtained from the horizontal proposals. Our calculations reveal that the angle range of these inscribed rotating rectangles is constrained by the aspect ratio of the horizontal proposals (as depicted in Fig. 5).

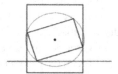

Fig. 5. Non-Inscribed Rotated Rectangle when the Rotation Angle Exceeds the Range.

During the generation of rotated proposals, our primary purpose is to exclude boxes with angles that fall outside a specific range. Among the rotated proposals that meet the angle range requirement, we design four inscribed rotated boxes with different angles and aspect ratios to optimize the recall rate for various objects [25] while minimizing the number of generated rotated proposals. By adopting this approach, we make the best use of prior information derived from the aspect ratio of the horizontal proposals, leading to the successful decoding of four rotated proposals. During the process of designing the four vertices of the rotated proposals based on the four vertices of the horizontal proposals, we adhere to the following three principles: 1. The four vertices of the rotated proposals lie on the four edges of the horizontal proposals; 2. The rotated proposals are in the form of rotated rectangles; 3. We strive to ensure the robustness of the rotated proposals as much as possible.

Figure 6 illustrates the precise process of decoding four rotated proposals from a given horizontal proposal. Initially, as depicted in Fig. 6. (a), we designate the longer side of the horizontal proposals as 'h', the shorter side as 'w', and the center point as 'x, y'. If the longer side 'h' corresponds to the left and right sides of the horizontal proposals, we select the top-left corner of the horizontal proposals as the first point, denoted as A1. On the other hand, if the longer side corresponds to the top and bottom sides of

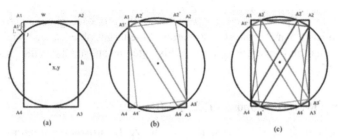

Fig. 6. Illustration of the DRPD Decoding Process.

the horizontal proposals, we choose the top-right corner of the horizontal proposals as the first point, also denoted as A1. Starting from point A1, we systematically assign the remaining three points in a clockwise direction, resulting in the construction of the rotated proposals, represented by the vertices A1, A2, A3, and A4, which form the quadrilateral A1A2A3A4. We determine the first vertex $A1\prime$ of the rotated proposals on the A1A4 side of the inscribed rotated rectangle. To ensure that the rotated proposals has an angle of 90 degrees and that all four vertices lie on the edges of the horizontal proposals, the distance L between $A1\prime$ and A1 must fall within a specific range known as the constrained range. The computation of the constrained range is as follows:

By considering the center point O of the horizontal proposals A1A2A3A4 as the center of a circle with a radius of h/2, we find the intersection point J between the circle and the A1A4 side. In order to satisfy the conditions of the rotated proposals having a 90 degree angle and all four vertices simultaneously lying on the edges of the horizontal proposals, the first vertex $A1\prime$ can only exist on the line segment between A1, J. Therefore, we can calculate the constrained range of the distance L. The constrained range of L is given by: $[0, (\frac{h}{2} - \sqrt{\left(\frac{h}{2}\right)^2 - \left(\frac{w}{2}\right)^2})/h] \cup [(\frac{h}{2} + \sqrt{\left(\frac{h}{2}\right)^2 - \left(\frac{w}{2}\right)^2})/h, 1]$. From the derived formula, it is evident that an increase in the aspect ratio of the horizontal proposals leads to a decrease in the constrained range. Consequently, the range for the existence of the first vertex of the rotated proposals becomes smaller, providing stronger prior information. The detailed procedure for generating rotated proposals utilizing this prior information is as follows:

In this paper, we set one vertex $A1\prime$ of the rotated proposals as the midpoint between points A1 and J. Taking O as the center and h/2 as the radius, as depicted in Fig. 6. (b), we find the points where the circle intersects with A1A2, A2A3, and A3A4, respectively. The intersection points between the circle and A1A2, along the A1 → A2 direction, are denoted as $A2\prime$ and $A2''$, while the lower intersection point between the circle and A2A3 is denoted as $A3\prime$. The intersection points between the circle and A3A4, along the A3 → A4 direction, are denoted as $A4\prime$ and $A4''$.

With these points, we obtain two rotated proposals with vertex coordinates $(A1\prime, A2\prime, A3\prime, A4\prime)$ and $(A1\prime, A2'', A3\prime, A4'')$. Next, as depicted in Fig. 6 (c), we horizontally mirror flip the two obtained rotated proposals with the center O, resulting in another two rotated proposals. Through these steps, we can directly convert one horizontal proposals into four rotated proposals. In this way, based on the high recall rate of horizontal suggestion boxes, we use dynamic decoding to obtain more accurate rotated proposals for final and accurate regression.

Once the four vertices of the rotated proposals are known, we calculate to obtain the parameters of the rotated proposals in the long-edge representation form $(x, y, w_j, h_j, \theta_j)$, where $j = 1, 2, 3, 4$ represents the four types of rotated proposals. Based on the label $(x^*, y^*, w^*, h^*, \theta^*)$, we can calculate the true offset values for the four types of rotated proposals:

$$L_x^*(j) = \frac{x^* - x}{w_j}, L_y^*(j) = \frac{y^* - y}{h_j} \tag{5}$$

$$L_w^*(j) = \log\left(\frac{w^*}{w_j}\right), L_h^*(j) = \log\left(\frac{h^*}{h_j}\right) \tag{6}$$

$$L_\theta^*(j) = \frac{\theta^* - \theta_j}{\pi} \tag{7}$$

Based on the parameters predicted by the Rotated ROI bbox head $L_{x'}(j), L_{y'}(j), L_{w'}(j), L_{h'}(j), L_{\theta'}(j)$, we can compute the loss function of the Rotated ROI bbox head:

$$L_{reg}\left(L_{n'}(j), L_n^*(j)\right) = smooth_{L1}\left(L_{n'}(j) - L_n^*(j)\right), n \in [x, y, w, h, \theta] \tag{8}$$

By augmenting the training with various angles, we can enhance the rotation sensitivity of the Rotated ROI bounding box head [26]. Simultaneously, the class labels assigned to the four types of rotated proposals remain consistent with those assigned to the horizontal boxes, thereby increasing the rotation invariance of the RROI classification head [27, 28].

3 Experiments

3.1 Datasets

This section introduces two datasets used in the experimental part: DOTA and HRSC2016. These datasets are widely employed in the field of object detection to assess the performance of algorithms on aerial remote sensing images.

DOTA Dataset: The DOTA (A Large-scale Dataset for Object Detection in Aerial Images) is a comprehensive dataset for aerial object detection. It comprises 2806 aerial images sourced from Google Earth satellite imagery. The dataset consists of a wide range of object categories, such as airplanes, ships, vehicles, basketball courts, and more. The images cover a wide range of scenes, such as urban areas, rural areas, and ports. DOTA provides meticulous bounding box annotations, rotated box coordinates, and object category information, enabling the evaluation of detection algorithms' accuracy and robustness in complex remote sensing images.

HRSC2016 Dataset: The HRSC2016 (High-Resolution Ship Dataset) is a specialized dataset specifically tailored for ship detection in high-resolution aerial images. It encompasses 1061 aerial images acquired from Google Earth, featuring various types of ships like cargo ships, cruise ships, and bulk carriers. Each image is accompanied by rotated bounding box annotations and object category information for the ships. The HRSC2016 dataset is characterized by its high resolution and complex object shapes, rendering it suitable for evaluating the performance of aerial image object detection algorithms in ship detection tasks.

3.2 Experimental Details

This section provides detailed information about the implementation. For the DOTA dataset, In order to achieve simplicity and efficiency, we employed a pre-trained ResNet-50 as the backbone network, which had been pre-trained on the ImageNet dataset. Unless explicitly stated, FPN was utilized as the neck network. The hyper parameter settings used were consistent with those of Rotated Faster R-CNN. Notably, the number of rotated proposals (RCNN) was set to 2048. We employed stochastic gradient descent (SGD) as the optimizer, with an initial learning rate of 0.0025, a momentum of 0.9, and a weight decay coefficient of 0.0001. To mitigate the risk of gradient explosion, we applied gradient clipping, which restricts the maximum norm of the gradients to 35.

For the HRSC2016 dataset, we resized the images to (800, 1333) while preserving their aspect ratio. The model was trained for 36 epochs, and the learning rate was reduced by a factor of 10 at the 24th and 33rd epochs. During both training and testing, the experiments were conducted using computing devices equipped with Tesla V100 GPUs. The batch size was set to 2. Mean average precision (mAP) with an IoU threshold of 0.5 was employed as the evaluation metric to measure the model's performance in the object detection task.

3.3 Ablation Study

In this section, we conducted a series of ablation experiments on the DOTA dataset to demonstrate the advantages of each proposed component in DRPDDET.

Table 1 presents the detailed results of the ablation experiments conducted on the DOTA1.0 dataset. The first row represents the performance of the baseline Rotated Faster R-CNN detector. By incorporating our Harmony RPN, as shown in the second row, the mAP score improved to 75.1.

Table 1. Ablation experiments and evaluations of our proposed method on the DOTA dataset.

	SMMF	DRPD	mAP(%)
baseline			73.4
Ours	✓		75.1
	✓	✓	75.5

Fig. 7. Example detection result of our method on DOTA.

Furthermore, as demonstrated in the third row, the application of the dynamic rotation proposals decoder (DRPD) strategy resulted in an additional increase of 1.8 in the mAP score. Additionally, the combination of Harmony RPN and the DRPD strategy achieved a 2.1 increase in mAP compared to the baseline model. Finally, when all strategies were combined, the mAP score reached 75.5, as indicated in the last row of Table 1. These experimental results provide substantial evidence of the effectiveness of our proposed approach, which integrates foreground information feature maps and utilizes the dynamic rotation proposals decoder, thereby significantly improving the performance of the object detection task. The visualization results are shown in Fig. 7.

To assess the generalizability and effectiveness of Harmony RPN, we evaluated its performance when integrated into the RoI Transformer and Gliding Vertex networks, as shown in Table 2. The inclusion of Harmony RPN in both networks leads to notable enhancements in performance. These experimental findings provide compelling evidence for the efficacy of Harmony RPN in improving proposals accuracy and recall rates.

Table 2. Demonstration of the effects of applying HarmonyRPN to RoI Transformer and Gliding Vertex Networks.

Method	RPN	HarmonyRPN	mAP (%)
RoI Transformer	√		69.56
		√	71.82
Gliding Vertex	√		75.02
		√	75.84

3.4 Contrast Test

Results on the DOTA dataset: Table 3 presents the results of 14 oriented detectors, including DRN [29], PIoU [30], G-Rep [31], Hou [32], CFA [33], SASM [34], AOPG [15], and

EARL [14]. Our DRPDDet achieved a mAP of 75.46% based on ResNet50-FPN and 75.97% based on ResNet101-FPN without employing any additional techniques. These performance scores surpass those of other state-of-the-art oriented detection methods.

Table 3. Comparison with state-of-the-art methods on the DOTA1.0 dataset.

Method	Backbone	PL	BD	BR	GTF	SV	LV	SH	TC	BC	ST	SBF	RA	HA	SP	HC	mAP
Single-stage																	
DRN	H104	88.91	80.22	43.52	63.35	73.48	70.69	84.94	90.14	83.85	84.11	50.12	58.41	67.62	68.60	52.50	70.70
PIoU	DLA-34	80.90	69.70	24.10	60.20	38.30	64.40	64.80	90.90	77.20	70.40	46.50	37.10	57.10	61.90	64.00	60.50
R3Det	R-101	88.76	83.09	50.91	67.27	76.23	80.39	86.72	90.78	84.68	83.24	61.98	61.35	66.91	70.63	53.94	73.79
RSDet	R-101	89.80	82.90	48.60	65.20	69.50	70.10	70.20	90.50	85.60	83.40	62.50	63.90	65.60	67.20	68.00	72.20
S2ANet	R-50	89.11	82.84	48.37	71.11	78.11	78.39	87.25	90.83	84.90	85.64	60.36	62.60	65.26	69.13	57.94	74.12
G-Rep	R101	88.89	74.62	43.92	70.24	67.26	67.26	79.80	90.87	84.46	78.47	54.59	62.60	66.67	67.98	52.16	70.59
Hou	R-101	89.32	76.05	50.33	70.25	76.44	79.45	86.02	90.84	82.80	82.50	58.17	62.46	67.38	71.93	45.52	72.63
Multi-stage																	
RoI Trans	R-101	88.64	78.52	43.44	75.92	68.81	73.68	83.59	90.74	77.27	81.46	58.39	53.54	62.83	58.93	37.67	69.56
SCRDet	R-101	89.98	80.65	52.09	68.36	68.36	60.32	72.41	90.85	87.94	86.86	65.02	66.68	66.25	68.24	65.21	72.61
G. Vertex	R-101	89.64	85.00	52.26	77.34	73.01	73.14	86.82	90.74	79.02	86.81	59.55	70.91	72.94	70.86	57.32	75.02
CFA	R-101	89.26	81.72	51.81	67.17	79.99	78.25	84.46	90.77	83.40	85.54	54.86	67.75	73.04	70.24	64.96	75.05
SASM	R-50	86.42	78.97	52.47	69.84	77.30	75.99	86.72	90.89	82.63	85.66	60.13	68.25	73.98	72.22	62.37	74.92
AOPG	R-101	89.14	82.74	51.87	69.28	77.65	82.42	88.08	90.89	86.26	85.13	60.60	66.30	74.05	66.76	58.77	75.39
EARL	R-50	89.76	78.79	47.01	65.20	80.98	79.99	87.33	90.74	79.17	86.23	49.09	65.87	65.75	71.86	55.21	72.87
DRPDDet	R-50	89.52	82.61	49.42	72.63	77.36	80.23	87.81	90.87	86.18	85.45	65.18	65.75	66.94	70.49	61.44	75.46
DRPDDet	R-101	**90.21**	83.05	**52.62**	71.88	77.20	80.72	**88.12**	90.90	87.64	85.89	64.75	68.79	73.63	69.95	54.20	75.97

Our model exhibited excellent performance in detecting relatively sparse objects, such as airplanes and bridges. Simultaneously, for denser objects like ships, our model demonstrated optimal detection results. This fully demonstrates the strong adaptability of our method in handling both sparse and dense scenarios.

Table 4. Performance comparison of different state-of-the-art methods on HRSC2016 dataset.

	RRPN	R2CNN	RoI Trans	G. Vertex	EARL	SASM	DRPDDet
mAP (VOC 07)	79.08	73.07	86.2.0	88.2.0	89.00	88.91	90.23
mAP (VOC 12)	85.64	79.73	*	*	93.00	*	95.41

Results on HRSC2016: The HRSC2016 dataset comprises ship targets with large aspect ratios. The experimental results on this dataset confirm the superiority of our method. As depicted in Table 4, our method outperforms other approaches in terms of object detection. The mAP of DRPDDet reaches 90.23% and 95.41% when we evaluated using VOC07 and VOC12 metrics, respectively.

4   Conclusion

In this study, we conducted a thorough analysis of the key challenges associated with remote sensing object detection, specifically focusing on the adaptability of horizontal and rotated proposals in dense and sparse scenes. To effectively address these challenges, we introduced an innovative two-stage strategy that intelligently combines the advantages of both horizontal and rotated proposals. Firstly, we employed the generation of horizontal proposals in the RPN stage to enhance recall, followed by the development of a dynamic rotated proposals decoder that adaptively generates rotated proposals based on the constraints of the horizontal proposals. Furthermore, we designed and implemented the Harmony RPN module and integrate it into the two-stage object detection network. The experimental results demonstrated that our approach significantly improves the precision and recall of proposals, leading to substantial enhancements in overall object detection performance. HormonyRPN employs a cross-scale hierarchical fusion strategy to seamlessly incorporate foreground scores into feature maps. While this approach enhances the richness of data or features, it inevitably results in an expansion of model parameters. In the ensuing phases, we have the capacity to encode foreground details into the feature space, acquiring feature representations of greater dimensionality. Interactively integrating these enhanced feature representations with feature maps can enhance the model's ability to distinguish and capture complex feature interactions.

References

1. Carion, N., Massa, F., Synnaeve, G., Usunier, N., Kirillov, A., Zagoruyko, S.: End-to-end object detection with transformers. In: Vedaldi, A., Bischof, H., Brox, T., Frahm, J.-M. (eds.) ECCV 2020. LNCS, vol. 12346, pp. 213–229. Springer, Cham (2020). https://doi.org/10.1007/978-3-030-58452-8_13
2. Lin, T.Y., Goyal, P., Girshick, R., et al.: Focal loss for dense object detection. In: Proceedings of IEEE International Conference on Computer Vision, pp. 2980–2988 (2017)
3. Tan, M., Pang, R., Le, Q.V.: EfficientDet: scalable and efficient object detection. In: Proceedings of IEEE/CVF Conference on Computer Vision and Pattern Recognition, pp. 10781–10790 (2020)
4. Lee, G., Kim, J., Kim, T., et al.:. Rotated-DETR: an end-to-end transformer-based oriented object detector for aerial images. In Proceedings of ACM/SIGAPP Symposium on Applied Computing, pp. 1248 1255 (2023)
5. Pu, Y., Wang, Y., Xia, Z., et al.: Adaptive rotated convolution for rotated object detection. arXiv preprint arXiv:2303.07820 (2023)
6. Liu, F., Chen, R., Zhang, J., Ding, S., Liu, H., Ma, S., Xing, K.: ESRTMDet: an end-to-end super-resolution enhanced real-time rotated object detector for degraded aerial images. IEEE J. Sel. Topics Appl. Earth Observ. Remote Sensing **16**, 4983–4998 (2023)
7. Wang, C.Y., Bochkovskiy, A., Liao, H.Y.M.: YOLOv7: trainable bag-of-freebies sets new state-of-the-art for real-time object detectors. In: Proceedings of IEEE/CVF Conference on Computer Vision and Pattern Recognition, pp. 7464–7475 (2023)
8. Li, F., Zhang, H., Xu, H., et al.: Mask Dino: towards a unified transformer-based framework for object detection and segmentation. In: Proceedings of IEEE/CVF Conference on Computer Vision and Pattern Recognition, pp. 3041–3050 (2023)
9. Liu, S., Zeng, Z., Ren, T., et al.: Grounding dino: marrying dino with grounded pre-training for open-set object detection. arXiv preprint arXiv:2303.05499 (2023)

10. Zhou, Q., Yu, C.: Object detection made simpler by eliminating heuristic NMS. IEEE Trans. Multimedia 1–10 (2023).
11. Han, J., Ding, J., Li, J., et al.: Align deep features for oriented object detection. IEEE Trans. Geosci. Remote Sens. **60**(1), 1–11 (2021)
12. Yang, X., Yan, J., Feng, Z., et al.: R3Det: refined single-stage detector with feature refinement for rotating object. In: Proceedings of AAAI Conference on Artificial Intelligence, pp. 3163–3171 (2021)
13. Qian, W., Yang, X., Peng, S., et al.: Learning modulated loss for rotated object detection. In: Proceedings of AAAI Conference on Artificial Intelligence, pp. 2458–2466 (2021)
14. Guan, J., Xie, M., Lin, Y., et al.: EARL: an elliptical distribution aided adaptive rotation label assignment for oriented object detection in remote sensing images. arXiv preprint arXiv:2301.05856 (2023)
15. Cheng, G., Wang, J., Li, K., et al.: Anchor-free oriented proposals generator for object detection. IEEE Trans. Geosci. Remote Sens. **60**(1), 1–11 (2022)
16. Xu, C., Ding, J., Wang, J., et al.: Dynamic coarse-to-fine learning for oriented tiny object detection. In: Proceedings of IEEE/CVF Conference on Computer Vision and Pattern Recognition, pp. 7318–7328 (2023)
17. Nabati, R., Qi, H.: RRPN: radar region proposals network for object detection in autonomous vehicles. In: Proceedings of IEEE International Conference on Image Processing, pp. 3093–3097 (2019)
18. Xu, Y., Fu, M., Wang, Q., et al.: Gliding vertex on the horizontal bounding box for multi-oriented object detection. IEEE Trans. Pattern Anal. Mach. Intell. **43**(4), 1452–1459 (2020)
19. Yang, X., Yang, J., Yan, J., et al.. SCRDet: towards more robust detection for small, cluttered and rotated objects. In: Proceedings of IEEE/CVF International Conference on Computer Vision, pp. 8232–8241 (2019)
20. Ding, J., Xue, N., Long, Y., et al.: Learning ROI transformer for oriented object detection in aerial images. In: Proceedings of IEEE/CVF Conference on Computer Vision and Pattern Recognition, pp. 2849–2858 (2019)
21. He, K., Zhang, X., Ren, S., et al.: Deep residual learning for image recognition. In: Proceedings of IEEE Conference on Computer Vision and Pattern Recognition, pp. 770–778 (2016)
22. Lin, T.Y., Dollár, P., Girshick, R., et al.: Feature pyramid networks for object detection. In: Proceedings of IEEE Conference on Computer Vision and Pattern Recognition, pp. 2117–2125 (2017)
23. Liu, W., Anguelov, D., Erhan, D., Szegedy, C., Reed, S., Cheng-Yang, Fu., Berg, A.C.: SSD: single shot multibox detector. In: Leibe, B., Matas, J., Sebe, N., Welling, M. (eds.) ECCV 2016. LNCS, vol. 9905, pp. 21–37. Springer, Cham (2016). https://doi.org/10.1007/978-3-319-46448-0_2
24. Zhu, C., He, Y., Savvides, M.: Feature selective anchor-free module for single-shot object detection. In: Proceedings of IEEE/CVF Conference on Computer Vision and Pattern Recognition, pp. 840–849 (2019)
25. Kong, T., Yao, A., Chen, Y., et al.: HyperNet: towards accurate region proposals generation and joint object detection. In: Proceedings of IEEE Conference on Computer Vision and Pattern Recognition, pp. 845–853 (2016)
26. Li, P., Zhao, H., Liu, P., Cao, F.: RTM3D: real-time monocular 3D detection from object keypoints for autonomous driving. In: Vedaldi, A., Bischof, H., Brox, T., Frahm, J.-M. (eds.) ECCV 2020. LNCS, vol. 12348, pp. 644–660. Springer, Cham (2020). https://doi.org/10.1007/978-3-030-58580-8_38
27. Kirillov, A., Wu, Y., He, K., et al.. PointRend: Image segmentation as rendering. In: Proceedings of IEEE/CVF Conference on Computer Vision and Pattern Recognition, pp. 9799–9808 (2020)

28. Pu, Y., Wang, Y., Xia, Z., et al.: Adaptive Rotated Convolution for Oriented Object Detection. arXiv preprint arXiv:2303.07820 (2023)
29. Pan, X., Ren, Y., Sheng, K., et al.: Dynamic refinement network for oriented and densely packed object detection. In: Proceedings of IEEE/CVF Conference on Computer Vision and Pattern Recognition, pp. 11207–11216 (2020)
30. Chen, Z., Chen, K., Lin, W., See, J., Hui, Yu., Ke, Y., Yang, C.: Piou loss: towards accurate oriented object detection in complex environments. In: Vedaldi, A., Bischof, H., Brox, T., Frahm, J.-M. (eds.) ECCV 2020. LNCS, vol. 12350, pp. 195–211. Springer, Cham (2020). https://doi.org/10.1007/978-3-030-58558-7_12
31. Hou, L., Lu, K., Yang, X., Li, Y., Xue, J.: Grep: Gaussian representation for arbitrary-oriented object detection. arXiv preprint arXiv:2205.11796 (2022)
32. Hou, L., Lu, K., Xue, J.: Refined one-stage oriented object detection method for remote sensing images. IEEE Trans. Image Process. **31**, 1545–1558 (2022)
33. Guo, Z., Liu, C., Zhang, X., et al.: Beyond bounding-box: convex-hull feature adaptation for oriented and densely packed object detection. In: Proceedings of IEEE/CVF Conference on Computer Vision and Pattern Recognition, pp. 8792–8801 (2021)
34. Hou, L., Lu, K., Xue, J., et al.: Shape-adaptive selection and measurement for oriented object detection. In: Proceedings of the AAAI Conference on Artificial Intelligence, pp. 923–932 (2022)

MFSFFuse: Multi-receptive Field Feature Extraction for Infrared and Visible Image Fusion Using Self-supervised Learning

Xueyan Gao and Shiguang Liu[✉]

College of Intelligence and Computing, Tianjin University, Tianjin 300350, China
lsg@tju.edu.cn

Abstract. The infrared and visible image fusion aims to fuse complementary information in different modalities to improve image quality and resolution, and facilitate subsequent visual tasks. Most of the current fusion methods suffer from incomplete feature extraction or redundancy, resulting in indistinctive targets or lost texture details. Moreover, the infrared and visible image fusion lacks ground truth, and the fusion results obtained by using unsupervised network training models may also cause the loss of important features. To solve these problems, we propose an infrared and visible image fusion method using self-supervised learning, called MFSFFuse. To overcome these challenges, we introduce a Multi-Receptive Field dilated convolution block that extracts multi-scale features using dilated convolutions. Additionally, different attention modules are employed to enhance information extraction in different branches. Furthermore, a specific loss function is devised to guide the optimization of the model to obtain an ideal fusion result. Extensive experiments show that, compared to the state-of-the-art methods, our method has achieved competitive results in both quantitative and qualitative experiments.

Keywords: Infrared and Visible Image · Image Fusion ·
Multi-receptive Field Feature Extraction · Self-supervised

1 Introduction

Image fusion aims to fuse images captured from different sensors or different shooting settings into an informative image to enhance the understanding of scene information [1,2]. In the field of image fusion, infrared and visible image fusion is the most widely used. Infrared images are sensitive to thermal radiation information and are not affected by the working environment, which can highlight significant targets, but infrared images have low spatial resolution and lack texture details. Compared with infrared images, visible images are easily affected by the working environment but have higher spatial resolution and richer texture details [3,4]. Fusing these two images with complementary properties can yield

---

This work was partly supported by the Natural Science Foundation of China under grants 62072328.

B. Luo et al. (Eds.): ICONIP 2023, LNCS 14452, pp. 118–132, 2024.
https://doi.org/10.1007/978-981-99-8076-5_9

a fusion image that contains both rich texture details and salient targets, which can be used for subsequent tasks such as image segmentation, target detection, tracking tasks, and military actions.

In the past years, infrared and visible image fusion has been greatly developed, and many infrared and visible image fusion methods have been proposed. The existing fusion methods mainly including traditional methods and deep learning-based methods. The traditional methods can be classified into five categories [5]: 1) multi-scale transform-based fusion methods, 2) sparse representation-based fusion methods, 3) sub-space-based fusion methods, 4) saliency-based fusion methods, and 5) hybrid methods. Although traditional methods have achieved great fusion results, there are still some shortcomings, such as 1) the fusion performance depends on the extraction of manual features and the design of fusion rules; 2) for complex source images, the extraction of manual features and the design of fusion rules tend to be complicated, which is time-consuming and difficult to implement.

In recent years, deep learning has seen rapid development, and its powerful feature extraction and data representation capabilities have drawn significant attention from researchers. A growing number of researchers have applied deep learning to image fusion. Due to the lack of ground truth for infrared and visible image fusion, most deep learning-based fusion methods are based on unsupervised learning to train the network to obtain the fusion image. According to the adopted network framework, it can be divided into three categories: CNN-based fusion methods [3,6–8], Auto-encoder based fusion methods [9,10], GAN-based fusion methods [11,12].

Although the fusion performance of existing fusion methods based on deep learning has been improved to some extent, there are still the following challenges:

a) **Limited performance of some unsupervised fusion methods:** Due to the lack of ground truth for infrared and visible image fusion, most methods constrain the fusion image and the source image by designing a loss function to generate fusion results. However, such fusion results may approximate the compromise of the source images, which can result in the loss of important features.

b) **Failure to fully leverage the correlation between features:** Some existing fusion methods use a single convolution kernel to extract features, but this will result in a relatively single extracted feature. Some use multiple different convolution kernels at the same level to obtain multi-scale features. However, each convolution kernel performs convolution operations independently, which may lead to redundancy in the extracted information, making the model fails to exploit the correlation between features effectively while simultaneously increasing computational complexity.

c) **Inadequate consideration of modality differences:** Many deep learning-based fusion algorithms do not adequately account for the inherent differences between infrared and visible image modalities. They often employ the same feature extraction strategy for both modalities, which may not effectively extract the most relevant features for optimal fusion. This limitation can lead to suboptimal fusion performance.

To address the challenges mentioned, we propose a novel approach named MFSFFuse, which utilizes self-supervised learning to fuse infrared and visible images by extracting features with multiple receptive fields. The key contributions of this paper can be summarized as follows:

1) **To address the limitations of certain unsupervised fusion methods, we propose a self-supervised fusion network.** Our approach utilizes the multi-receptive field dilated convolution block to extract features from both infrared and visible images. We then concatenate the features from both branches along the channel dimension to generate a fusion image. Next, we decompose the fusion image into infrared and visible images, and apply constraints to these decomposed images and the source images to achieve self-supervision.

2) **To address the limitation that the correlation between features cannot be fully exploited, we introduce a novel convolutional module with multiple receptive fields.** This module utilizes dilated convolutions with varying dilation factors to construct a multi-receptive field convolution module. It overcomes the problems of information redundancy and extensive computations caused by traditional multi-receptive field convolution. Moreover, it provides a wider range of context information, enabling the model to effectively capture long-range dependencies in the image, extract multi-scale features better, and improve the fusion performance of the model.

3) **Considering the modality difference of infrared and visible image, in the branches of feature extraction for both infrared and visible, we introduce two distinct attention modules to enhance the performance of fusion.** The first is an intensity attention module, which selectively focuses on the most important features for the fusion results. The second is a detail attention module, which further enhances the fusion results by attending to the most significant details in the extracted features. These attention mechanisms allow us to selectively attend to the most relevant features in each modality and produce fused images with superior quality.

2   Proposed Method

In this section, we will introduce MFSFFuse, a novel method for fusing infrared and visible images. We will start by presenting the overall framework of our proposed approach, followed by a detailed explanation of the network structure. Finally, we will describe the loss function employed in our method.

2.1   Overall Framework

Since infrared and visible images belong to different modalities, we use $ir - path$ and $vi - path$ to extract multi-scale features from the source images respectively

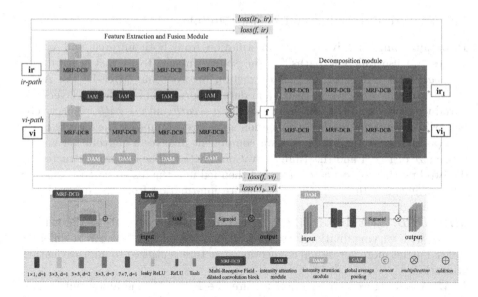

Fig. 1. The overall network framework of the proposed method. d is the dilation factor in the dilated convolution. $loss(x, y)$ represents the loss between x and y. *ir-path* and *vi-path* represent infrared and visible image branches, respectively. ir and vi are the infrared and visible images, respectively. f denotes the fusion image, ir_1 and vi_1 represent the reconstructed infrared and visible images after decomposition. The self-supervised pattern formed between the decomposed and reconstructed images and their corresponding source images.

to obtain more comprehensive features of the images in different modalities. We are aware of the valuable capabilities of the infrared sensor in highlighting significant targets, while visible sensor excels in capturing intricate texture details. These distinctive characteristics provide fusion images with prominent target information and detailed texture information, respectively. To maximize the utilization of meaningful features from both image types during the fusion process, we incorporate two attention modules into the two branches: an intensity attention module and a detail attention module. These modules allow us to focus on the features that are most relevant for the fusion image. The features extracted by the two branches are then concatenated along the channel dimension and passed through a convolutional layer to obtain the fusion image.

We know that the fusion of infrared and visible image lacks ground truth, and the fusion result obtained by designing the loss function to constrain the fusion image and the source image is close to the compromise of the source images, which easily leads to the loss of important features. To address this, we reconstruct the source image in a self-supervised manner in the decomposition module. This helps to ensure that the fusion image contains more important features from the source images. The overall framework of our proposed method is shown in Fig. 1.

2.2 Feature Extraction and Fusion Module

To obtain comprehensive feature information from both infrared and visible images, we use different paths for the source images to extract features. In $ir - path$ and $vi - path$, feature extraction involves four MRF-DCBs. Additionally, to fully preserve intensity and detail information in different layers, we introduce four cascaded IAMs and DAMs in $ir - path$ and $vi - path$, respectively. Meanwhile, we use a residual connection structure to improve the feature reuse rate (as shown in Fig. 1). The fusion component involves a convolution with a kernel size of 1×1 and the Tanh activation function. The features extracted by the two branches are concatenated and then fed into the fusion part to generate a fusion image with comprehensive feature information from the infrared and visible images.

MRF-DCB: Each MRF-DCB comprises three dilated convolutions (kernel size 3×3) with varying dilation factors ($d = 1, 2, 3$) to extract features at different scales. Furthermore, a residual connection structure is utilized to improve feature utilization.

IAM: Assuming we have an input feature $F \in \mathbb{R}^{H \times W \times C}$ and an output feature $X \in \mathbb{R}^{H \times W \times C}$, H and W are the height and width of the feature map, and C is the number of channels. The calculation process of the IAM can be described as follows:

$$X_{ir} = \mathrm{Sigmoid}(\mathrm{conv}(\mathrm{GAP}(F))) \odot F \tag{1}$$

where GAP stands for global average pooling, conv refers to 1-D convolution operation, Sigmoid is a sigmoid function, and $\odot$ means Hadamard Product.

DAM: The calculation process of the DAM can be described as follows:

$$X_{vi} = \mathrm{Sigmoid}(\mathrm{BN}(\mathrm{conv}\,7(\mathrm{Re}(\mathrm{BN}(\mathrm{conv}\,7(\,F)))))) \odot F \tag{2}$$

where BN represents Batch Normalization, conv7 represents a convolution operation with the kernel size of 7×7, and Re represents the Relu activation function.

2.3 Decomposition Module

Infrared and visible images are images of different modalities, and their features have certain differences, and the fusion of the two lacks ground truth. Optimizing the model to generate fusion images by constraining the relationship between the fusion image and the source images may cause the loss of important features. Therefore, we use a decomposition module to decompose and reconstruct the fusion image into its corresponding source image, thereby achieving self-supervision by constraining the decomposition image and the source image. This module has two branches, one branch reconstructs the infrared image, and the other branch reconstructs the visible image, each branch includes three MRF-DCBs, and finally, the 1×1 convolution and tanh activation function are applied to generate the reconstructed image.

In our approach, we consider the source image as the reference image, leveraging a loss function to enforce consistency between the decomposed image and the reference image. This process facilitates the reconstruction of the source image, effectively achieving self-supervision, thus enabling the fusion image to obtain more information from the source image.

2.4   Loss Function

To improve the visual quality of fusion image and preserve important features such as texture details and salient objects, we design a loss function denoted as L_{loss} to guide the optimization of the model, written as

$$L_{loss} = L_{\text{content}} + \alpha L_{ssim} + L_s, \tag{3}$$

where $L_{content}$ represents the content loss, L_{ssim} denotes the structural similarity loss, L_s is the loss used to train the self-supervised network. By optimizing the model using this loss function, we can produce high-quality fusion images that accurately capture the most important features of the source images. α is the trade-off parameter to control the balance between three terms.

The content loss $L_{content}$ forces the fusion image to have richer texture details and preserve salient target information, it is defined as:

$$L_{content} = L_{\text{int}} + \beta L_{detail}, \tag{4}$$

where L_{int} denotes the intensity loss, L_{detail} represents the detail loss, β is a trade-off parameter. In the intensity loss we use mean squared error (MSE) as the loss function, defined as follows:

$$L_{int} = \gamma_1 MSE\left(I_f, I_{vi}\right) + \gamma_2 MSE\left(I_f, I_{ir}\right), \tag{5}$$

where I_f, I_{ir} and I_{vi} stand for the fusion image, the infrared image, and the visible image, respectively. γ_1 and γ_2 are two weight parameters utilized to regulate the balance of loss values.

In the detail loss function, we incorporate a maximum gradient operation. We assume that the texture details present in the fused image are the maximum combination of textures from the infrared and visible images. The detail loss function is defined as follows:

$$L_{detail} = \frac{\left\| |\nabla I_f| - \max\left(|\nabla I_{ir}|, |\nabla I_{vi}|\right) \right\|_1}{HW}, \tag{6}$$

where ∇ represents the Sobel gradient operator, $\| \cdot \|_1$ is the l_1-norm, I_f, I_{ir} and I_{vi} stand for the fusion image, the infrared image, and the visible image, respectively. H and W denote the height and width of the image, respectively. $|\cdot|$ denotes the absolute value. $\max(\cdot)$ refers to the element-wise maximum selection.

To force the fusion image and the source image to have similar structures, we add a modified structural similarity [13] loss L_{ssim} to the loss function L_{loss}, which is defined as

$$\text{SSIM} = \begin{cases} \text{SSIM}\left(I_{vi}, I_f\right) \\ \quad \text{if } \sigma^2(I_{vi}) > \sigma^2(I_{ir}) \\ \text{SSIM}\left(I_{ir}, I_f\right) \\ \quad \text{if } \sigma^2(I_{ir}) >= \sigma^2(I_{vi}) \end{cases}, \tag{7}$$

where $L_{SSIM} = 1 - SSIM$. I_f denotes the fusion image, I_{vi} stands for the visible image, I_{ir} is the infrared image. σ^2 represent variance. The SSIM aims to calculate the structural similarity of two images. The larger the value of SSIM, the more similar the two images. Therefore, we take $(1 - SSIM)$ when calculating the loss.

For the training of self-supervised network, we adopt the standard mean square error (MSE) as the loss function, defined as follows:

$$L_s = MSE\left(I_{vi1}, I_{vi}\right) + MSE\left(I_{ir1}, I_{ir}\right), \tag{8}$$

where I_{vi1} and I_{ir1} are the results of fusion image decomposition and reconstruction.

3   Experiments

3.1   Datasets and Training Details

In this study, we randomly chose 32 image pairs from the TNO dataset[1] for training our model, while reserving the remaining images for testing. To ensure a comprehensive assessment of the fusion performance and generalization capabilities of our proposed method, we also added quantitative and qualitative analysis on the LLVIP dataset [14], MSRS dataset[2], and M3FD dataset [15]. The image pairs used in our experiment have all been preregistered. During training, since the ideal model cannot be obtained when the amount of data is too small, we use a cropping strategy to expand the data to obtain more data. We crop the source images into image patch pairs of size 120×120 for training, and the cropping stride is set to 12. We use the Adam optimizer to update the parameters, the batch size is set to 16, and the learning rate is set to 1e−4. According to the extensive experiments, we set the weight parameters in the loss function as $\alpha = 100$, $\beta = 50$, $\gamma_1 = 10$ and $\gamma_2 = 30$. Our method is implemented on Pytorch framework and trained on a computer with 3.10-GHz Intel Core i9-9900 CPU, 32 GB RAM, and GPU NVIDIV RTX 2060.

3.2   Fusion Metrics

To further prove the performance of the proposed method, in addition to qualitative experiments, we also select several representative evaluation indicators to quantitatively evaluate the fusion results [20], i.e., standard deviation (SD), spatial frequency (SF), average gradient (AG), entropy (EN). Larger values of these metrics stand for better fusion results. SD can reflect the distribution and contrast of fusion image. SF can be used to measure the gradient distribution of an image and evaluate image texture and details. AG is used to quantify the gradient information of the fusion image, reflecting the details and textures in the fusion image. EN can be used to measure the amount of information contained in the fusion image.

---

[1] https://figshare.com/articles/TN_Image_Fusion_Dataset/1008029.
[2] https://github.com/Linfeng-Tang/MSRS.

3.3   Comparative Experiment

To demonstrate the effectiveness of our proposed method, we conducted a comparison with several state-of-the-art image fusion methods, including two traditional methods (GTF [16] and LatLRR [17]) and ten deep learning methods (FusionGAN [11], PMGI [7], U2Fusion [1], SDNet [8], RFN-Nest [10], Super-Fusion [18], SwinFuse [19], SeAFusion [3], TarDAL [15], and AT-GAN [20]). The parameters of the contrasting methods were set according to their original reports.

1) Results on TNO dataset

Qualitative Comparison. We test the performance of MFSFFuse and twelve other state-of-the-art fusion methods on the TNO dataset, and the results are presented in Fig. 2(a). It is evident from the figure that most of the fusion methods were successful in completing the task of fusion. The results of GTF, Fusion-GAN, PMGI, SwinFuse, SDNet, and SuperFusion are more inclined to infrared images. The fusion results of LatLRR, U2Fusion, and RFN-Nest with rich texture information, but the contrast is low, which impacts the visual quality. In contrast, SeAFusion, TarDAL, AT-GAN, and our method produced fusion images that had prominent targets and rich texture details, which aligned well with the human visual perception. As can be seen from the leaves in the red box in the figure, our fusion results are the most visually effective and the most informative of all methods.

Quantitative Comparison. Qualitative evaluation of image fusion results can be subjective, especially when the differences between images are subtle. To overcome this, we also conducted a quantitative analysis of our proposed fusion methods. As shown in Table 1(a), our method achieves the highest values in terms of SD, AG, and EN metrics, and the second-best value in SF. A larger SD value indicates a higher contrast in our fusion result, while a larger AG value indicates that our fusion result contains rich texture and detail information. The EN value reflects the amount of information contained in our fusion result. Overall, our fusion method effectively preserves important information from the source images.

2) Results on LLVIP dataset

Qualitative Comparison. The LLVIP dataset comprises a large number of nighttime scene image pairs. A well-performing fusion algorithm should generate fusion images with rich texture details and salient objects, even under low-light conditions. Given the limited information provided by both infrared and visible images in nighttime scenes, it is crucial to integrate their complementary features to enhance the visual quality of fusion images. To demonstrate the efficacy of our proposed method in improving the visual quality of fusion images and integrating the complementary features of infrared and visible images, we evaluated it on the LLVIP dataset and compared its performance with state-of-the-art fusion algorithms, as shown in Fig. 2(b). All algorithms complete the fusion task to some extent, but the quality of the fusion results varies. Apart from our method, other methods introduce irrelevant information in the fusion process, resulting in the loss of texture details and a significant reduction in

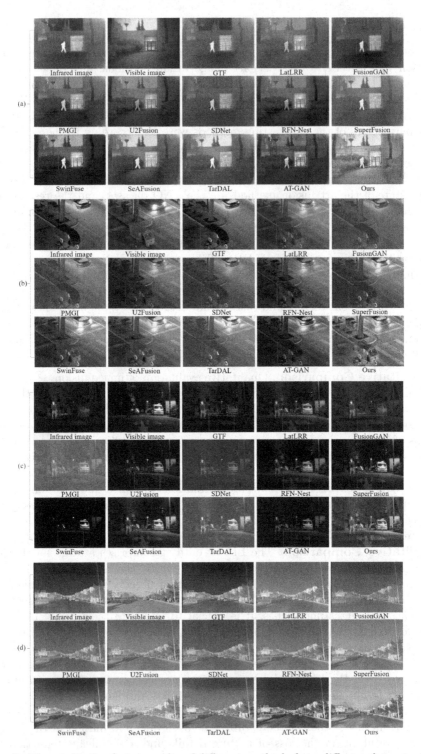

Fig. 2. The qualitative fusion results of different methods from different datasets. (a) TNO dataset, (b) LLVIP dataset, (c) MSRS dataset, (d) M^3FD dataset.

object contrast. In contrast, our method extracts multi-scale features and uses the synergistic effects of intensity attention and detail attention modules, resulting in fusion images with rich texture details and salient objects. These results demonstrate the superiority of our approach in fusing low-light infrared and visible images and highlight its potential applications (such as image enhancement, target detection, wilderness rescue, intelligent transportation, etc.) in various fields.

Table 1. The average quantitative results on the different datasets. The best and second-best fusion results for each evaluation metric are highlighted in bold and italic.

(a) TNO dataset				(b) LLLVIP dataset			
Methods	SD ↑	SF ↑	AG ↑	EN ↑	Methods	SD ↑	SF ↑	AG ↑	EN ↑
GTF	9.4559	0.0315	3.2319	6.7854	GTF	*9.6182*	0.0530	3.7378	7.3782
LatLRR	8.6825	0.0297	3.1506	6.5109	LatLRR	9.1372	0.0425	2.9960	7.0506
FusionGAN	8.7266	0.0240	2.4364	6.5691	FusionGAN	8.6127	0.0299	2.1484	6.5675
PMGI	*9.6512*	0.0333	3.6963	7.0228	PMGI	9.5219	0.0339	2.7424	7.0431
U2Fusion	9.4991	0.0455	5.1333	7.0189	U2Fusion	8.8186	0.0499	3.9600	6.8095
SDNet	9.1310	0.0444	4.6136	6.7019	SDNet	9.0090	0.0555	4.0928	6.9491
STDFusionNet	9.2626	0.0462	4.4936	6.9977	STDFusionNet	7.5675	0.0550	3.6277	5.9250
SuperFusion	9.1478	0.0329	3.3971	6.7996	SuperFusion	9.4570	0.0462	3.0551	7.2985
SwinFuse	9.3388	0.0497	4.6344	7.0279	SwinFuse	8.0158	0.0553	3.5657	6.4294
SeAFusion	9.6354	0.0465	4.9622	7.1300	SeAFusion	9.5177	0.0593	4.3825	*7.4603*
TarDAL	9.5234	0.0462	4.3277	7.1418	TarDAL	9.5219	0.0541	3.6567	7.3903
AT-GAN	9.3015	**0.0722**	*6.9538*	*7.1659*	AT-GAN	8.6333	*0.0652*	*4.7407*	6.9863
RFNNest	9.4242	0.0222	2.7181	6.9977	RFNNest	9.3305	0.0274	2.4266	7.1356
Ours	**10.1428**	*0.0573*	**6.9847**	**7.1854**	Ours	**10.1600**	**0.0715**	**6.1103**	**7.5873**
(c) MSRS dataset				(d) M^3FD dataset			
Methods	SD ↑	SF ↑	AG ↑	EN ↑	Methods	SD ↑	SF ↑	AG ↑	EN ↑
GTF	6.2803	0.0311	2.4389	5.4009	GTF	9.6401	0.0633	5.7226	7.2623
LatLRR	7.9810	0.0311	2.7031	6.3026	LatLRR	8.4141	0.0424	3.9724	6.6165
FusionGAN	5.7659	0.0169	1.4342	5.3356	FusionGAN	9.3450	0.0404	3.7267	6.8787
PMGI	8.0309	0.0330	3.0396	6.2800	PMGI	8.8045	0.0463	4.4801	6.9000
U2Fusion	7.2116	0.0380	3.1761	5.7681	U2Fusion	9.2830	0.0706	7.0247	6.9991
SDNet	5.9641	0.0350	2.8178	5.3023	SDNet	9.0745	0.0716	6.7906	6.9704
STDFusionNet	7.5460	0.0424	3.1407	5.7876	STDFusionNet	9.8151	0.0717	6.4913	6.9519
SuperFusion	8.8779	*0.0438*	3.6098	6.7303	SuperFusion	9.3339	0.0556	5.1721	6.8633
SwinFuse	5.3636	0.0389	2.1488	4.5381	SwinFuse	9.7908	0.0785	7.3444	**7.5165**
SeAFusion	*8.9295*	0.0450	*3.9098*	*6.7882*	SeAFusion	*10.1821*	0.0737	6.8173	7.0007
TarDAL	8.4195	0.0415	3.4116	6.5950	TarDAL	9.8938	0.0596	5.5534	7.2459
AT-GAN	6.2844	0.0339	2.5816	5.3576	AT-GAN	9.0916	*0.1016*	**9.7448**	7.2955
RFNNest	8.1426	0.0246	2.2309	6.2605	RFNNest	8.8753	0.0399	4.0053	6.9086
Ours	**9.3541**	**0.0631**	**6.3545**	**7.1927**	Ours	**10.8800**	**0.1024**	*9.7097*	*7.3753*

Quantitative Comparison. We select 96 pairs of images from the LLVIP dataset for quantitative analysis. Table 1(b) presents the corresponding quantitative indicators. As shown in the table, our method achieved the best results on SD, SF, AG and EN. These metrics also reflect that our method has certain advantages in preserving texture information as well as salient objects. Overall, our method is effective in preserving source images information.

3) Results on MSRS dataset

Qualitative Comparison. To ensure the credibility of our results, we conduct generalization experiments on the MSRS dataset. Figure 2(c) shows qualitative results of different methods on the MSRS dataset. The figures reveal that methods such as GTF, LatLRR, FusionGAN, PMGI, and SDNet retain the essential target information, but fail to capture texture details. On the other hand, RFN-Nest, U2Fusion, SwinFuse, SuperFusion, SeAFusion, TarDAL and AT-GAN not only preserve critical target information but also reserve some texture details in the fusion result. However, these results have low contrast, leading to a less visually appealing image. Overall, our method achieve a balance between retaining rich information and producing visually pleasing results.

Quantitative Comparison. Table 1(c) presents the quantitative results of the MSRS dataset, and it shows that our method achieve the optimal values in SF, AG, SF and EN. This reflects the effectiveness of our method in retaining features.

4) Results on M^3FD dataset

Qualitative Comparison. The application scenarios of infrared and visible image fusion are complex and diverse. For this reason, we choose to verify on the M^3FD dataset. Figure 2(d) is the qualitative results of different methods on the M^3FD dataset. However, some fusion methods such as GTF, LatLRR, FusionGAN, PMGI, SDNet, RFN-Nest, and SuperFusion have failed to preserve the texture information of trees, leading to suboptimal fusion results. In contrast, our method and a few others, such as TarDAL and AT-GAN, have successfully maintained rich texture details while avoiding over-reduction in contrast. As can be seen from the leaves in the red boxes, our method preserves rich texture details. Overall, our experimental results demonstrate the effectiveness and superiority of our method on the M^3FD dataset.

Quantitative Comparison. We select 76 pairs of images from the M^3FD dataset for quantitative experimental analysis, and the results are presented in Table 1(d). Our method outperforms all other methods in terms of SD and SF, and ranks second in AG and EN. From the results, it can be seen that our method achieves the overall best performance.

3.4   Ablation Study

To verify the rationality of different components in the proposed method, we conduct ablation experiments. We design different structures for experiments: 1) *w/o MRF-DCB:* Use single convolutions instead of MRF-DCB; 2) *w/o D:* without decomposition module; 3) *w/o DAM:* without detail attention module; 4) *w/o IAM:* without intensity attention module; 5) *w/o DAM+IAM:* without detail attention module and intensity attention module; 6) *Ours:* the proposed method. We conducted qualitative and quantitative experiments to validate the effectiveness of different modules. The role of structure 1) is to prove the effectiveness of the proposed multi-receptive field convolution module; structure 2) is to prove that the self-supervised network can promote the fusion results; structure 3), structure 4) and structure 5) are to illustrate the effectiveness of the intensity attention and the detail attention module. The qualitative results are shown in Fig. 3, and the quantitative results are shown in Table 2.

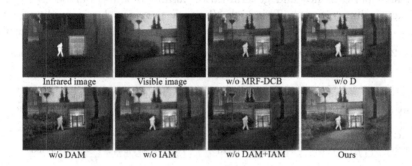

Fig. 3. Fusion images with different structures.

Qualitative Analysis: In our method, the cascaded DAM and IAM play a crucial role in preserving rich texture details and salient object information in the fusion image. Qualitative experiments have demonstrated the efficacy of these two modules in optimizing the model. In the absence of DAM, the fusion result tends to lose certain information and exhibit reduced contrast. Similarly, without the IAM, the overall brightness of the fused image decreases. If these two modules are not available, the fusion result may suffer from artifacts and exhibit an overall reduction in visual quality. The decomposition module in this paper is actually a module related to self-supervision and is designed to promote the optimization of the model through the interaction between the decomposed and reconstructed image and the source image. In the ablation experiment, we further validated the efficacy of this module, which demonstrated a significant promotion effect on the model's performance. To evaluate the effectiveness of MRF-DCB, we design a single convolution structure for comparison. The experimental results demonstrate that our proposed method outperforms the single convolution structure in terms of detail preservation and visual quality.

Quantitative Analysis: From the quantitative results shown in Table 2, it can be seen that the proposed method achieves the optimal value on SD, SF, and AG, and obtains the third best value on EN. These outcomes demonstrate that each module in the proposed method has contributed positively towards enhancing the overall performance of the model.

3.5 Execution Time

In addition to the generalization performance, the efficiency of fusion algorithms is also an important evaluation index, especially when fusing infrared and visible images in advanced vision tasks, where real-time performance is crucial. Table 3 presents the average fusion time of different fusion methods on TNO dataset. As shown in the table, our method demonstrates high processing efficiency, which is only second to AT-GAN. In summary, our method not only achieves superior fusion performance and generalization ability but also demonstrates high efficiency in image fusion. This high efficiency provides the possibility for the algorithm to be applied to a wide range of advanced vision tasks in the future.

Table 2. The quantitative results of different structures. The best and second-best results for each evaluation metric are highlighted in bold and italic.

Methods	w/o MRF-DCB	w/o D	w/o DAM	w/o IAM	w/o DAM+IAM	Ours
SD ↑	9.8270	10.0524	9.4767	*10.0584*	9.6089	**10.1428**
SF ↑	0.0456	0.0385	0.0498	0.0483	*0.0563*	**0.0573**
AG ↑	5.0309	4.2922	5.3681	5.1951	*5.8155*	**6.9847**
EN ↑	7.1223	*7.2276*	7.1735	**7.2758**	7.0244	7.1854

Table 3. The average runtime for different methods.

Methods	GTF	LatLRR	FusionGAN	PMGI		U2Fusion	SDNet	RFN-Nest
time (s)	2.850	51.718	0.478	0.237		3.425	0.181	1.771
Methods	AT-GAN	SuperFusion	SwinFuse	SeAFusion	TarDAL	Ours		
time (s)	0.028	0.220	0.727	0.100	0.780	0.056		

4 Conclusion

In this paper, we propose a novel method, MFSFFuse, for fusing infrared and visible images. Firstly, we introduce MRF-DCB, which comprehensively captures multi-scale features from the source images. Secondly, considering the disparities between infrared and visible image modalities and characteristics, we design two branches for feature extraction. In $ir - path$, we incorporate an IAM to preserve the crucial target information, while in $vi - path$, a DAM is employed to retain texture details. These modules ensure that the fusion result contains the most meaningful information from both modalities. To tackle the issue of important feature loss in unsupervised fusion methods, we adopt a self-supervised approach by treating the source image as the ground truth, and utilize constrained decomposition reconstructed images and corresponding source images to optimize the model. Experimental results on diverse datasets demonstrate that our fusion results surpass the performance of state-of-the-art methods in terms of qualitative, quantitative, and generalization experiments. In our future work, we will continue refining the network and explore its integration with high-level vision tasks, such as object detection and image segmentation, to further enhance its practical applicability.

References

1. Xu, H., Ma, J., Jiang, J., Guo, X., Ling, H.: U2Fusion: a unified unsupervised image fusion network. IEEE Trans. Pattern Anal. Mach. Intell. **44**(1), 502–518 (2020)
2. Liu, S., Wang, M., Song, Z.: WaveFuse: a unified unsupervised framework for image fusion with discrete wavelet transform. In 28th International Conference on Neural Information Processing (ICONIP), pp. 162–174 (2021)
3. Tang, L., Yuan, J., Ma, J.: Image fusion in the loop of high-level vision tasks: a semantic-aware real-time infrared and visible image fusion network. Inf. Fus. **82**, 28–42 (2022)
4. Gao, X., Liu, S.: DAFuse: a fusion for infrared and visible images based on generative adversarial network. J. Electron. Imaging **31**(4), 043023 (2022)
5. Han, M., et al.: Boosting target-level infrared and visible image fusion with regional information coordination. Inf. Fus. **92**, 268–288 (2023)
6. Ma, J., Tang, L., Xu, M., Zhang, H., Xiao, G.: STDFusionNet: an infrared and visible image fusion network based on salient target detection. IEEE Trans. Instrum. Meas. **70**, 1–13 (2021)
7. Zhang, H., Xu, H., Xiao, Y., Guo, X., Ma, J.: Rethinking the image fusion: a fast unified image fusion network based on proportional maintenance of gradient and intensity. Proc. AAAI Conf. Artif. Intell. **34**(7), 12797–12804 (2020)
8. Zhang, H., Ma, J.: SDNet: a versatile squeeze-and-decomposition network for real-time image fusion. Int. J. Comput. Vis. **129**, 2761–2785 (2021)
9. Li, H., Wu, X.J.: DenseFuse: a fusion approach to infrared and visible images. IEEE Trans. Image Process. **28**(5), 2614–2623 (2018)
10. Li, H., Wu, X.J., Kittler, J.: RFN-Nest: an end-to-end residual fusion network for infrared and visible images. Inf. Fus. **73**, 72–86 (2021)
11. Ma, J., Yu, W., Liang, P., Li, C., Jiang, J.: FusionGAN: a generative adversarial network for infrared and visible image fusion. Inf. Fus. **48**, 11–26 (2019)
12. Ma, J., Zhang, H., Shao, Z., Liang, P., Xu, H.: GANMcC: a generative adversarial network with multiclassification constraints for infrared and visible image fusion. IEEE Trans. Instrum. Meas. **70**, 1–14 (2020)
13. Rao, D., Xu, T., Wu, X.J.: TGFuse: An infrared and visible image fusion approach based on transformer and generative adversarial network. IEEE Trans. Image Process. (2023). https://doi.org/10.1109/TIP.2023.3273451
14. Jia, X., Zhu, C., Li, M., Tang, W., Zhou, W.: LLVIP: a visible-infrared paired dataset for low-light vision. In: Proceedings of the IEEE/CVF International Conference on Computer Vision, pp. 3496–3504 (2021)
15. Liu, J., et al.: Target-aware dual adversarial learning and a multi-scenario multi-modality benchmark to fuse infrared and visible for object detection. In Proceedings of the IEEE/CVF Conference on Computer Vision and Pattern Recognition, pp. 5802–5811 (2022)
16. Ma, J., Chen, C., Li, C., Huang, J.: Infrared and visible image fusion via gradient transfer and total variation minimization. Inf. Fus. **31**, 100–109 (2016)
17. Li, H., Wu, X.J.: Infrared and visible image fusion using latent low-rank representation. arXiv preprint arXiv:1804.08992 (2018)
18. Tang, L., Deng, Y., Ma, Y., Huang, J., Ma, J.: SuperFusion: a versatile image registration and fusion network with semantic awareness. IEEE/CAA J. Automatica Sinica **9**(12), 2121–2137 (2022)

19. Wang, Z., Chen, Y., Shao, W., Li, H., Zhang, L.: SwinFuse: a residual Swin transformer fusion network for infrared and visible images. IEEE Trans. Instrum. Meas. **71**, 1–12 (2022)
20. Rao, Y., et al.: AT-GAN: a generative adversarial network with attention and transition for infrared and visible image fusion. Inf. Fus. **92**, 336–349 (2023)

Progressive Temporal Transformer for Bird's-Eye-View Camera Pose Estimation

Zhuoyuan Wu[✉], Jiancheng Cai, Ranran Huang, Xinmin Liu,
and Zhenhua Chai

Meituan, 7 Rongda Road, Chaoyang, Beijing 100012, China
{wuzhuoyuan02,caijiangcheng,huangranran,
liuxinmin,chaizhenhua}@meituan.com

Abstract. Visual relocalization is a crucial technique used in visual odometry and SLAM to predict the 6-DoF camera pose of a query image. Existing works mainly focus on ground view in indoor or outdoor scenes. However, camera relocalization on unmanned aerial vehicles is less focused. Also, frequent view changes and a large depth of view make it more challenging. In this work, we establish a Bird's-Eye-View (BEV) dataset for camera relocalization, a large dataset contains four distinct scenes (*roof, farmland, bare ground,* and *urban area*) with such challenging problems as frequent view changing, repetitive or weak textures and large depths of fields. All images in the dataset are associated with a ground-truth camera pose. The BEV dataset contains 177242 images, a challenging large-scale dataset for camera relocalization. We also propose a Progressive Temporal transFormer (dubbed as PTFormer) as the baseline model. PTFormer is a sequence-based transformer with a designed progressive temporal aggregation module for temporal correlation exploitation and a parallel absolute and relative prediction head for implicitly modeling the temporal constraint. Thorough experiments are exhibited on both the BEV dataset and widely used handheld datasets of 7Scenes and Cambridge Landmarks to prove the robustness of our proposed method.

Keywords: Camera Pose Estimation · Birds-Eye-View · Transformer

1 Introduction

Camera relocalization aims to regress the 6-DoF pose of a given image relative to a scene, which is a crucial technique widely applied in such fields as robot navigation, augmented reality, and autonomous driving. Recently some learning-based research [19,26] have shown impressive performance in visual relocalization on several widely used datasets such as Cambridge Landmarks [17], Oxford Robotcar [21], 7Scenes [27]. However, these datasets are mostly recorded by handheld devices or ground robots with limited perspective changes and a small depth of field. Also, they concentrate less on weak or repetitive textures, which boosts the need for building a more challenging dataset. The drone has recently improved

B. Luo et al. (Eds.): ICONIP 2023, LNCS 14452, pp. 133–147, 2024.
https://doi.org/10.1007/978-981-99-8076-5_10

efficiency in mining, sea ports, oil, and other industrial facilities because of its limitless aerial perspective. Practically view changes for the aerial images are more remarkable, and the immense depth of field makes it more sensitive to texture changes. As such aerial images collected by drones are suitable for the mentioned challenges. Consequently, we build a bird's eye view dataset for camera pose estimation to stimulate future research on BEV camera relocalization. The dataset contains four scenes, i.e., *roof*, *farmland*, *bare ground*, and *urban area*. *Roof* and *urban area* are two scenes collected in a city, while *farmland* and *bare ground* are collected in a rural district. The whole dataset constitutes 177242 images in total.

Methods for camera relocalization can be divided into structure-based and regression ones. Structure-based methods [23,24,31] predict the camera pose hierarchically. Given a query image, any image retrieval algorithm is applied to find similar images, followed by a feature extraction and matching to get numerous 2D-2D matches. Combined with the information on depth and camera intrinsic, 2D image features are mapped to their 3D counterpart coordinates. Camera pose is obtained via Perspective-n-Point (PnP) [14] and RANSAC [12]. Although geometry-based methods achieve state-of-the-art performance, they are relatively time-consuming because of the iterative optimization process and also need to store dense keyframes. Recent years have witnessed the vigorous development of deep learning, and studies on deep learning solve the problem of camera relocalization through the absolute pose regression network (APRs) [17]. Training on multiple collected images with 6-DoF pose, the network can infer the query image with one forward pass. Although the performance of APRs is less accurate than structure-based methods, they have irreplaceable advantages in terms of faster speed and robustness on repetitive or weak textures.

However, no tailored method is proposed for BEV camera relocalization. Because small movements may cause a significant change in the visual field, we rely on sequence-based methods to introduce temporal constraints. In addition, we design the temporal aggregation module (TAM) to exploit the temporal correlation and a parallel APR and RPR prediction head to draw into the relative pose restraint in feature representation. Our contribution can be summarized below:

- We build a BEV dataset collected by 6-rotor drones with ground truth poses for camera relocalization. The dataset with 177242 images in total is challenging regarding movements in 3D space, frequent view changing, and variability of textures in different scenes.
- We propose Progressive Temporal Transformer (PTFormer) with three techniques: a progressive temporal aggregation module (TAM), a parallel absolute prediction regression (APR) and relative prediction regression (RPR) head to exploit temporal correlation, and inner attention integrated in the original multi-head attention.
- We give thorough experiments on the built BEV dataset and two public datasets 7Scenes, and Cambridge Landmarks. PTFormer achieves the

best performance on the BEV dataset and a comparable result with SOTA approaches on two public datasets.

2   Related Work

2.1   Dataset for Camera Relocalization

Existing datasets for camera pose estimation are usually obtained via a Structure-from-motion (Sfm) reconstruction or differential GPS. 7Scenes [27] is an indoor dataset with all scenes recorded from a handheld Kinect RGB-D camera. Cambridge Landmarks dataset is an outdoor dataset [17] which is collected by pedestrians holding a Google LG Nexus 5 smartphone containing different lighting and weather conditions. Also, numerous datasets are recorded by car-mounted cameras like Oxford Robotcar [21]. DAG [33] and VPAIR [25] are most related to our dataset. However, DAG targets visual place recognition and localization by retrieving the closest database aerial image given a street-level query image, while ours focuses on localization only given BEV images. VPAIR is a low frame-rate dataset collected by a single flight route, while ours is a high frame-rate one with dozens of flight routes in each scene.

2.2   Camera Relocalization

Camera relocalization can be categorized into structure-based methods and regression-based methods. Structure-based methods [6,20,24,30] solve the problem of camera relocalization by matching features descriptors of the query image to descriptors of 3D points of an Sfm model. Due to the weak representation of handcrafted descriptors, recently, many methods rely on the convolutional network [9,10] or transformer [29] for the sake of extracting features more robust to view changes. Besides matching feature descriptors, scene coordinate regression methods [2–4,27] predict matching from the coordinate of the query image to 3D scene space, which performs decently in the small-scale environments while less accurately in large-scale ones. Although structure-based methods usually have more accurate predictions, facts like computation-heavy in large-scale environments and the need for known camera intrinsic or depth sensors restrict its development.

The regression-based methods train a deep model to predict the pose from a given image. PoseNet [17] is one of the earliest works which attaches an MLP after a GooLeNet. Compared with structured-based methods, PoseNet has shown the advantages of the regression-based methods in terms of robustness towards view changes and free of matching. A sequence of works is modified on top of PoseNet. [15] proposes the Bayesian PoseNet to estimate the uncertainty of the predicted pose. [35] introduces an LSTM after the FC layers to relieve the overfitting problem. [16] designs geometrically inspired pose losses to improve orientation accuracy. Some works focus on refining the architecture.

[22] proposes Hourglass, which replaces the original GooLeNet with an encoder-decoder architecture implemented with a ResNet34. BranchNet [37] designs a dual branch to regress the pose and orientation separately.

Besides, a group of works predicts camera poses via consecutive sequences. VidLoc [8] predicts camera poses recurrently and also estimates the trajectories, which can reduce localization error. MapNet [5] introduces geometric constraints by using additional inputs of visual odometry and GPS to provide extra supervision for model training. LsG [38] use conLSTM to regress relative poses between consecutive frames, trying to exploit the temporal correlation to boost the performance. Sequence-based methods usually outperform image-based ones because of additional temporal information. Such temporal correlation can give model auxiliary information to restrain performance flicker caused by view changes. So in this work, we focus on the sequence-based method.

Recently, Transformer [34] has been widely used in diverse vision tasks. A recent work [26] proposes MS-Transformer and achieves SOTA performance on camera relocalization. MS-Transformer is a work inspired by DETR [7]. It learns scene-specific queries with the Transformer decoder to regress the camera poses in different scenes. Compared with MS-Transformer, with fewer parameters, our method is tailored for BEV localization with sequence input, also the proposed temporal aggregation module and a parallel APR, RPR prediction head introduce temporal restriction while learning high dimensional features.

3   Method

In this section, we give details of our proposed PTFormer as shown in Fig. 1. Given a consecutive sequence of frames $\{\mathbf{F}_{t+i}\}_{i=-L:L}$, PTFormer regress the camera pose $\mathbf{p} = <\mathbf{x}, \mathbf{q}>$ of the intermediate frame $\mathbf{F}_t$, where $\mathbf{F}_t \in \mathbb{R}^{H \times W \times C}$, $\mathbf{x} \in \mathbf{R}^3$ is the position of the camera in the world and $\mathbf{q} \in \mathbf{R}^4$ is the corresponding 3D orientation in quaternion form. First, a pre-trained CNN extracts visual features from given images. Then a transformer encoder is followed to augment spatial features. Afterward, PTFromer applies a progressive temporal aggregation module to make feature interaction to better model the temporal correspondence. Finally, besides the APR module regressing the camera pose of each frame, the RPR module is aimed to predict the relative camera pose of all frame pairs, trying to impose a temporal restriction on augmented features in latent space.

3.1   Network Architecture

Convolutional Backbone. Following [26], we use a pre-trained CNN backbone to extract visual features. However, we use the same activation map to regress both position and orientation.

Sequential Representation. In order to make the extracted visual features compatible with Transformer, the activation map $\mathbf{M}_t \in \mathbb{R}^{H_m \times W_m \times C_m}$ is converted to a sequential representation $\tilde{\mathbf{M}}_t$ through a 1×1 convolution and flattening. As mentioned in [26], two one-dimensional encodings are separately learned for the X and Y axes for the sake of reducing the parameters of learned parameters. Specifically, for a certain position (i, j), $i = 1, ..., H_m, j = 1, ..., W_m$, the positional embedding is defined as $\mathbf{E}_{\text{pos}}^{ij} = [\mathbf{E}_u^j, \mathbf{E}_v^i]$ where $\mathbf{E}_u \in \mathbb{R}^{W_m \times C_m/2}$ and $\mathbf{E}_v \in \mathbb{R}^{H_m} \times C_m/2$. So the input for Transformer is given by:

$$\mathbf{h}_t = \tilde{\mathbf{M}}_t + \mathbf{E}. \tag{1}$$

This process is applied for each frame.

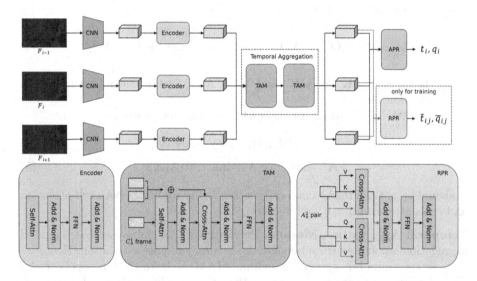

Fig. 1. Overview of our proposed PTFormer. Given a sequence of frames $\mathbf{F}_i, i \in \{1, ..., n\}$, a pre-trained CNN extracts the visual features from each image. Then a weight-shared Transformer is used to augment visual representation. Two layers of TAM are followed to exploit all groups of temporal correlation (C_n^1) among features of a certain frame with the other ones. Finally, given augmented features, the APR head predicts the camera pose of each frame, while the RPR head predicts the relative pose of all frame pairs. Note that the RPR head only exists in the training phase.

Attention with Inner Attention. Transformer Encoder is composed of m identical layers, with each comprising multi-head attention (MHA) and multi-layer perception (MLP) layers, where each part is followed with LayerNorm and residual connection. Different from the standard Transformer encoder in [7], the architecture of the Transformer encoder in our work is integrated with attention in attention module (AiA) as mentioned in [13]. Given a set of queries, keys, and values as $\mathbf{Q}, \mathbf{K}, \mathbf{V} \in \mathbb{R}^{H_m W_m \times C_m}$, the vanilla multi-head attention (VaniMHA) is formulated as:

$$\text{VaniMHA}(\mathbf{Q}, \mathbf{K}, \mathbf{V}) = (\text{Softmax}(\frac{\hat{\mathbf{Q}}\hat{\mathbf{K}}^\top}{\sqrt{C_m}})\hat{\mathbf{V}})\mathbf{W_o}, \tag{2}$$

where $\hat{\mathbf{Q}} = \mathbf{Q}\mathbf{W}_q$, $\hat{\mathbf{K}} = \mathbf{K}\mathbf{W}_k$ and $\hat{\mathbf{V}} = \mathbf{V}\mathbf{W}_v$ are linear transformations while $\mathbf{W}_q$, $\mathbf{W}_k$, $\mathbf{W}_v$ and $\mathbf{W}_o$ are learnable weights.

One can observe that the correlation among query-key pairs is ignored in conventional attention. Intuitively introducing interaction among query-key pairs can relieve the side effect caused by imperfect representation of attention score, i.e., imposing second-order attention on one-order attention score can restrain the noise and help relevant query-key pairs augment each other. Denote the attention map as $\mathbf{A} = (\text{Softmax}(\frac{\hat{\mathbf{Q}}\hat{\mathbf{K}}^\top}{\sqrt{C_m}})\hat{\mathbf{V}})$. The inner multi-attention (Inner-MHA) is formulated as follows:

$$\text{InnerMHA}(\mathbf{Q}', \mathbf{K}', \mathbf{V}') = (\text{Softmax}(\frac{\mathbf{Q}'\mathbf{K}'^\top}{\sqrt{D}})\mathbf{V}')(1 + \mathbf{W_o'}), \tag{3}$$

where $\mathbf{Q}' = \mathbf{A}\mathbf{W}_q'$, $\mathbf{K}' = \mathbf{A}\mathbf{W}_k'$ and $\mathbf{V}' = \mathbf{A}\mathbf{W}_v'$ is the linear transformation from attention score $\mathbf{A}$.

Finally, the second-order multi-head attention mechanism is given by:

$$\text{SecOrdMHA}(\mathbf{Q}, \mathbf{K}, \mathbf{V}) = \hat{\mathbf{A}}\hat{\mathbf{V}}\mathbf{W_o}, \tag{4}$$

$$\hat{\mathbf{A}} = \text{Softmax}(\mathbf{A} + \text{InnerMHA}(\mathbf{A})) \tag{5}$$

It should be noticed that second-order attention is applied in both self-attention and cross-attention in our architecture.

Temporal Aggregation Module. Previous work [28,38] prove that temporal aggregation among sequence benefits the accuracy of pose estimation. Hence we propose the temporal aggregation module (TAM) to learn the temporal correlation. Considering a sequence of feature $\{\mathbf{h}'_{t+i}\}_{i=-L:L}$ obtained after Transformer encoders, the coarse temporal aggregation between current feature $\mathbf{h}'_t$ and the concatenation of other features $\mathbf{h}'^c_t$ is formulated as:

$$\mathbf{h}''_t = \text{TAM}_1(\mathbf{h}'_t, \mathbf{h}'^c_t) = \text{SecOrdMHA}(\hat{\mathbf{Q}}, \hat{\mathbf{K}}, \hat{\mathbf{V}}), \tag{6}$$

where $\hat{\mathbf{Q}} = \mathbf{h}'_t\hat{\mathbf{W}}_q$, $\hat{\mathbf{K}} = \mathbf{h}'^c_t\hat{\mathbf{W}}_k$ and $\hat{\mathbf{V}} = \mathbf{h}'^c_t\hat{\mathbf{W}}_v$. Further, the fine aggregation is the same procedure:

$$\mathbf{h}^o_t = \text{TAM}_2(\mathbf{h}''_t, \mathbf{h}''^c_t). \tag{7}$$

Absolute Pose Regression Head. The APR head is one hidden layer MLP followed by a gelu activation function to regress the camera pose. Concretely, features of corresponding frames obtained after the temporal aggregation module are followed by an APR head to predict the camera pose of each frame.

Relative Pose Regression Head. As mentioned in [11,18], learning relative camera poses and absolute ones simultaneously help the model to learn better representations for features and improve the accuracy of absolute pose estimation. In this work, we design a parallel regression head for the aforementioned reasons. Following [11], the relative pose is given in the reference system of the second camera. Given a pair of frames $(\mathbf{F}_1, \mathbf{F}_2)$, their corresponding rotation matrix and translations from world coordinates to camera ones are $(\mathbf{R}_1, \mathbf{x}_1, \mathbf{R}_2, \mathbf{x}_2)$. Denotes the rotation quaternions as $(\mathbf{q}_1, \mathbf{q}_2)$, the relative camera pose $(\mathbf{q}_{1,2}, \mathbf{x}_{1,2})$ is formulated as:

$$\mathbf{q}_{1,2} = \mathbf{q}_2 \times \mathbf{q}_1^*, \tag{8}$$

$$\mathbf{x}_{1,2} = \mathbf{R}_2(-\mathbf{R}_1^\top \mathbf{x}_1) + \mathbf{x}_2, \tag{9}$$

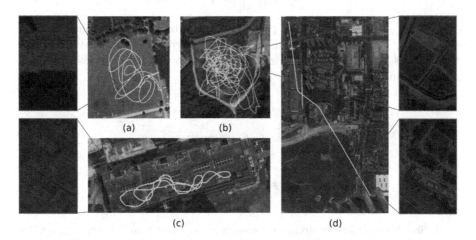

(a)   (b)

(c)   (d)

Fig. 2. Trace and a sample of our proposed BEV dataset. (a) is *bare ground*, (b) is *farmland*, (c) is *roof*, (d) is *urban area*. For visually pleasant, in each scene, we select only one trace for visualization. The red boxes are samples of collected data in each scene. (Color figure online)

where $\mathbf{q}_1^*$ is the conjugate of $\mathbf{q}_1$ and $\times$ is the multiplication of quaternions. Then we will give the details about how to predict pairwise relative camera pose. Suppose $\mathbf{Q}_1, \mathbf{K}_1, \mathbf{V}_1$ and $\mathbf{Q}_2, \mathbf{K}_2, \mathbf{V}_2$ is the linear mapping from feature $\mathbf{h}_1^o$ and $\mathbf{h}_2^o$ respectively, the relative pose is obtained via the MLP on the concatenation of features $[\mathrm{Softmax}(\mathbf{Q}_1\mathbf{K}_2^\top)\mathbf{V}_2, \mathrm{Softmax}(\mathbf{Q}_2\mathbf{K}_1^\top)\mathbf{V}_1]$. Given n frames, all A_n^2 feature permutation pairs are sent to the RPR head to predict relative camera poses.

3.2 Loss Function

PTFormer is trained in an end-to-end fashion, guided by a joint loss function to supervise both absolute poses and relative poses:

$$\mathcal{L} = \sum_i (||\mathbf{x}_i - \overline{\mathbf{x}}_i|| \exp^{-s_x} + s_x$$

$$+ ||\mathbf{q}_i - \frac{\overline{\mathbf{q}}_i}{||\overline{\mathbf{q}}_i||}|| \exp^{-s_q} + s_q)$$

$$+ \sum_{ij} (||\mathbf{x}_{ij} - \overline{\mathbf{x}}_{ij}|| \exp^{-s_{rx}} + s_{rx}$$

$$+ ||\mathbf{q}_{ij} - \frac{\overline{\mathbf{q}}_{ij}}{||\overline{\mathbf{q}}_{ij}||}|| \exp^{-s_{rq}} + s_{rq}), \tag{10}$$

where $(\mathbf{x}_i, \mathbf{q}_i)$ and $(\overline{\mathbf{x}}_i, \overline{\mathbf{q}}_i)$ is the prediction and ground-truth absolute camera pose, $(\mathbf{x}_{ij}, \mathbf{q}_{ij})$ and $(\overline{\mathbf{x}}_{ij}, \overline{\mathbf{q}}_{ij})$ is the prediction and ground-truth relative camera pose. s_x, s_q, s_{rx}, s_{rq} are learned parameters to balance different parts in the loss function as suggested in [16] (Fig. 3).

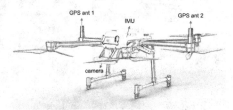

Fig. 3. Sketch of our 6-rotor drone and sensors location. BEV images are captured by a down-view camera. GPS antennas are on both sides of the drone.

3.3   Implementation Details

Our model is implemented in PyTorch with a single NVIDIA Tesla A100 GPU. Adam optimizer is used to optimize parameters with $\beta_1 = 0.9, \beta_2 = 0.999, \epsilon = 10^{-10}$. The batch size is 8, and the initial learning rate is set as 1×10^{-4} with a weight decay of 1×10^{-4} every 200 epochs out of 600 epochs. A pre-trained EfficientNet [32] is integrated to extract visual features. In order to make the model capable of regressing poses in different orientations, we supply an angle augmentation. During training, all images are rescaled to 256×256 pixels and a center crop with 224×224. The dimension of the feature in the Transformer is set as $C_m = 256$. The hidden dimension in transformer blocks is set as 256 with a dropout of $p = 0.1$. The head number of MHA is 4, followed by 2-layer MLP layers.

4   Aerial Visual Localization Dataset

4.1   Sensor Setup

The data is collected by a Meituan-developed 6-rotor fixed-wing drone with a nominal cruise speed of 10 km/h. The drone is equipped with a binocular camera with a resolution of 1280×800 at $10\,\mathrm{Hz}$. For convenience, our BEV dataset is constructed only with the left camera. The 6-DoF poses are provided by the self-developed navigation system with an expected uncertainty of $0.05°$ in rotation and less than $80\,\mathrm{cm}$ in position.

4.2 BEV Dataset

BEV dataset contains 4 scenes. The *roof* sequences consist of 16 sequences collected on the roof of a building. The *farmland* sequences comprise 12 sequences recorded on top of farmland. The *bare ground* sequences are composed of 9 sequences collected in a sub-urban area with bare ground and trees. Sequences in these scenes cover the area of 3.6×10^4, 1×10^4, and 8.2×10^4 m^2, respectively. The *urban area* sequences are 6 sequences from an urban area. Due to aerial regulations, drones can only fly in a fixed route, and we select a 3 km one. The BEV dataset contains 177242 images in total with diverse scenes. All images have a ground-truth camera pose with a pose error of less than $80\,\mathrm{cm}$. Figure 2 gives the sample of our dataset.

Table 1. Comparative analysis on the Cambridge Landmarks dataset (outdoor localization). We report the median position/orientation error in meters/degrees for each method. Bold highlighting indicates better performance.

Scene	Roof	Farmland	Bare ground	Urban area	Avg
Scene scale	3.6×10^4 m^2	1×10^4 m^2	8.2×10^4 m^2	3×10^3 m	
AtLoc [36]	4.159 m, 6.002°	7.611 m, 9.810°	7.499 m, 6.417°	24.731 m, 7.166°	11.000 m, 7.349°
MS-Transformer [26]	2.417 m, 4.259°	4.950 m, 10.981°	3.368 m, 8.636°	7.718 m, 1.353°	4.613 m, 6.307°
PTFormer (ours)	0.869 m, 1.710°	1.706 m, 3.018°	2.100 m, 1.931°	5.041 m, 1.132°	2.429 m, 1.948°

Table 2. Ablation study of model configuration on *roof*

Model config	Avg
baseline	1.361 m, 2.272°
+TAF × 1	1.193 m, 2.041°
+TAF × 2	1.178 m, 1.850°
+aia	1.107 m, 1.763°
+RPR head	0.869 m, 1.710°

Table 3. Ablation study of dimension on *roof*

Dimension	Avg
64	0.940 m, 2.217°
128	0.907 m, 1.787°
256	0.869 m, 1.710°
512	0.876 m, 1.773°

Table 4. Ablation study of encoder layers on *roof*

layers	Avg
2	0.986 m, 1.860°
4	0.900 m, 1.736°
6	0.869 m, 1.710°

Table 5. Ablation study of CNN backbone on *roof*

backbone	Avg
ResNet18	1.100 m, 2.556°
ResNet50	0.962 m, 1.978°
Efficient-net	0.869 m, 1.710°

5   Experiments

Besides evaluate on our dataset, we also evaluate our method on the 7Scenes [27], and the Cambridge Landmarks Dataset [17] to verify PTFormer is robust on diverse views and scenes. The evaluation matrices are the median error of camera rotation (°) and translation (m).

BEV Dataset. *Farmland* and *Bare ground* are scenes with weak and repetitive textures. *Roof* and *Urban area* are both in the city scene, while the *urban area* are more sophisticated sequences containing moving underground objects and scene changes. Due to limited open-source methods and reproduction difficulties, in this part, we select AtLoc [36] and MS-Transformer [26] for comparison, as shown in Table 1. When facing weak or repetitive scenes, i.e., *farmland* and *bare ground*, our method shows a more robust capability to capture the global feature. Our method significantly outperforms the other two methods in urban scenes, especially in *urban area*.

Ablation Study. In this section, we give step-by-step experiments to demonstrate the effectiveness of our designed module, shown in Table 2. The baseline model is the one only with a feature extractor, a Transformer encoder, and an APR head. First, we validate the efficiency of the two-step temporal aggregation module. The experiments show that the two-step fusion makes a 0.183 m and 0.422° decrease of pose error. Then the experiments of the aia module replacing original multi-head attention show that inner attention can better exploit the correlation of query-key pairs. Finally, the additional RPR head can further improve the accuracy of APR, with a decrease of 0.238 m and 0.053°.

Table 3 discusses the impact of hidden dimensions in Transformer. We choose the dimension of 256 as a compromise result. Table 4 analyses the effect of encoder layers. Considering the memory cost, we set the encoder layers to 6. Table 5 shows the influence of different pre-trained CNN backbones, and we select efficient-net as our pre-trained model.

7Scenes. 7Scenes is collected by a handheld Kinect camera, covering indoor scenes composed of RGB-D sequences with a spatial extent of $1 \sim 10\,m^2$. Many scenes have repetitive or weak textures, making it challenging for camera pose estimation. The comparison between our method and the recent state-of-the-art

Table 6. Experiment results on the 7Scenes Dataset [27]. Results are cited directly, the best results are **highlighted**.

Scene	Chess	Fire	Heads	Office	Pumpkin	Kitchen	Stairs	Avg.
Scene scale	$3 \times 2\,\mathrm{m}^2$	$2.5 \times 1\,\mathrm{m}^2$	$2 \times 0.5\,\mathrm{m}^2$	$2.5 \times 2\,\mathrm{m}^2$	$2.5 \times 2\,\mathrm{m}^2$	$4 \times 3\,\mathrm{m}^2$	$2.5 \times 2\,\mathrm{m}^2$	
PoseNet15 [17]	0.32 m, 8.12°	0.47 m, 14.4°	0.29 m, 12.0°	0.48 m, 7.68°	0.47 m, 8.42°	0.59 m, 8.64°	0.47 m, 13.8°	0.44 m, 10.4°
PoseNet16 [15]	0.37 m, 7.24°	0.43 m, 13.7°	0.31 m, 12.0°	0.48 m, 8.04°	0.61 m, 7.08°	0.58 m, 7.54°	0.48 m, 13.1°	0.47 m, 9.81°
PoseNet17 [16]	0.14 m, 4.50°	0.27 m, 11.80°	0.18 m, 12.10°	0.20 m, 5.77°	0.25 m, 4.82°	0.24 m, 5.52°	0.37 m, 10.60°	0.24 m, 7.87°
LSTM+Pose [35]	0.24 m, 5.77°	0.34 m, 11.9°	0.21 m, 13.7°	0.30 m, 8.08°	0.33 m, 7.00°	0.37 m, 8.83°	0.40 m, 13.7°	0.31 m, 9.85°
RelocNet [1]	0.12 m, 4.14°	0.26 m, 10.4°	0.14 m, 10.5°	0.18 m, 5.32°	0.26 m, 4.17°	0.23 m, 5.08°	0.28 m, 7.53°	0.21 m, 6.73°
Hourglass [22]	0.15 m, 6.17°	0.27 m, 10.84°	0.19 m, 11.63°	0.21 m, 8.48°	0.25 m, 7.01°	0.27 m, 10.15°	0.29 m, 12.46°	0.23 m, 9.53°
BranchNet [37]	0.18 m, 5.17°	0.34 m, 8.99°	0.20 m, 14.15°	0.30 m, 7.05°	0.27 m, 5.10°	0.33 m, 7.40°	0.38 m, 10.26°	0.29 m, 8.30°
VMLoc [40]	0.10 m, 3.70°	0.25 m, 10.5°	0.15 m, 10.80°	0.16 m, 5.08°	0.20 m, 4.01°	0.21 m, 5.01°	0.24 m, 10.00°	0.19 m, 7.01°
VidLoc [8]	0.18 m, –	0.26 m, –	0.14 m, –	0.26 m, –	0.36 m, –	0.31 m, –	0.26 m, –	0.25 m, –
LsG [38]	0.09 m, 3.28°	0.26 m, 10.92°	0.17 m, 12.70°	0.18 m, 5.45°	0.20 m, 3.69°	0.23 m, 4.92°	0.23 m, 11.3°	0.19 m, 7.47°
MapNet [5]	0.08 m, 3.25°	0.27 m, 11.69°	0.18 m, 13.25°	0.17 m, 5.15°	0.22 m, 4.02°	0.23 m, 4.93°	0.30 m, 12.08°	0.21 m, 7.77°
GL-Net [39]	0.08 m, 2.82°	0.26 m, 8.94°	0.17 m, 11.41°	0.18 m, 5.08°	0.15 m, 2.77°	0.25 m, 4.48°	0.23 m, 8.78°	0.19 m, 6.33°
MS-Transformer [26]	0.11 m, 4.66°	0.24 m, 9.6°	0.14 m, 12.19°	0.17 m, 5.66°	0.18 m, 4.44°	0.17 m, 5.94°	0.26 m, 8.45°	0.18 m, 7.28°
PTFormer (Ours)	0.10 m, 3.12°	0.26 m, 9.27°	0.13 m, 12.10°	0.18 m, 5.46°	0.19 m, 4.10°	0.20 m, 4.41°	0.24 m, 8.87°	0.19 m, 6.76°

ones is listed in Table 6. Existing methods can be roughly categorized into three types: i) image-based APRs, ii) sequence-based APRs, and iii) Transformer-based APRs. Our method can be served as a combination of sequence-based and Transformer-based approaches. Basically, the sequenced-based methods perform better than image-based ones because of temporal correlation. Our method shows comparable performance with SOTA methods, which indicated that our method is robust on indoor scenes (Table 7).

Cambridge Landmarks. The Cambridge Landmarks dataset is collected by mobile phone camera at Cambridge University, containing six outdoor scenes with moving pedestrians and weather changes. Following [26], we use four scenes that are commonly benchmarked by APRs. The categories of existing methods

Table 7. Experiment results on the Cambridge Dataset [17]. Evaluations are cited directly. The average is taken on the first four datasets. The best results are **highlighted**.

Scene	College	Shop	Church	Hospital	Avg.
Scene scale	$5.6 \times 10^3\,\mathrm{m}^2$	$8.8 \times 10^3\,\mathrm{m}^2$	$4.8 \times 10^3\,\mathrm{m}^2$	$2.0 \times 10^3\,\mathrm{m}^2$	
PoseNet15 [17]	1.66 m, 4.86°	1.41 m, 7.18°	2.45 m, 7.96°	2.62 m, 4.90°	2.04 m, 6.23°
PoseNet16 [15]	1.74 m, 4.06°	1.25 m, 7.54°	2.11 m, 8.38°	2.57 m, 5.14°	1.92 m, 6.28°
LSTM+Pose [35]	0.99 m, 3.65°	1.18 m, 7.44°	1.52 m, 6.68°	1.51 m, 4.29°	1.30 m, 5.52°
PoseNet17 [16]	0.99 m, 1.06°	1.05 m, 3.97°	1.49 m, 3.43°	2.17 m, 2.94°	1.43 m, 2.85°
PoseNet17+ [16]	0.88 m, 1.04°	0.88 m, 3.78°	1.57 m, 3.32°	3.20 m, 3.29°	1.63 m, 2.86°
GL-Net [39]	0.59 m 0.65°	0.50 m, 2.87°	1.90 m, 3.29°	1.88 m, 2.78°	1.12 m, 2.40°
MapNet [5]	1.07 m, 1.89°	1.49 m, 4.22°	2.00 m, 4.53°	1.94 m, 3.91°	1.63 m, 3.64°
MS-Transformer [26]	0.83 m, 1.47°	0.86 m, 3.07°	1.62 m, 3.99°	1.81 m, 2.39°	1.28 m, 2.73°
PTFormer (Ours)	0.71 m, 1.32°	0.66 m, 2.80°	1.17 m, 3.31°	1.60 m, 2.69°	1.04 m, 2.53°

keep the same with Sect. 5. Our methods have a minor improvement compared with MS-Transformer and GL-Net, which proves that PTFormer can also perform well on car-mounted data.

6   Conclusion

In this work, we build a novel challenging BEV dataset for camera relocalization, which includes diverse scenes with large translation and rotation changes. We hope to provide a drone-specific benchmark for further research. At the same time, we provide a baseline method named PTFormer, a sequence-based transformer for camera pose estimation with inner attention, temporal progressive aggregation module, and parallel absolute and relative pose regression head. Experiments show the effectiveness of our designed modules. PTFormer performs best on the BEV dataset and achieves comparable performance both on the indoor 7Scenes dataset and the outdoor Cambridge Landmarks dataset compared with SOTA methods. In future work, we will continually expand the BEV dataset to contain more abundant cases such as data with illumination change, the same scene collected every several months, more expansive flying areas, and so on. Also, altitude change is a challenging problem, and we aim to synthesize images from different altitudes to make the model robust to altitude.

References

1. Balntas, V., Li, S., Prisacariu, V.: RelocNet: continuous metric learning relocalisation using neural nets. In: Ferrari, V., Hebert, M., Sminchisescu, C., Weiss, Y. (eds.) Computer Vision – ECCV 2018. LNCS, vol. 11218, pp. 782–799. Springer, Cham (2018). https://doi.org/10.1007/978-3-030-01264-9_46
2. Brachmann, E., et al.: DSAC-differentiable RANSAC for camera localization. In: Proceedings of the IEEE Conference on Computer Vision and Pattern Recognition, pp. 6684–6692 (2017)
3. Brachmann, E., Michel, F., Krull, A., Yang, M.Y., Gumhold, S., et al.: Uncertainty-driven 6d pose estimation of objects and scenes from a single RGB image. In: Proceedings of the IEEE Conference on Computer Vision and Pattern Recognition, pp. 3364–3372 (2016)
4. Brachmann, E., Rother, C.: Learning less is more-6d camera localization via 3d surface regression. In: Proceedings of the IEEE Conference on Computer Vision and Pattern Recognition, pp. 4654–4662 (2018)
5. Brahmbhatt, S., Gu, J., Kim, K., Hays, J., Kautz, J.: Geometry-aware learning of maps for camera localization. In: Proceedings of the IEEE Conference on Computer Vision and Pattern Recognition, pp. 2616–2625 (2018)
6. Cao, S., Snavely, N.: Minimal scene descriptions from structure from motion models. In: Proceedings of the IEEE Conference on Computer Vision and Pattern Recognition, pp. 461–468 (2014)

7. Carion, N., Massa, F., Synnaeve, G., Usunier, N., Kirillov, A., Zagoruyko, S.: End-to-end object detection with transformers. In: Vedaldi, A., Bischof, H., Brox, T., Frahm, J.-M. (eds.) ECCV 2020. LNCS, vol. 12346, pp. 213–229. Springer, Cham (2020). https://doi.org/10.1007/978-3-030-58452-8_13

8. Clark, R., Wang, S., Markham, A., Trigoni, N., Wen, H.: VidLoc: a deep spatio-temporal model for 6-DoF video-clip relocalization. In: Proceedings of the IEEE Conference on Computer Vision and Pattern Recognition, pp. 6856–6864 (2017)

9. DeTone, D., Malisiewicz, T., Rabinovich, A.: SuperPoint: self-supervised interest point detection and description. In: Proceedings of the IEEE Conference on Computer Vision and Pattern Recognition Workshops, pp. 224–236 (2018)

10. Dusmanu, M., et al.: D2- Net: a trainable CNN for joint detection and description of local features. In: CVPR 2019-IEEE Conference on Computer Vision and Pattern Recognition (2019)

11. En, S., Lechervy, A., Jurie, F.: RPNet: an end-to-end network for relative camera pose estimation. In: Leal-Taixé, L., Roth, S. (eds.) ECCV 2018. LNCS, vol. 11129, pp. 738–745. Springer, Cham (2019). https://doi.org/10.1007/978-3-030-11009-3_46

12. Fischler, M.A., Bolles, R.C.: Random sample consensus: a paradigm for model fitting with applications to image analysis and automated cartography. Commun. ACM **24**(6), 381–395 (1981)

13. Gao, S., Zhou, C., Ma, C., Wang, X., Yuan, J.: AiATrack: attention in attention for transformer visual tracking. In: Avidan, S., Brostow, G., Cissé, M., Farinella, G.M., Hassner, T. (eds.) ECCV 2022. LNCS, vol. 13682, pp. 146–164. Springer, Cham (2022). https://doi.org/10.1007/978-3-031-20047-2_9

14. Horn, B.K.: Closed-form solution of absolute orientation using unit quaternions. Josa a **4**(4), 629–642 (1987)

15. Kendall, A., Cipolla, R.: Modelling uncertainty in deep learning for camera relocalization. In: 2016 IEEE International Conference on Robotics and Automation, pp. 4762–4769. IEEE (2016)

16. Kendall, A., Cipolla, R.: Geometric loss functions for camera pose regression with deep learning. In: Proceedings of the IEEE Conference on Computer Vision and Pattern Recognition, pp. 5974–5983 (2017)

17. Kendall, A., Grimes, M., Cipolla, R.: PoseNet: a convolutional network for real-time 6-DoF camera relocalization. In: Proceedings of the IEEE International Conference on Computer Vision, pp. 2938–2946 (2015)

18. Laskar, Z., Melekhov, I., Kalia, S., Kannala, J.: Camera relocalization by computing pairwise relative poses using convolutional neural network. In: Proceedings of the IEEE International Conference on Computer Vision Workshops, pp. 929–938 (2017)

19. Li, X., Ling, H.: GTCaR: graph transformer for camera re-localization. In: Avidan, S., Brostow, G., Cisé, M., Farinella, G.M., Hassner, T. (eds.) ECCV 2022. LNCS, vol. 13670, pp. 229–246. Springer, Cham (2022). https://doi.org/10.1007/978-3-031-20080-9_14

20. Li, Y., Snavely, N., Huttenlocher, D.P., Fua, P.: Worldwide pose estimation using 3D point clouds. In: Zamir, A.R.R., Hakeem, A., Van Van Gool, L., Shah, M., Szeliski, R. (eds.) Large-Scale Visual Geo-Localization. ACVPR, pp. 147–163. Springer, Cham (2016). https://doi.org/10.1007/978-3-319-25781-5_8

21. Maddern, W., Pascoe, G., Linegar, C., Newman, P.: 1 year, 1000 km: the oxford robotcar dataset. Int. J. Robot. Res. **36**(1), 3–15 (2017)

22. Melekhov, I., Ylioinas, J., Kannala, J., Rahtu, E.: Image-based localization using hourglass networks. In: Proceedings of the IEEE International Conference on Computer Vision Workshops, pp. 879–886 (2017)
23. Sarlin, P.E., Cadena, C., Siegwart, R., Dymczyk, M.: From coarse to fine: Robust hierarchical localization at large scale. In: Proceedings of the IEEE/CVF Conference on Computer Vision and Pattern Recognition, pp. 12716–12725 (2019)
24. Sattler, T., Leibe, B., Kobbelt, L.: Efficient & effective prioritized matching for large-scale image-based localization. IEEE Trans. Pattern Anal. Mach. Intell. **39**(9), 1744–1756 (2016)
25. Schleiss, M., Rouatbi, F., Cremers, D.: Vpair-aerial visual place recognition and localization in large-scale outdoor environments. arXiv preprint arXiv:2205.11567 (2022)
26. Shavit, Y., Ferens, R., Keller, Y.: Learning multi-scene absolute pose regression with transformers. In: Proceedings of the IEEE/CVF International Conference on Computer Vision, pp. 2733–2742 (2021)
27. Shotton, J., Glocker, B., Zach, C., Izadi, S., Criminisi, A., Fitzgibbon, A.: Scene coordinate regression forests for camera relocalization in RGB-D images. In: Proceedings of the IEEE Conference on Computer Vision and Pattern Recognition, pp. 2930–2937 (2013)
28. Stenborg, E., Sattler, T., Hammarstrand, L.: Using image sequences for long-term visual localization. In: 2020 International Conference on 3d Vision, pp. 938–948. IEEE (2020)
29. Sun, J., Shen, Z., Wang, Y., Bao, H., Zhou, X.: LoFTR: detector-free local feature matching with transformers. In: Proceedings of the IEEE/CVF Conference on Computer Vision and Pattern Recognition, pp. 8922–8931 (2021)
30. Svärm, L., Enqvist, O., Kahl, F., Oskarsson, M.: City-scale localization for cameras with known vertical direction. IEEE Trans. Pattern Anal. Mach. Intell. **39**(7), 1455–1461 (2016)
31. Taira, H., et al.: InLoc: indoor visual localization with dense matching and view synthesis. In: Proceedings of the IEEE Conference on Computer Vision and Pattern Recognition, pp. 7199–7209 (2018)
32. Tan, M., Le, Q.: EfficientNet: rethinking model scaling for convolutional neural networks. In: Proceedings of International Conference on Machine Learning, pp. 6105–6114. PMLR (2019)
33. Vallone, A., Warburg, F., Hansen, H., Hauberg, S., Civera, J.: Danish airs and grounds: a dataset for aerial-to-street-level place recognition and localization. IEEE Robot. Autom. Lett. **7**(4), 9207–9214 (2022)
34. Vaswani, A., et al.: Attention is all you need. In: Advances in Neural Information Processing Systems, vol. 30 (2017)
35. Walch, F., Hazirbas, C., Leal-Taixe, L., Sattler, T., Hilsenbeck, S., Cremers, D.: Image-based localization using LSTMs for structured feature correlation. In: Proceedings of the IEEE International Conference on Computer Vision, pp. 627–637 (2017)
36. Wang, B., Chen, C., Lu, C.X., Zhao, P., Trigoni, N., Markham, A.: AtLoc: attention guided camera localization. In: Proceedings of the AAAI Conference on Artificial Intelligence, vol. 34, pp. 10393–10401 (2020)
37. Wu, J., Ma, L., Hu, X.: Delving deeper into convolutional neural networks for camera relocalization. In: 2017 IEEE International Conference on Robotics and Automation, pp. 5644–5651. IEEE (2017)

38. Xue, F., Wang, X., Yan, Z., Wang, Q., Wang, J., Zha, H.: Local supports global: deep camera relocalization with sequence enhancement. In: Proceedings of the IEEE/CVF International Conference on Computer Vision, pp. 2841–2850 (2019)
39. Xue, F., Wu, X., Cai, S., Wang, J.: Learning multi-view camera relocalization with graph neural networks. In: 2020 IEEE/CVF Conference on Computer Vision and Pattern Recognition, pp. 11372–11381. IEEE (2020)
40. Zhou, K., Chen, C., Wang, B., Saputra, M.R.U., Trigoni, N., Markham, A.: VMLoc: variational fusion for learning-based multimodal camera localization. In: Proceedings of the AAAI Conference on Artificial Intelligence, vol. 35, pp. 6165–6173 (2021)

Adaptive Focal Inverse Distance Transform Maps for Cell Recognition

Wenjie Huang[1], Xing Wu[1(✉)], Chengliang Wang[1], Zailin Yang[2],
Longrong Ran[2], and Yao Liu[2]

[1] Chongqing University, Chongqing 400000, China
{20172957,wuxing,wangcl}@cqu.edu.cn
[2] Department of Hematology-Oncology, Chongqing Key Laboratory of Translational
Research for Cancer Metastasis and Individualized Treatment, Chongqing University
Cancer Hospital, Chongqing 400000, China
{zailinyang,liuyao77}@cqu.edu.cn

Abstract. The quantitative analysis of cells is crucial for clinical diagnosis, and effective analysis requires accurate detection and classification. Using point annotations for weakly supervised learning is a common approach for cell recognition, which significantly reduces the labeling workload. Cell recognition methods based on point annotations primarily rely on manually crafted smooth pseudo labels. However, the diversity of cell shapes can render the fixed encodings ineffective. In this paper, we propose a multi-task cell recognition framework. The framework utilizes a regression task to adaptively generate smooth pseudo labels with cell morphological features to guide the robust learning of probability branch and utilizes an additional branch for classification. Meanwhile, in order to address the issue of multiple high-response points in one cell, we introduce Non-Maximum Suppression (NMS) to avoid duplicate detection. On a bone marrow cell recognition dataset, our method is compared with five representative methods. Compared with the best performing method, our method achieves improvements of 2.0 F1 score and 3.6 F1 score in detection and classification, respectively.

Keywords: Cell recognition · Point annotation · Distance transform ·
Multi-task learning · Proposal matching

1 Introduce

Diffuse large B-cell lymphoma (DLBCL) is a major form of human blood cancer. The proportion of DLBCL cells in bone marrow cells is an important basis for cancer clinical diagnosis. However, microscopic imaging often involves a large number of cells, and manual cell counting using the naked eye can be burdensome task for doctors. With the increasing quality of microscopic imaging, utilizing computer-aided techniques to analyze digital cytology images for cancer screening [14] can significantly alleviate the workload of doctors and improve the accuracy of diagnosis. However, both object detection and semantic segmentation

B. Luo et al. (Eds.): ICONIP 2023, LNCS 14452, pp. 148–160, 2024.
https://doi.org/10.1007/978-981-99-8076-5_11

rely on strong labels, which require a significant amount of labeling workload. Moreover, due to the high density of cells, strong labels may introduce additional interference information. Therefore, in the field of cell recognition, utilizing point annotations is a more common and practical approach.

The most common method based on point annotations is to generate intermediate density maps to simulate the probability density of cells. Although this approach can significantly enhance the robustness of model, cells often exhibit diverse morphological features, and using the fixed encodings may regress cells into approximate circular shapes. Some studies [21–23] have indicated that these density map-based methods have limitations. Moreover, traditional density map-based methods can only count the number of objects and cannot precisely locate their positions.

Some methods treat point detection as a set prediction problem [9–11], where a bipartite graph algorithm is used to match point proposals with optimal learning targets. However, these methods essentially involve non-smooth regression of the point proposals. Although they can achieve good results when the object features are prominent, in case the input images are high resolution and exhibit numerous morphological and textural features, if the labels are not smooth, they may face challenges in robust learning.

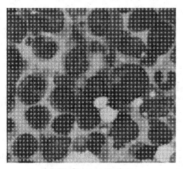
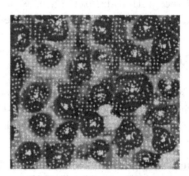

(a) Reference Points   (b) Predicted Proposal Points

Fig. 1. Regression demonstration in P2PNet. The proposal points near the cell center tend to converge towards the cell center after extracting meaningful features.

The FIDT map [13] demonstrated excellent performance of crowd counting in dense scenes. While it addressed the issue of object localization, the FIDT map assigns probability values to each pixel based solely on the Euclidean distance from the ground truth labels. In essence, it is a pseudo label that contains a significant amount of noise and regresses the objects into approximate circular shapes without incorporating the information from the image. If we can find a method to correct the coordinate of each pixel, the generated intermediate map would be more accurate. P2PNet [9] performs open set prediction based on preset reference points on the original image. We observed that P2PNet regresses the coordinate offsets between the reference points and the ground truth points (see Fig. 1), as a result, if a proposal point extracts the features of cells, it tends to instinctively shift towards the nearest ground truth point, and the Euclidean distance from the point proposals to the ground truth points can be used to measure the similarity between them.

In this paper, based on the above considerations, we extend the set prediction framework by introducing a probability branch for cell detection and a classification branch for cell classification. Specifically, our research makes the following contributions:

- We propose a multi-task cell recognition framework that achieves both cell localization and cell classification. The framework is based on set prediction and utilizes a regression task to adaptively generate AFIDT map with cell morphological features to guide the robust learning of probability branch, thereby achieving robust cell detection.
- We redesign the process of local maximum detection. Specifically, we apply non-maximum suppression (NMS) to the set of predicted points based on probability scores. This approach avoids duplicate detection of one cell and improves the accuracy of cell detection.
- To validate the effectiveness of our method, we construct a bone marrow cell recognition dataset. On this dataset, our method is compared with five representative methods and achieves the best performance.

2   Related Work

In general, deep learning-based cell recognition methods are primarily categorized into three types: segmentation-based methods, intermediate map-based methods, and detection-based methods.

Segmentation-Based Methods. Cell segmentation methods based on traditional image techniques include threshold-based methods [28] and watershed-based methods [29], however, these non-data-driven methods lack robustness. Deep leaning-based methods utilize masks that contain category information to perform cell segmentation and classification. With the rapid development of convolutional neural networks (CNN), using CNN for cell segmentation has become a hot research topic [2,15]. Recently, some studies aimed to design better feature extraction networks using self-attention structure [18]. Chen et al. [25] integrated the self-attention structure into the U-Net architecture. Ji et al. [24] used the self-attention structure to fuse multi-scale features. However, segmentation-based methods require complex manual annotations. Furthermore, due to the density and similarity of cells, the boundaries of cell segmentation are often ambiguous. Therefore, segmentation-based methods cannot directly help us accurately count the number of different types of cells.

Intermediate Map-Based Methods. Density map is a type of intermediate map that simulate the probability density of objects. Lempitsky et al. [12] first introduced the density map method into the task of object counting. Some researchers aimed to design better network architectures to improve the usability of density map. Li et al. [16] designed a network architecture based on dilated convolution to understand highly congested scenes and generate high quality density map. Wan et al. [1] proposed a density estimation network with an adaptive density map generator. However, density map-based methods still cannot precisely locate the positions of objects. Some studies attempted to use intermediate maps for precise localization of objects. Liang et al. [4] proposed the repel encoding method to differentiate adjacent cells in crowded regions. Liang et al. [13] proposed the inverse distance transform maps method that exhibits excellent robustness in extremely dense scenes. Sugimoto et al. [17] designed an interactive network architecture for cell recognition. Zhang et al. [5] adopted a multi-task strategy to generate intermediate probability maps as additional supervision signals.

Detection-Based Methods. Cell detection methods based on traditional image techniques include hand-designed feature-based methods [26] and traditional machine learning-based methods [19]. In recent years, the most common approach in deep learning based object detection algorithms is to regress bounding boxes [6–8]. But in cell recognition task, it is not necessary to outline the entire cells. Instead, the focus is on locating the cells and classifying them. Some end-to-end detection methods directly perform cell recognition on point annotations. Zhou et al. [3] adopted a multi-task interactive framework to optimize both the detection and classification tasks. Some methods treat point localization and classification as set prediction problems. In contrast to traditional object detection, these methods only regress the coordinates of objects without regressing the entire bounding box. Song et al. [9] first proposed the set prediction framework based on point annotations. Liang et al. [10] used the extracted features and trainable embedding as inputs to the transformer decoder for set prediction. Shui et al. [11] introduced this framework into the field of cell recognition and proposed a pyramid feature aggregation strategy to aggregate multi-level features.

3   Methods

In this chapter, we provide a detailed description of our method. In Sect. 3.1, we explain how our framework works during the training and inference stages. In Sects. 3.2–3.4, we explain the internal implementations of three modules. In Sect. 3.5, we explain how we construct loss functions.

3.1   Overall Framework

From the Fig. 2, during training, in match module, to assign the optimal learning target for each point proposal, we adopt detection task and regression task for

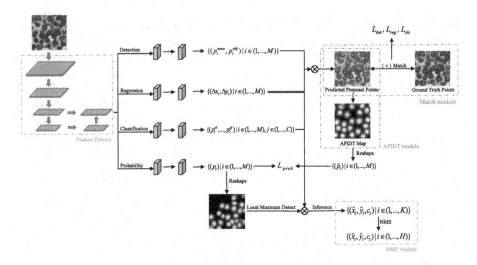

Fig. 2. The whole framework based on multi-task learning. The cell image first passes through a deep feature extraction network, then we adopt four tasks to make predictions for the reference points. For all point proposals, the detection task predicts detection score set, the regression task predicts offset set, the classification task predicts category set, and the probability task predicts probability set.

proposal matching. Then, the $L_{det}, L_{reg}, L_{cls}$ are calculated based on the matching results. Meanwhile, in AFIDT module, the regression task is used to regress the coordinates of all point proposals, and the predicted coordinates are used to generate the AFIDT map, then, the L_{prob} is calculated using the probability set and the reshaped AFIDT map.

During inference, the detection branch is inactive. The probability set is reshaped into a 2-dimension probability map, then, a local maximum detection algorithm [13] is applied on the probability map to obtain the predicted set of point proposals. Finally, in NMS module, the NMS algorithm is used to suppress point proposals with lower probability scores, resulting in the final output.

3.2 AFIDT Module

Our method does not require restoring the feature map to the original image size. Instead, it generates the AFIDT map based on the coordinates of all predicted point proposals. Specifically, we add the regression values to the preset reference points to get the predicted coordinates of all point proposals, then the AFIDT map is generated using the coordinates (Fig. 3). $\hat{p}_i$ is the i-th probability pseudo label in the AFIDT map, which is defined as follow:

$$\hat{x}_i = x_i + k\Delta x_i, \hat{y}_i = y_i + k\Delta y_i \tag{1}$$

$$\hat{p}_i = \frac{1}{D(\hat{x}_i, \hat{y}_i)^{(\alpha D(\hat{x}_i, \hat{y}_i) + \beta)} + C} \tag{2}$$

Where (x_i, y_i) is the i-th original coordinate of the i-th reference point, $(\Delta x_i, \Delta y_i)$ is the predicted regression value for the i-th reference point, $(\hat{x}_i, \hat{y}_i)$ is the coordinate of the i-th point proposal, k is a balance coefficient, and $D(\hat{x}_i, \hat{y}_i)$ is the Euclidean distance between the i-th point proposal and its nearest ground truth point.

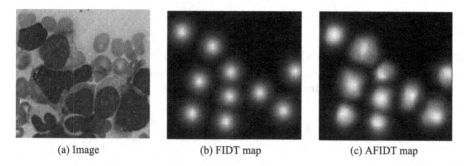

      (a) Image   (b) FIDT map   (c) AFIDT map

Fig. 3. The differences between AFIDT map and FIDT map. Compared with the FIDT map, our AFIDT map incorporates morphological features of cells such as shape and size. Moreover, the boundaries of cells in the AFIDT map are more precise and realistic.

3.3 Match Module

The purpose of proposal matching is to assign the optimal learning target to each point proposal, enabling more effective learning and improving the accuracy of the classification and regression tasks. In our matching method, we consider the detection score of each point proposal and the Euclidean distance between each proposal point and each ground truth point to construct the cost matrix for one-to-one matching. The cost matrix D is defined as follows:

$$loc_i = (x_i + \Delta x_i, y_i + \Delta y_i) \tag{3}$$

$$D = (\mu \| loc_i - loc_j^* \|_1 - p_i^{det})_{i \in M, j \in N} \tag{4}$$

Where loc_i and loc_j^* are the coordinates of the i-th point proposal and the j-th ground truth point, respectively. p_i^{det} is the predicted detection score for the i-th point proposal, μ is a balancing coefficient, M is the total number of point proposals, and N is the total number of ground truth points, which is also equal to the number of positive proposals.

3.4 NMS Module

Although our method can generate AFIDT maps adaptively, the process of generation does not strictly adhere to the principle of maximizing the center probability. As a result, there may be more than one high-response points in one cell.

when local maximum detection radius is small, all the high-response points may be detected, meaning that a cell may be detected multiple times. To address this problem, we use Non-Maximum Suppression (NMS) algorithm at the end of our framework. Specifically, we perform NMS on the set of all predicted effective points based on the probability values, and the suppression distance is set to 12 pixels.

3.5 Loss Function

Probability Loss. The AFIDT map generated by the coordinates of point proposals is reshaped to supervise the probability branch. We adopt the mean squared error (MSE) loss for probability training:

$$L_{prob} = \frac{1}{M} \sum_{i=1}^{M} ||\hat{p}_i - p_i||_2 \tag{5}$$

Where p_i is the predicted probability value for the i-th point proposal, and $\hat{p}_i$ is the probability pseudo label for the i-th point proposal.

Regression Loss. To calculate the regression loss between the predicted coordinates and the ground truth coordinates, we adopt the MSE loss for regression training:

$$loc_i = (x_i + \lambda \Delta x_i, y_i + \lambda \Delta y_i) \tag{6}$$

$$L_{reg} = \frac{1}{N} \sum_{i=1}^{N} ||loc_i - loc_i^*||_2 \tag{7}$$

Where loc_i is the coordinate of the i-th positive proposal, and loc_i^* is the coordinate of the i-th corresponding ground truth point. It is important to note that the coefficients in formulas 1, 3 and 6 may not necessarily be the same.

Detection Loss. In bone marrow smear images, stained cells are often not distributed too densely, resulting in a significant number of negative proposals. In order to enhance detection of cells, we reduce the proportion of negative proposals in loss computation. For the detection loss, we adopt weighted cross-entropy (CE) loss for training:

$$L_{det} = -\frac{1}{M} (\sum_{i=1}^{N} \log(p_i^{obj}) + \omega \sum_{i=N}^{M} \log(p_i^{none})) \tag{8}$$

Where ω is a balance coefficient, p_i^{obj} is the foreground score of the i-th positive proposal, and p_i^{none} is the background score of the $(i\text{-}N)$-th negative proposal.

Classification Loss. Due to the relatively balanced number of two cell types in these experiments, we adopt CE loss for classification training:

$$L_{cls} = -\frac{1}{N} \sum_{i=1}^{N} \log(p_i^{cls}) \tag{9}$$

Where p_i^{cls} is the classification score of the i-th positive proposal.

Total Loss. The overall loss function is constructed as follows:

$$L_{total} = L_{prob} + \rho_c L_{cls} + \rho_r L_{reg} + \rho_d L_{det} \tag{10}$$

Where ρ_c, ρ_r, ρ_d are the balance coefficients.

4   Experiments

4.1   Dataset

The diagnosis of DLBCL requires the exclusion of other cells that are pathologically similar to DLBCL cells [20]. We performed experiments on bone marrow smear stained images, which were collected using the most advanced equipment under different conditions. First, we used a sliding window of size 1800×1600 to extract 704 patches from the regions of interest (ROI) in high resolution images. Subsequently, the images were downsampled to a resolution of 896×800 for training. The training set and test set were divided into a ratio of 8:2. To determine the proportion of DLBCL cells in bone marrow cells, we invited two pathology experts to label all stained cells. The stained cells are labeled as either DLBCL or non-DLBCL types, and the annotations were repeatedly checked to ensure their accuracy. A total of 12,164 DLBCL cells and 20,083 non-DLBCL cells were labeled in this dataset.

4.2   Implementation Details

In these experiments, We compared our method with five representative methods. Except for CSRNet [16], all five networks utilized VGG-16_bn [27] for feature extraction. Adam optimizer was employed with a momentum of 0.9, learning rate of 1e−4, and the batch size is set to 4. The hyper-parameters are set as follows, $\mu = 0.05$, $\rho_c = 1$, $\rho_r = 0.01$, $\rho_d = 1$, $\alpha = 0.02$, $\beta = 0.02$, $C = 1$, $\omega = 0.5$, $k = 0.5$, $\lambda = 1$. All algorithms used random flipping and cropping as data augmentation during training and our training was performed using a single NVIDIA 2080TI GPU.

Due to the high resolution of bone marrow cell images, after an upsampling path, all tasks are performed on the feature map that is 1/8 the size of the original image. Additionally, reference points are placed every 8 pixels on the original image. For calculating metrics, the effective matching range was defined as the radius of 12 pixels around the ground truth points. The intermediate map-based methods used local maximum detection to locate positions of cells, with detection radius set to 2 points. In terms of evaluation metrics, we adopted precision(P), recall(R), and F1 scores.

4.3   Experimental Results

Comparative Experiments. Our method was compared with five representative methods in the fields of crowd counting and cell recognition, including three set prediction-based methods [9–11] and two intermediate map-based methods [13,16].

Table 1. Comparison of cell recognition with different methods

Method	Detection			Classification		
	P	R	F1	P	R	F1
P2PNet [9]	87.2	89.7	88.4	78.8	81.0	79.9
Method in [11]	87.0	87.3	87.1	78.3	78.5	78.4
CLTR [10]	85.6	84.1	84.8	77.6	78.4	78.0
CSRNet [16]	87.9	90.2	89.1	76.5	78.6	77.5
FIDT [13]	89.8	90.2	90.0	79.5	79.9	79.7
Ours	**92.2**	**91.9**	**92.0**	**83.7**	**82.3**	**83.5**

From the Table 1, On the bone marrow cell recognition dataset, our method demonstrates significantly better performance in both detection and classification compared with the other five methods. Specifically, compared with FIDT, our method improved the F1 score for detection and classification by 2.0 and 3.8, respectively. We can see that the intermediate map-based methods have an advantage in terms of detection performance. This is because they utilize smooth pseudo labels, which enhances the robustness of the model for detection. However, set prediction-based methods can even achieve a reverse advantage in terms of classification performance. This is because the proposal matching can assign the optimal learning targets for each point proposal, thereby they can achieve better classification performance. Our multi-task framework retains the classification advantage of proposal matching and benefits from the superior detection robustness of AFIDT maps, resulting in optimal performance.

Ablation Experiments. To verify the role of NMS in post-processing, we compared the effects of using NMS with different local maximum detection radii on the F1 score for classification.

From the Table 2, NMS can effectively improve the F1 score, and when the detection radius is smaller, the improvement becomes more obvious. When the

Table 2. Results of ablation experiments

Method	R = 3	R = 3 & NMS	R = 1	R = 1 & NMS	R = 2	**R = 2 & NMS**
F1	83.2	83.2	80.9	83.3	83.4	**83.5**

detection radius is 1, using NMS can significantly increase the F1 score by 2.4. When the detection radius is 2, using NMS can achieves the best F1 score. When the detection radius is 3, NMS does not work, but it also does not decrease the F1 score. This result can be explained as the detection radius increases, there will be fewer instances of duplicate detection, and when the detection radius is large enough, there will be no duplicate detection.

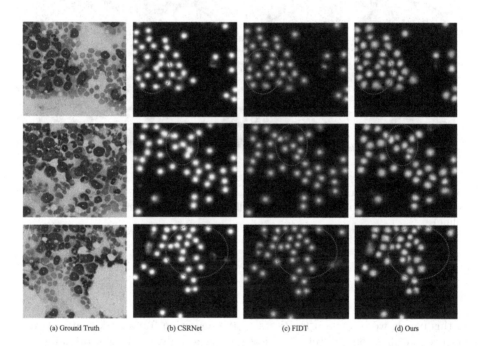

(a) Ground Truth   (b) CSRNet   (c) FIDT   (d) Ours

Fig. 4. Visualization results of probability maps generated by three intermediate map-based methods.

4.4 Visualization Analysis

From the Fig. 4, both FIDT [13] and CSRNet [16] regress all cells into approximate circular shapes, while our method can reflect the size and shape of cells on the probability map. When faced dense cells with irregular shapes, both FIDT and CSRNet perform poorly. They fail to correctly identify individual cells or mistakenly merge multiple cells into one. But our method accurately recognizes dense cells with irregular shapes, and the boundaries between cells are more defined (Fig. 5).

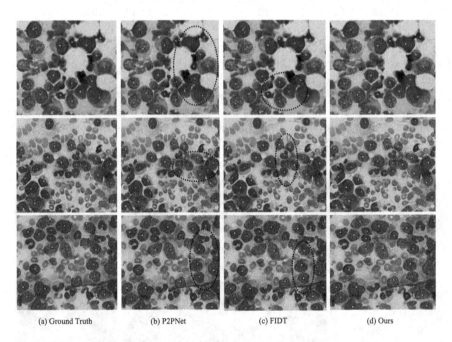

(a) Ground Truth (b) P2PNet (c) FIDT (d) Ours

Fig. 5. Detection visualization results of three representative methods. We mark DLBCL cells and non-DLBCL cells in yellow and blue, respectively. (Color figure online)

5 Conclusion

In this paper, we propose a multi-task cell recognition framework that achieves both cell localization and cell classification. Specifically, our framework adopts a multi-task learning approach, utilizing the extracted morphological features from the regression branch to construct AFIDT map, then, we utilize the generated AFIDT map to supervise the probability branch. Unlike traditional intermediate map-based methods, our method allows the probability map to reflect various morphological features of cells and utilizes an additional branch for cell classification, which effectively improves the detection and classification performance. Additionally, we redesigned the post-processing method by introducing NMS to avoid duplicate detection of one cell. Our method achieves the best performance on the bone marrow cell recognition dataset. Finally, we visualized and analyzed the output of the different networks, validating the effectiveness of our method.

Acknowledgements. This work is supported by the Fundamental Research Funds for the Central Universities (No. 2022CDJYGRH-001) and the Chongqing Technology Innovation & Application Development Key Project (cstc2020jscx-dxwtBX0055; cstb2022tiad-kpx0148).

References

1. Wan, J., Chan, A.: Adaptive density map generation for crowd counting. In: Proceedings of the IEEE/CVF International Conference on Computer Vision, pp. 1130–1139 (2019)

2. Zhou, Z., Rahman Siddiquee, M.M., Tajbakhsh, N., Liang, J.: UNet++: a nested U-Net architecture for medical image segmentation. In: Stoyanov, D., et al. (eds.) DLMIA/ML-CDS -2018. LNCS, vol. 11045, pp. 3–11. Springer, Cham (2018). https://doi.org/10.1007/978-3-030-00889-5_1

3. Zhou, Y., Dou, Q., Chen, H., et al.: SFCN-OPI: detection and fine-grained classification of nuclei using sibling FCN with objectness prior interaction. In: Proceedings of the AAAI Conference on Artificial Intelligence, vol. 32, no. 1 (2018)

4. Liang, H., Naik, A., Williams, C.L., et al.: Enhanced center coding for cell detection with convolutional neural networks. arXiv preprint arXiv:1904.08864 (2019)

5. Zhang, S., Zhu, C., Li, H., et al.: Weakly supervised learning for cell recognition in immunohistochemical cytoplasm staining images. In: 2022 IEEE 19th International Symposium on Biomedical Imaging (ISBI), pp. 1–5. IEEE (2022)

6. Ren, S., He, K., Girshick, R., et al.: Faster R-CNN: towards real-time object detection with region proposal networks. Adv. Neural Inf. Process. Syst. **28** (2015)

7. Tian, Z., Shen, C., Chen, H., et al.: FCOS: fully convolutional one-stage object detection. In: Proceedings of the IEEE/CVF International Conference on Computer Vision, pp. 9627–9636 (2019)

8. Carion, N., Massa, F., Synnaeve, G., Usunier, N., Kirillov, A., Zagoruyko, S.: End-to-end object detection with transformers. In: Vedaldi, A., Bischof, H., Brox, T., Frahm, J.-M. (eds.) ECCV 2020, Part I. LNCS, vol. 12346, pp. 213–229. Springer, Cham (2020). https://doi.org/10.1007/978-3-030-58452-8_13

9. Song, Q., Wang, C., Jiang, Z., et al.: Rethinking counting and localization in crowds: a purely point-based framework. In: Proceedings of the IEEE/CVF International Conference on Computer Vision, pp. 3365–3374 (2021)

10. Liang, D., Xu, W., Bai, X.: An end-to-end transformer model for crowd localization. In: Avidan, S., Brostow, G., Cissé, M., Farinella, G.M., Hassner, T. (eds.) ECCV 2022, Part I. LNCS, vol. 13661, pp. 38–54. Springer, Cham (2022). https://doi.org/10.1007/978-3-031-19769-7_3

11. Shui, Z., Zhang, S., Zhu, C., et al.: End-to-end cell recognition by point annotation. In: Wang, L., Dou, Q., Fletcher, P.T., Speidel, S., Li, S. (eds.) MICCAI 2022, Part IV. LNCS, vol. 13434, pp. 109–118. Springer, Cham (2022). https://doi.org/10.1007/978-3-031-16440-8_11

12. Lempitsky, V., Zisserman, A.: Learning to count objects in images. Adv. Neural Inf. Process. Syst. **23** (2010)

13. Liang, D., Xu, W., Zhu, Y., et al.: Focal inverse distance transform maps for crowd localization. IEEE Trans. Multimedia, 1–13 (2022)

14. Jiang, H., Zhou, Y., Lin, Y., et al.: Deep learning for computational cytology: a survey. Med. Image Anal., 102691 (2022)

15. Ronneberger, O., Fischer, P., Brox, T.: U-Net: convolutional networks for biomedical image segmentation. In: Navab, N., Hornegger, J., Wells, W.M., Frangi, A.F. (eds.) MICCAI 2015, Part III. LNCS, vol. 9351, pp. 234–241. Springer, Cham (2015). https://doi.org/10.1007/978-3-319-24574-4_28

16. Li, Y., Zhang, X., Chen, D.: CSRNet: dilated convolutional neural networks for understanding the highly congested scenes. In: Proceedings of the IEEE Conference on Computer Vision and Pattern Recognition, pp. 1091–1100 (2018)

17. Sugimoto, T., Ito, H., Teramoto, Y., et al.: Multi-class cell detection using modified self-attention. In: Proceedings of the IEEE/CVF Conference on Computer Vision and Pattern Recognition, pp. 1855–1863 (2022)

18. Vaswani, A., Shazeer, N., Parmar, N., et al.: Attention is all you need. Adv. Neural. Inf. Process. Syst. **30** (2017)

19. Tikkanen, T., Ruusuvuori, P., Latonen, L., et al.: Training based cell detection from bright-field microscope images. In: 2015 9th International Symposium on Image and Signal Processing and Analysis (ISPA), pp. 160–164. IEEE (2015)

20. Li, D., Bledsoe, J.R., Zeng, Y., et al.: A deep learning diagnostic platform for diffuse large B-cell lymphoma with high accuracy across multiple hospitals. Nat. Commun. **11**(1), 6004 (2020)

21. Bai, S., He, Z., Qiao, Y., et al.: Adaptive dilated network with self-correction supervision for counting. In: Proceedings of the IEEE/CVF Conference on Computer Vision and Pattern Recognition, pp. 4594–4603 (2020)

22. Ma, Z., Wei, X., Hong, X., et al.: Bayesian loss for crowd count estimation with point supervision. In: Proceedings of the IEEE/CVF International Conference on Computer Vision, pp. 6142–6151 (2019)

23. Wang, B., Liu, H., Samaras, D., et al.: Distribution matching for crowd counting. Adv. Neural. Inf. Process. Syst. **33**, 1595–1607 (2020)

24. Ji, Y., et al.: Multi-compound transformer for accurate biomedical image segmentation. In: de Bruijne, M., et al. (eds.) MICCAI 2021, Part I. LNCS, vol. 12901, pp. 326–336. Springer, Cham (2021). https://doi.org/10.1007/978-3-030-87193-2_31

25. Chen, J., Lu, Y., Yu, Q., et al.: TransUNet: transformers make strong encoders for medical image segmentation. arXiv preprint arXiv:2102.04306 (2021)

26. Arteta, C., Lempitsky, V., Noble, J.A., Zisserman, A.: Learning to detect cells using non-overlapping extremal regions. In: Ayache, N., Delingette, H., Golland, P., Mori, K. (eds.) MICCAI 2012, Part I. LNCS, vol. 7510, pp. 348–356. Springer, Heidelberg (2012). https://doi.org/10.1007/978-3-642-33415-3_43

27. Simonyan, K., Zisserman, A.: Very deep convolutional networks for large-scale image recognition. arXiv preprint arXiv:1409.1556 (2014) 5

28. Otsu, N.: A threshold selection method from gray-level histograms. IEEE Trans. Syst. Man Cybern. **9**(1), 62–66 (1979)

29. Chalfoun, J., Majurski, M., Dima, A., et al.: FogBank: a single cell segmentation across multiple cell lines and image modalities. BMC Bioinform. **15**, 1–12 (2014)

Stereo Visual Mesh for Generating Sparse Semantic Maps at High Frame Rates

Alexander Biddulph[1,2(✉)] [ID], Trent Houliston[2] [ID], Alexandre Mendes[1] [ID], and Stephan Chalup[1] [ID]

[1] The University of Newcastle, Callaghan, Australia
Alexander.Biddulph@uon.edu.au,
{Alexandre.Mendes,Stephan.Chalup}@newcastle.edu.au
[2] 4Tel Pty Ltd., Newcastle, Australia
{abiddulph,thouliston}@4tel.com.au
https://www.newcastle.edu.au, https://www.4tel.com.au

Abstract. The Visual Mesh is an input transform for deep learning that allows depth independent object detection at very high frame rates. The present study introduces a Visual Mesh based stereo vision method for sparse stereo semantic segmentation. A dataset of simulated 3D scenes was generated and used for training to show that the method is capable of processing high resolution stereo inputs to generate both left and right sparse semantic maps. The new stereo method demonstrated better classification accuracy than the corresponding monocular approach. The high frame rates and high accuracy may make the proposed approach attractive to fast-paced on-board robot or IoT applications.

Keywords: Deep Learning · Stereo Vision · Semantic Segmentation

1 Introduction

For the task of detecting objects of interest in highly structured environments, such as autonomous driving [6,32], robotic soccer [7,34], or navigation in marine environments [5,24], dense depth and semantic predictions are not required. If the sizes of the objects of interest are known and their boundaries can be accurately determined then there exist simple and efficient algorithms to calculate the distances between the camera and the objects. For example, the formula $D = r/\sin \delta$ provides the distance to a spherical object, like a soccer ball, that has an angular radius, δ, and an actual radius, r. The present study focuses on high-speed semantic classification and leaves the determination of distances as a separate task.

The Visual Mesh input transform [12] achieves object detection at exceptionally high frame rates by warping the input space in a way that allows sampling

A. Biddulph was supported by an Australian Government Research Training Program scholarship and a top-up scholarship through 4Tel Pty Ltd. S. Chalup was supported by ARC DP210103304.

B. Luo et al. (Eds.): ICONIP 2023, LNCS 14452, pp. 161–178, 2024.
https://doi.org/10.1007/978-981-99-8076-5_12

of the target object using a fixed number of pixels, independently of the object's distance to the camera. The convolutional neural network (CNN) with Visual Mesh proposed in [12] was only trained to detect a single class of objects from a single view of the scene, leaving open the question of whether multi-class classifications, both from a single view and from multiple views, would be possible with this type of input transformation.

The main contributions of the present paper are:

1. Training the Visual Mesh to show that it is capable of detecting multiple classes.
2. Expanding the concept of the Visual Mesh to include sparse stereo semantic segmentation, receiving a stereo image pair as input and simultaneously producing a sparse stereo semantic segmentation map (SSM) for both the left and right views.
3. Show that the proposed Stereo Visual Mesh can achieve better classification accuracy than the mono version at comparable computational expense.

2   Related Work

There are very few works which deal with stereo semantic segmentation. In the context of the present work a "stereo semantic segmentation system" is defined as a system which directly computes a SSM, or a pair of SSMs, from an input pair of stereo images. While there are many works which compute a SSM in a stereo context, these works typically only compute a SSM on a single input image, and then either use this SSM to refine a disparity map, as is done by [22,38], or use a disparity map to refine the SSM, as is done by [2,26] who use multiple U-Net [31] models to compute 12 SSMs. The SSM are then refined with the aid of a disparity map obtained from PSMNet [3].

A disparity map is used by [4] to refine a SSM. Visible and near-infrared (VNIR) stereo images are used to compute the disparity map. Single RGB and multi-spectral images are used to compute semantic features which are then fused together with the output of a disparity fusion segmentation network which fuses disparity features with stereo semantic features. The fused SSMs are post-processed with the computed disparity map.

A 3D bounding box detection methods is proposed by [17]. A ResNet38 [11] encoder is used to extract image features which are then used to compute a SSM and bounding box detection and regression. A disparity map is computed and used to convert the 2D bounding box proposals into 3D bounding boxes.

Fan et al. [9] use a depth image to compute surface normals which are then fed into a DenseNet [15] and ResNet [11] inspired network to reinforce semantic features for the purposes of estimating free space in the scene.

The following works have been identified as performing stereo semantic segmentation, where either a single SSM or a pair of SSMs are generated as output:

- [25,36] use Graph-Cut and energy minimisation techniques to perform foreground/background segmentation with the aid of disparity maps and user input.

- [4,8,40] use stereo image features combined with depth features to compute a single SSM.
- Finally, [23] use the encoder portion of the ResNet50 [11] network to compute features on both the left and right input images, and a custom decoder network to compute a single SSM.

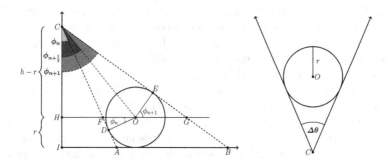

Fig. 1. Left: Depicted is the location of the observation point, $C = (0,0,0)$, and its height, h, above the horizontal observation plane at the bottom. I is the vertical projection of C. The relationship between ϕ_n and ϕ_{n+1} is also shown that determine concentric "ϕ-rings" around I in the observation plane via their intersection points A and B. Right: A top-down view of the observation plane shows how $\Delta\theta$ is measured and provides a radial subdivision into pizza slice style sections.

3   The Visual Mesh

The Visual Mesh [12,13] method calculates a graph that overlays the input image and serves as a preprocessing stage for CNNs. Initially designed as a single-object detector, the Visual Mesh aims to ensure a constant pixel-sampling density over an input image in such a way that no matter where the object of interest appears in the image, provided the object is below the horizon and closer than a predefined maximum distance, d, it will always intersect the Visual Mesh at a fixed predefined number of k points. Figure 1 shows the construction of the Visual Mesh in lateral view (left) and top view (right) where a sphere of radius r is used as the target object. A number of rays are sent out, from the camera's observation point, C, to the observation plane where the intersections A and B draw ϕ rings around I that are subdivided by radial θ sections. The final construction takes a number of parameters into account, including the distance of the ϕ rings from the origin, the number of intersections with the target, k, the maximum projected distance, d, and the size of the target, as explained in more detail in [12,13].

For example, for the task of detecting a soccer ball on a soccer field, a sphere with the same radius as the ball, r, can be used as the geometric model and the soccer field can be used as the observation plane. From there simple trigonometry, as depicted in Fig. 1, can be used to solve for the coordinates of camera rays, expressed as unit vectors, that will result in the specified number of intersections, k, with the sphere. These rays are then joined to their nearest neighbours in order to create the Visual Mesh graph. The original Visual Mesh [12] used a 6-neighbour ring graph structure and this same graph structure is adopted in this work to facilitate a more direct comparison with the original work.

While the Visual Mesh itself is lens-agnostic, in order to project the 3D unit vectors into the image plane to obtain 2D pixel coordinates, it requires knowledge of the camera's lens parameters; focal length, field of view, optical centre offset, projection type, distortion parameters, and image resolution.

The Visual Mesh has been shown [12] to allow for much smaller CNNs to be used, resulting in much fewer parameters and much higher inference speeds than other state-of-the-art object detectors, in the order of 450 fps on an Nvidia GTX1080Ti on 1280 × 1024 RGB images. Furthermore, due to the constant sampling density of the Visual Mesh graph, scale invariance is achieved for objects conforming to the geometric model, resulting in consistent detections over a large range of distances, something that other state-of-the-art networks, such as YOLO [28–30] and MobileNet [14], failed to achieve, see [12, Fig. 5(c)]. Finally, sampling pixels via the graph structure of the Visual Mesh results in convolution-like operations, with the underlying structure of the graph allowing for non-square, in our case hexagonal, filter shapes.

4   The Stereo Visual Mesh

The Visual Mesh should be able to improve its classification performance by incorporating data from a different viewpoint. In this way, objects that appear partially occluded in one viewpoint may appear less occluded in the other viewpoint allowing the Visual Mesh to still see and classify these objects.

Given two Visual Meshes, each one centred underneath a camera of a stereo system, the question of how to link these two meshes together and how much information to share between the two views of the scene arises.

Regarding linking the two Visual Meshes together, two options are immediately apparent. Every node in the Visual Mesh graph has a unique index. If we are using the same graph structure for both Visual Meshes we could simply link the same index in both Visual Meshes. Unfortunately, if the relative position or orientation of the two cameras with respect to the observation plane differ too much it is possible for a large portion of the indices that are on-screen in one view to be entirely off-screen in the other view, resulting in a lot of dead links. As a result, this linking scheme will preclude the possibility of a multi-view setup, and will also preclude the option of using different graph structures for each camera. However, if these constraints are not too restrictive, this linking scheme is very simple to implement, although sampling the same pixel coordinate in both images may not necessarily result in useful information being shared between the views.

Alternatively, linking the Visual Mesh nodes in the left and right mesh that appear visually closest to each other will resolve most of the issues that are present in the first option, but with extra computational overhead. Different Visual Mesh graph structures, different camera positioning, and more than two cameras are possible with this linking scheme, provided there is a large enough overlap in the visual fields of all cameras. The steps in the implementation of this linking scheme are as follows

1. Given a ray corresponding to a node in the left Visual Mesh, project it into the view of the right camera
   (a) Find the intersection between the Visual Mesh ray and the observation plane
   (b) Shift by the translation between the two cameras
   (c) Re-normalise the ray with respect to the right camera
2. Search the right Visual Mesh to find the node that is closest to the projected ray
3. Repeat steps 1 and 2 for every node in the left Visual Mesh

Steps 1–3 describe how to find a matching node in the right Visual Mesh for every node in the left Visual Mesh. To allow for different mesh geometries for each camera and arbitrary camera baselines, to find a matching node in the left Visual Mesh for every node in the right Visual Mesh steps 1–3 need to be repeated for the right camera. Sections 4.1 to 4.3 detail the algorithms involved in performing step 2.

Regarding how much information to share between the two views at least two options present themselves. Figure 2 provides a graphical depiction of the options. Given nodes $L_0 \cdots L_6$ from the left Visual Mesh, we can share only the node from the right Visual Mesh that is visually closest to L_0, R_0 in this case. Alternatively, we can share the visually closest node, R_0, along with that node's immediate neighbours, $R_1 \cdots R_6$. In this paper we only consider sharing R_0, as it is anticipated that the extra overhead in including $R_1 \cdots R_6$ will outweigh any performance benefits that may be gained.

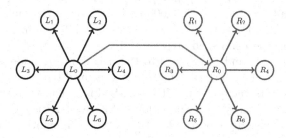

Fig. 2. Given a 6-neighbour graph structure on both the left and right either the nearest link, R_0 (in red), can be shared between views, or the nearest link and its neighbours, $R_0 \cdots R_6$ (the red link plus the blue links) can be shared. (Color figure online)

Figure 3 depicts the relationship between the ground truth labels in the dataset, Sect. 6. Given a pair of left and right images from the dataset and their ground truth labels a Visual Mesh was projected on to both images and the stereo links were computed. For every L_0 in the dataset, Fig. 2, the relationship between the labels for L_0 and R_0 was determined and is shown in Fig. 3. The figure shows, e.g., that when L_0 had the label of "Ball", 85.1% of the time R_0 also had the "Ball" label and 8.8% of the time R_0 had the "Field" label.

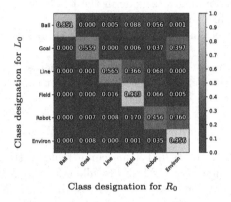

Fig. 3. Dataset-specific quantitative relationships between stereo link ground truth labels. How frequently L_0 (y-axis) has the same ground truth label as R_0 (x-axis). See Fig. 2 for definitions of L_0 and R_0.

4.1   Linking Two Visual Meshes

Given a Visual Mesh for each camera, finding the visually closest node in the right Visual Mesh for a given node in the left Visual Mesh can take a number of forms. The naive option would involve a brute force search of every node in the right Visual Mesh. However, depending on the exact parameters of the Visual Meshes, each could contain over 100,000 nodes. This search would be very slow and ignores the connected structure of the Visual Mesh graph.

By leveraging the connected structure of the Visual Mesh we can select a random starting node and then move to the neighbouring node which has the shortest distance to the query ray and repeat this until the search arrives at the node that is closest to the query ray. This method is listed as "Graph" in Table 1.

By employing a binary space partition (BSP) to partition the Visual Mesh the search for the visually closest Visual Mesh node can start from a randomly selected Visual Mesh node from the BSP leaf node that encloses the query ray. From this starting node the search algorithm proceeds as was described for the "Graph" method. We call the just described third method "BSP+Graph".

Table 1 provides an overview of the search times for the three different search methods discussed. The table shows that the overhead associated with setting up and maintaining the BSP provides an almost 4000-fold decrease in search time when finding the visually closest node.

Table 1. Search algorithm run times. Tests were performed with a pair of Visual Meshes, each containing 98,900 nodes and using the 6-neighbour graph structure. Total time is the time taken to find the visually closest node in the right Visual Mesh for each node in the left Visual Mesh. The average search time is how long, on average, it would take to find the visually closest node for a single node in the left Visual Mesh.

Method	Avg. Time (μs)	Total Time (ms)
BSP+Graph	6.24	617
Graph	207.89	20,560
Brute Force	23,692.66	2,343,204

4.2 Binary Space Partition

The BSP is created during construction of the Visual Mesh. Each node in the BSP tree stores the ϕ and θ ranges that enclose the Visual Mesh rays spanned by the current tree node. Figure 1 depicts these quantities. Each BSP node also stores a minimum bounding cone that contains all the Visual Mesh rays spanned by the current tree node.

The root node of the BSP tree has a cone axis of $(0, 0, -1)$ and the internal angle is set to the largest z-component of all Visual Mesh rays, since Visual Mesh rays are also unit vectors this corresponds to the angle between the z-axis and the ray. The Visual Mesh rays are then partitioned into two sets based on the sign of the y-component of the rays. These two sets form the two children of the BSP root node. This partitioning scheme simplifies the calculation and comparison of ϕ and θ later on.

For each child node in the BSP tree a cone, with vertex at the observation point, is created with an internal angle just large enough to ensure all Visual Mesh rays spanned by the current tree node are encompassed by the cone. The range of ϕ and θ values for the current set of Visual Mesh rays is determined, and the largest range is found. Finally, the current set of rays are partitioned using the average value of the largest range as the partition point. The average is used as it is simpler to calculate than the median as the average does not require the data to be sorted.

By representing each ray in the Visual Mesh by a unit vector, we can calculate ϕ as the z-component of the ray. To avoid using trigonometric functions, θ is calculated as the normalised x-component of the ray, $\theta = \vec{v}x/\sqrt{1-\vec{v}z^2}$. This introduces the possibility of θ becoming infinite when $\vec{v} = (0, 0, -1)$. Since this calculation is only used for partitioning the Visual Mesh rays this has the effect of always forcing the origin point into the second half of the partition.

Traversing the BSP tree is implemented as a recursive algorithm, terminating when a leaf node of the BSP tree is reached. As with the creation of the BSP tree the first decision point is treated specially, where the sign of the y-component of the target ray is used to determine which child node to traverse. For every subsequent decision point, we calculate the ϕ and θ values of the target ray. If both ϕ and θ are contained within the ϕ and θ ranges of the first child node then we traverse down that branch, otherwise, down the other branch.

4.3   Visual Mesh Graph Traversal

To traverse the Visual Mesh graph, we measure the distance between each neighbour of the current Visual Mesh node and our target ray. We then move to the neighbour that has the shortest distance to the target, and then repeat. To prevent the possibility of infinite searches, a distance threshold is introduced. If the distance to all the neighbours of the current node does not decrease the distance to the target node by more than this threshold then the search terminates. Empirically, we found that k^{-1}, where k is the number of intersections used at Visual Mesh construction, works well.

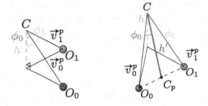

Fig. 4. The relationship between h and h', ϕ_0 and ϕ_0', and ϕ_1 and ϕ_1'. C represents the observation position, O_0 and O_1 represent the two objects that we wish to measure the distance between, and $\vec{v0}^p$ and $\vec{v1}^p$ represent the vectors pointing from the observation point to the centres of the objects.

While a Euclidean distance metric could be used for calculating the distance between Visual Mesh nodes, a graph-based distance metric is more appropriate. Given the observation height, h, angle from the z-axis, ϕ, and the height of the object's centre above the observation plane, r, it is possible to calculate how many objects, $n(h, \phi, r)$, can fit between the object that would be pointed to by our target ray and the origin using Eq. (1).

$$n(h, \phi, r) \overset{\text{def}}{=} \frac{\operatorname{arcsinh}(\tan(-\phi))}{\ln\left(1 - \frac{2r}{h}\right)} \tag{1}$$

However, in order to use Eq. (1) to calculate the object distance between two different objects a consistent observation height, h', for both objects is required. If $\vec{v0}$ and $\vec{v1}$ are unit vectors pointing to centre of each object, we obtain their projection to a plane that is mutually orthogonal to both objects,

$\vec{v}0^p$ and $\vec{v}1^p$. The consistent observation height, h', can then be calculated using $h' = \|\vec{v}0^p \times \vec{v}1^p\|/\|\vec{v}0^p - \vec{v}1^p\|$. From the projected vectors, $\vec{v}0^p$ and $\vec{v}1^p$, and consistent observation height, h', we can calculate the new angles

$$\phi'_0 = \arccos\left(\frac{h'}{\|\vec{v}0^p\|}\right) \text{ and } \phi'_1 = \arccos\left(\frac{h'}{\|\vec{v}1^p\|}\right). \tag{2}$$

Figure 4 depicts the relationship between these computed quantities.

Finally, Eq. (3) is used to obtain a count, d_o, of the number of objects that would fit between $\vec{v}0$ and $\vec{v}1$ when observed from a height of h'.

$$d_o = |n(h', \phi'_0, r) - n(h', \phi'_1, r)|. \tag{3}$$

5   CNN Architecture

To reduce the variability in training, a standard CNN architecture was chosen for all Visual Mesh networks. The chosen architecture is large enough to produce satisfactory classification results on the chosen dataset without being overly large to maintain the fast inference speed of the Visual Mesh [12].

Table 2 provides the layer details. Each layer is preceded by a Visual Mesh gathering step which samples the pixels corresponding to the Visual Mesh graph structure. For the first layer, the gathering step samples the raw pixel values from the input image, while all succeeding layers sample the outputs of the previous layer. All layers, except for the last layer, use a SELU [16] activation function, while the last layer uses a softmax [10, Chapter 6.2.2.3] activation function. All layer weights are initialised using a truncated normal distribution, $\mathcal{N}\left(0, \sqrt{\frac{1}{n}}\right)$, where n is the number of inputs to the layer and any weights that are further than two standard deviations from the mean are discarded and redrawn [16,18].

Table 2. Network architecture for all trained networks. Dotted rows indicate a row that is identical to the previous non-dotted row.

Layer	Width	Parameters	
		Mono	Stereo
1 (input)	16	352	400
2	16	1808	2064
...	...	...	...
7	16	1808	2064
8	8	904	1032
9	8	456	520
...	...	...	...
12	8	456	520
13 (output)	6	342	390
Total Parameters		14,270	16,286

The parameter count for each layer is computed as $P = N_i N_g N_o + N_o$, where N_i and N_o are the number of inputs and outputs to/from the layer, respectively, and N_g is the number of nodes in the graph structure. For the mono network $N_g = 7$, comprising $L_0, \cdots, L_6$ shown in Fig. 2, and for the stereo network $N_g = 8$, comprising $L_0, \cdots, L_6$ and the stereo link, R_0, discussed in Sect. 4. The Linked Mono network, discussed in Sect. 7, has the same number of parameters as the Stereo network.

To further reduce the variability in training, a standard set of training hyperparameters was chosen for all Visual Mesh networks. Table 3 shows the settings used for all experiments in this paper. During training the height and orientation of the camera with respect to the observation plane are augmented. A small perturbation is added to the height. For the orientation, a rotation matrix is formed by rotating a small amount around a randomly constructed unit vector. This rotation matrix is then applied to the orientation of the camera.

Table 3. Visual Mesh parameters for all trained networks. Maximum distance is the furthest distance that the Visual Mesh is projected from the camera in all directions.

Hyperparameter	Setting
Graph Structure	6-neighbour graph
Maximum Distance	20 m
Intersections	6
Geometric Model	Sphere - Radius 0.095 m
Data Augmentations	Height $\sim \mathcal{N}(0, 0.05)$
	Rotation $\sim \mathcal{N}(0, 0.08727)$

6   Dataset

To the knowledge of the authors there are no publicly available stereo datasets that provide semantic segmentation ground truth labels for both the left and right views while also providing the camera lens parameters and extrinsics that the Visual Mesh requires, facilitating the need to create a custom dataset. The created dataset consists of stereo images taken from a humanoid robot model with a stereo camera setup in various poses in a robotic soccer simulation using the Webots simulation environment [21].

The segmentation masks identify 9 classes (Table 4). In actuality, the 9 classes are reduced to 6 classes by taking the multiple classes for goals and robots and combining them each to a single class for goals and a single class for robots.

All images have a resolution of 640×480 and are taken with a rectilinear camera lens with a $90°$ field of view and a 1.98 mm focal length. Input images are stored as 3-channel JPEGs while the ground truth segmentation masks are stored as 4-channel RGBA PNGs. The alpha channel in the mask images allows the

Table 4. The classes of objects in the dataset and the labels assigned to them.

Index	Class	Colour
0	Ball	red
1	Goal (large)	yellow
1	Goal (small)	olive
2	Field line	white
3	Field	green
4	Robot (self)	gray
4	Robot (red team)	magenta
4	Robot (blue team)	cyan
4	Robot (other)	charcoal
5	Environment	black

Visual Mesh to ignore those pixel labels, effectively making any pixel with a zero alpha channel unlabelled. However, this dataset does not have any unlabelled pixels. Figure 5 gives an example of the images in the dataset.

Fig. 5. Sample left input image with corresponding ground truth segmentation mask.

In total, the dataset contains 65,138 stereo image pairs and this is split 45%/10%/45% for training/validation/testing. This results in 29,132 stereo image pairs in each of the training and testing sets and 6514 stereo image pairs in the validation set.

7   Linked Mono Network

Table 2 shows that the stereo network has 2012 extra trainable parameters over the mono network. The source of these extra parameters is due to the stereo link effectively causing a widening of the layers in the network. Due to this, it is possible that any increase in performance over the mono network could be the result of these extra parameters and not because of the extra stereo data. To account for this, a mono network with an equivalent number of parameters was trained. This was achieved by feeding the centre node, L_0 in Fig. 2, a second time in lieu of the stereo link, R_0. In this way, the network had the extra parameters available for learning without any new data being presented to the network.

8   Training

Training was performed in TensorFlow [1] on a number of different GPUs. All training was performed over 500 epochs with 1000 batches per epoch. A validation epoch was performed at the end of every training epoch. Batch size and learning rate parameters are summarised in Table 5.

Table 5. Summary of batch size and learning rate parameters across the different GPUs used for training. Only training batch size is shown, validation batch size is twice as large as the training batch size, and the testing batch size is twice as large as the validation batch size. Batch accumulation was used when necessary, see column 3.

GPU	RAM	Batch Size	LR
Nvidia GTX1070	8 GB	15 ($\times 4$)	8e$-$4
Nvidia GTX1080Ti	11 GB	30 ($\times 2$)	5e$-$4
Nvidia V100	32 GB	60 ($\times 1$)	2e$-$04

A batch size of 60 was empirically determined to provide good training results. Batch accumulation was implemented to allow the batches of 60 to fit into the available RAM of the GPUs. Results from sub-batches were summed to give the result on the complete batch. Learning rates for each GPU are dependent on the batch size and the number of sub-batches needed. To accommodate this a learning rate finding algorithm was used to find a good learning rate. The learning rate finder [35] is a variation on the algorithm presented by [33].

The learning rate finder increases the learning rate per-batch from some predefined minimum, $\mathrm{lr_{min}}$, to a predefined maximum, $\mathrm{lr_{max}}$, over n batches, with the i-th batch having learning rate

$$\mathrm{lr}_i = \mathrm{lr_{min}} \sqrt[n]{\left(\frac{\mathrm{lr_{max}}}{\mathrm{lr_{min}}}\right)^i}. \tag{4}$$

Figure 6 shows a plot of loss against learning rate, and the learning rate, 2.6158×10^{-3}, that resulted in the lowest loss. It is suggested [33,35] that a good learning rate is an order of magnitude lower than the learning rate that resulted in the lowest loss as this is where the loss is still decreasing. However, in practice, the loss may not be decreasing at the chosen learning rate, in this case a learning rate from the negative loss slope just before the lowest loss should be chosen.

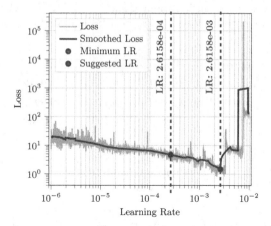

Fig. 6. Results of the learning rate finder algorithm on the Nvidia GTX1070.

Network weights were optimised using the Ranger algorithm [37]. The lookahead mechanism in Ranger was set to have a synchronisation period of 6 and a step size for the slow weights of 0.5. Focal loss [19] was used for the loss function.

When training the mono and linked mono networks only the left image from each stereo pair was used.

9   Results

Table 6 gives an overview of the training results. The Matthews correlation coefficient (MCC) [20] is used as the metric for comparison as it takes into account the sizes of the classes as well as all four confusion matrix categories. Furthermore, unlike most confusion matrix metrics, the MCC has been generalised to the multi-class case, allowing a single metric to be calculated that will give an overall view of how well the trained network has performed across all classes. Valid values for the MCC lie in the range $[-1, 1]$ with a value of 1 corresponding to perfect predictions, a value of -1 corresponding to total disagreement, and a value of 0 indicates that the classifier is indistinguishable from random guessing.

As can be seen in Table 6, the Stereo Visual Mesh has a higher MCC than the mono Visual Mesh indicating that the Stereo Visual Mesh has learnt to classify elements in the dataset better than the mono Visual Mesh. Furthermore, we see that the Stereo Visual Mesh also has a higher MCC than the Linked Mono Visual Mesh indicating that the increase in classifier performance is the result of more than the increased number of parameters available for learning.

To ascertain the robustness of the Stereo Visual Mesh MCC result the order of the test set was randomised and divided into 10 chunks of 2913 elements and tested the Stereo Visual Mesh on each of these chunks. The resulting MCC from each chunk was recorded, and the mean was found to be

Table 6. Summary of Visual Mesh training results. Training and validation loss are the minimum losses achieved during training and validation, respectively. MCC is the multi-class Matthews correlation coefficient obtained from the multi-class confusion matrix for each network.

Network	Train Loss	Valid Loss	MCC
Mono	0.1001	0.1653	0.9919
Linked Mono	0.0667	0.1267	0.9925
Stereo	0.0877	0.1045	0.9930

0.992964 (SD = 0.0001276) showing that random selection of test data had minimal impact on the classification outcome.

Table 7. Difference between true positive, true negative, false positive, and false negative rates for the mono and Stereo Visual Meshes. Calculated as $X_{stereo} - X_{mono}$, where X is one of TPR, TNR, FPR, or FNR. All values are relative to 1×10^{-5}.

Class	TPR	TNR	FPR	FNR
Ball	−5.15	3.61	−3.60	5.15
Goal	187.04	20.43	−20.50	−187.04
Line	0.32	12.97	−12.98	−0.32
Field	26.62	−40.31	40.31	−26.62
Robot	114.89	34.51	−34.50	−114.89
Environ	44.12	42.31	−42.30	−44.12

With respect to the mono Visual Mesh, the Stereo Visual Mesh shows an improvement to true positive, TN, FP, and FN rates across all classes apart from the ball and field classes, Table 7. The ball class experiences a slight decrease in true positive rate and a corresponding increase in false negative rate. Similarly, for the TN and FP rates for the field class. This is due to a Visual Mesh node that should be classified as a ball being linked, via the stereo link, to a Visual Mesh node that should be classified as field. This can be seen in Fig. 3 where there is an 8.8% chance of this type of link occurring. Although Fig. 3 shows high chances of confusing links occurring between the robot class and all other classes, and with the goal and environment classes, the network has learnt to adequately deal with the potential source of confusion, likely because the size of the objects in these classes is helping the network to stay on track. Overall, however, the Stereo Visual Mesh outperforms the mono Visual Mesh.

One of the most remarkable properties of the Visual Mesh is its extremely short inference time. In a benchmarking experiment we recorded inference times of approximately 610 fps for the mono Visual Mesh on an Nvidia GTX1080Ti and approximately 275 fps for the Stereo Visual Mesh. This results in an approximately 2.2-fold increase in inference time over the mono Visual Mesh. This increase in inference time can be explained by the more than 2-fold increase in the amount of data that needs to be processed on the GPU.

Finally, a ResNet-18 [11] variant of StreoScenNet [23] was trained on the dataset in Sect. 6. StreoScenNet achieved a MCC of 0.9990 and an inference speed of 32.49 fps. This increase in classification performance, and decrease in inference speed, compared to the Stereo Visual Mesh is due to the 3500-fold increase in the parameter count of StreoScenNet, with StreoScenNet having 56,205,462 parameters compared to the Stereo Visual Mesh's 16,286.

10   Limitations

The Visual Mesh requires knowledge of the position and orientation of the camera with respect to an observation plane. Training is typically performed with data augmentation performed on the camera's position and orientation to allow for some inaccuracies in the measurement of these quantities, however, this is not foolproof. The Visual Mesh also requires camera calibration to be performed. This also means that any dataset that is used for training must have camera intrinsics and extrinsics available. The position of the camera relative to some origin is not needed, merely the height above, and orientation with respect to, an observation plane is needed. The observation plane might be the road surface in a driving scenario, the ground in a soccer scenario, or the surface of the ocean in a marine navigation task.

The algorithm for finding the visually closest Visual Mesh node is dependent upon the translation between the two cameras. If a custom stereo camera setup is being employed and there is the potential for the two cameras to move relative to each other, this will introduce a source of error in the classifications.

11   Conclusion

In this paper we have presented a new method for sparse stereo semantic segmentation that simultaneously generates a sparse SSM for both the left and right views. Although the proposed method is approximately two times slower than the corresponding mono method, 275 fps is still an exceptionally fast inference speed for processing high resolution stereo images. Furthermore, the proposed method provides an increase in classification accuracy over its mono counterpart.

Further work based on the Stereo Visual Mesh could investigate two directions of study. First, the feasibility of the Visual Mesh input transformation for stereo disparity estimation, both independently of and in conjunction with semantic segmentation could be investigated, to achieve sparse disparity maps at similar inference speeds. Second, the Stereo Visual Mesh could be adapted to a MultiView Visual Mesh where there are potentially more than two cameras and where the cameras are not necessarily on the same baseline. The WoodScape dataset [27, 39] could be a good dataset for this branch of work.

References

1. Abadi, M., et al.: TensorFlow: Large-Scale Machine Learning on Heterogeneous Distributed Systems, November 2021. https://www.tensorflow.org/
2. Bosch, M., Foster, K., Christie, G., Wang, S., Hager, G.D., Brown, M.: Semantic stereo for incidental satellite images. In: 2019 IEEE Winter Conference on Applications of Computer Vision (WACV), pp. 1524–1532, January 2019. https://doi.org/10.1109/WACV.2019.00167
3. Chang, J.R., Chen, Y.S.: Pyramid stereo matching network. In: Proceedings of the IEEE Conference on Computer Vision and Pattern Recognition, pp. 5410–5418 (2018). http://openaccess.thecvf.com/content_cvpr_2018/html/Chang_Pyramid_Stereo_Matching_CVPR_2018_paper.html
4. Chen, H., et al.: Multi-level fusion of the multi-receptive fields contextual networks and disparity network for pairwise semantic stereo. In: IGARSS 2019–2019 IEEE International Geoscience and Remote Sensing Symposium, pp. 4967–4970, July 2019. https://doi.org/10.1109/IGARSS.2019.8899306
5. Chen, X., Liu, Y., Achuthan, K.: WODIS: water obstacle detection network based on image segmentation for autonomous surface vehicles in maritime environments. IEEE Trans. Instrum. Meas. **70**, 1–13 (2021). https://doi.org/10.1109/TIM.2021.3092070
6. Cordts, M., et al.: The cityscapes dataset for semantic urban scene understanding. In: Proceedings of the IEEE Conference on Computer Vision and Pattern Recognition (CVPR), June 2016
7. van Dijk, S.G., Scheunemann, M.M.: Deep learning for semantic segmentation on minimal hardware. In: Holz, D., Genter, K., Saad, M., von Stryk, O. (eds.) RoboCup 2018. LNCS (LNAI), vol. 11374, pp. 349–361. Springer, Cham (2019). https://doi.org/10.1007/978-3-030-27544-0_29
8. Durner, M., Boerdijk, W., Sundermeyer, M., Friedl, W., Marton, Z.C., Triebel, R.: Unknown Object Segmentation from Stereo Images. arXiv:2103.06796 [cs], March 2021. http://arxiv.org/abs/2103.06796
9. Fan, R., Wang, H., Cai, P., Liu, M.: SNE-RoadSeg: incorporating surface normal information into semantic segmentation for accurate freespace detection. In: Vedaldi, A., Bischof, H., Brox, T., Frahm, J.-M. (eds.) ECCV 2020. LNCS, vol. 12375, pp. 340–356. Springer, Cham (2020). https://doi.org/10.1007/978-3-030-58577-8_21
10. Goodfellow, I., Bengio, Y., Courville, A.: Deep Learning. MIT Press, Cambridge (2016)
11. He, K., Zhang, X., Ren, S., Sun, J.: Deep residual learning for image recognition. In: Proceedings of the IEEE Conference on Computer Vision and Pattern Recognition, pp. 770–778 (2016). https://openaccess.thecvf.com/content_cvpr_2016/html/He_Deep_Residual_Learning_CVPR_2016_paper.html
12. Houliston, T., Chalup, S.K.: Visual mesh: real-time object detection using constant sample density. In: Holz, D., Genter, K., Saad, M., von Stryk, O. (eds.) RoboCup 2018. LNCS (LNAI), vol. 11374, pp. 45–56. Springer, Cham (2019). https://doi.org/10.1007/978-3-030-27544-0_4
13. Houliston, T.J.: Software architecture and computer vision for resource constrained robotics. Ph.D. thesis, University of Newcastle (2018). http://hdl.handle.net/1959.13/1389336
14. Howard, A.G., et al.: MobileNets: efficient convolutional neural networks for mobile vision applications. arXiv: 1704.04861 [cs], April 2017. http://arxiv.org/abs/1704.04861

15. Huang, G., Liu, Z., v. d. Maaten, L., Weinberger, K.Q.: Densely connected convolutional networks. In: 2017 IEEE Conference on Computer Vision and Pattern Recognition (CVPR), pp. 2261–2269, July 2017. https://doi.org/10.1109/CVPR.2017.243

16. Klambauer, G., Unterthiner, T., Mayr, A., Hochreiter, S.: Self-normalizing neural networks. In: Advances in Neural Information Processing Systems, vol. 30 (2017). https://papers.nips.cc/paper/2017/hash/5d44ee6f2c3f71b73125876103c8f6c4-Abstract.html

17. Königshof, H., Salscheider, N.O., Stiller, C.: Realtime 3D object detection for automated driving using stereo vision and semantic information. In: 2019 IEEE Intelligent Transportation Systems Conference (ITSC), pp. 1405–1410, October 2019. https://doi.org/10.1109/ITSC.2019.8917330

18. LeCun, Y.A., Bottou, L., Orr, G.B., Müller, K.-R.: Efficient BackProp. In: Montavon, G., Orr, G.B., Müller, K.-R. (eds.) Neural Networks: Tricks of the Trade. LNCS, vol. 7700, pp. 9–48. Springer, Heidelberg (2012). https://doi.org/10.1007/978-3-642-35289-8_3

19. Lin, T.Y., Goyal, P., Girshick, R., He, K., Dollár, P.: Focal loss for dense object detection. In: 2017 IEEE International Conference on Computer Vision (ICCV), pp. 2999–3007 (2017). https://doi.org/10.1109/ICCV.2017.324

20. Matthews, B.W.: Comparison of the predicted and observed secondary structure of T4 phage lysozyme. Biochimica et Biophysica Acta (BBA) - Protein Struct. **405**(2), 442–451 (1975). https://doi.org/10.1016/0005-2795(75)90109-9

21. Michel, O.: Cyberbotics Ltd., Webots^{TM}: Professional mobile robot simulation. Int. J. Adv. Robot. Syst. **1**(1), 5 (2004). https://doi.org/10.5772/5618

22. Miclea, V.C., Nedevschi, S.: Real-time semantic segmentation-based stereo reconstruction. IEEE Trans. Intell. Transp. Syst. **21**(4), 1514–1524 (2020). https://doi.org/10.1109/TITS.2019.2913883

23. Mohammed, A., Yildirim, S., Farup, I., Pedersen, M., Hovde, Ø.: StreoScenNet: surgical stereo robotic scene segmentation. In: Medical Imaging 2019: Image-Guided Procedures, Robotic Interventions, and Modeling, vol. 10951, pp. 174–182, March 2019. https://doi.org/10.1117/12.2512518

24. Peng, H., et al.: An adaptive coarse-fine semantic segmentation method for the attachment recognition on marine current turbines. Comput. Electr. Eng. **93**, 107182 (2021). https://doi.org/10.1016/j.compeleceng.2021.107182. https://www.sciencedirect.com/science/article/pii/S004579062100183X

25. Peng, J., Shen, J., Li, X.: High-order energies for stereo segmentation. IEEE Trans. Cybernet. **46**(7), 1616–1627 (2016). https://doi.org/10.1109/TCYB.2015.2453091

26. Qin, R., Huang, X., Liu, W., Xiao, C.: Pairwise stereo image disparity and semantics estimation with the combination of U-Net and pyramid stereo matching network. In: IGARSS 2019–2019 IEEE International Geoscience and Remote Sensing Symposium, pp. 4971–4974, July 2019. https://doi.org/10.1109/IGARSS.2019.8900262

27. Ramachandran, S., Sistu, G., McDonald, J.B., Yogamani, S.K.: Woodscape fisheye semantic segmentation for autonomous driving - CVPR 2021 OmniCV workshop challenge. CoRR abs/2107.08246 (2021). https://arxiv.org/abs/2107.08246

28. Redmon, J., Divvala, S., Girshick, R., Farhadi, A.: You only look once: unified, real-time object detection. arXiv: 1506.02640 [cs], May 2016. http://arxiv.org/abs/1506.02640

29. Redmon, J., Farhadi, A.: YOLO9000: better, faster, stronger. arXiv: 1612.08242 [cs], December 2016. http://arxiv.org/abs/1612.08242

30. Redmon, J., Farhadi, A.: YOLOv3: an incremental improvement. arXiv: 1804.02767 [cs], April 2018. http://arxiv.org/abs/1804.02767

31. Ronneberger, O., Fischer, P., Brox, T.: U-Net: convolutional networks for biomedical image segmentation. In: Navab, N., Hornegger, J., Wells, W.M., Frangi, A.F. (eds.) MICCAI 2015. LNCS, vol. 9351, pp. 234–241. Springer, Cham (2015). https://doi.org/10.1007/978-3-319-24574-4_28

32. Ros, G., Sellart, L., Materzynska, J., Vazquez, D., Lopez, A.M.: The synthia dataset: a large collection of synthetic images for semantic segmentation of urban scenes. In: The IEEE Conference on Computer Vision and Pattern Recognition (CVPR), June 2016

33. Smith, L.N.: Cyclical learning rates for training neural networks. In: 2017 IEEE Winter Conference on Applications of Computer Vision (WACV), pp. 464–472, March 2017. https://doi.org/10.1109/WACV.2017.58

34. Szemenyei, M., Estivill-Castro, V.: Real-time scene understanding using deep neural networks for RoboCup SPL. In: Holz, D., Genter, K., Saad, M., von Stryk, O. (eds.) RoboCup 2018. LNCS (LNAI), vol. 11374, pp. 96–108. Springer, Cham (2019). https://doi.org/10.1007/978-3-030-27544-0_8

35. Tanksale, N.: Finding Good Learning Rate and The One Cycle Policy, May 2019. https://towardsdatascience.com/finding-good-learning-rate-and-the-one-cycle-policy-7159fe1db5d6

36. Tasli, H.E., Alatan, A.A.: User assisted stereo image segmentation. In: 2012 3DTV-Conference: The True Vision - Capture, Transmission and Display of 3D Video (3DTV-CON), pp. 1–4, October 2012. https://doi.org/10.1109/3DTV.2012.6365447

37. Wright, L.: Ranger - a synergistic optimizer (2019). https://github.com/lessw2020/Ranger-Deep-Learning-Optimizer

38. Wu, Z., Wu, X., Zhang, X., Wang, S., Ju, L.: Semantic stereo matching with pyramid cost volumes. In: Proceedings of the IEEE/CVF International Conference on Computer Vision, pp. 7484–7493 (2019). https://openaccess.thecvf.com/content_ICCV_2019/html/Wu_Semantic_Stereo_Matching_With_Pyramid_Cost_Volumes_ICCV_2019_paper.html

39. Yogamani, S., et al.: WoodScape: a multi-task, multi-camera fisheye dataset for autonomous driving. In: Proceedings of the IEEE/CVF International Conference on Computer Vision (ICCV), October 2019

40. Zhou, L., Zhang, H.: 3SP-Net: semantic segmentation network with stereo image pairs for urban scene parsing. In: Geng, X., Kang, B.-H. (eds.) PRICAI 2018. LNCS (LNAI), vol. 11012, pp. 503–517. Springer, Cham (2018). https://doi.org/10.1007/978-3-319-97304-3_39

Micro-expression Recognition Based on PCB-PCANet+

Shiqi Wang, Fei Long$^{(\boxtimes)}$, and Junfeng Yao

Center for Digital Media Computing, School of Film, Xiamen University,
Xiamen 361005, China
flong@xmu.edu.cn

Abstract. Micro-expressions (MEs) have the characteristics of small motion amplitude and short duration. How to learn discriminative ME features is a key issue in ME recognition. Motivated by the success of PCB model in person retrieval, this paper proposes a ME recognition method called PCB-PCANet+. Considering that the important information of MEs is mainly concentrated in a few key facial areas like eyebrows and eyes, based on the output of shallow PCANet+, we use a multiple branch LSTM networks to separately learn the local spatio-temporal features for each facial ROI region. In addition, in the stage of multiple branch fusion, we design a feature weighting strategy according to the significances of different facial regions to further improve the performances of ME recognition. The experimental results on the SMIC and CASME II datasets validate the effectiveness of the proposed method.

Keywords: Micro-expression recognition · PCANet+ · PCB

1 Introduction

Micro-expressions (MEs) can reflect people's true emotions due to its spontaneity and irrepressibility. So, ME recognition has a wide range of applications in many fields such as psychotherapy, criminal interrogation and business negotiation. In recent years, the study of automatic ME spotting and recognition has attracted increasing attentions in the field of computer vision.

Early ME recognition methods used manually designed descriptors to encode the features of ME image sequences, and many of which were based on Local Binary Patterns (LBP). In 2011, Pfister et al. [1] utilized LBP-TOP [2] to extract dynamic features of MEs on the Spontaneous Micro-expression Database (SMIC) [3] and proposed a benchmark framework for automatic ME recognition. In order to solve the problem of duplicate encoding in LBP-TOP, Wang et al. [4] proposed LBP-SIP, which reduces computational complexity and redundant features by removing six intersections of duplicate encoding. Zong et al. [5] extended the granularity of the LBP operator through layered STLBP-IP features and used sparse learning to reduce the feature dimension.

Due to the promising performances of deep learning methods achieved in macro-expression recognition, researchers try to apply deep learning to ME recognition tasks in recent years. In the ME recognition framework proposed by

B. Luo et al. (Eds.): ICONIP 2023, LNCS 14452, pp. 179–190, 2024.
https://doi.org/10.1007/978-981-99-8076-5_13

Verma et al. [6], the dynamic image synthesis method [7] is first used to encode a ME image sequence into a single image instance, and then a 2D LEARNet model is used to learn the specific features of MEs. Thuseethan et al. [8] proposed a ME recognition framework called Deep3DCANN. In which, a 3D CNN was first used to learn spatio-temporal features from facial image sequences. At the same time, a deep network was used to track the visual associations between different sub-regions of the faces, and the learned facial features were combined with the semantic relationships between regions to recognize MEs. Xu et al. [9] proposed a feature disentanglement learning method based on asymmetric adversarial properties, which uses residual network to learn the emotional and domain features of MEs from two branches, and exhibits excellent performance in cross dataset ME recognition.

Deep learning methods rely heavily on a large amount of available training data, however, the sizes of existing ME datasets are almost all small, which limits the performances of deep learning methods in ME recognition. By applying transfer learning to ME recognition, the impact of insufficient training data on performances can be reduced to some extent. Mayya et al. [10] used a CNN model pre-trained on ImageNet [11] to extract features, and then used Support Vector Machine (SVM) in ME classification. Xia et al. [12] proposed a knowledge transfer framework from macro-expressions to MEs. In which, MiNet and MaNet were trained using ME datasets and macro-expression datasets, respectively. Then, MaNet was used to guide MiNet from the spatio-temporal domain for fine-tuning, enabling the model to better complete ME recognition tasks.

The Part based Convolutional Baseline (PCB) model [13] was initially proposed for the task of person retrieval. PCB provides more fine-grained information for pedestrian image description by learning part-informed features, and therefore can effectively boost the performances of person retrieval. Motivated by the idea of PCB model, in this paper we propose a ME recognition algorithm called PCB-PCANet+. Which uses PCANet+ as the backbone network for feature extraction, then divides the output feature tensor into multiple parts based on facial ROI region segmentation, and uses a separate LSTM network for each part to learn local spatiotemporal features, and finally fuses the features of all parts for ME recognition. The experimental results demonstrate the effectiveness of the algorithm.

The rest of this paper is organized as follows. In Sect. 2, we provide a detailed introduction to our PCB-PCANet+ model. Section 3 presents the experimental results and discussions, and the conclusion will be given in Sect. 4.

2   Proposed Method

In this section, we provide a detailed description of the proposed ME recognition method. PCB takes existing CNNs as the backbone networks, divides the feature tensor output into several parts after convolution layer and pooling layer, and uses each part to train a separate classifier, so as to achieve more effective feature learning. The framework of our proposed PCB-PCANet+ model for ME recognition is shown in Fig. 1.

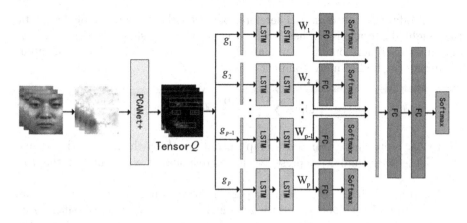

Fig. 1. Framework of proposed PCB-PCANet+ model for ME recognition

2.1 PCANet+ Feature Extraction

For a ME video clip, data preprocessing is required before feature extraction. First, 68 facial landmarks are detected using the Active Shape Model (ASM), and then the face area is aligned and cropped to remove the background in the image and also reduce the influence of head offset on ME recognition. Then we use TIM algorithm to normalize the frame number of ME video clip.

We perform optical flow calculation on the preprocessed video clip to obtain the horizontal and vertical optical flow sequence $\mathbf{U}, \mathbf{V} \in \mathbb{R}^{N \times M \times (L-1)}$, where N and M are the height and width of the image, and L is the number of frames. We use a sliding window (length T, step size s) to sample $\mathbf{U}$ and $\mathbf{V}$ simultaneously and stack these sampled horizontal and vertical optical flow components to obtain a multi-channel image set $\mathcal{I} = \{\mathbf{I}_1, \mathbf{I}_2, \ldots, \mathbf{I}_K\}$ with the number of channels of $2T$, and input them into PCANet+ for feature extraction.

In our algorithm, we use a two-layer PCANet+ network, where the number of filters in the first layer is F_1 and the filter size is $K_1 \times K_1$, the number of filters in the second layer is F_2, and the filter size is $K_2 \times K_2$. For a multi-channel image $\mathbf{I}_i$, after passing through the two-layer PCANet+, a tensor Q consisting of F_2 feature maps $\{\mathbf{O}_1, \mathbf{O}_2, \ldots, \mathbf{O}_{F_2}\}$ will be obtained.

In the original PCANet+, histogram features will be calculated with hash coding for each feature map in Q. In order to further model the temporal information of MEs on the basis of PCANet+ features, this paper directly uses the feature map output from the second layer of PCANet+ network as the input for the subsequent LSTM networks.

2.2 Multi Branch LSTMs Based on Facial ROIs

According to the learning process in PCANet+, the filters in the second layer corresponds to the eigenvectors computed in that layer, and generally the larger the corresponding eigenvalue of the filter, the more important information contained in the produced feature map. Therefore, based on the eigenvalues of the filters, the weighted average is applied to the feature maps, as shown in Formula 1, to obtain a two-dimensional feature map $\mathcal{O}_i$. By this processing we make the input to subsequent LSTM more compact.

$$\mathcal{O}_i = \sum_{j=1}^{F_2} \frac{\exp(\lambda_j)}{\sum_{k=1}^{F_2} \exp(\lambda_k)} \mathbf{O}_j \tag{1}$$

where $\mathbf{O}_j$ represents the feature map output by the j-th filter in the second PCANet+ layer, and λ_j represents the corresponding eigenvalue of the j-th filter.

The original PCB model used for person retrieval uniformly divided feature tensors into multiple blocks, and then trained the corresponding classifier based on each block. In this paper, for the task of ME recognition, we divide the feature map based on facial ROI regions. The illustration of facial ROI region segmentation is shown in Fig. 2.

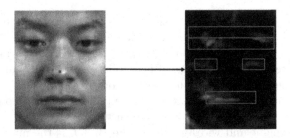

Fig. 2. Illustration of facial ROI region segmentation

Then, the features of each region are transformed into a feature vector g_i, which serves as input for subsequent part based classifiers. In order to better incorporate PCB model into ME recognition task, the local classifier we designed consists of a two-layer LSTM, a fully connected layer, and a Softmax function.

2.3   Multi-branch Fusion and Classification

To further enhance the ME recognition performance, we use a feature weighting strategy to take into account of significances of different ROI regions. The weight of each branch is calculated according to the accuracy of the classifier or the prediction score. The two methods of feature weighting are given in Formula 2 and Formula 3 respectively.

$$w_i = \frac{\exp(\text{accuracy}(i))}{\sum_{j=1}^{p} \exp(\text{accuracy}(j))} \tag{2}$$

$$w'_i = \sum_{j=1}^{n} \frac{\exp\left(\text{score}_j^i(y_j)\right)}{\sum_{k=1}^{c} \exp(\text{score}_j^i(k))} \tag{3}$$

where w_i and w'_i represent the weights calculated based on accuracy and prediction score for the i-th branch, p represents the number of branches, and accuracy(i) represents the accuracy of the validation set in the i-th branch. n represents the number of samples in the validation set, c represents the number of ME classes in the dataset, $\text{score}_j^i(k)$ represents the score of the k-th class predicted by the i-th classifier for the j-th sample, and y_j represents the real class corresponding to the j-th sample.

Finally, the features learned from each branch are weighted and concatenated, and then are input to a global classifier composed of two full connection layers and a Softmax function to complete ME classification. So the model can ensemble part-informed features to improve ME recognition performance.

3 Experimental Results and Analysis

To evaluate the proposed ME recognition method and also to analyze the impact of model settings and model parameters on performance, we conducted extensive experiments on SMIC and CASME II datasets.

3.1 Dataset

SMIC. The SMIC [3] dataset is the first spontaneous ME dataset and contains three subsets, namely SMIC-HS, SMIC-NIR, and SMIC-VIS. Among them, SMIC-NIR and SMIC-VIS subsets were captured by cameras with 25 fps. Due to the rapid and brief facial movements of MEs, cameras with low frame rates can not capture the dynamic information of MEs. Therefore, SMIC-HS will be used to evaluate our approach, which was captured by a high-speed camera with a frame rate of 100 fps and included 164 ME samples from 16 volunteers. The samples were divided into three ME categories: Positive, Negative and Surprise.

CASME II. The samples in CASME II [14] dataset were captured by a high-speed camera with frame rate of 200 fps, involving the MEs of 26 volunteers, including 255 video clips. The CASME II dataset contains seven ME categories: Happiness, Surprise, Sadness, Fear, Disgust, Repression, and Others. However, since there are only nine samples of sadness and fear in the data set, which is not conducive to network training, only 246 samples of the other five categories were used in our experiments.

3.2 Performance Metric

In ME recognition experiments, we use Accuracy, Macro-F1, and Macro-recall as performance metrics.

$$Accuracy = \frac{\sum_{i=1}^{C} TP_i}{\sum_{i=1}^{C} TP_i + \sum_{i=1}^{C} FP_i} \tag{4}$$

$$P_i = \frac{TP_i}{TP_i + FP_i} \tag{5}$$

$$R_i = \frac{TP_i}{TP_i + FN_i} \tag{6}$$

$$Macro - F1 = \frac{1}{C} \sum_{i=1}^{C} \frac{2 \times P_i \times R_i}{P_i + R_i} \tag{7}$$

$$Macro - recall = \frac{1}{C} \sum_{i=1}^{C} \frac{TP_i}{TP_i + FP_i} \tag{8}$$

where C represents the number of classes of MEs. TP_i, FP_i and FN_i represents the number of true positive samples, false positive samples and false negative samples of class i respectively.

3.3   Different Methods of Feature Tensor Segmentation

In the proposed ME recognition method, the output of PCANet+ is divided into several parts by key facial regions. To verify the validity of this segmentation method, in this section we compare it with the other two methods of segmentation. The first method, as shown in Fig. 3, evenly divides the feature tensor into four parts along the horizontal direction, which is consistent with the method proposed in [13].

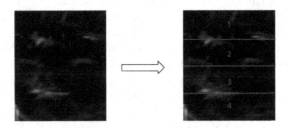

Fig. 3. Horizontal segmentation of tensors

The second method of segmentation is shown in Fig. 4. The feature tensor is divided into four parts along both horizontal and vertical directions. As shown in Fig. 5, our method divides the feature tensor into several parts based on ROI regions like eyebrows, eyes, nasal wings and mouth.

This section compares three segmentation methods on SMIC dataset and CASME II dataset, and the experimental results are shown in Table 1 and Table 2. From the experimental results, it can be seen that the recognition performances of the feature segmentation method based on facial ROI regions are superior to the other two simple segmentation methods. Which indicates that the motion information contained in key facial regions is more important to ME recognition.

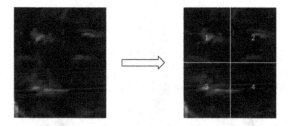

Fig. 4. Horizontal and vertical segmentation of tensor

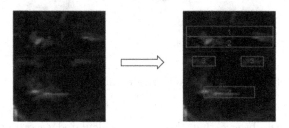

Fig. 5. Tensor segmentation based on facial key regions

Table 1. Comparison of different tensor segmentation methods on SMIC

Tensor segmentation methods	Accuracy	Macro-F1	Macro-recall
Horizontal	0.6524	0.6535	0.6557
Horizontal+Vertical	0.6463	0.6479	0.6505
Key facial regions	**0.6768**	**0.6797**	**0.6875**

Table 2. Comparison of different tensor segmentation methods on CASME II

Tensor segmentation methods	Accuracy	Macro-F1	Macro-recall
Horizontal	0.5528	0.5563	0.5420
Horizontal+Vertical	0.5447	0.5342	0.5103
Key facial regions	**0.5691**	**0.5754**	**0.5636**

3.4   The Influence of Feature Weight

In order to verify the effectiveness of feature weighting for multiple branches, this section compares recognition performances of models with and without this feature weighting process. The experimental results are shown in Fig. 6.

It can be observed from Fig. 6 that the recognition accuracies without feature weighting are inferior to that using feature weighting. Which indicates that feature weighting for multiple branches we used can boost performances by taking into account the contributions of different facial regions to ME recognition. At

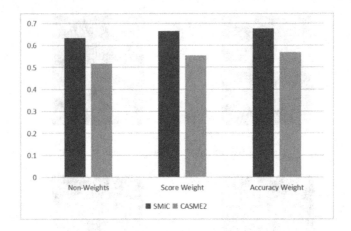

Fig. 6. Influence of feature weighting on model recognition accuracy

the same time, we observe that the weights calculated according to accuracy works slightly better than the weights calculated based on prediction score.

3.5   Model Parameter Optimization

In this section, we carry out experiments on SMIC dataset to analyze the impacts of model parameters on recognition performances, such as the size of feature tensor, and the number of partitions by tensor segmentation.

The Size of Feature Tensor. In the experiment, the spatial size of input video was normalized to 139×170. The spatial size of feature tensor can be changed by using the different step size in pooling in the second layer of PCANet+. Table 3 shows the relationships between the pooling stride and the spatial size of feature tensor. Figure 7 shows the model performances with different spatial sizes of feature tensor, and it can be seen that the best recognition results are achieved when the size is 46×56.

Table 3. Relationships between pooling stride and the spatial size of feature tensor

Pooling stride	Spatial size of feature tensor
1	139×170
2	69×85
3	46×56
4	34×42

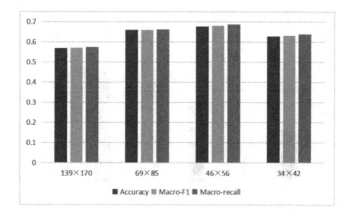

Fig. 7. Model performances with different tensor sizes

Number of Tensor Partitions. The number of tensor partitions can be changed by different ways of combinations of key facial regions, and their corresponding relationships are shown in Table 4. In the left column, the numbers 1, 2, 3, and 4 represent facial regions such as eyebrows, eyes, nasal wings and mouth, respectively. [1, 2] represents the combination of eyebrows and eye regions. [3, 4] represents the combination of nasal wings and eye mouth.

Table 4. Relationships between the different combinations of facial regions and the number of tensor partitions

Ways of facial region combinations	Number of tensor partitions
[1, 2][3, 4]	2
[1, 2][3][4]	3
[1][2][3][4]	4

Figure 8 shows the model performances with different number of tensor partitions. It can be seen that model performances increase with more fine granularity of facial region segmentations. Which indicates that the fine-grained features learned from local facial regions are benificial to ME recognition.

3.6 Comparison with Other Methods

In order to verify the effectiveness of our proposed method, we compare it with some existing methods including both of hand-crafted features and deep learning methods. In comparison, our method uses the optimal parameter configuration obtained from the above experiments. In addition, for better comparison, we re-implement LBP-TOP and STLBP-IP based on the same data pre-processing

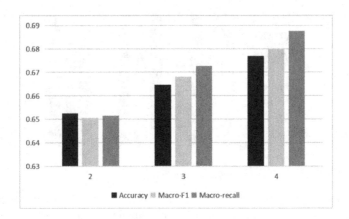

Fig. 8. Model performances with different number of tensor partitions

Table 5. Comparison of different methods on SMIC

Method	Accuracy	Macro-F1	Macro-recall
LBP-TOP [14]	0.4207	0.4266	0.4429
STLBP-IP [15]	0.4329	0.4270	0.4241
Selective [16]	0.5366	N/A	N/A
3D-FCNN [17]	0.5549	N/A	N/A
FR [18]	0.5790	N/A	N/A
OF-PCANet+ [19]	0.6280	0.6309	0.6369
Ours(PCB-PCANet+)	**0.6463**	**0.6467**	**0.6514**

Table 6. Comparison of different methods on CASME II

Method	Accuracy	Macro-F1	Macro-recall
LBP-TOP [14]	0.4390	0.4297	0.4259
STLBP-IP [15]	0.4173	0.4026	0.4282
Selective [16]	0.4575	N/A	N/A
ELRCN [20]	0.5244	0.5000	0.4396
3D-FCNN [17]	0.5911	N/A	N/A
OF-PCANet+ [19]	0.5325	0.5493	0.5241
Ours(PCB-PCANet+)	**0.5569**	**0.5574**	**0.5543**

and evaluation indicators, and their parameters are consistent with the optimal settings in the original paper [14,15]. For deep learning based algorithms, we directly use the results from the original papers. Table 5 and Table 6 respectively provide the comparison results of Accuracy, Macro-F1 and Macro-recall for different methods on the SMIC and CASME II, where N/A indicates that the corresponding performance metrics were not provided in the original paper.

From Table 5 and Table 6, it can be seen that deep learning based methods perform significantly better than traditional hand-crafted methods on the SMIC and CASME II. Meanwhile, our method produces competitive results compared to other deep learning based algorithms. The experimental results indicate that the proposed PCB-PCANet+ method can improve ME recognition performance effectively by learning part-informed features from facial ROI regions.

4   Conclusion

In this paper we propose a ME recognition method called PCB-PCANet+. By dividing the output of PCANet+ into several parts based on facial ROI regions, we use a multiple branch LSTM networks to learn part-informed spatio-temporal features from different parts. To further enhance the performance, we present a feature weighting strategy to fuse the features of different facial regions before ME classification. The experimental results on the SMIC and CASME II datasets indicate that PCB-PCANet+ can effectively improve ME recognition performances by ensembling part-level features of key facial regions.

References

1. Pfister, T., Li, X., Zhao, G., Pietikäinen, M.: Recognising spontaneous facial micro-expressions. In: Proceedings of International Conference on Computer Vision, pp. 1449–1456 (2011)
2. Zhao, G., Pietikainen, M.: Dynamic texture recognition using local binary patterns with an application to facial expressions. IEEE Trans. Pattern Anal. Mach. Intell. **29**(6), 915–928 (2007)
3. Li, X., Pfister, T., Huang, X., Zhao, G., Pietikäinen, M.: A spontaneous micro-expression database: Inducement, collection and baseline. In: Proceedings of 10th IEEE International Conference and Workshops on Automatic Face and Gesture Recognition, pp. 1–6 (2013)
4. Wang, Y., See, J., Phan, R.C.-W., Oh, Y.-H.: LBP with six intersection points: reducing redundant information in LBP-TOP for micro-expression recognition. In: Cremers, D., Reid, I., Saito, H., Yang, M.-H. (eds.) ACCV 2014. LNCS, vol. 9003, pp. 525–537. Springer, Cham (2015). https://doi.org/10.1007/978-3-319-16865-4_34
5. Zong, Y., Huang, X., Zheng, W., Cui, Z., Zhao, G.: Learning from hierarchical spatiotemporal descriptors for micro-expression recognition. IEEE Trans. Multimedia **20**(11), 3160–3172 (2018)
6. Verma, M., Vipparthi, S.K., Singh, G., Murala, S.: Learnet: dynamic imaging network for micro expression recognition. IEEE Trans. Image Process. **29**, 1618–1627 (2020)

7. Bilen, H., Fernando, B., Gavves, E., Vedaldi, A.: Action recognition with dynamic image networks. IEEE Trans. Pattern Anal. Mach. Intell. **40**(12), 2799–2813 (2018)
8. Thuseethan, S., Rajasegarar, S., Yearwood, J.: Deep3DCANN: a deep 3DCNN-ANN framework for spontaneous micro-expression recognition. Inf. Sci. **630**, 341–355 (2023)
9. Xu, S., Zhou, Z., Shang, J.: Asymmetric adversarial-based feature disentanglement learning for cross-database micro-expression recognition. In: Proceedings of the 30th ACM International Conference on Multimedia, pp. 5342–5350 (2022)
10. Mayya, V., Pai, R.M., Manohara Pai, M.: Combining temporal interpolation and DCNN for faster recognition of micro-expressions in video sequences. In: 2016 International Conference on Advances in Computing, Communications and Informatics (ICACCI), pp. 699–703 (2016)
11. Krizhevsky, A., Sutskever, I., Hinton, G.E.: Imagenet classification with deep convolutional neural networks. Commun. ACM **60**(6), 84–90 (2017)
12. Xia, B., Wang, S.: Micro-expression recognition enhanced by macro-expression from spatial-temporal domain. In: Proceedings of the 30th International Joint Conference on Artificial Intelligence (IJCAI), pp. 1186–1193 (2021)
13. Sun, Y., Zheng, L., Yang, Y., Tian, Q., Wang, S.: Beyond part models: Person retrieval with refined part pooling (and a strong convolutional baseline). In: Proceedings of the European conference on computer vision (ECCV), pp. 480–496 (2018)
14. Yan, W.J., et al.: CASME II: an improved spontaneous micro-expression database and the baseline evaluation. PLoS ONE **9**(1), 1–8 (2014)
15. Huang, X., Wang, S., Zhao, G., Piteikainen, M.: Facial micro-expression recognition using spatiotemporal local binary pattern with integral projection. In: Proceedings of IEEE International Conference on Computer Vision Works, Los Alamitos, CA, USA, pp. 1–9 (2015)
16. Patel, D., Hong, X., Zhao, G.: Selective deep features for micro-expression recognition. In: 2016 23rd International Conference on Pattern Recognition (ICPR), pp. 2258–2263 (2016)
17. Li, J., Wang, Y., See, J., Liu, W.: Micro-expression recognition based on 3d flow convolutional neural network. Pattern Analysis and Applications **22**(4), 1331–1339 (2019)
18. Zhou, L., Mao, Q., Huang, X., Zhang, F., Zhang, Z.: Feature refinement: an expression-specific feature learning and fusion method for micro-expression recognition. Pattern Recogn. **122**, 108275 (2022)
19. Wang, S., Guan, S., Lin, H., Huang, J., Long, F., Yao, J.: Micro-expression recognition based on optical flow and pcanet+. Sensors **22**(11), 4296 (2022)
20. Khor, H.Q., See, J., Phan, R.C.W., Lin, W.: Enriched long-term recurrent convolutional network for facial micro-expression recognition. In: 2018 13th IEEE International Conference on Automatic Face Gesture Recognition (FG 2018), pp. 667–674 (2018)

Exploring Adaptive Regression Loss and Feature Focusing in Industrial Scenarios

Mingle Zhou[1,2] , Zhanzhi Su[1,2] , Min Li[1,2] , Delong Han[1,2] ,
and Gang Li[1,2(✉)]

[1] Key Laboratory of Computing Power Network and Information Security, Ministry
of Education, Shandong Computer Science Center (National Supercomputer Center in
Jinan), Qilu University of Technology (Shandong Academy of Sciences), Jinan, China
Lig@qlu.edu.cn
[2] Shandong Provincial Key Laboratory of Computer Networks, Shandong
Fundamental Research Center for Computer Science, Jinan, China

Abstract. Industrial defect detection is designed to detect quality
defects in industrial products. However, the surface defects of different industrial products vary greatly-for example, the variety of texture
shapes and the complexity of background information. A lightweight
Focus Encoder-Decoder Network (FEDNet) is presented to solve these
problems. Specifically, the novelty of FEDNet is as follows: First, the
feature focusing module (FFM) is designed to focus the attention on
defect features in complex backgrounds. Secondly, a lightweight texture
extraction module (LTEM) is proposed to lightly extract the texture and
relative location information of shallow network defect features. Finally,
the AZIoU, an adaptive adjustment loss function, is reexamined in the
prediction box's specific circumference and length-width bits. Experiments on two industrial defect datasets show that FEDNet achieves the
accuracy of Steel at 42.86% and DeepPCB at 72.19% using only 15.3
GFLOPs.

Keywords: Industrial Quality Detection · Feature Focusing ·
Adaptive Adjustment Loss · Lightweight Texture Extraction

1 Introduction

In order to promote the urgent need for deep integration of information technology and industrialization, industrial surface defect detection is one of the essential technologies to ensure product quality and improve production efficiency.
Recently, defect detection methods based on computer vision have been widely
used in industrial fields. Among them, product quality testing in the industrial
field is critical.

Industrial defect detection methods are divided into traditional methods and deep
learning-based methods. Traditional methods can be divided into three categories.

B. Luo et al. (Eds.): ICONIP 2023, LNCS 14452, pp. 191–205, 2024.
https://doi.org/10.1007/978-981-99-8076-5_14

For example, a clustering algorithm based on image fusion [1], a method based on Hoff transformation [2] Fourier Image Reconstruction Method [3]. Most of these methods depend on prior knowledge and handcrafted defect descriptions. Recently, due to the rise of Convolutional Neural Network (CNN), surface detection algorithms based on CNN have developed rapidly. By the invariance of translation of convolution and strong correlation of local features, the accuracy and robustness of model detection are enhanced. New methods such as dynamic label assignment [4], attention mechanism, and edge feature thinning have emerged, focusing more on solving the coupling relationship between features, achieving feature thinning and focus extraction. However, images in the industrial world differ from those in the natural scene dataset. As shown in Fig. 1, shape textures are diverse, background information is complex, and small samples are fuzzy. The above methods are often tested for conventional defects, and they cannot be well promoted to industrial defect detection in complex scenarios. Especially for the small defect objects in the industrial scene, the texture shape is changeable, etc., and the generalization of the detection is poor. When deployed, industrial surface defect detection requires high-precision detection and ensures low computational complexity, aiming to achieve leak-free and real-time detection.

For the above challenges, we design a novel FENDet network. Specifically, the defective objects are considered to be tiny. We design the FFM module, which can focus on the tiny defective features in the complex background and learn the semantic information of the tiny defective features in a more targeted way. Considering that the shape texture is diverse and the background information is complex. We design LTEM module, which aims to extract texture and relative position information of shallow network defect features in a lightweight way, while using depth-separable convolution to build lightweight modules. In this paper, the perimeter factor of the prediction box is reviewed again, and an adaptive adjustment loss calculation paradigm is designed to accelerate the convergence of the model.

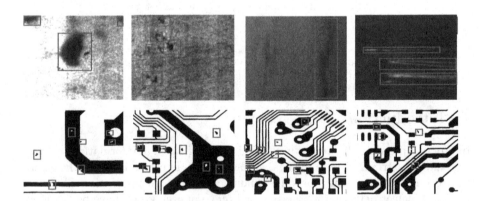

Fig. 1. The first row is the Steel dataset, where shape and texture are diverse, and background information is complex. The second row is the DeepPCB dataset, where you can see that the defect is minor and the regression task is difficult.

The core contributions of this methodology are as follows:

1. This paper presents a FFM module to focus attention on defect features in complex backgrounds, to learn more specifically the semantic information of defect features in complex backgrounds, and to improve the detection accuracy of the network.
2. This paper presents an LTEM module and introduces a deep detachable convolution to build a lightweight module. At the same time, extract and fuse features from multiple channels, aiming at lightly extracting texture and relative location information of shallow network defect features.
3. In this paper, an adaptive adjustment loss function, AZIoU, is designed to reexamine the validity of the prediction box under specific circumference and length-width bits and to accelerate the convergence of the model. The problem of complex regression for small objects is solved.

The other section in this paper are as follows: The second section reviews the previous work and contributions of object detection tasks. The third section describes the core modules and architecture of this model. The fourth section makes an experimental demonstration and analysis of the model. The fifth section comprehensively summarizes this article and its prospects for future work.

2   Related Work

2.1   Universal Object Detection

In recent years, with the rise of convolution neural network, object detection algorithm based on deep learning has also achieved rapid development. Currently, the main algorithms are divided into one-stage and two-stage object detection. One-stage object detection predicts the class and location of objects directly by the features extracted from the network. The mainstream algorithms include YOLO [4–8], YOLOX [9], and YOLOP [10]. The two-stage object detection algorithm first creates a suggestion box that is more likely to contain the object to be detected and then classify and locate it. The mainstream algorithms include SPP-Net [11] and Faster-RCNN [12]. Researchers have introduced attention mechanisms in response to problems, such as complex defect backgrounds and difficult detect detection. Attention is a mechanism for focusing on global or local information. The models of soft attention mainly include CA [13], ResNeXt [14], GAM [14], and DANet [15]; Models involving self-attention include the multi-headed self-attention MSA proposed in Transformer [16], the multi-headed attention SW-MSA based on sliding windows proposed in Swing Transformer [17]. There are also several excellent algorithms in the field: Xu et al. [18] proposed the PP YOLOE algorithm for anchor-free detection, introducing advanced technologies such as the dynamic label allocation algorithm TAL. Yu et al. [19] used a two-stage FCN based algorithm, Li et al. [20] adopted a lightweight backbone network MobileNet and a single stage continuous detector SSD, Song et al. [21] proposed a significance detection method based on Encoder-Decoder residual network structure.

2.2   Regression Loss Function for Object Detection

The regression loss function plays a critical role in object detection. The GIoU proposed by [22] and others solves the problem where the prediction box and label box do not intersect with an IoU value of 0; Zheng et al. [23] added the concepts of center point distances and minimum circumscribed rectangles between prediction boxes and label boxes on this basis, and subsequently discovered the aspect ratio factor of the two boxes, making positional regression more accurate. Zhora et al. [24] again considered the factor of vector angle in the regression process, once again improving the speed of model training and the accuracy of reasoning.

However, regression is difficult for data sets with minor industrial defects and drastic scale changes. When the center points of the prediction box and the label box coincide, the penalty terms in most regression losses will fail.

3   Method

3.1   Overview

This paper proposes an end-to-end focused Encoder-Decoder network. Figure 2 shows that the feature map is first processed by a data enhancement method and sent to Encoder for feature code extraction. Secondly, this article uses the Encoder's P1, P2, and P3 feature layers as input to the Decoder and performs serial decoding and fusion for multi-level feature information. Finally, the fully decoded and fused P4, P5, and P6 feature information in the Decoder network is transmitted to the detector for classification, regression, and confidence detection of defective objects.

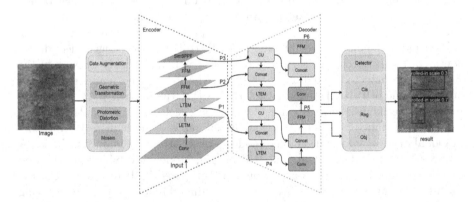

Fig. 2. The overall architecture of the FEDNet model proposed in this paper, the surface defect images are firstly enhanced by the data; then they are decoded and fused by the Encoder module for feature encoding extraction and the Decoder module for multi-layer feature information; finally the prediction results are output by the detection head.

3.2   Encoder

In this paper, we stack LETM and FFM convolution modules to continuously expand the receptive image field and extract rich multi-level feature information. Finally, the robustness of the feature is guaranteed through the SimSPPF [7] module.

In industrial defect images, most defects are relatively small, resulting in an imbalance in the number of defective and non-defective samples. For this reason, the FFM module focuses on the features of defective objects and highlights the semantic information of defects. As shown in Fig. 3, in the feature layer $F_F = X_c(i,j) \in R^{H \times W \times C}$, the FFM module captures channel information and spatial location awareness information through two branches. It divides the feature map evenly into two branches F_1, F_2 and on the channel as $C/2 \times H \times W$. The normalized weight factor of the sample features in the LayerNorm [25] regularization function is used in the channel branch to generate focus weights. First, F_1 undergoes the convolution of 5×5 to obtain a large receptive field. Then it adjusts the feature to $H \times W \times C/2$ and inputs it into LayerNorm by calculating the mean and standard deviation within the sample feature channel by channel. After linear mapping $g(\alpha)$, the normalized features are mapped to the new feature distribution F_e through learnable parameters α and β. Finally, use the focus weight regeneration function $g(\alpha)$ to calculate the score of each channel sample and multiply by F_e to obtain the channel focus weight G_c; The spatial branch first undergoes the convolution of 3×3 to obtain the characteristic information of the receptive field different from the channel branch. After encoding the spatial position information in the spatial H and W directions through the pooling layer, the dimensions are changed to $C \times H \times 1$ and $C \times W \times 1$, and the feature information F_s in these two directions is obtained. The two feature information is fused through the convolution layer of 1×1 and decomposed into two separate feature layers along the spatial dimensions H and W. Use the Sigmoid [26] activation function to obtain the branch ultimately focus weights $G_s^h(i)$ and $G_s^w(j)$ for rows and columns. Multiply the three weights to obtain the focus weight, then multiply them by the input feature to obtain F_F the focus feature map F'. The formula involved in FFM:

$$f(F_1) = \frac{F_1 - mean}{\sqrt{std + \epsilon}} \times \alpha + \beta \tag{1}$$

$$g(\alpha_i) = \ln^{\frac{\alpha_i}{\sum_i^n \alpha_i + \epsilon}} \tag{2}$$

$$G_c = \sigma\big(Fe \otimes \mathbf{g}(\alpha_i)\big) \otimes F_1 \tag{3}$$

where mean represents the average value of each sample in the channel; std represents the variance of each sample in the channel, and ϵ represents the coding channel in the horizontal coordinate direction; α is a normalized weight factor parameter that represents learnable; β is a learnable paranoid parameter, $\otimes$ is multiplied element by element, and σ represents the sigmoid activation function, i represents the encoding channel in the horizontal direction of the coordinate; j is the encoding channel representing the vertical direction of coordinates.

$$F_s = Add\big(MaxPooling(F_F), AvgPooling(F_F)\big) \tag{4}$$

$$G_s^h(i), G_s^w(j) = \sigma\big(Conv\big(Split\big(BN(Conv(F_s))\big)\big)\big) \tag{5}$$

$$F' = \big(G_c \otimes G_s^h(i) \otimes G_s^w(j)\big) \otimes F_F \tag{6}$$

where σ represents the sigmoid activation function; Conv is the convolution of 1×1. Such as Eq. (5), j represents the coding channel in the horizontal direction of coordinates; i is a coding channel representing the vertical direction of coordinates; F_F is the initial input feature, and $\otimes$ is multiplied element by element.

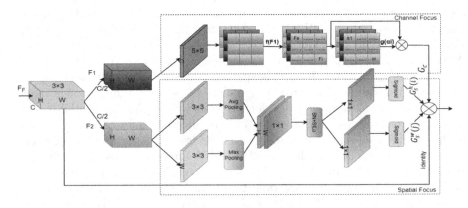

Fig. 3. FFM module, mainly through the fusion of channel focus branch and spatial focus branch, can assist the model to better focus on the shallow texture shape information and the deep important feature information.

LETM is designed to encode and extract the characteristics of external networks and decode and fuse the characteristics of deep networks. As shown in Fig. 4, the cascade convolution module LETM continuously adjusts the size of the receptive field and the number of lightweight convolutions N for different branches as the model width w and depth d increase. Define input $F_L \in R^{H \times W \times C}$, and each time a feature passes through a layer, the definition feature layer $\{f_d, d = 1, 2, 3\}$ will contain two branches. Contains a lightweight branch definition $\{f_d^o \times N^d, N = 1, 3\}$ and different receptive field branches $\{f_d^i = DaConv \in H \times W \times \frac{c}{r}d^i, r = 1, 2, 3, i = 1, 2, 3\}$. The feature layer undergoes different amounts of deep separable convolution processing, committed to filtering non important information in the original image to reduce noise in the image, and reducing the amount of parameters; At the same time, common convolutions of different receptive fields are used to obtain complex and variable edge texture features. Finally, the feature information from multiple branches is spliced and fused along the channel direction, with the purpose of assisting the feature extraction network to better focus on the spatial location information of the target. The validity of the LETM module is demonstrated in Table 3.

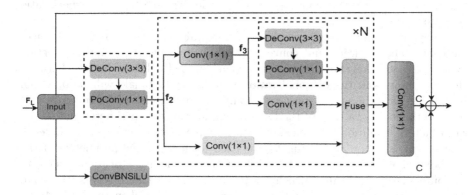

Fig. 4. The LETM module structure, adjusting the number of depth-separable convolutions and branching structure in different widths and depths, is dedicated to filtering the non-important information in the original image to reduce the noise in the image, while reducing the number of model parameters to ensure light-weight extraction of features.

3.3 Decoder

In Decoder Network, we perform serial decoding and fusion of multi-level feature information, sequentially using LTEM and FFM modules to recover features while using the idea of BiFPN [27] to efficiently fuse shallow texture edge information and deep semantic information. Different weights are assigned to different feature layers for fusion through ω_i learnable weight a, thereby improving the model detection head for regression and classification.

The $BiFPN_{Concat}$ fusion formula mentioned in this article is as follows.

$$Out = \sum_i \frac{\omega_i}{\varepsilon + \Sigma_j \omega_j} \cdot I_i \tag{7}$$

where ω_i is a learnable weight parameter, which is normalized to ensure stable; ε is a non zero random parameter, and the weights from different feature layers are adjusted to be the same or zero; I_i represents the feature information of different feature layers to be fused.

3.4 Loss Function

Although current regression loss functions for target detection consider a variety of factors, they mainly focus on the centroid distance and edge aspect ratio consistency between the prediction frame and the true frame. As in Fig. 5-(a) when the centroid of the prediction frame does not coincide with the centroid of the true frame, gradient optimization proceeds efficiently regardless of which loss function (GIoU [22] or CIoU [23]) is used. However, as in Fig. 5-(b), when the centroids of the prediction frame coincide exactly with the centroids of the real frame and have the same aspect ratio, the distance loss and aspect ratio consistency loss as well as IoU fail.

Many small and defect targets have large-scale changes in industrial defect data sets. The accuracy and convergence rate of model regression is affected. When the center points of the two boxes coincide, the Euclidean distance between the two points and the diagonal ratio of the outer rectangle will be 0. When both boxes have the same aspect ratio, only the IoU loss is valid. Therefore, this paper rethinks the validity of the two boxes under specific perimeter and aspect ratio factors conditions. Specifically, the new perimeter factor is considered a supplement, and the arctan function is used to measure the perimeters of the two boxes. For tiny target defective objects, the perimeter factor is considered to compute a larger IoU value, which can produce a minor loss to regress the small defective objects. As shown in Fig. 5-(c), we found that when the centroids of the two boxes overlap there will be a variety of cases of the prediction box does not accurately return to the labeled box. But at this time the distance between the perimeter of the two boxes is present, and at this time the distance and the diagonal distance with the outside rectangle there is a gap, you can continue to guide the model to continue to converge to this part of the distance than the direction of loss reduction, to ensure more accurate so that the prediction box wirelessly close to the real box. At the same time, in order to accelerate the magnitude of gradient update between different factors, a balance function is designed as the new weight calculation paradigm. Finally, AZIoU segmented adaptive loss function formula is as follows:

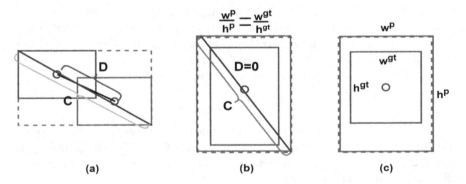

Fig. 5. Example diagrams of real and predicted boxes in different cases. Where red represents the real box, green represents the target box, the blue line segment represents the minimum outer rectangle diagonal distance, and the black line segment represents the distance between the center points of the two boxes. (Color figure online)

$$Loss_{re.g.} = 1 - \omega_1 IoU(B, B_{gt}) + AZIoU \tag{8}$$

$$\omega_i = \frac{e^a/e^b/e^c}{e^a + e^b + e^c + epsilon} * \frac{a/b/c}{a+b+c} \tag{9}$$

$$AZIoU = \begin{cases} \omega_2 * \frac{\rho^2(B,B_{gt})}{c^2} + \omega_3 * \alpha\frac{4}{\pi^2}\left(\tan^{-1}\frac{w^{gt}}{h^{gt}} - \tan^{-1}\frac{w^p}{h^p}\right)^2 \\ \omega_2 * \frac{\rho^2(B,B_{gt})}{c^2} + \omega_3 * \alpha\frac{4}{\pi^2}\left(2\tan^{-1}(h^{gt} + w^{gt}) - 2\tan^{-1}(h^p + w^p)\right)^2 \end{cases} \tag{10}$$

where B is the center point of the target frame; B_{gt} is the center point of the true box; ρ is the Euclidean distance; c represents the minimum outer rectangular diagonal length; α is the balance factor; h^{gt}, w^{gt}, w^p, h^p are the height and width of the real box and the prediction box, respectively; a, b and c are learnable parameters.

4   Experiments and Analysis

4.1   Parameter Settings and Experimental Environment

The IDE used in this experiment is Pycharm2021 Professional Edition. The PyTorch version is 1.9.1; The CUDA version is 11.6. Model training and reasoning are performed on NVIDIA A100-SMX with 40GB GPU memory and 16GB CPU memory.

During training, the size of the input image is adjusted to 640 × 640, epoch 300, initial learning rate 0.001, weight decay 0.0001, Momentum 0.9. At the same time, the Adam optimization algorithm is used, and the learning rate is adjusted by cosine annealing. In addition, IoU, mAP, GFLOPs, Recall, Precision, and F1 Score were selected for the evaluation index.

4.2   Experimental Datasets

Steel is a dataset published by Northeast University that contains six typical surface defects of hot rolled steel strips: pitted surface, crazing, rolled-in scale, patches, inclusion, and scratches. The shape and texture in the defect image change dramatically. In addition, the brightness and background complexity of the defect image is highly different. Before the experiment started, the data set was preprocessed: Firstly, photometric distortion was added to the data set, and the brightness of some pictures was adjusted; Next, the processed image was processed by panning and flipping horizontally. We selected 1400 more complex pictures of defect types as training sets. Finally, the images are divided into the training, test, and verification sets according to the 6:2:2 ratio.

The DeepPCB dataset is a semiconductor chip manufactured using linear scanning microelectronics technology with 1500 images. Image contains six defect categories: open, short, mouse bite, spur, pinhole, and spurious copper. According to the ratio of 7:1:2, the set is divided into the training, test, and validation sets.

4.3   Comparative Experiment

This paper selects the traditional model, new industrial detection model, and two-stage model with higher accuracy for comparison on two datasets. As shown in Table 1, traditional industrial defect detection models do not perform well on both datasets. We also compared the new model of YOLO series, FED-Net's mAP@.5-.95 And F1 scores are higher than models such as YOLOv7 and YOLOR. To verify the robustness of the model, we compared the new industrial detection model to see the FEDNet on the Steel dataset. mAP@.5 4% higher than PPYOLOE and AirDet, F1 up to 0.78; Both recall, and precision indices on DeepPCB datasets were higher than AirDet 3%. It is worth noting that when the experimental results of FEDNet on two datasets are similar to the two-stage model Faster-Rcnn, the GFLOPs are 10 times more petite. Finally, we show the detection results for the two datasets, as shown in Fig. 6.

4.4   Ablation Experiment

As shown in Table 2, this paper chooses YOLOv5s as the baseline to verify the ability of FFM modules to focus on defect features. The second line included the CA attention module in the C3 module of the baseline model, and F1 was 1% higher than the baseline. Add the NAM attention module, mAP@.5 1% increase and 1% increase in recall rate; Choose GAM attention regardless of cost. You can see mAP@.5 And F1 indicators increased by 2%. Finally, the FFM Focus module was added, the precision was increased by 3%, and the other indicators reached the best.

Figure 7 shows a visual thermogram of the FFM module, which can be visually observed to have a higher focus on defective objects.

LETM ablation experiment: As shown in Table 3, this article continues to conduct ablation experiments on the LETM module using the YOLOv5s model

Table 1. Comparison tests on two representative datasets of industrial surface defects.

Approach	GFLOPs	Steel					DeepPCB			
–	–	mAP@.5	mAP@.5-.95	precision	recall	F1	mAP@.5	mAP@.5-.95	recall	F1
Faster-Rcnn	201.3	76.3%	40.91%	54.14%	82.25%	0.64	97.95%	72.91%	97.86%	0.98
YOLOv3	154.9	68.92%	37.41%	72.25%	65.94%	0.69	95.86%	66.78 %	92.68%	0.92
YOLOv4	120.2	68.73%	39.62%	85.61%	62.54%	0.72	95.38%	65.18%	94.53%	0.93
YOLOv5s	15.8	74.92%	40.29%	70.97%	76.03%	0.73	97.21%	67.72%	95.89%	0.95
YOLOv7	103.2	74.52%	39.21%	80.31%	67.46%	0.72	97.53%	72.23%	97.11%	0.97
Efficientdet-d3	24.9	65.65%	37.11%	84.76%	48.44%	0.61	85.78%	63.91%	78.44%	0.84
Centernet	109.8	67.24%	36.51%	81.51%	51.33%	0.63	82.54%	61.57%	78.33%	0.76
Retinanet	191.6	64.24%	31.93%	76.73%	49.62%	0.64	86.49%	65.85%	72.22%	0.78
YOLOR-P6	57	76.52%	41.32%	65.56%	83.37%	0.74	98.14%	72.12%	97.42%	0.97
PPYOLOE-s	17.4	75.42%	40.89%	73.75%	76.74%	0.75	97.34%	70.23%	94.36%	0.95
AirDet-s	28	74.52%	40.38%	74.62%	73.58%	0.74	96.32%	69.81%	94.78%	0.94
FEDNet	15.3	78.68%	42.86%	76.23%	80.52%	0.78	98.42%	72.19%	97.55%	0.97

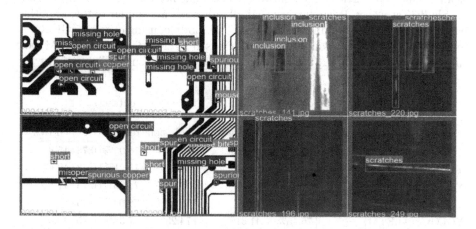

Fig. 6. Detection results of Steel and DeepPCB surface defects.

Table 2. Ablation experiment of FFM module on Steel dataset.

Approach and improvement	mAP@.5	mAP@.5-.95	precision	recall	F1
YOLOv5s(+C3)	74.92%	40.29%	70.97%	76.03%	0.73
YOLOv5s(+C3+CA)	75.38%	40.62%	71.43%	76.72%	0.74
YOLOv5s(+C3+NAM)	75.85%	40.86%	72.21%	77.42%	0.74
YOLOv5s(+C3+GAM)	76.18%	41.06%	72.83%	77.92%	0.75
YOLOv5s(+FFM)	76.54%	41.25%	73.12%	78.21%	0.75

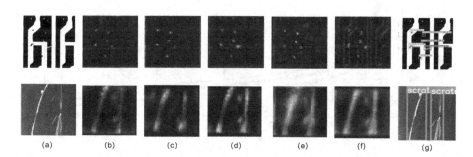

Fig. 7. (a) Original image; (b) Thermal diagram of C3 module; (c) Thermal diagram of CA attention; (d) A thermal map of NAM attention; (e) Thermal diagram of GAM; (f) Thermal diagram of FFM module; (g) Test result diagram.

as a baseline. This article uses lightweight backbone networks such as MobileNet
v3, ShuffleNet v2 ect. It can be seen that all indicators are similar to the C3 of the
baseline model. The fifth line shows the LETM module, using only 11.5 GFLOPs
to make mAP@.5 Increased by 2%, F1 increased by 2%. Therefore, LETM can
effectively extract complex edge texture features while ensuring lightweight.

Table 3. Ablation experiment of LETM on Steel dataset.

Approach and improvement	mAP@.5	mAP@.5-.95	precision	recall	F1	GFLOPs
YOLOv5s(+C3)	74.92%	40.29%	70.97%	76.03%	0.73	15.8
YOLOv5s(+MobileNet-v3)	74.12%	39.32%	71.21%	75.21%	0.73	10.4
YOLOv5s(+ShuffleNet-v2)	74.43%	38.72%	72.16%	74.36%	0.73	11.5
YOLOv5s(+CSPDarkNet53)	74.92%	40.29%	70.97%	76.03%	0.73	13.8
YOLOv5s(+LETM)	76.45%	41.31%	74.23%	77.34%	0.75	11.5

AZIoU ablation experiment: The FEDNet model was chosen as the baseline
for the experiment. The first five rows of data were the experimental results of
DIoU, CIoU, EIoU, and SIoU loss functions, respectively. The result of the SIoU
loss function is the best. mAP@.5-.95 and F1 were 41.29% and 0.77, respectively.
The sixth row of data is the AZIoU proposed in this article. You can see that
recall is 1% higher than SIoU. mAP@.5 Increase by 1%. It is worth noting that,
as shown in Fig. 8, AZIoU converges the fastest on steel datasets (Table 4).

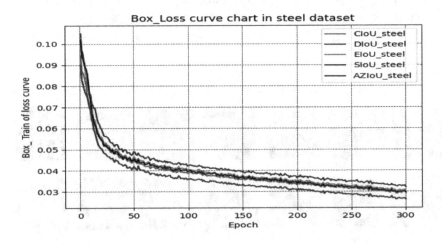

Fig. 8. Detection results of Steel and DeepPCB surface defects.

Table 4. Ablation experiment of Loss on Steel dataset.

Approach and improvement	mAP@.5	mAP@.5-.95	precision	recall	F1
FEDNet (DIoU)	76.35%	40.08%	71.19%	76.25%	0.74
FEDNet (CIoU)	77.42%	40.41%	74.25%	78.94%	0.76
FEDNet (EIoU)	76.73%	39.62%	72.61%	78.54%	0.75
FEDNet (SIoU)	77.92%	41.29%	75.17%	79.03%	0.77
FEDNet(AZIoU)	78.68%	42.86%	76.23%	80.52%	0.78

5 Conclusions

This paper presents a lightweight focused Encoder-Decoder network for defect detection in the industry. Specifically, the FFM structure can focus on defect features in a complex background to distinguish the different features of the defect. LTEM module dedicated to lightly extracting texture and relative location information of shallow network defect features. Finally, an adaptive adjustment loss function AZIoU for perimeter and aspect ratio factors is designed, accelerating the model's convergence speed. In the future, our network will be optimized to meet the robustness of different industrial defect detection.

Acknowledgements. This research was funded by Key R&D Program of Shandong Province, China (2022RZB02018), the Taishan Scholars Program (NO. tscy20221110).

References

1. Li, Y., Liu, M.: Aerial image classification using color coherence vectors and rotation and uniform invariant LBP descriptors. In: 2018 IEEE 3rd Advanced Information Technology, Electronic And Automation Control Conference (IAEAC), pp. 653–656 (2018)
2. Long, X., et al.: PP-YOLO: an effective and efficient implementation of object detector. ArXiv Preprint ArXiv:2007.12099 (2020)
3. Tsai, D., Huang, C.: Defect detection in electronic surfaces using template-based Fourier image reconstruction. IEEE Trans. Comp. Pack. Manuf. Technol. **9**, 163–172 (2018)
4. Ge, Z., Liu, S., Li, Z., Yoshie, O., Sun, J.: Ota: optimal transport assignment for object detection. In: Proceedings of the IEEE/CVF Conference on Computer Vision and Pattern Recognition, pp. 303–312 (2021)
5. Redmon, J., Farhadi, A.: Yolov3: an incremental improvement. ArXiv Preprint ArXiv:1804.02767 (2018)
6. Bochkovskiy, A., Wang, C., Liao, H.: Yolov4: optimal speed and accuracy of object detection. ArXiv Preprint ArXiv:2004.10934 (2020)
7. Li, C., et al.: YOLOv6: a single-stage object detection framework for industrial applications. ArXiv Preprint ArXiv:2209.02976 (2022)

8. Wang, C., Bochkovskiy, A., Liao, H. YOLOv7: trainable bag-of-freebies sets new state-of-the-art for real-time object detectors. In: Proceedings of the IEEE/CVF, pp. 7464–7475 (2023)

9. Ge, Z., Liu, S., Wang, F., Li, Z., Sun, J.: Yolox: Exceeding yolo series in 2021. ArXiv Preprint ArXiv:2107.08430 (2021)

10. Wu, D., et al.: YOLOP: you only look once for panoptic driving perception. Mach. Intell. Res. **19**, 550–562 (2022)

11. He, K., Zhang, X., Ren, S., Sun, J.: Spatial pyramid pooling in deep convolutional networks for visual recognition. IEEE Trans. Pattern Anal. Mach. Intell. **37**, 1904–1916 (2015). https://doi.org/10.1109/TPAMI.2015.2389824

12. Ren, S., He, K., Girshick, R., Sun, J.: Faster R-CNN: towards real-time object detection with region proposal networks. In: Advances In Neural Information Processing Systems, vol. 28 (2015)

13. Hou, Q., Zhou, D., Feng, J.: Coordinate attention for efficient mobile network design. In: Proceedings of the IEEE/CVF Conference on Computer Vision and Pattern Recognition, pp. 13713–13722 (2021)

14. Xie, S., Girshick, R., Dollár, P., Tu, Z., He, K.: Aggregated residual transformations for deep neural networks. In: Proceedings of the IEEE Conference on Computer Vision And Pattern Recognition, pp. 1492–1500 (2017)

15. Fu, J., et al.: Dual attention network for scene segmentation. In: Proceedings of the IEEE/CVF Conference on Computer Vision and Pattern Recognition, pp. 3146–3154 (2019)

16. Vaswani, A., et al.: Attention is all you need. In: Advances In Neural Information Processing Systems, vol. 30 (2017)

17. Liu, Z., et al.: Swin transformer: Hierarchical vision transformer using shifted windows. In: Proceedings of the IEEE/CVF International Conference on Computer Vision, pp. 10012–10022 (2021)

18. Xu, S., et al.: PP-YOLOE: an evolved version of YOLO. ArXiv Preprint ArXiv:2203.16250 (2022)

19. Long, Z., Wei, B., Feng, P., Yu, P., Liu, Y.: A fully convolutional networks (FCN) based image segmentation algorithm in binocular imaging system. In: 2017 International Conference on Optical Instruments and Technology: Optoelectronic Measurement Technology and Systems, vol. 10621, pp. 568–575 (2018). oelectronic Measurement Technology and Systems. SPIE, p 106211W

20. Li, K., Tian, Y., Qi, Z.: Carnet: a lightweight and efficient encoder-decoder architecture for high-quality road crack detection. ArXiv Preprint ArXiv:2109.05707 (2021)

21. Song, G., Song, K., Yan, Y.: EDRNet: encoder-decoder residual network for salient object detection of strip steel surface defects. IEEE. **69**, 9709–9719 (2020)

22. Rezatofighi, H., Tsoi, N., Gwak, J., Sadeghian, A., Reid, I., Savarese, S.: Generalized intersection over union: a metric and a loss for bounding box regression. In: Proceedings of the IEEE/CVF Conference On Computer Vision And Pattern Recognition, pp. 658–666 (2019)

23. Zheng, Z., Wang, P., Liu, W., Li, J., Ye, R., Ren, D.: Distance-IoU loss: faster and better learning for bounding box regression. In: Proceedings of the AAAI Conference on Artificial Intelligence, vol. 34, pp. 12993–13000 (2020)

24. Lin, T., Goyal, P., Girshick, R., He, K., Dollár, P.: Focal loss for dense object detection. In: Proceedings of the IEEE International Conference On Computer Vision, pp. 2980–2988 (2017)

25. Ba, J., Kiros, J., Hinton, G.: Layer normalization. ArXiv:1607.06450 (2016)

26. Elfwing, S., Uchibe, E., Doya, K.: Sigmoid-weighted linear units for neural network function approximation in reinforcement learning. Neural Netw. **107**, 3–11 (2018)
27. Tan, M., Pang, R., Le, Q.: Efficientdet: scalable and efficient object detection. In: Proceedings of the IEEE/CVF Conference on Computer Vision and Pattern Recognition, pp. 10781–10790 (2020)

Optimal Task Grouping Approach in Multitask Learning

Reza Khoshkangini[1,3]([✉]), Mohsen Tajgardan[2], Peyman Mashhadi[3], Thorsteinn Rögnvaldsson[3], and Daniel Tegnered[4]

[1] Internet of Things and People Research Center (IoTap) Department of Computer Science and Media Technology, Malmö University, Malmö, Sweden
reza.khoshkangini@mau.se
[2] Faculty of Electrical and Computer Engineering, Qom University of Technology, Qom, Iran
tajgardan.m@qut.ac.ir
[3] Center for Applied Intelligent Systems Research (CAISR), Halmstad University, Halmstad, Sweden
{peyman.mashhadi,reza.khoshkangini,thorsteinn.rognvaldsson}@hh.se
[4] Volvo Group Connected Solutions, Gothenburg, Sweden
daniel.tegnered@volvo.com

Abstract. Multi-task learning has become a powerful solution in which multiple tasks are trained together to leverage the knowledge learned from one task to improve the performance of the other tasks. However, the tasks are not always constructive on each other in the multi-task formulation and might play negatively during the training process leading to poor results. Thus, this study focuses on finding the optimal group of tasks that should be trained together for multi-task learning in an automotive context. We proposed a multi-task learning approach to model multiple vehicle long-term behaviors using low-resolution data and utilized gradient descent to efficiently discover the optimal group of tasks/vehicle behaviors that can increase the performance of the predictive models in a single training process. In this study, we also quantified the contribution of individual tasks in their groups and to the other groups' performance. The experimental evaluation of the data collected from thousands of heavy-duty trucks shows that the proposed approach is promising.

Keywords: Machine Learning · Vehicle Usage Behavior · Multitask learning

1 Introduction

Today's automotive manufacturers are becoming more motivated to investigate how vehicles behave during their operations. This is because understanding the long-term behavior of the vehicle support manufacturers' maintenance strategy to lower the costs and increase the performance of their fleets based on their

B. Luo et al. (Eds.): ICONIP 2023, LNCS 14452, pp. 206–225, 2024.
https://doi.org/10.1007/978-981-99-8076-5_15

capabilities, needs, and customers' demands. Although using high-resolution time-series data such as GPS position or other sensory data can be potentially employed to extract vehicle behavior patterns, such data is costly to transfer from connected assets and, most importantly, has privacy issues. In our previous study [10], we demonstrated and acknowledged that modeling vehicle behavior with low-resolution data can be made by building a complex ensemble multi-task deep neural network. The finding could indeed support overcoming the concern highlighted above. In this study (a continuation of our previous work), we investigate how different vehicle behavior chosen from high-resolution can be optimally trained together in a multi-task formulation using low-resolution data to enhance the predictive model's performance. We aim to find out–in one training process–multiple groups of tasks that outperform the multi-task model when all tasks are trained together.

Analysis of truck operation is critical to understand customers' real requirements to improve future truck design and developments. Vehicle long term-behaviors such as night stop behavior, whether the vehicle usually stops at a single home base, several distant bases, or is irregular, or driver class, whether only one driver or multiple drivers drove a vehicle, can be utilized to inform the design of future trucks and their infrastructure.

We could observe multiple studies in the automotive sector that have investigated short-term vehicle behavior and attempt to correlate the behaviors such as driver's lane changing patterns [16], fatigue [4], aggressiveness [21] to energy consumption, breakdowns [1,11], and CO2 emission [17,19,25] to decrease safety issues and enhance the performance of the vehicle. Many of these studies formulated the practice as a single-task problem and employed machine learning and deep neural networks to map such patterns to the particular performance factor. For example, [11] built a deep neural network model by feeding multiple sensor data such as engine speed, vehicle speed, and engine load to predict safe and unsafe driving behaviors. In [20], researchers focused on a single driving behavior–aggressive driving– and tried to measure and correlate Nox emission to different levels of aggressive driving. A similar investigation has been done by Cohi et al. in [5] to understand how aggressive driving impacts fuel consumption. Various environmental factors such as weather, time, and road conditions have been used to model vehicle behavior and correlate to vehicle performance [3].

Although the studies briefly reviewed vary widely in terms of what kinds of vehicle demeanor are evaluated and what approaches are built and utilized to handle the problem, AI and machine learning have proved to be successful in modeling vehicle behavior in the automotive sector [1,11]. However, studies are lacking in modeling multiple long-term vehicle behaviors. In this context, multiple patterns of operations can occur in a particular time window. Thus, we believe such modeling operations can be improved by incorporating AI and advanced multi-task deep neural network approaches.

Motivated by the above, in this study, we propose a deep neural network approach to map vehicle usage to multiple vehicle behaviors and focus on finding the optimal group of tasks that should be trained together to enhance the predictive model performance. Two main phases construct the proposed method; the first phase is *Data pre-processing*, concerning the extraction of hidden information from low-frequency data, where this information is later combined with vehicle behaviors obtained from high-resolution data (also serves as prediction variables)–here there are four different behaviors/tasks for each vehicle; In the second phase, we developed a multi-task deep neural network by constructing a specific head layer for each individual task that enables the network to train and add more behaviors with a low amount of data without training the whole network. In this phase, we utilized the gradient descent of each task to find the optimal group of tasks that should be trained together to boost the predictive model's performance to forecast different vehicle behaviors. We computed the gradient descent of each task giving the shared layer to assess how the models make similar mistakes to predict different tasks. In this way, we could quantify the transferred knowledge between each task leading to finding the optimal groups. In addition, we quantify the contribution of each task in group training to understand how individual tasks can behave when coupled with other tasks in group training.

Considering our previous study [10] in this context, this work is the first one to investigate the optimization of task grouping in modeling and predicting multiple vehicle long-term behaviors through one single training process. This allows us to quantify the positive and negative contributions of each task. The following research questions (RQs) further elaborate the investigative objectives of our proposed approach:

- **RQ1- Vehicle Task Grouping:** To what extent could the task of vehicle behaviors be grouped optimally?
- **RQ2- Vehicle Behavior Transference Quantification:** To what extent could the vehicle behavior transference be quantified?
  - **SbQ2.1-**Quantifying the contribution of each task in its group?
  - **SbQ2.2-**Quantifying the contribution of each task onto different groups?

Taken together, with these research questions, our work aims to counter the practice noted above by utilizing the vehicle's low-resolution operational usage to find the optimal group of tasks (long-term behaviors) to be trained together wherein; First, we concentrate on modeling and predicting multiple behaviors simultaneously via a multi-task learning fashion as the baseline. Then, to answer RQ1 we develop a predictive multi-task model where gradient descent is used from the shared layers to extract which tasks might have similar errors over the training process. A dot-product function was designed to acknowledge which of those gradients descend points to a similar direction over the training process. The reported figures revealed how this approach could find the optimal clusters of tasks leading to enhanced performance without training all combinations. To answer RQ2, we employed a pairwise comparison between the inner join value of multiple groups and explored each individual task contribution. The result of RQ2 allows us to understand which tasks have positive and negative contributions in its cluster and on other groups of tasks, and most importantly, we could find out the task that plays an important role in the multi-task predictive model.

The development of this multi-task and gradient approach in the automotive sector – particularly – vehicle long-term behavior modeling our research outlines and constitutes this paper's main contribution. The contributions in this study are underlined below:

- We developed a multi-task network and utilized a gradient descent algorithm to efficiently find the optimal tasks that should be trained together.
- We could quantify an inter-task transference associated with our multi-task network to extract and reveal the contribution and transference of the tasks in task grouping.
- We could quantify each task's positive and negative contribution in each cluster that shows tasks behave differently when trained in different groups.

The remainder of the paper is organized as follows; Section 2 presents the related studies in this context. In Sect. 3, we describe the data used in this study. Section 4 formulates the problem; The proposed approach is described in Sect. 5.2. Section 6 covers the experimental evaluation and results. Section 7 gives a discussion and summary of the work.

2   Related Work

This study introduces an efficient approach to finding the optimal clusters of tasks that should be trained together in a multi-task learning problem. The approach is applied to the automotive context where long-term vehicle behavior modeling was concerned with multi-task deep neural networks based on transferring knowledge from high-resolution to low-resolution data. Therefore, before we dive into the details of our proposed approach, we review some relevant solutions in multi-task learning, vehicle behavior modeling, and position our own work concerning those areas.

Driving behavior modeling has become an essential field of study among automotive sector researchers seeking to handle different challenges. Earlier studies concentrated on developing machine learning models to build a self-driving system [2,18], and later we could observe investigations focused more on modeling drivers' maneuver patterns [13,14,28]. For example, in [9], an ensemble unsupervised machine learning approach is used to extract vehicle usage patterns in different seasons and assess their performances by focusing on vehicle breakdowns and fuel consumption.

Yet, many studies have concentrated on signal processing and utilized single or few sensors data to estimate immediate behaviors such as driver frustration or lane changing [16] to forecasting fuel consumption [26] or build up ADAS systems [15,22]. This differs from our unique study, which presents vehicles' long-term behavior modeling approach based on transferring knowledge from high-resolution to low-resolution data and using gradient descent to find the optimal cluster of tasks.

Multi-task learning approaches are used to gain the knowledge shared between related tasks. The aim of multi-task learning is that the shared representation has more generalization power since it needs to learn a more general representation useful for multiple tasks. Multi-task learning has gained much attention from the automotive sector, where researchers used them in various domains, such as predictive maintenance, autonomous driving, drivers' behavior reasoning [6,23,24,27]. For instance, Chowdhuri et al. in [6] introduced a multi-task learning approach for drivers' behavior recognition. They developed a CNN-based model as the encoder and utilized multiple decoders, including fully connected and LSTM layers heads to different tasks. In a similar context of driving behavior detection, we could observe a study in [27], where multi-task learning is used for three related tasks: illegal driver detection, legal driver identification, and driving behavior evaluation. In this study, sequential modeling–LSTM–is first used to extract the common representation of data, then injected into an SVDD model to detect illegal driver behavior and two feed-forward neural networks for legal driver identification and behavior evaluation.

In the autonomous driving domain, in [7], a multi-task network is designed to model and predict three short-term behavioral modes: Direct mode, follow mode, and furtive mode, where they used self-driving model cars for driving in unstructured environments such as sidewalks and unpaved roads to evaluate the proposed approach. In [12], a graph-based multi-task network is developed to predict trajectories of traffic actors with interactive behaviors. This study considered Trajectory predictions, 3-D bounding box prediction, and interactive events recognition as related tasks in building the MLT network.

We found the closest study to our works in [8], where they tried to efficiently find the task grouping using inter-task-affinity. They attempted to quantify the effect of one task on another by calculating the loss value of each task before and after updating the shared layers in the multi-task network. Although the approach can find the optimal groups in one training process, the network should be trained once more in each epoch to quantify the impact of one task on another, which computationally is expensive. This can be intensified when the number of tasks gets more.

All in all, three main limitations can be observed in earlier studies: 1) in terms of vehicle behavior modeling, most are limited to using only a few input parameters to model short-term vehicle behavior; 2) they suffer from the lack of available real-life data and high prediction accuracy; 3) in terms of application and multi-task learning, we have not seen any work in this domain to find the optimal clusters of tasks using gradient descent.

3   Data Representation

This section describes the two data sets used for the proposed vehicle behavior modeling approach: Logged Vehicle Data (LVD), which is low-resolution (infrequent) but high-dimensional (many features) data that is aggregated in a cumulative fashion; and Dynafleet data, consisting of high-resolution (frequent) but more low-dimensional (fewer features) data.

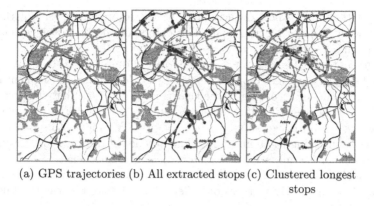

(a) GPS trajectories (b) All extracted stops (c) Clustered longest stops

Fig. 1. Part of the data mining pipeline to determine the night stop class. First, the stops are extracted from the GPS trajectories. Then, the stop with the longest duration per day is selected. These stops are then clustered geographically. The class is then determined based on the number of clusters and how often they are visited.

Low Resolution Data – Logged Vehicle Data (LVD). The logged vehicle data (LVD) were collected from commercial trucks over 2019. The LVD holds the aggregated sensor information for a fleet of heavy-duty trucks operating worldwide. The values of the features are collected using telematics and through manual inspection when a vehicle visits an authorized workshop for repairs and service. In general, two types of features are logged in this dataset. The first type of feature shows the operational sensor values for the vehicle in the form of scalars and histograms during its operation. This data is continuously aggregated and includes several features such as oil temperature, fuel consumption, compressor usage, gears used, cargo load, etc. The scalars are commonly "life of vehicle" values, meaning they represent the cumulative value so far in the vehicle's life, such as the total mileage, fuel consumption or engine time. The histograms represent the total time spent in a certain bin defined by the histogram axes. For instance, the histogram describing the engine speed and engine torque has bins that contain the total time spent in a certain interval in engine speed and engine torque. The readout frequency of these parameters will vary depending on importance, from everyday to occasional workshop visits (Fig. 1).

High Resolution Data-Dynafleet Data. The Dynafleet database contains fleet management data related to the services provided by Dynafleet (Volvo Trucks) and Optifleet (Renault trucks). Depending on the services provided, the vehicles have tracking events logged with intervals of 1 to 10 min, and in connection with certain events such as turning the engine on or off. The tracking events include information such as GPS position, speed, odometer, and accumulated fuel consumption. Using this data, it is possible to estimate the average vehicle behaviors described below. The labels rely on positional data and can not be directly computed from the aggregated data in the LVD dataset.

- **Geopattern:** Classification according to the approximate size of the area traveled during the desired year. The extent of the geographical cloud of positional data during a year is used to determine the classes.
- **Longest stop:** Classification according to the duration of the daily longest stop (typically the night stop). The classes are determined by segmenting the trucks into classes according to the yearly median of their daily longest stops.
- **Night stop class:** Classification according to if the vehicle has a common home base or an irregular night stop pattern. The night stop locations are found by clustering the night stops. The number of resulting clusters needed to account for the majority of the stops determines the class.
- **Driver class:** Classification according to the number of regular drivers of the vehicle (one main driver or several different ones), based on the total number of unique drivers and the distance driven by each driver.

All of these labels are calculated by an internal Volvo Group framework used for vehicle utilization analysis from Dynafleet data. The labels have been used, and their precision has been verified in several projects. The framework can output labels for different time frames but only the yearly labels are used in this study.

4 Problem Formulation

This section presents the formulations determined to tackle vehicle behavior modeling. Our approach considers low-resolution data (LVD), which are labeled (Geopattern, Longest stop, Night stop class, and Driver class) with high-resolution data as the system's input, and employs multi-task deep neural networks. The conceptual view of the proposed approach is illustrated in Fig. 2. The intention of this vehicle behavior modeling and prediction investigation can be formulated as follows:

- The design of a deep neural multi-task learning approach is studied, formulating the task as a supervised learning problem in a multi-task fashion, to model and predict vehicle usage. Given the tasks representing vehicle behavior, the approach tries to find the optimal group of tasks to be trained together for increasing the predictive performance.

5 Proposed Methodology

5.1 Data Preparation

Data preparation is an integral part of any machine learning problem, and our multi-task learning practice in the automotive sector is not an exception. Cleaning and transferring histograms to the right format have been done in this phase. However, the most essential step in this phase was to merge the LVD and the labels calculated from the Dynafleet data to create an integrated dossier with the usage and behavior designations. These two data sources are combined based on the vehicles' "Chassis id" and "Date of readout". The labels from dynafleet data are estimated yearly, and the LVD data are therefore also labeled yearly. The LVD readouts can be available weekly, which means that every vehicle has a maximum of 52 weeks of LVD data assigned with the same labels.

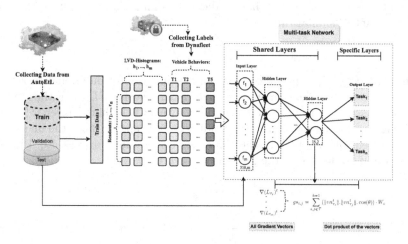

Fig. 2. The conceptual view of the proposed task/vehicle behavior grouping for multi-task learning in the automotive sector.

5.2   Multitask Network

Multi-task Learning (MTL) refers to the idea of learning several related tasks at the same time with the same algorithm to learn representations that are more general, lead to improved generalization, and possibly also faster learning [29]. In our specific case of modeling vehicle behaviors, the different behaviors/tasks are denoted τ_i and $\mathcal{T} = \{\tau_1, \tau_2, \ldots, \tau_m\}$ is the set of all task. The total loss vector is:

$$min_\theta L(\theta) = (L_{\tau_1}(\theta), L_{\tau_2}(\theta), \ldots, L_{\tau_m}(\theta)) \tag{1}$$

where $L_{\tau_i}(\theta)$ is the loss function of the i_{th} task (each task refers to a particular behavior mentioned in Sect. 1). A multi-task learning aims to perform joint learning at the same time and optimize all the tasks by utilizing $D = \{x_j^i, y_j^i\}_{j=1}^{n_i}$ The models learn from the data points D, and takes advantage of θ_s as the shared layers to calculate a loss for each task $L(\tau_i|\theta_s, \theta_i)$. Let's assume a multitask loss function parameterized by $\{\theta_s\} \cup \{\theta_i|\tau_i \in \mathcal{T}\}$, where "$\theta_s$" indicates the shared parameters in the shared layers and τ_i is the task i. Thus, given a batch of samples X, the total loss function of the MTL for vehicle behavior prediction is calculated by Eq. 2.

$$L_{all}(X, \theta_s, \theta_{i=1,..,m}) = \sum_{i=1}^{m} L(\mathcal{T}_i; \theta_s, \theta_i) \tag{2}$$

5.3 Training Multi-task Network

Given the shared parameters θ_s and set of tasks defined in $\mathcal{T}$, we aim to find the cluster of tasks that are more optimal to be trained together. Indeed, we aim to train the network in a way that gradients of similar tasks—tasks in the same cluster—point in similar direction.

$$\|v_{\tau_i}^t\| \leftarrow \nabla_{\theta_{s_L}^t} L_{\tau_i}(X^t, \theta_s^t, \theta_{\tau_i}^t), \quad \forall \tau_i \in \mathcal{T} \tag{3}$$

In Eq. 3, $L_{\tau_i}(X^t, \theta_s^t, \theta_{\tau_i}^t)$ indicates the loss function of individual task given X^t as input and shared parameters θ_s^t at the macro level t. $\nabla L_{\tau_i}(.)$ is the gradient descent of task τ_i with respect to the last layer of the shared layer $(\theta_{s_L}^t)$ and $\|v_{\tau_i}^t\|$ refers to the vector of gradient holding at each epoch. Since the magnitude of the gradient of each task might differ over the training process, Eq. 4 is used to find a unit to normalize the gradient of each task.

$$u_{\tau_i} = max(\|v_{\tau_i}^1\|, \ldots, (\|v_{\tau_i}^T\|) \quad i = 1, 2, \ldots, m \tag{4}$$

$$\|vn_{\tau_i}^t\| \leftarrow \frac{\|v_{\tau_i}^t\|}{u_{\tau_i}} \quad i = 1, 2, \ldots, m \tag{5}$$

In order to calculate the similarity between the normalized gradients of two tasks i and j—$(gs_{i,j})$—, Eq. 6 is used.

$$gs_{i,j} = \sum_{i,j \in \mathcal{T}}^{k=1} (\|vn_{\tau_i}^t\|.\|vn_{\tau_j}^t\|.\cos(\theta)) \cdot W_c \tag{6}$$

$$W_c = \frac{t}{T} \quad t = 1, 2, \ldots, T \cdot \tag{7}$$

In Eq. 6, θ is the angle between the two gradient vectors. W_c is used as an important weight over the training process to inject more weight into the inner value of the later epochs. In fact, $InnerVal$ illustrates whether the two or more vectors point to the same direction when they inner join together.

5.4 Individual Task Transference Quantification into Group

To gain insights into the impact of the tasks on each other as well as on a group of tasks during the training process, we adapt the inter-task affinity to vehicle behavior impact, introduced in our previous study [10]. We define the quantity $\theta_{s|i}^{t+1}$ to indicate the model with the updated shared parameters towards the task τ_i.

$$\theta_{s|i}^{t+1} = \theta_s^t - \zeta \Delta_{\theta_s^t} L_{\tau_i}(X^t, \theta_s^t, \theta_{\tau_i}^t) \tag{8}$$

Using $\theta_{s|i}^{t+1}$ in Eq. 8, we could measure the impact of task τ_i on the performance of the other tasks defined in $\mathcal{T} = \{\tau_1, \tau_2, \ldots, \tau_m\}$. Thus, given the input $X^t \in D$ we can measure the loss for each task by taking the updated shared parameter θ_s as well as the specific task parameters θ_i. Indeed, we assess the impact IM of the gradient update of task τ_i on a given set of tasks (S_τ) which in our case are chosen to be tuple, and triple. Then we can compare the ratio between the average loss value of tasks in a given set (S_τ) before and after conducting the gradient update from task τ_i towards the shared parameters as follows:

$$IM_{\tau_i \to L_{S_\tau}} = 1 - \frac{L_{S_\tau}(X^t, \theta_{s|i}^{t+1}, (\theta_{S_\tau}))}{L_{S_\tau}(X^t, \theta_s^t, (\theta_{S_\tau}))} \cdot W_c \tag{9}$$

$$W_c = \frac{t}{T} \quad t = 1, 2, \ldots, T \tag{10}$$

where θ_{S_τ} represent parameters of the tasks in a given set. Thus, we translate $IM_{\tau_i \to L_{S_\tau}}$ as a measure of transference from meta-train task τ_i to tasks in S_τ. A positive value of $IM_{\tau_i \to L_{S_\tau}}$ shows the update on the shard parameters θ_s led to a lower loss value on the set of tasks with respect to the original parameters. This basically expresses the positive effect of task i to generalize the predictive model on the set of tasks (S_τ), while the negative value of $IM_{\tau_i \to L_{S_\tau}}$ describes the destructive impact of task i on that set over the training process. We adapted the overall inter-task affinity measure onto the group of tasks by incrementally adding certain weights, defined in Eq. 10 on each iteration over the training process. t refers to the current epoch number, and T is the maximum number of epochs that the multi-task network should be iterated. This is due to the fact that at the beginning of the training, the weights are randomly generated, so it is not expected the earlier loss have the same impact w.r.t to the parameters at the end of the training.

Thus, in Eq. 11 we calculate the transference over all epochs from task i to the group of tasks:

$$\hat{IM}_{\tau_i \to L_{S_\tau}} = \frac{1}{T} \sum_{e=1}^{T} IM_{\tau_i \to L_{S_\tau}}^e \tag{11}$$

$IM_{\tau_i \to L_{S_\tau}}$ can be employed in different levels of granularity, such as per-epoch level or even micro-level/batch level. In this study, we measure the transference at the epoch level, where T refers to the number of epochs. Given Eq. 9, we calculate the effect of each task on a group of tasks to quantify the behavior of each individual task in a group. This basically reflects the hypothesis that each task might have different contributions when they are coupled with other tasks.

6   Experimental Evaluation and Results

To facilitate the implementation of the approach, we recall the two research questions, introduced in Sect. 1, on which we based the evaluation of the proposed approach as follows: RQ1) To what extent could the task of vehicle behaviors be grouped optimally? and RQ2) To what extent could the vehicle behavior transference be quantified to find the contribution of each task in different groups and on different groups?

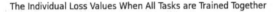

The Individual Loss Values When All Tasks are Trained Together

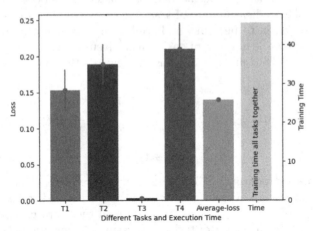

Fig. 3. The individual tasks loss values when they are trained all together with a shared and specific layer.

6.1 RQ1: Vehicle Task Grouping Results

Before directly answering the research questions, we formulate the prediction task as a multi-task learning problem wherein, in one single training process, we simultaneously trained all the tasks, having the shared and their specific parameters using Eq. 2. This implementation has been done to understand to what extent–as a baseline–a multi-task deep neural network can handle the complex problem of mapping low-resolution data to vehicle behavior obtained from high-resolution data.

Figure 3 shows the results of conducting Eq. 2 when all four tasks are considered and trained together. To get a reliable result, we have iterated the training phase 5 times and reported the overall average.

Within the reported values, we could observe Task 3 performed very well with a loss value close to zero 0.0027. Concerning the complex problem of modeling vehicle long-term behavior with low-resolution data, we obtained relatively poor results with $Task1 = 0.153$, $Task2 = 0.189$, and $Task4 = 0.210$. By taking into account all tasks together, we reached the average loss value=0.139, which leads us to the point that there is a need to improve the predictive model to lower the errors. Thus, we consider these numbers as a baseline for the rest of our study and attempt to increase the performance–reduce the loss value– by finding the optimal cluster (by extracting gradient descent in the shared layer) of tasks that should be trained together.

To answer the RQ1, we considered only the shared layers and used Eq. 3 to calculate the gradient descent of each task over the training process. Since we might face the gradient with a highly different magnitude, Eq. 4 is used to find a unit that we could inject into Eq. 5 to normalize the gradient vectors for all

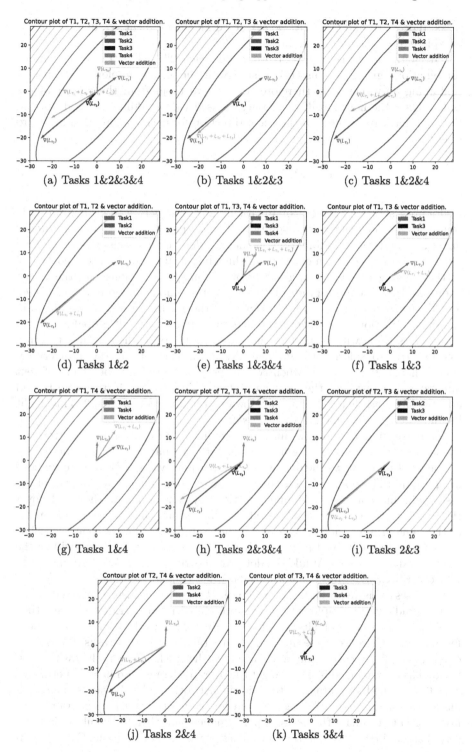

Fig. 4. Gradient and cross product of all combinations.

tasks. Indeed, to find the unit, we extracted all the gradient vectors, then Eq. 4 is exploited to calculate the unit.

Given the four tasks $m = 4$, we could reach $2^m - 1 - (m) = 11$ various combinations/groups or clusters of tasks (Note: the second m refers to the groups containing only one task; therefore, we remove them from the cluster of the tasks). Once the normalization is done, we used vector addition (Va) defined in Eq. 12 to intuitively show the gradient direction for each combination at macro level t.

$$Va = \sum_{i=1}^{m} \nabla(L_{\tau_i})^t \quad t = 1, 2, \ldots, T \tag{12}$$

Figure 4 shows all eleven combinations' at one epoch. It is clear from some combinations that the vector addition of tasks points to the same or similar directions, and some combinations point to different directions. However, these plots are constructed from one epoch, which might differ in the later epochs due to the training process. The green arrows in all plots demonstrate the vector addition of two or more gradient vectors. For example, the gradient of $Task2$ and $Task3$ point to the same direction, and the Va of the vectors also head in the same direction. This means training these two tasks will probably lead to a lower loss value with respect to the $Task2$ and $Task4$ when trained together (see Fig. 4j). Considering the gradient descent in each epoch, the most critical and challenging situation is when two or more task gradients point in different directions. For instance, the gradient of $Task1$ and $Task2$ illustrated in Fig. 4d are opposite, and their Va is inclined towards the direction of $Task2$. This might be due to the higher magnitude of $Task2$.

To better understand how different combinations might perform better over the whole training process, we utilized Eq. 6 to calculate the dot product of the gradients. Table 1 shows the dot-product (inner join) values of those productions in 11 groups. The positive values express the combination of the gradient of the tasks pointing more-less to similar directions by considering all the epochs over the training process. The negative values demonstrate that training those tasks in the same group is not optimal when the overall loss value is concerned. The figures reported in Table 1 show G1-G7 are the optimal groups by providing the positive values in theirs. While the dot products of the gradient of groups G8 to G11 describe, the groups are not optimal when trained together.

To assess this hypothesis and evaluate the optimal cluster obtained based on the gradient, we employed the network constructed by Eq. 2 with multiple combinations covering all clusters of tasks. Table 2 shows the results of the above formulation, where we came up with 11 varieties or groups of tasks. We trained only those tasks in each group and froze the other tasks. Each multi-task combination has been trained 5-times to get the overall average, and in all experiments were 60% of the LVD data used to train the networks and 30% was used as a validation set, and finally, 10% was held out for testing the multitask network. The loss is used to evaluate the network performance, and then the combinations are sorted ascendingly based on their average loss values. For example, G1 (T1&T3)

and G2 (T2&T3) have the lowest loss values by 0.0008 and 0.01, respectively. Concerning the computational time reported in the table, it is clear that the clusters with more tasks took relatively more time than those with fewer tasks.

Taking into account the numbers reported in Table 1, where the groups are sorted based on dot product value (from positive to negative), and figures illustrated in Table 2, where the groups are sorted based on the overall loss values, we could notice that most of the groups are located in the similar order with minor differences showing the proposed approach is promising. However, it is fair to remark G2, which is placed in the second order in terms of the dot product; it is located in the 9th place when the loss value was considered (see Table 2).

Table 1. The dot product of the gradient of the tasks calculated at the shared layers. Here, the more positive value shows that the more similar the gradient and negative represents the vectors are different.

Groups	Tasks Combinations	Inner join Score
G1	**T2&T3**	0.005606
G2	**T2&T4**	0.003383
G3	**T1&T3**	0.003016
G4	**T1&T2&T3**	0.00145
G5	**T1&T3&T4**	0.00063
G6	**T1&T4**	0.000168
G7	**T2&T3&T4**	0.00014
G8	**T1&T2&T3&T4**	9.25568e−05
G9	**T1&T2**	−0.000587
G10	**T1&T2&T4**	−0.00081
G11	**T3&T4**	−0.00153

Concerning individual task improvement in different group training, numbers obtained and reported in Table 2 shows how dramatically the loss values decreased when they were trained in different groups. In Fig. 5, we illustrated an A/B test comparison between individual task loss values obtained in different groups and the baseline loss values in G10. For instance, the loss value of T2 in G10 has dropped by 13.2% from 0.189 to 0.025 in G2. A similar improvement was achieved for T3 in G2 and T1 and T4 in G7.

Table 2. All combinations of the task with their individual and average loss values. * means the particular task was not included in the group and consequently in the training process.

Groups #	Task Combinations Trained Together	T1	T2	T3	T4	AVG Tasks Loss	Time Spent
G1	T1&T3	0.016 ± 0.002	*	$0.0008 \pm 3.01e-5$	*	0.00862	40.006
G2	T2&T3	*	0.0255 ± 0.0031	$0.0007 \pm 7.63e-5$	*	0.01315	41.38
G3	T3&T4	*	*	0.00138 ± 0.0001	0.0251 ± 0.0029	0.0132	39.84
G4	T1&T2&T3	0.0548 ± 0.019	0.0617 ± 0.0157	0.0009 ± 0.0002	*	0.039	45.99
G5	T1&T3&T4	0.0671 ± 0.014	*	$0.0019 \pm 2.196e-05$	0.081 ± 0.013	0.050	44.57
G6	T2&T3&T4	*	0.079 ± 0.008	0.001 ± 0.0002	0.078 ± 0.012	0.053	45.33
G7	T1&T4	0.060 ± 0.007	*	*	0.058 ± 0.01	0.059	39.90
G8	T1&T2	0.085 ± 0.005	0.086 ± 0.014	*	*	0.085	39.98
G9	T2&T4	*	0.111 ± 0.005	*	0.093 ± 0.019	0.10	41.85
G10	T1&T2&T3 &T4	0.153 ± 0.029	0.189 ± 0.028	0.0027 ± 0.0005	0.210 ± 0.036	0.139	45.62
G11	T1&T2&T4	0.192 ± 0.008	0.284 ± 0.0132	*	0.373 ± 0.020	0.283	41.67

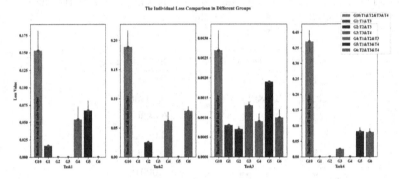

Fig. 5. The individual tasks improvement in different groups. The loss value of individual tasks in different groups is compared vs. the loss value of individual tasks in Group 10 as a baseline, where all tasks are trained together.

6.2   RQ2: Vehicle Behavior Transference Quantification Results

To answer RQ2, we defined two sub-research questions and carried out two sorts of implementations. **SbQ2.1**) Given the multi-task network, first, we aim to quantify the individual task behavior in its group by utilizing the gradient descent of each task using Eq. 6. This is because the behavior or effect of each task might differ in different groups, and one task could positively or negatively contribute to its group performance or multi-task training. To quantify the contribution, we took G4, G5, G7, G8, and G10 from all combinations (Table 1) since their groups hold more than two tasks. Considering G4, we employed three analyses where every time we took one task out, e.g., ($T3$), and measured the inner join value, and compared it with the value of the inner join when it was only two tasks ($T1$ and $T2$). We iterated the same analysis for the other tasks in the same group and other groups as well. Table 3 illustrated the figures obtained for the task contribution analysis. We observed that different tasks contribute differently when they are in various groups. For instance, $Task1$ in $G4$ has a negative contribution with -0.0041 in G4 (highlighted by orange), while it has a positive impact in G5 with $+0.0021$ (highlighted by green). It can be seen from the table the contributions of the tasks in $G10$ are all negative, and that is the reason G10 performed worst compared to the other groups, both in dot product values shown in Table 1 and in loss evaluation represented in Table 2. We could also observe that $Task3$ plays a positive role in most groups, such as G4, G5, and G8. However, the negative contribution of $Taks3$ is obtained in G7, which highlights that even if one task performed well in most cases, there is a possibility to play negatively in one group. This case shows the importance of obtaining the optimal group of tasks to be trained together. It needs to be noted such a negative contribution in one group does not mean that the performance of that group is considerably poor. Still, it suggests we could perform better by finding an optimal combination.

Table 3. The contribution of each task in group training.

Groups	Tasks Combinations (TC)	T1	T2	T3	T4
G4	T1&T2&T3	*	*	+0.002	-
	T1&T2&T3	*	-0.0015	*	-
	T1&T2&T3	-0.0041	*	*	-
G5	T1&T3&T4	+0.0021	-	*	*
	T1&T3&T4	*	-	+0.0004	*
	T1&T3&T4	*	-	*	-0.0023
G7	T2&T3&T4	-	+0.001		
	T2&T3&T4	-	*	-0.001	*
	T2&T3&T4	-	*	*	-0.004
G8	T1&T2&T3&T4	-0.00013	*	*	*
	T1&T2&T3&T4	*	-0.0006	*	*
	T1&T2&T3&T4	*	*	+0.00081	*
	T1&T2&T3&T4	*	*	*	-0.001
G10	T1&T2&T4	-0.004	*	-	*
	T1&T2&T4	*	-0.0009	-	*
	T1&T2&T4	*	*	-	-0.001

In the second evaluation **SbQ2.2)**, we aim to understand the individual task transfer or contribution onto the group of tasks over the training process by utilizing the shared and specific layers. To acknowledge this, we utilized Eq. 9, which we introduced in [10] and adapted to the below equation.

$$IM_{\tau_i \to L_{S_\tau}} = \frac{avg(L_{S_\tau}(X_{val}^t, X_{tr}^t, \theta_{s|i}^{t+1}, \theta_{S_\tau})) - avg(L_{\tau_j, \tau_k}(X_{val}^t, X_{tr}^t, \theta_s^t, \theta_{S_\tau})}{avg(L_{S_\tau}(X_{val}^t, X_{tr}^t, \theta_s^t, \theta_{S_\tau})} .W_c$$

$$(13)$$

where, X_{val}^t and X_{tr}^t are the independent vehicles usage predictors for training and validating, respectively. $avg(L_{S_\tau}(X_{val}^t, X_{tr}^t, \theta_{s|i}^{t+1}, \theta_{S_\tau}))$ refers to the average loss value of tasks in the set S_τ after the network was updated and trained with the shared parameters for specific task τ_i $(\theta_{s|i}^{t+1})$, and $avg(L_{S_\tau}(X_{val}^t, X_{tr}^t, \theta_s^t, \theta_{S_\tau})$ points to the average loss value before the update. This intuitively reveals to what extent one task can transfer positive or negative knowledge onto the performance of the other group of tasks. Note: here, the main question of transferability is on the modeling aspect and not the effect of different data representations. This means different forms of data representation could have various consequences on the multi-task learning results.

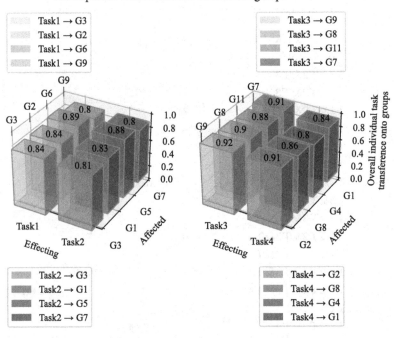

Fig. 6. The overall individual tasks inter-task transference onto a group of tasks. In these two plots, we could observe the impact of each task into a group obtained by Eq. 9.

Figure 6 shows how the individual task can impact the performance of the group of tasks. In this experiment, we evaluate the impact of individual tasks on 10 different groups (we ignore G10 since this group contains all the tasks so we could not quantify the impact of one task on this group). The figures illustrate that overall, all tasks positively impact the groups of tasks with $Task1 = 0.63.5$, $Task2 = 0.65$, $Task3 = 0.58$, and $Task4 = 0.62$. In contrast with the previous analysis, in this experiment, we obtained a less contribution of $Task3$ on the performance of the other groups w.r.t to the other tasks. The negative contribution of Task1 and Task2 in their groups (G4, G8, and G10) changed to positive when quantifying the impact of other groups' performance. The numbers show $Task2$ transfered more positive knowledge ($> 2\%$) to the other group compared to the other tasks. This suggests one task may negatively contribute to its group and positively to other groups.

7   Discussion and Conclusion

This study presents a multi-task deep neural network approach for finding the optimal group of tasks and ultimately quantifying vehicle behavior transference in multitask settings. The experimental evaluation of thousands of heavy-duty Volvo trucks' logged real data show that the proposed approach can sufficiently find the optimal group of tasks that should be trained together, leading to better performance. In addition, we could quantify knowledge transferred within and onto the groups of tasks.

Considering the first objective (RQ1), figures obtained in task grouping show a significant difference between the performance of the groups found by the proposed gradient descent approach and the baseline (when all tasks were trained together). The figures obtained in the individual tasks comparison show the potential of finding the optimal group and building a general multi-task forecasting model for this complex behavioral problem. It described how the optimal groups could reduce the overall loss values of multi-task training. Taking the figures obtained in this experiment into account, we, therefore, can state that *the proposed gradient decent approach could provide an equal and, in most cases, optimal group of tasks leading to better predictive models than when all the tasks are trained together for the multi-task formulations.*

Considering the second objective (RQ2), the inter-task transference evaluations explained how the individual tasks contributed to the performance of their group over the training process. Assessing the overall impact and the results on each group of tasks shown in the bar plots and the table, it is clear that *given the shared model, the individual tasks have different positive and negative contributions in the tasks grouping, and given the specific layer, tasks have a constructive impact on all groups performance in the multi-task formulation.*

The findings of this work furthermore indicate limitations of the proposed approach, which present new directions for future investigation. In this study, we only focused on the individual impact of one task on the group task performance. To understand which group of tasks can lead to behaving the other tasks in vehicle behavior modeling, we need to investigate quantifying the impact of a cluster of tasks on each task.

The second limitation pertains to the generality assessment. In this study, we applied the approach in the automotive domain to find the optimal cluster of tasks that should be trained together for vehicle behavior modeling. Further investigation is needed to employ the approach on the datasets with multiple target values to evaluate how the approach can perform in other contexts. An interesting example would be computer vision. In particular, the scene understanding over time, where multiple objects in the scene need to be recognized, and more importantly, the impact of a group of objects can be quantified to understand the scene better.

References

1. Alizadeh, M., Rahimi, S., Ma, J.: A hybrid Arima-WNN approach to model vehicle operating behavior and detect unhealthy states. Expert Syst. Appl. 116515 (2022)
2. Badue, C., et al.: Self-driving cars: a survey. Expert Syst. Appl. **165**, 113816 (2021)
3. Bousonville, T., Dirichs, M., Krüger, T.: Estimating truck fuel consumption with machine learning using telematics, topology and weather data. In: 2019 International Conference on Industrial Engineering and Systems Management (IESM), pp. 1–6. IEEE (2019)
4. Chen, J., Wang, S., He, E., Wang, H., Wang, L.: Two-dimensional phase lag index image representation of electroencephalography for automated recognition of driver fatigue using convolutional neural network. Expert Syst. Appl. **191**, 116339 (2022)
5. Choi, E., Kim, E.: Critical aggressive acceleration values and models for fuel consumption when starting and driving a passenger car running on lpg. Int. J. Sustain. Transp. **11**(6), 395–405 (2017)
6. Chowdhuri, S., Pankaj, T., Zipser, K.: Multinet: Multi-modal multi-task learning for autonomous driving. In: 2019 IEEE Winter Conference on Applications of Computer Vision (WACV), pp. 1496–1504. IEEE (2019)
7. Chowdhuri, S., Pankaj, T., Zipser, K.: Multinet: multi-modal multi-task learning for autonomous driving. In: 2019 IEEE Winter Conference on Applications of Computer Vision (WACV), pp. 1496–1504 (2019). https://doi.org/10.1109/WACV.2019.00164
8. Fifty, C., Amid, E., Zhao, Z., Yu, T., Anil, R., Finn, C.: Efficiently identifying task groupings for multi-task learning. In: Advances in Neural Information Processing Systems, vol. 34 (2021)
9. Khoshkangini, R., Kalia, N.R., Ashwathanarayana, S., Orand, A., Maktobian, J., Tajgardan, M.: Vehicle usage extraction using unsupervised ensemble approach. In: Arai, K. (ed.) IntelliSys 2022, vol. 542, pp. 588–604. Springer, Cham (2023). https://doi.org/10.1007/978-3-031-16072-1_43
10. Khoshkangini, R., Mashhadi, P., Tegnered, D., Lundström, J., Rögnvaldsson, T.: Predicting vehicle behavior using multi-task ensemble learning. Expert Syst. Appl. **212**, 118716 (2023). https://doi.org/10.1016/j.eswa.2022.118716, https://www.sciencedirect.com/science/article/pii/S0957417422017419
11. Lattanzi, E., Freschi, V.: Machine learning techniques to identify unsafe driving behavior by means of in-vehicle sensor data. Expert Syst. Appl. **176**, 114818 (2021)
12. Li, Z., Gong, J., Lu, C., Yi, Y.: Interactive behavior prediction for heterogeneous traffic participants in the urban road: a graph-neural-network-based multitask learning framework. IEEE/ASME Trans. Mechatron. **26**(3), 1339–1349 (2021)
13. Lin, N., Zong, C., Tomizuka, M., Song, P., Zhang, Z., Li, G.: An overview on study of identification of driver behavior characteristics for automotive control. Math. Probl. Eng. **2014** (2014)
14. Liu, P., Kurt, A., Özgüner, Ü.: Trajectory prediction of a lane changing vehicle based on driver behavior estimation and classification. In: 17th international IEEE conference on intelligent transportation systems (ITSC), pp. 942–947. IEEE (2014)

15. Marina Martinez, C., Heucke, M., Wang, F.Y., Gao, B., Cao, D.: Driving style recognition for intelligent vehicle control and advanced driver assistance: a survey. IEEE Trans. Intell. Transp. Syst. **19**(3), 666–676 (2018). https://doi.org/10.1109/TITS.2017.2706978

16. Miyajima, C., Takeda, K.: Driver-behavior modeling using on-road driving data: a new application for behavior signal processing. IEEE Signal Process. Mag. **33**(6), 14–21 (2016). https://doi.org/10.1109/MSP.2016.2602377

17. Mondal, S., Gupta, A.: Evaluation of driver acceleration/deceleration behavior at signalized intersections using vehicle trajectory data. Transp. Lett. **15**, 350–362 (2022)

18. Pentland, A., Liu, A.: Modeling and prediction of human behavior. Neural Comput. **11**(1), 229–242 (1999)

19. Powell, S., Cezar, G.V., Rajagopal, R.: Scalable probabilistic estimates of electric vehicle charging given observed driver behavior. Appl. Energy **309**, 118382 (2022)

20. Prakash, S., Bodisco, T.A.: An investigation into the effect of road gradient and driving style on NOX emissions from a diesel vehicle driven on urban roads. Transp. Res. Part D: Transp. Environ. **72**, 220–231 (2019)

21. Shahverdy, M., Fathy, M., Berangi, R., Sabokrou, M.: Driver behavior detection and classification using deep convolutional neural networks. Expert Syst. Appl. **149**, 113240 (2020)

22. Wang, Z., et al.: Driver behavior modeling using game engine and real vehicle: a learning-based approach. IEEE Trans. Intell. Veh. **5**(4), 738–749 (2020). https://doi.org/10.1109/TIV.2020.2991948

23. Xie, J., Hu, K., Li, G., Guo, Y.: CNN-based driving maneuver classification using multi-sliding window fusion. Expert Syst. Appl. **169**, 114442 (2021). https://doi.org/10.1016/j.eswa.2020.114442, https://www.sciencedirect.com/science/article/pii/S0957417420311003

24. Xing, Y., Lv, C., Cao, D., Velenis, E.: A unified multi-scale and multi-task learning framework for driver behaviors reasoning. arXiv preprint arXiv:2003.08026 (2020)

25. Xu, Y., Zheng, Y., Yang, Y.: On the movement simulations of electric vehicles: a behavioral model-based approach. Appl. Energy **283**, 116356 (2021). https://doi.org/10.1016/j.apenergy.2020.116356, https://www.sciencedirect.com/science/article/pii/S0306261920317360

26. Xu, Z., Wei, T., Easa, S., Zhao, X., Qu, X.: Modeling relationship between truck fuel consumption and driving behavior using data from internet of vehicles. Comput.-Aided Civil Infrastruct. Eng. **33**(3), 209–219 (2018). https://doi.org/10.1111/mice.12344, https://onlinelibrary.wiley.com/doi/abs/10.1111/mice.12344

27. Xun, Y., Liu, J., Shi, Z.: Multitask learning assisted driver identity authentication and driving behavior evaluation. IEEE Trans. Industr. Inf. **17**(10), 7093–7102 (2020)

28. Yao, W., Zhao, H., Davoine, F., Zha, H.: Learning lane change trajectories from on-road driving data. In: 2012 IEEE Intelligent Vehicles Symposium, pp. 885–890. IEEE (2012)

29. Zhang, Y., Yang, Q.: A survey on multi-task learning. IEEE Trans. Knowl. Data Eng. **34**, 5586–5609 (2021)

Effective Guidance in Zero-Shot Multilingual Translation via Multiple Language Prototypes

Yafang Zheng[1,2], Lei Lin[1,2], Yuxuan Yuan[1,2], and Xiaodong Shi[1,2(✉)]

[1] Department of Artificial Intelligence, School of Informatics, Xiamen University, Xiamen, China
{zhengyafang,linlei}@stu.xmu.edu.cn
[2] Key Laboratory of Digital Protection and Intelligent Processing of Intangible Cultural Heritage of Fujian and Taiwan, Ministry of Culture and Tourism, Xiamen, China
mandel@xmu.edu.cn

Abstract. In a multilingual neural machine translation model that fully shares parameters across all languages, a popular approach is to use an artificial language token to guide translation into the desired target language. However, recent studies have shown that language-specific signals in prepended language tokens are not adequate to guide the MNMT models to translate into right directions, especially on zero-shot translation (i.e., *off-target translation* issue). We argue that the representations of prepended language tokens are overly affected by its context information, resulting in potential information loss of language tokens and insufficient indicative ability. To address this issue, we introduce multiple language prototypes to guide translation into the desired target language. Specifically, we categorize sparse contextualized language representations into a few representative prototypes over training set, and inject their representations into each individual token to guide the models. Experiments on several multilingual datasets show that our method significantly alleviates the off-target translation issue and improves the translation quality on both zero-shot and supervised directions.

Keywords: Zero-Shot Multilingual Machine Translation · Off-Target Issue · Language Tag Strategy

1 Introduction

Unlike traditional neural machine translation (NMT) models that focus on specific language pairs, the many-to-many multilingual neural machine translation (MNMT) models aim to translate between multiple source and target languages using a single model [1,4,7,10,12]. Parameter sharing across different languages

© The Author(s), under exclusive license to Springer Nature Singapore Pte Ltd. 2024
B. Luo et al. (Eds.): ICONIP 2023, LNCS 14452, pp. 226–238, 2024.
https://doi.org/10.1007/978-981-99-8076-5_16

Table 1. Illustration of the analysis of the off-target translation issue with German → Chinese zero-shot translations using a multilingual NMT model with prepended tokens in the source sentences. The baseline multilingual NMT model fails to maintain the Chinese translation when introducing lexical variations to the source sentence, resulting in English.

Source	__zh__ Das ist ein ernstzunehmendes, klinisches Problem.
Reference	这可是个严重的临床问题。
Hypothesis	这是个严肃的临床问题。
Source	__zh__ Es handelt sich um ein ernstzunehmendes, klinisches Problem.
Reference	这是一个严重的临床问题。
Hypothesis	It's about a 重大、临床问题。

in the MNMT model makes it benefit from the transferring ability of intermediate representations among languages, thereby achieving better translation quality between low-resource and even zero-resource language directions [2,3,8,19] than bilingual models.

Since it can greatly reduce the MT system's deployment cost and benefit low-resource languages, MNMT has been gaining increasing attention. One line of research on MNMT focuses on partial parameter sharing models with language-specific components such as separate encoders, separate decoders or separate cross-attention networks [7,13,22]. However, partial sharing faces the challenge of a rapid increase in the number of parameters as the number of languages grows. Johnson et al. [12] propose to train full sharing models for MNMT with a prepended token to guide the translation direction. Despite its deployment efficiency and the transferring ability of multilingual modeling, the full sharing models still suffer from the *off-target translation* issue [8,27] where a model translates into a wrong language. Specifically, Zhang et al. [27] identify this issue as the major source of the inferior zero-shot performance. Since then, numerous researchers have been paying attention to solve the off-target problem from different perspectives. Some of them [8,27] attribute this issue to the lack of zero-shot directional data, Jin and Xiong [11] suggest that the efficacy of the language tag diminishes as the translation information propagates through deeper layers, while Chen et al. [6] tackle the problem by considering the aspect of dictionary sharing.

In this work, we conduct a comprehensive analysis of the off-target issue from a new perspective. A consensus is that the prepended language tokens help models distinguish the target language that should be translated to, playing a significant role in language-specific knowledge learning. However, we argue that the prepended tokens in sentence pairs in full-sharing models are influenced by their context information (the content of the sentence pairs), potentially hindering their ability to guide the MNMT models in translating to the right languages.

We illustrate this issue with a specific example, as shown in Table 1, where the initial German source sentence "Das ist ein ernstzunehmendes, klinishes Problem." ("It is a serious clinical problem." in English) and its corresponding Chinese reference result is taken from the TED-59 zero-shot test sets [11]. In our experiment, we introduce slight lexical variations to the sentence while preserving its underlying semantic. However, when employing the baseline model, which is trained with a prepending special token (__zh__) at the source sentence, instead of producing the expected generation in Chinese, the model outputs several unexpected English words. The same phenomenon occurs when the prepended token is added to the target sentence.

We hypothesize that this phenomenon is due to the entanglement of the indicative information (prepended language tokens) and translation information (content of sentence pairs) after encoding. As a result, the representations of prepended language tokens are overly affected by their context information, which hinders the indicative function of the prepended tokens, leading to the off-target problem. Building upon this, an intuitive way to solve it is to alleviate the excessive reliance of language representations on context information and enhance their indicative function by injecting the indicative representations into each individual token in the sentence. Given the fact that the representations of language tags encode typological properties of languages, we categorize sparse contextualized language representations into a few representative prototypes over training instances, and make use of them to enrich indicative representations to guide the models.

To be specific, we propose a two-stage approach. In the first stage, we initialize a Transformer model, which is trained with prepended tokens added to the source sentences or target sentences. This initialization enables the model to generate reasonable language-indicative representations, serving as a foundation for subsequent stages. In the second stage, we utilize the pre-initialized model to categorize contextualized representations of the prepended tokens from the training corpus for each language. Specifically, we perform clustering techniques, such as K-Means, to obtain Target Language Prototypes (TLP) that capture the essential characteristics of each language to guide the translation direction. Then, we propose an extension by integrating the TLP into the encoder or decoder using a Lang-Attention module. The extended model is trained iteratively until convergence.

Our empirical experiments prove that our method improves the language accuracy from 70.54% to 93.72% and increases the BLEU score of zero-shot translation directions by 5.28 points on IWLST17. In the large-scale setting, such as the TED-59 dataset, our method demonstrates a significant enhancement of the BLEU score of zero-shot translation directions by 3.74 points on the average of 3306 translation directions and improves the language accuracy from 60.67% to 76.21%. Extensive analyses demonstrate that fusing TLP appropriately leads to better language accuracy and improves the translation quality of zero-shot directions while preserving the performance of the supervised directions.

2   Background and Motivation

Early studies on multilingual Neural Machine Translation mainly focused on extending the standard bilingual model to MNMT by incorporating language-specific components [7,13,22]. However, these approaches encounter a rapid growth in the number of parameters as the number of languages increases. In contrast, Johnson et al. [12] successfully trained a single NMT model for multilingual translation with prepending an artificial target language token to source sentences without modifying the model architecture, which is parameter efficient and beneficial for zero-shot translation directions. Therefore, it becomes the dominant approach to many-to-many MNMT [1,3,24]. However, it usually suffers from off-target translation issue [27] on zero-shot translation directions. This issue indicates that the MNMT model tends to translate input sentences to the wrong languages, which leads to low translation quality. To alleviate it, pioneering studies propose approaches such as generating pseudo sentence pairs for zero-shot directions [27], increasing the model cardinality [25], exploring the language tag strategies [24], adding the language tag embedding to the model during training [11], modifying the vocabulary sharing [6] and so on.

In our practical investigation, we have discovered a novel perspective for exploring the underlying factors that contribute to the off-target issue. For instance, as demonstrated in Table 1, when introducing multiple lexical variations to sentence pairs while maintaining the fundamental semantic structure, the off-target problem arises. This demonstrates that the representations of prepended language tokens are overly affected by its context information. Therefore, the indicative information of language tokens potentially loses during training and inference, leading to insufficient indicative ability of language tags. In other words, the contextual representation of the prepended tokens significantly diminishes the indicative function of the prepended tokens, which is detrimental to both supervised and zero-shot translation. Drawing inspiration from this, we propose a method to introduce multiple language prototypes to guide translation into the desired target language. To be specific, we design a two-stage method to first obtain the multiple language prototypes and then integrate them into the multilingual neural machine translation models through a Lang-Attention modules.

3   Method

In this section, we provide a detailed explanation of our two-stage method. First, we briefly introduce the baseline strategies, which are also employed in our method in the first stage, aiming to enable the model to generate reasonable indicative representations. Then, we describe the second stage in two separate sections. The first section discusses how we leverage the basic MNMT model to obtain TLP (target language prototypes). In the second section, we explain how these TLP are utilized to train MNMT models expanding with additional Lang-Attention modules.

3.1 Baseline Strategies

Previous studies [10,12,24] demonstrated that placing the target language tags (TLT) on the source side or target side can alleviate the off-target issue and improve the quality of zero-shot translation, which have become one of the fundamental strategies for MNMT. The specific examples of two strategies are shown in Table 2.

Table 2. Examples of modified input data by different language tag strategies. The bold tokens is the target language tag (__zh__). T-ENC means adding the target language tag (TLT) to the encoder side. T-DEC means placing the TLT on the decoder side of model.

Strategy	Source sentence	Target sentence
Original	`Hello World!`	你好，世界！
T-ENC	`__zh__ Hello World!`	你好，世界！
T-DEC	`Hello World!`	`__zh__` 你好，世界！

In the first stage, we train a base Transformer model for N epochs with the basic T-ENC or T-DEC strategy until it is able to generate reasonable representations of language tags to guide the translation. Given the training corpus $\mathcal{D} = \{(X,Y)\}$, where X and Y denote a source sentence and target sentence respectively, the model is optimized by minimizing cross-entropy loss. After training for N epochs we obtain a pre-initialized model $\theta^{(N)}$.

3.2 Language Representative Prototypes

In the second stage, we first utilize the pre-initialized model to obtain target language prototypes (TLP). Building upon insights from previous studies, we introduce two types of TLP for the subsequent training procedure.

Language Token Representation (LTR). Prior studies reveal that the embedding of the prepended tokens encode typological properties of languages [15–17], which can serve as guiding information for translation direction. However, this feature is susceptible to contextual influence from the input sentence pair based on our observation. As demonstrated in Fig. 1, the language representation in the encoder output may appear disorderly. However, it still exhibits a detectable distribution that can be categorized into distinct clusters. Inspired by this observation, we categorize sparse contextualized language representations into several representative prototypes over training set.

To be specific, using the pre-initialized model $\theta^{(N)}$, we build an language representative prototypes lookup table C. We iterate through the training corpus and aggregate contextualized representations $\{H_1^{lang_i}, ..., H_{M(lang_i)}^{lang_i}\}$ of each language $lang_i$, where $M(lang_i)$ is the number of contextualized representations of the language tag for $lang_i$. Next, for each language $lang_i$, we use K-Means [14] to cluster the contextualized representations due to its efficiency for large number of samples in high dimensions:

$$C^{lang_i} = \text{K-Means}(H_1^{lang_i}, ..., H_{M(lang_i)}^{lang_i}) \tag{1}$$

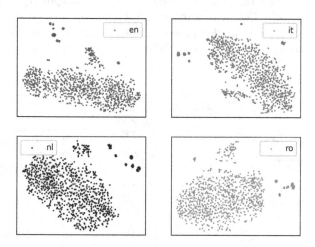

Fig. 1. The language representation, obtained through encoder visualization using t-SNE over a randomly selected set of 1000 sentences from the training corpus, is generated from the Baseline models trained on IWSLT17.

Mean-Pooling Sentence Representation (MPSR). Inspired by the application in the classification task of pre-trained language models [21], we further consider the sentence embeddings in the training corpus of each language to be the indicative representations. For simplicity and efficiency, we utilize the mean-pooling technique to generate the sentence embedding, which involves computing the average representation across all tokens in the sentence. Then, we use the clustering approach described above to gain k target language prototypes.

3.3   Integration of Multiple TLP

For each language $lang_i$, we gain the TLP and packed as a matrix $C^{lang_i} \in \mathbb{R}^{d \times k}$, where d is the dimension of the pre-initialized model and k is the number of TLP. Then, we extend the model $\theta^{(N)}$ with a Lang-Attention module on the top of the self-attention module in each encoder layer or on the top of the cross-attention module in each decoder layer, which is inspired by Yin et al. [26]. The Lang-Attention aggregates the language representative prototypes and refines each token by the Multi-Head Attention (MHA) mechanism [23]:

$$H_{pa}^l = \text{MHA}(H_{sa}^l, C, C) \tag{2}$$

$$H_{pa}^l = \text{MHA}(H_{ca}^l, C, C) \tag{3}$$

where C is the TLP of the target language, H_{sa}^l denotes the output of the self-attention in the corresponding encoder layer and H_{ca}^l denotes the output of the cross-attention in the corresponding decoder layer. The output H_{pa}^l is fed into the feed-forward network. The overall process of enhancing the decoder is depicted in Fig. 2, and it shares a similar process of enhancement with the encoder. In our experimentation, we set $k = 3$ for default.

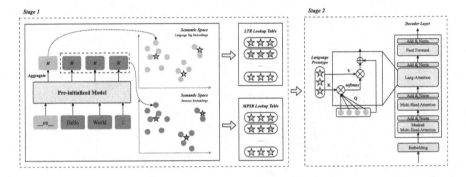

Fig. 2. The overall process of enhancing the decoder with the T-DEC strategy. The small circles denote the representation of tokens which is outputted by the last layer of the decoder.

4   Experiments and Analysis

4.1   Experimental Settings

Dataset. We conduct experiments on the following three multilingual datasets: IWSLT17 [5], OPUS-7 [9] and TED-59 [11]. We choose four different languages for IWSLT17, seven languages from OPUS-100 [27] for its zero-shot translation test sets and all languages for TED-59. All the training data are English-centric parallel data, which means either the source-side or target-side of the sentence pairs is English. Table 3 shows the detail statistics of the three datasets.

Evaluation. All the results are generated with beam search = 5. The translation quality is evaluated using the case-sensitive detokenized sacreBLEU[1] [20]. BLEU scores were averaged over test sets. Following Zhang et al. [27], we used LangAcc as a complementary evaluation metric for the off-target issue, which calculates the proportion of the translations in the desired target language[2]. Additionally, we use Win Rate (WR) to denote the proportion of translation directions that outperform the *ref* system in terms of BLEU.

Model and Training. We implement our MNMT models based on Fairseq [18]. We set the dimension of word embeddings and FFN layer to 512/2048. Embeddings were shared for the encoder, decoder and the output projection. As IWSLT17 is smaller, we use a 5-layer encoder and 5-layer decoder variation of Transformer-based model and train the models for 35 epochs [23] for IWSLT17. For OPUS-7 and TED-59, we follow the detail experiment settings in [11]. We train the models for 30 epochs, and use the last 5 epochs to be averaged for testing [11]. For all experiments, we set the default value of k to three.

Table 3. The detail statistics of the datasets.

Dataset	Type	#Language	#Supervised	#Zero-shot	#Training sentences
IWSLT17	English-centric	4	6	6	0.87M
OPUS-7	English-centric	7	12	30	12M
TED-59	English-centric	59	116	3306	5M

4.2   Overall Performance

We mainly carried out experiments on the IWSLT17 and OPUS-7 datasets to examine the effectiveness of our methods. Table 4 shows the experimental results on the two datasets. It demonstrates that our methods can significantly improve the translation performance, especially on zero-shot translation directions. While for both T-ENC and T-DEC strategies, our methods can gain consistently improvement. It proves that the integration of target language prototypes can eliminate the off-target phenomenon.

We further evaluated T-ENC+LTR and T-ENC+MPSR (the relative best systems in terms of average BLEU on all zero-shot translation directions on the IWSLT17 dataset) on the TED-59 dataset in order to compare our methods with Jin and Xiong [11] in the massively multilingual neural machine translation

---

[1] For all datasets, the signature is: BLEU+case.mixed+nrefs.1+smooth.exp+tok.{13a, zh,ja-mecab-0.996}+version.2.3.1, tok.zh and tok.ja-mecab-0.996 are only for Chinese and Japanese respectively.

[2] We employed langid.id toolkit for language identification.

scenario as shown in Table 5. Our main comparisons are made against LAA$_{enc.self}$ and Adapter$_{enc}$, which also incorporate additional parameters in the encoder. Furthermore, we consider the best-performing experiments for the LEE methods. The experimental results strongly demonstrate the effectiveness of our proposed methods.

4.3   Ablation Study

Effect of Multiple Target Language Prototypes. Previous studies generally use only one indicative representation to guide the models. Based on the visualization of language tokens' representations obtained by baseline model, as shown in Fig. 1, we use three as the default number of the target language representative prototypes. In this part, we consider the effectiveness of multiple indicative representations.

We conduct the experiment on IWSLT17 and OPUS-7 by average the three TLP value obtained through clustering technique as the sole TLP to guide the model for efficiency, while maintaining the entire framework for comparison. Table 6 reveals that the method using the average representation consistently yields inferior results compared to the corresponding methods utilizing three representations, particularly in zero-shot translation directions. It demonstration substantiates the effectiveness of incorporating multiple indicative features.

Effect of Lang-Attention Modules. As our methods introduce Lang-Attention modules into the baseline models to incorporate multiple target

Table 4. Experiment results on IWSLT17 and OPUS-7. T-DEC and T-ENC denote the addition of the target language tags to the decoder side and the encoder side, respectively. LTR and MPSR denote the use of language tag representation and the mean-pooling sentence representation for clustering to gain TLP, as explained in Sect. 3.2.

Dataset	Method	En → XX	XX → EN	Supervised		Zero-shot		
		BLEU	BLEU	BLEU	WR	BLEU	LangAcc	WR
IWSLT17	T-DEC	27.75	30.30	29.03	ref	10.78	70.54%	ref
	+LTR	28.18	30.63	29.41	50.00%	15.69	92.81%	100.00%
	+MPSR	27.74	30.30	29.02	33.33%	16.06	93.72%	100.00%
	T-ENC	28.18	30.63	29.41	83.33%	15.69	91.61%	100.00%
	+LTR	27.81	30.65	29.23	66.67%	**16.60**	**94.12%**	100.00%
	+MPSR	27.92	30.82	29.37	83.33%	16.39	93.77%	100.00%
OPUS-7	T-DEC	28.16	31.89	30.03	ref	13.03	84.78%	ref
	+LTR	28.37	32.15	30.26	100.00%	**15.16**	87.65%	**100.00%**
	+MPSR	28.23	31.97	30.10	58.33%	15.12	87.34%	93.33%
	T-ENC	28.11	31.77	29.94	33.33%	13.47	79.49%	53.33%
	+LTR	28.21	32.12	30.16	75.00%	15.11	86.99%	76.67%
	+MPSR	28.39	31.90	30.15	75.00%	14.88	86.33%	80.00%

Table 5. Experiment results on TED-59. The results for $LEE_{2,5}$, $LAA_{enc.self}$ and $Adapter_{enc}$ are taken from the experiments conducted by Jin and Xiong [11].

Dataset	Method	#Param	En → XX	XX → EN	Supervised	Zero-shot	
			BLEU	BLEU	BLEU	BLEU	LangAcc
TED-59	$LEE_{2,5}$	77M	21.26	23.79	22.53	9.82	74.44%
	$LAA_{enc.self}$	92M	20.11	21.13	20.62	6.30	71.92%
	$Adapter_{enc}$	92M	21.04	22.91	21.97	8.20	74.44%
	T-ENC	77M	21.28	23.34	22.21	8.00	60.67%
	+LTR	83M	21.78	23.75	22.77	**11.74**	**76.21%**
	+MPSR	83M	21.82	23.87	22.85	11.36	75.96%

Table 6. Experimental results on the exploration about effect of multiple target language prototypes. "+AVG" means the TLP gained by average the corresponding three target language prototypes obtained by the clustering techniques.

Dataset	Method	En → XX	XX → EN	Supervised	Zero-shot	
		BLEU	BLEU	BLEU	BLEU	LangAcc
IWSLT17	T-DEC	27.75	30.30	29.03	10.78	70.54%
	+LTR	28.18	30.63	29.41	15.69	92.81%
	+AVG	27.80	30.43	29.11	14.87	89.92%
	+MPSR	27.74	30.30	29.02	**16.06**	**93.72%**
	+AVG	27.83	30.64	29.23	15.03	89.11%
	T-ENC	28.18	30.63	29.41	15.69	91.61%
	+LTR	27.81	30.65	29.23	**16.60**	**94.12%**
	+AVG	27.91	30.54	29.23	16.40	93.03%
	+MPSR	27.92	30.82	29.37	16.39	93.77%
	+AVG	28.11	30.82	29.47	16.27	93.40%
OPUS-7	T-DEC	28.16	31.89	30.03	13.03	84.78%
	+LTR	28.37	32.15	30.26	**15.16**	**87.65%**
	+AVG	27.76	31.52	29.64	14.39	87.25%
	+MPSR	28.23	31.97	30.10	15.12	87.34%
	+AVG	27.77	31.54	29.66	14.58	87.05%
	T-ENC	28.11	31.77	29.94	13.47	79.49%
	+LTR	28.21	32.12	30.16	**15.11**	**86.99%**
	+AVG	28.39	31.99	30.19	14.72	84.26%
	+MPSR	28.39	31.90	30.15	14.88	86.33%
	+AVG	28.22	31.93	30.07	14.38	83.45%

language prototypes, we conduct experiments on IWSLT17 to further analyse the effect of additional Lang-Attention modules. First, we consider the perfor-

Table 7. Experimental results on the effect of additional attention modules. "DECi" and "ENCi" means only adding the Lang-Attention (LA) module in the i-th Decoder Layer and the i-th Encoder Layer, respectively. "ALL" means adding the Lang-Attention modules in all decoder layers or all encoder layers. "Decoder Layer" means extend the baseline architecture with an additional decoder layer. "Encoder Layer" means extend the baseline architecture with an additional encoder layer.

Dataset	Method	#Param	En → XX	XX → EN	Supervised	Zero-shot	
			BLEU	BLEU	BLEU	BLEU	LangAcc
IWSLT17	T-DEC	46M	27.75	30.30	29.03	10.78	70.54%
	+ Decoder Layer	51M	28.06	30.31	29.19	12.08	74.93%
	+ LA in DEC1	47M	27.88	30.15	29.01	13.62	84.93%
	+ LA in DEC2	47M	27.83	30.43	29.13	14.05	87.87%
	+ LA in DEC3	47M	27.72	30.37	29.04	14.65	89.20%
	+ LA in DEC4	47M	27.98	30.24	29.11	15.26	90.46%
	+ LA in DEC5	47M	27.83	30.51	29.17	15.10	90.04%
	+ LA in ALL	52M	28.18	30.63	29.41	**15.69**	**92.81%**
	T-ENC	46M	28.18	30.63	29.41	15.69	91.61%
	+ Encoder Layer	54M	26.97	28.69	27.82	15.37	95.16%
	+ LA in ENC1	47M	27.89	30.36	29.12	16.22	93.76%
	+ LA in ENC2	47M	27.92	30.55	29.24	16.29	93.76%
	+ LA in ENC3	47M	28.05	30.57	29.31	16.29	93.74%
	+ LA in ENC4	47M	27.95	30.32	29.14	16.29	**94.22%**
	+ LA in ENC5	47M	27.97	30.50	29.23	16.08	94.05%
	+ LA in ALL	52M	27.81	30.65	29.23	**16.60**	94.12%

mance is effected by the increasing amount of parameters. Thus, we compare the deeper Transformer with 6 encoder layers or 6 decoder layers (the Baseline models on IWSLT17 only contain 5 encoder layers and 5 decoder layers). Second, we further analyses the additional attention modules of different locations contribute to the performance. Therefore, we introduce the module to different encoder or decoder layers. The experimental results shown in Table 7. It demonstrates that with only one additional Lang-Attention module, our methods can perform better than the Baseline model and the deeper one, while maintaining less parameters. Another conclusion is that the Lang-Attention adding to the top layer (e.g. in fourth or fifth layer) contributes more to alleviate the off-target phenomenon.

5   Conclusion

In this work, we focus on enhancing the language accuracy of fully shared multilingual neural machine translation models to improve their zero-shot translation performance. Based on our practice, we argue that the representations of prepended language tokens are overly affected by their context information, resulting in potential information loss of language tokens and leading to insufficient indicative ability. Therefore, the language-specific signals in prepended language tokens are not adequate to guide the MNMT models to translate into right directions. Start from this inspiration, we further propose a two-stage approach to introduce multiple target language prototypes into the baseline models to guide the translation direction. The experimental results demonstrate that our method consistently improves translation quality across diverse multilingual datasets. Further analyses show the effectiveness of our method in alleviating the off-target translation issue and improving the translation quality in both zero-shot and supervised directions.

Acknowledgement. This work was supported by the Key Support Project of NSFC-Liaoning Joint Foundation (No. U1908216), and the Project of Research and Development for Neural Machine Translation Models between Cantonese and Mandarin (No. WT135-76). We thank all anonymous reviewers for their valuable suggestions on this work.

References

1. Aharoni, R., Johnson, M., Firat, O.: Massively multilingual neural machine translation. In: Proceedings of the NAACL (2019)
2. Al-Shedivat, M., Parikh, A.: Consistency by agreement in zero-shot neural machine translation. In: Proceedings of the NAACL (2019)
3. Arivazhagan, N., Bapna, A., Firat, O., Aharoni, R., Johnson, M., Macherey, W.: The missing ingredient in zero-shot neural machine translation (2019)
4. Arivazhagan, N., et al.: Massively multilingual neural machine translation in the wild: findings and challenges (2019)
5. Cettolo, M., et al.: Overview of the IWSLT 2017 evaluation campaign. In: Proceedings of the 14th International Conference on Spoken Language Translation (2017)
6. Chen, L., Ma, S., Zhang, D., Wei, F., Chang, B.: On the off-target problem of zero-shot multilingual neural machine translation. In: Proceedings of the ACL Findings (2023)
7. Firat, O., Cho, K., Bengio, Y.: Multi-way, multilingual neural machine translation with a shared attention mechanism. In: Proceedings of the NAACL (2016)
8. Gu, J., Wang, Y., Cho, K., Li, V.O.: Improved zero-shot neural machine translation via ignoring spurious correlations. In: Proceedings of the ACL (2019)
9. Gu, S., Feng, Y.: Improving zero-shot multilingual translation with universal representations and cross-mapping. In: Proceedings of the EMNLP Findings (2022)
10. Ha, T.L., Niehues, J., Waibel, A.: Toward multilingual neural machine translation with universal encoder and decoder. In: Proceedings of the 13th International Conference on Spoken Language Translation (2016)
11. Jin, R., Xiong, D.: Informative language representation learning for massively multilingual neural machine translation. In: Proceedings of the COLING (2022)
12. Johnson, M., et al.: Google's multilingual neural machine translation system: enabling zero-shot translation. Trans. Assoc. Comput. Linguist. **5**, 339–351 (2017)
13. Kong, X., Renduchintala, A., Cross, J., Tang, Y., Gu, J., Li, X.: Multilingual neural machine translation with deep encoder and multiple shallow decoders. In: Proceedings of the EACL (2021)

14. Lloyd, S.: Least squares quantization in PCM. IEEE Trans. Inf. Theor. **28**, 129–137 (1982)
15. Malaviya, C., Neubig, G., Littell, P.: Learning language representations for typology prediction. In: Proceedings of the EMNLP (2017)
16. Oncevay, A., Haddow, B., Birch, A.: Bridging linguistic typology and multilingual machine translation with multi-view language representations. In: Proceedings of the EMNLP (2020)
17. Östling, R., Tiedemann, J.: Continuous multilinguality with language vectors. In: Proceedings of the EACL (2017)
18. Ott, M., et al.: fairseq: a fast, extensible toolkit for sequence modeling. In: Proceedings of the NAACL (2019)
19. Pham, N.Q., Niehues, J., Ha, T.L., Waibel, A.: Improving zero-shot translation with language-independent constraints. In: Proceedings of the Fourth Conference on Machine Translation (Volume 1: Research Papers) (2019)
20. Post, M.: A call for clarity in reporting BLEU scores. In: Proceedings of the Third Conference on Machine Translation: Research Papers (2018)
21. Reimers, N., Gurevych, I.: Sentence-BERT: sentence embeddings using Siamese BERT-networks. In: Proceedings of the EMNLP (2019)
22. Sachan, D., Neubig, G.: Parameter sharing methods for multilingual self-attentional translation models. In: Proceedings of the Third Conference on Machine Translation: Research Papers (2018)
23. Vaswani, A., et al.: Attention is all you need. In: Proceedings of the NeurIPS (2017)
24. Wu, L., Cheng, S., Wang, M., Li, L.: Language tags matter for zero-shot neural machine translation. In: Proceedings of the ACL Findings (2021)
25. Xu, H., Liu, Q., van Genabith, J., Xiong, D.: Modeling task-aware MIMO cardinality for efficient multilingual neural machine translation. In: Proceedings of the ACL (2021)
26. Yin, Y., Li, Y., Meng, F., Zhou, J., Zhang, Y.: Categorizing semantic representations for neural machine translation. In: Proceedings of the COLING (2022)
27. Zhang, B., Williams, P., Titov, I., Sennrich, R.: Improving massively multilingual neural machine translation and zero-shot translation. In: Proceedings of the ACL (2020)

Extending DenseHMM with Continuous Emission

Klaudia Balcer[(✉)] and Piotr Lipinski

Computational Intelligence Research Group, Institute of Computer Science,
University of Wrocław, Wrocław, Poland
{klaudia.balcer,piotr.lipinski}@cs.uni.wroc.pl

Abstract. Traditional Hidden Markov Models (HMM) allow us to discover the latent structure of the observed data (both discrete and continuous). Recently proposed DenseHMM provides hidden states embedding and uses the co-occurrence-based learning schema. However, it is limited to discrete emissions, which does not meet many real-world problems. We address this shortcoming by discretizing observations and using a region-based co-occurrence matrix in the training procedure. It allows embedding hidden states for continuous emission problems and reducing the training time for large sequences. An application of the proposed approach concerns recommender systems, where we try to explain how the current interest of a given user in a given group of products (current state of the user) influences the saturation of the list of recommended products with the group of products. Computational experiments confirmed that the proposed approach outperformed regular HMMs in several benchmark problems. Although the emissions are estimated roughly, we can accurately infer the states.

Keywords: HMM · embedding · co-occurrence · emission discretization

1 Introduction

Hidden Markov Models (HMMs) continue to appeal to scientists, although their history dates back to the 20th century. They are appreciated for their simplicity, solid theoretical understanding, reliability, and ease of interpretation, especially in terms of explainable artificial intelligence. Recently reported applications concern motion recognition [10], finance [11], medical engineering [1], and others.

Such a wide range of applications also exposes the limitations of the traditional model, such as unacceptably long learning time for large datasets (caused by the quadratic complexity of the Baum-Welch learning procedure) and the assumption of emission distribution coming from a known parametrized family. A number of extensions and enhancements of the standard model have been proposed to address those issues [5,8]. One of the recent is DenseHMM [9], which introduces continuous dense representations (embeddings) of discrete observations and states, and applies an efficient learning algorithm based on direct co-occurrence optimization. Embedding provides a dense representation of the discrete hidden states and values which might be interpreted geometrically. However, this novelty limits to discrete emission, which is not suitable for many applications. Our work focused mainly on extending the idea of dense representations of hidden states on continuous observations.

In this paper, we extend DenseHMMs for continuous observations with Gaussian distributions and propose a suitable learning algorithm that, first, endeavors to cluster observations, and next, evaluates a region-based co-occurrence matrix in a similar way to DenseHMMs. One of the interesting applications (which is not feasible for the original DenseHMMs because of its limitation to discrete emission) of the proposed approach may concern recommender systems, where we try to explain how the current state of the interests of a given user in a given group of products influences the saturation of recomendations, i.e. saturation of the list of recommended products with the group of products.

This paper is structured in the following manner: Sect. 2 introduces the Gaussian Dense Hidden Markov Model. Sections 3 and 4 propose two learning algorithms: one based on a regular EM algorithm and one using on a region-based co-occurrence matrix, along with the computational experiments. Section 5 concludes our research.

2   Gaussian Dense Hidden Markov Model

A Hidden Markov Model (HMM) is defined by two stochastic processes: discrete hidden states $\{X_t\}_{t\in\mathbb{N}}$ and observations $\{Y_t\}_{t\in\mathbb{N}}$.

The Markov process $\{X_t\}_{t\in\mathbb{N}}$ is a discrete stochastic process of n hidden states $(q_1, q_2, \ldots, q_n)$. The first hidden state has the starting distribution $X_1 \sim \pi$. At each next timestamp, the random variable depends only on the one in the previous timestamp: $\mathbb{P}(X_t|X_{t-1}, X_{t-2}, \ldots, X_1) = \mathbb{P}(X_t|X_{t-1})$ (the process follows the Markov assumption). The probabilities of transiting from q_i to q_j are gathered in a transition matrix $A_{(i,j)} = \mathbb{P}(X_t = q_j|X_{t-1} = q_i)$ for each $i, j \in \{1, 2, \ldots, n\}$.

The distribution of observed random variables $\{Y_t\}_{t\in\mathbb{N}}$ depends on current state: $\mathbb{P}(Y_t|Y_{t-1}, Y_{t-2}, \ldots, Y_1; X_t, X_{t-1}, \ldots, X_1) = \mathbb{P}(Y_t|X_t)$; this property is called output independence assumption. In a basic HMM, observations are discrete. We will consider GaussianHMM, a model with emission coming from the (multivariate) normal distribution $Y_t|X_t = q_i \sim \mathcal{N}(\mu_i, \Sigma_i)$ with probability distribution function $\mathbb{P}(Y_t = y|X_t = q_i) = \phi_i(y)$. The standard learning algorithm for HMMs is the commonly used Baum-Welch (also called Foward-Backward) algorithm, a special case of Expectation-Maximization [4].

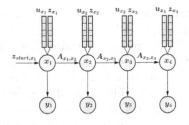

Fig. 1.   Schema   of   Gaussian-DenseHMM

Recently, Sicking et al. [9] came up with embedding basic HMM and defined DenseHMM. They enriched a basic HMM with dense representations of discrete states and observations. Each possible value was represented by two vectors (one incoming and one outgoing). They adopted two learning schemas for the purpose of training the model: the Expectation-Maximization algorithm (with a gradient-based part in the M-step) and the co-occurrence-based learning algorithm (unconstrained and faster).

In this paper, we propose to combine the continuous emission of GaussianHMMs with the dense representation of hidden states in DenseHMMs. Our model, called **Gaussian Dense Hidden Markov Model**, has continuous emission and embedded hidden states. We will use in-coming embedding vectors u_i and outgoing embedding vectors z_i of each hidden state x_i and starting vector z_{start}, as presented at the schema in Fig. 1, to obtain transitions probabilities via softmax kernelization:

$$\pi_i := \mathbb{P}(X_1 = q_i) = \frac{exp\langle z_{start}, u_i \rangle}{\sum_{k=1}^n exp\langle z_{start}, u_k \rangle}, \tag{1a}$$

$$A_{(i,j)} := \mathbb{P}(X_t = q_j | X_{t-1} = q_i) = \frac{exp\langle z_i, u_j \rangle}{\sum_{k=1}^n exp\langle z_i, u_k \rangle}. \tag{1b}$$

We adopt both mentioned learning algorithms to our model. The modified Forward-Backward algorithm is presented in Sect. 3. The benefit of embedding GaussianHMM is using continuous representation in a fast, unconstrained, co-occurrence-based learning algorithm. In Sect. 4, we propose to use discretization and adapt the co-occurrence-based learning algorithm to our model by defining the region-based co-occurrence.

3   Basic Learning Algorithm

One of the basic but also very powerful concepts of parameter estimation is maximizing the likelihood function. In GaussianDenseHMMs we assume the distribution of the process. However, we do not know the values of the hidden states. For the case of missing values, the **E**xpectation-**M**aximization (EM) algorithm was provided [2]. The idea behind it is to iteratively: provide expectations of the missing values using the current parameter estimation (E) and find new parameters maximizing the likelihood function (fed with the expectations from the previous step). EM adapted to the case of HMMs is called the Baum-Welch algorithm.

Loss. For simplicity of the notation we will denote the whole paramater set as $\Theta = \left(A, \pi, (\phi_i)_{i=1}^n \right)$. The loss function maximized in the Baum-Welch algorithm is not directly the likelihood, but the **E**vidence **L**ower **BO**und (ELBO) of the logarithm of the likelihood function. We present it in the form of three summands

$\mathcal{L}(Y,\Theta) = \mathcal{L}_1(Y,\Theta) + \mathcal{L}_2(Y,\Theta) + \mathcal{L}_3(Y,\Theta)$:

$$\mathcal{L}_1(Y,\Theta) = \sum_{i=1}^{n}\sum_{j=1}^{n}\sum_{t=2}^{T} \mathbb{P}(X_t = i, X_{t-1} = j) \cdot log\left(\frac{exp\langle z_j, u_i\rangle}{\sum_{k=1}^{n} exp\langle z_j, u_k\rangle}\right) \qquad (2a)$$

$$\mathcal{L}_2(Y,\Theta) = \sum_{i=1}^{n}\sum_{t=1}^{T} \mathbb{P}(X_t = i) \cdot log\left(\phi_i(Y_t)\right) \qquad (2b)$$

$$\mathcal{L}_3(Y,\Theta) = \sum_{i=1}^{n} \mathbb{P}(X_1 = i) \cdot log\left(\frac{exp\langle z_{start}, u_i\rangle}{\sum_{k=1}^{n} exp\langle z_{start}, u_k\rangle}\right) \qquad (2c)$$

Please note that $\mathbb{P}(X_t = i, X_{t-1} = j)$ and $\mathbb{P}(X_t = i)$ are unknown, as we do not observe $(X_t)_{t\in\mathbb{N}}$.

Step E. In the first step of each iteration, we calculate how probable each state and transition is, assuming the current parameter value. To run those calculations, first, we need to introduce two recursive formulas to calculate the likelihood of a sentence:

– forward probability:

$$\alpha_1(i) = \frac{exp\langle z_{start}, u_i\rangle}{\sum_{k=1}^{n} exp\langle z_{start}, u_k\rangle} \cdot \phi_i(Y_1) \qquad (3a)$$

$$\alpha_t(i) = \sum_{j=1}^{n} \alpha_{t-1}(j) \cdot \frac{exp\langle z_j, u_i\rangle}{\sum_{k=1}^{n} exp\langle z_j, u_k\rangle} \cdot \phi_i(Y_t) \qquad (3b)$$

$$\mathbb{P}(Y) = \sum_{i=1}^{n} \alpha_T(i) \qquad (3c)$$

– backward probability:

$$\beta_T(i) = 1 \qquad (4a)$$

$$\beta_t(i) = \sum_{j=1}^{n} \frac{exp\langle z_i, u_j\rangle}{\sum_{k=1}^{n} exp\langle z_i, u_k\rangle} \cdot \phi_j(Y_{t+1}) \cdot \beta_{t+1}(j) \qquad (4b)$$

$$\mathbb{P}(Y) = \sum_{j=1}^{n} \frac{exp\langle z_{start}, u_j\rangle}{\sum_{k=1}^{n} exp\langle z_{start}, u_k\rangle} \cdot \phi_j(Y_1) \cdot \beta_1(j) \qquad (4c)$$

To estimate the current expectations, we will use the probabilities:

$$\gamma_t^{(k)}(i) = \mathbb{P}(X_t = i|\Theta^{(k-1)}) = \frac{\alpha_t(i)\beta_t(i)}{\sum_{j=1}^{n} \alpha_t(j)\beta_t(j)} \qquad (5a)$$

$$\xi_t^{(k)}(i,j) = \mathbb{P}(X_t = j, X_{t-1} = i|\Theta^{(k-1)}) = \frac{\alpha_{t-1}(i) \cdot \frac{exp\langle z_i, u_j\rangle}{\sum_{k=1}^{n} exp\langle z_i, u_k\rangle} \cdot \phi_j(Y_t) \cdot \beta_t(j)}{\sum_{j=1}^{n} \alpha_t(j)\beta_t(j)}$$
$$(5b)$$

Step M. In the M step, we inject the calculated expectations into the ELBO (Eq. 2) and find parameters maximizing the modified function:
$$z^{(k)}, u^{(k)}, \mu^{(k)}, \Sigma^{(k)} = \arg\max_{z,u,\mu,\Sigma} \mathcal{L}_1^*(Y, \Theta^{(k)}) + \mathcal{L}_2^*(Y, \Theta^{(k)}) + \mathcal{L}_3^*(Y, \Theta^{(k)})$$

New emission distribution parameters can be calculated analytically. Embedding vectors are updated through a gradient-based procedure like SGD [7] (used in the implementation) or Adam [6].

Algorithm 1. EM

Initialize $\mu^{(0)}$, $\Sigma^{(0)}$, $u^{(0)}$, $z^{(0)}$
k = 1
while not converged **do**
    for $t = 1 \rightarrow T$, $i = 1 \rightarrow n$ **do**
        calculate $\alpha_t^{(k)}(i)$ using Eq. 3
    end for
    for $t = T \rightarrow 1$, $i = 1 \rightarrow n$ **do**
        calculate $\beta_t^{(k)}(i)$ using Eq. 4
    end for
    for $t = 1 \rightarrow T$, $i = 1 \rightarrow n$ **do**
        calculate $\gamma_t^{(k)}(i)$ using Eq. 5a
        for $j = 1 \rightarrow n$ **do**
            calculate $\xi_t^{(k)}(i, j)$ using Eq. 5b
        end for
    end for
    Update $\mu^{(k)}$, $\Sigma^{(k)}$ analytically, $z^{(k)}$, $u^{(k)}$ using SGD (in *max_iter* steps)
    k += 1
end while

Algorithm. First, we initialize $\mu^{(0)}$, $\Sigma^{(0)}$ (usually using k-means like in the standard implementation of HMMs[1]) and $u^{(0)}$, $z^{(0)}$ (randomly, for example using standard normal distribution). Then we can start the iterative learning process. In each iteration, we do the E step consisting of calculating forward probabilities $\alpha_t^{(k)}(i)$, backward probabilities $\beta_t^{(k)}(i)$, and current state and transition probabilities estimate $\gamma_t^{(k)}(i)$ and $\xi_t^{(k)}(i, j)$, respectively. Then, we can update the model parameters estimates in the M step. We update $\mu^{(k)}$ and $\Sigma^{(k)}$ using the analytical formulas and run an iterative, gradient-based procedure to update the embedding vectors $z^{(k)}$ and $u^{(k)}$. We repeat the E and M steps until we meet the convergence criterion, which can be the maximum number of iterations or a small improvement in likelihood.

4 Region-Based Co-occurrence Learning Algorithm

In this section, we propose a learning algorithm based on region-based **CO-OC**currence (rCOOC). A major drawback of the EM algorithm is that it goes

---

[1] https://hmmlearn.readthedocs.io/en/latest/, Last accessed 31 Mar 2023.

through all the observations at each iteration. Thus, its complexity grows with the data size. We address this issue with the rCOOC-based learning algorithm. The main idea is to provide a summary of the data (the co-occurrence matrix) and look for consistent parameters. It has been shown to work well for HMMs and DenseHMMs which are both models with discrete emissions. Now, we use the discretization of continuous emission to adopt it for our model.

Region-Based Co-occurrence Matrix. First, we need to provide the initial values of the parameters similar to the EM algorithm (emission parameters from the k-means algorithm and randomly selected embedding). Calculating the co-occurrences of observations may be seen as part of the initialization process. To specify the co-occurrences for GaussianDenseHMM we need to establish the space division into regions $r_1, \ldots, r_m$. Then, we can provide region-based emission probability matrix $B_{i,j}^r = \int_{r_j} \phi_i(y) dy$ (Fig. 2).

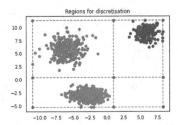

Fig. 2. Regions obtained in the discretization procedure and from the true model.

The regions can be selected ambiguous. On the one hand, we want to estimate the parameters as accurately as possible. On the other hand, fine-grained division makes the task unnecessarily complicated by distinguishing many separate regions within the same distribution. We decided to keep the division simple and make it maximally informative. First, we divide only the minimal hyper cuboid containing all observations (with some small predefined margin). We reuse the k-means model from initializing the emission parameters and build a simple decision tree classifier on the labels obtained. We use the division rules from nodes to divide the observation space.

Estimating the precise emission parameters is part of the further learning procedure, so we are satisfied with this inaccurate but reasonable space division. However, for specific data, one could propose a custom, more optimal discretizing technique.

Now, we are able to provide the definition of the rCOOC matrix. It is a matrix describing probabilities of co-occurring observations from given regions. For a collection $r_1, \ldots, r_m \subset \mathbb{R}^p$ such that for each Y_t exists only r_i containing Y_t, we define the **region-based co-occurrence matrix** (rCOOC matrix) as a probability matrix $\Omega_{i,j} = \mathbb{P}(Y_t \in r_i, Y_{t+1} \in r_j)$ for $i, j = 1, \ldots, m$.

Loss. We will consider the ground truth (empirical) rCOOC matrix Ω^{GT} summarising the training sequences:

$$\Omega_{i,j}^{GT} = \#\{t : Y_t \in r_i, Y_{t+1} \in r_j\}/(T-1) \tag{6}$$

and one calculated based on current parameter estimation $\Omega^{(k)}$, where the upper index (k) denotes the current iteration. The aim of the algorithm is to find parameters resulting in a rCOOC matrix possibly close to the empirical one.

To provide a formula to derive the rCOOC matrix from model parameters, we need to first define the state-co-occurrence matrix T, which is a stochastic matrix $T_{k,l} = \mathbb{P}(X_t = q_k, X_{t+1} = q_l)$. Using the above definition we can present the matrix $\Omega = (B^r)^T T B^r$. If we assume that starting probability is the stationary distribution of the transition matrix ($\pi_i = \sum_j A_{ji}\pi_j$), we can provide an analytical formula $T_{k,l} = A_{kl}\pi_k$. From now on, we do not parametrize the starting probability. Inserting the formula for T in the definition of Ω, we obtain the following result:

$$\Omega_{i,j} = \sum_{k=1}^{n} \sum_{l=1}^{n} \pi_k b'_{i,k} a_{k,l} b'_{l,j} \tag{7}$$

Using the model parameter estimates from k-th iteration we get the matrix $\Omega^{(k)}$. To learn the distribution we minimize the difference between the empirical and estimated the rCOOC matrix:

$$\mathcal{L}^{rCOOC}(Y, \Theta^{(k)}) = \|\Omega^{GT} - \Omega^{(k)}\|^2 \tag{8}$$

using a gradient-descent procedure like SGD [7] or Adam [6].

Algorithm 2. Co-occurrence based learning

---

model = k-means trained on Y
Initialize $\mu^{(0)}$, $\Sigma^{(0)}$, $u^{(0)}$, $z^{(0)}$, k = 0
labels = Y clustering from k-means
tree = Decision Tree trained on labels
nodes = partition rules from tree
Y_{disc} = regions of observations splitted on nodes
Ω^{gt} = the rCOOC matrix obtained using Eq. 6
while convergence **do**
    Calculate $\Omega^{(k)}$, $\mathcal{L}^{rCOOC}(Y, \Theta^{(k)})$ according to Eq. 7, 8
    Calculate the derivatives of $\mathcal{L}^{rCOOC}(Y, \Theta^{(k)})$ with respect to $\Theta^{(k)}$
    Update parameters $\Theta^{(k)}$ following a gradient-based procedure; k += 1
end while
Calculate matrix A and vector π from embedding according to Eq. 1b, 1a

---

Algorithm. First, we cluster the observations using k-means. We use the model not only for $\mu^{(0)}$, and $\Sigma^{(0)}$ initialization, but also for building a decision tree. The classifier rules are then propagated as observation region boundaries. Stating the partition, we can transform the training data into discrete vector Y_{disc} and calculate the region co-occurrence matrix Ω^{GT}. We also need to initialize the embedding vectors $u^{(0)}$, $z^{(0)}$. After this extensive preprocessing, we learn the parameters straight-forward by iteratively updating the parameters in a gradient-based procedure. The convergence criterion is met after a specified number of iterations or when the loss improvement is small.

We investigate our model in its properties and its application in recommendation saturation. In the first experiment, we intend to show the results quality of the proposed approach with respect to the time consumption for different numbers of hidden states. The goal is to check whether we speed up the learning while getting sufficiently good results. After that, we confront our approach with real-world-related task. We will model genre saturation in recommendations. We refer to regular GaussianHMM, implemented in the hmmlearn library, as the baseline solution.

4.1 Properties Study

To study the capabilities of GaussianDenseHMM, we used synthetic data, as shown in Fig. 3, and compare the model to regular GaussianHMM. We repeated the run 10 times and compare the results obtained in likelihood, accuracy, co-occurrence, and time consumption. The goal is to check whether we can speed the learning up for big datasets while getting good results.

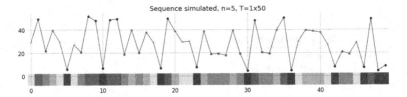

Fig. 3. Data sampled from GaussianHMM with $n = 5$ hidden states. The color marks the hidden states.

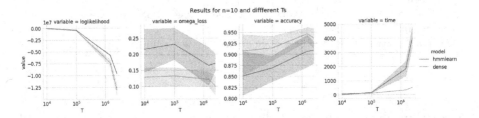

Fig. 4. Results (loglikelihood, region-co-occurrence loss, accuracy, time consumption) for n = 10 and T = 10 × 10000, 10 × 100000, 10 × 1500000, 10 × 2500000 for standard GaussianHMM and GaussianDenseHMM.

The time consumed by the rCOOC-based learning algorithm clearly depends on the hyper-parameters set. In this experiment, we used a non-optimized parameter set (learning rate = 0.003, number of iterations = $n \cdot 10000$, where n is the number of hidden states). Also, the HMM implementation requires specifying

Table 1. Results of synthetic experiment for $n = 5, 7, 10, 20$ and $T = 10 \times 10000, 10 \times 10000, 10 \times 100000, 10 \times 100000$, respectively

n		Standard GaussianHMM	GaussianDenseHMM
5	loglikelihood	$(-3.02 \pm 0.13) \times 10^4$	$(-3.24 \pm 0.16) \times 10^4$
	region-co-occur.	$(1.26 \pm 0.10) \times 10^{-2}$	$(6.31 \pm 0.07) \times 10^{-2}$
	accuracy	$(10.00 \pm 0.00) \times 10^{-1}$	$(9.90 \pm 0.01) \times 10^{-1}$
	time	$(5.00 \pm 0.60) \times 10^{-1}$	$(3.91 \pm 0.11) \times 10^1$
7	loglikelihood	$(-3.42 \pm 0.15) \times 10^4$	$(-3.90 \pm 0.53) \times 10^4$
	region-co-occur.	0.10 ± 0.10	$(1.70 \pm 0.03) \times 10^{-2}$
	accuracy	$(9.39 \pm 0.64) \times 10^{-1}$	$(9.84 \pm 3.10) \times 10^{-1}$
	time	1.91 ± 0.80	$(5.60 \pm 0.23) \times 10^1$
10	loglikelihood	$(-3.95 \pm 0.12) \times 10^6$	$(-5.08 \pm 0.76) \times 10^6$
	region-co-occur.	$(2.29 \pm 0.78) \times 10^{-1}$	$(1.22 \pm 0.45) \times 10^{-1}$
	accuracy	$(8.60 \pm 0.06) \times 10^{-1}$	$(9.32 \pm 0.39) \times 10^{-1}$
	time	$(3.674 \pm 0.053) \times 10^3$	$(2.449 \pm 0.063) \times 10^2$
20	loglikelihood	$(-4.73 \pm 0.44) \times 10^6$	$(-6.10 \pm 0.80) \times 10^6$
	region-co-occur.	$(3.00 \pm 0.30) \times 10^{-1}$	$(8.67 \pm 0.05) \times 10^{-2}$
	accuracy	$(7.70 \pm 4.10) \times 10^{-1}$	$(9.08 \pm 0.65) \times 10^{-1}$
	time	$(1.59 \pm 0.77) \times 10^4$	$(1.45 \pm 0.15) \times 10^3$

the number of iterations parameter. However, the learning can be exited earlier if the convergence condition is met (we used $2000 \cdot n$ iterations and tolerance $= 0.01$). Data size was denoted as $T = s \times t$, where s is the number of sequences and t is the length of each sequence.

Figure 4 shows the differences between the models for a fixed number of states and a different size of training data for 10 replications of the experiment. Table 1 presents the exact results for a different number of states. Each of the models beats one another in its loss function. For small data sizes, our implementation is relatively slow. However, when the complexity of the task grows (greater number of states or bigger data), the regular GaussianHMM using the EM algorithm becomes very time-consuming.

4.2   Recommendation Saturation Simulation

The practical importance of GaussianDenseHMM was studied in the context of recommender systems (RS). RS tries to suggest to a given user the most accurate products from a given set of available products. In some sense, an RS is a type of black box, because it is usually unclear how different types of user activities affect the results of the RS. Particularly: how the RS reflects the real (unknown, hidden) user interests. In order to study such a research question, we consider a dataset prepared on the basis of the MovieLens20M dataset [3].

We focus on a selected group of users and their interests in a particular group of products (we consider a group of users/products only in order to avoid over-fitting issues and/or insufficient data problems, so our studies may also consider single user/product if a sufficient amount of data is available). In some periods, these products may be additionally advertised, which increases interest in them. In some periods, these products may be unpopular, e.g. due to seasonality or weather reasons, which decreases interest in them.

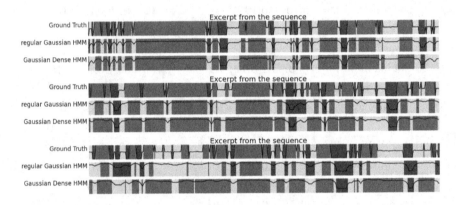

Fig. 5. Ground truth states with unprocessed data and results obtained from the baseline model and GaussianDenseHMM with preprocessed data for different lengths of the period considered in the experiment $\Delta T = 3, 5, 7$. The similarity to the unprocessed data is blurred with the growth of ΔT, which results in worse consistency to baseline states (80.88–84.02, 69.00–74.60, 60.35–60.60 percent accuracy, respectively).

In each time t, the users may be active in the e-commerce system and strongly interested in the products, e.g. due to some advertising of the products (state $x_t = 2$); active and regularly interested (state $x_t = 1$) or inactive/weakly interested (state $x_t = 0$). Let y_t denote the number of positive interactions with the products by the users at time t (e.g. the number of product page clicks, the number of product buys, or the number of product positive ratings).

In order to measure how the RS reflects the number of positive interactions in the list of recommended products, we transform y_t by the Hill saturation curve $h(y) = \frac{y^p}{k + y^p}$ into the saturation of the list of recommended products by the particular group of products.

In our experiments, the number y_t of positive interactions came from a 3-state HMM, and the Hill saturation curve was estimated on the Movielens20M dataset: after selecting randomly 10 users, we have provided recommendations for them 1000 times while injecting high ratings for items from the specified genre (simultaneously avoiding the mean converging to 5). Using the recommendations

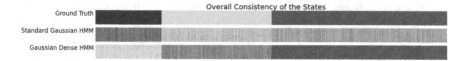

Fig. 6. True states in comparison to states obtained from regular GaussianHMM and GaussianDenseHMM, omitting the sequence order, for $\Delta T = 7$. The picture shows the overall state consistency. As the data is blurred, the trained models seem to distinguish low and high interest while mixing the medium state with one of the extreme states. Differences between models results occur due to convergence to local optima of both learning algorithms.

obtained from FunkSVD[2], we calculated the saturation of recommendations and fitted the Hills equation to the average saturation sequence. In order to denoise the user interest data, we preprocessed y_t with a moving average of length $\Delta T = 3, 5,$ or 7 days.

Our approach was used to discover the hidden states x_t of the users' interests in the particular group of products. Figure 5 and 6 as well as Table 2 and 3 present the results of comparison GaussianDenseHMMs with regular GaussianHMMs for 3 different lengths ΔT of moving average in the preprocessing. GaussianDenseHMMs outperformed regular GaussianHMMs in all cases, in terms of accuracy as well as computing time.

Table 2. Detailed results of recommendation saturation modeling experiment for $\Delta T = 3$: confusion matrix (true states in rows, predicted states in columns), sensitivity and specificity (for state indicators), accuracy, loglikelihood, and model training time

state		Standard Gaussian HMM			GaussianDenseHMM		
		0	1	2	0	1	2
confusion matrix	0	597086	597086	0	458191	365425	112565
	1	139716	1423215	0	0	1342466	220465
	2	1633	475022	2021233	0	100243	2397645
sensitivity		63.78%	91.06%	80.92%	48.94%	85.89%	95.99%
specificity		96.52%	76.29%	100.0%	100.0%	86.44%	86.67%
accuracy		80.88%			84.02%		
loglikelihood		−9553362.99			−47334974.89		
time [s]		300.24			147.45		

[2] https://sifter.org/simon/journal/20061211.html. Last accessed 2 Dec 2022.

Table 3. Summary of recommendation saturation modeling experiment

ΔT		accuracy	loglikelihood	time [s]
3	regular GaussianHMM	80.88%	−9553362.99	300.24
	GaussianDenseHMM	84.02%	−47334974.89	147.45
5	regular GaussianHMM	69.00%	−10089991.59	313.05
	GaussianDenseHMM	74.60%	−57123400.16	122.54
7	regular GaussianHMM	60.60%	−10352963.04	246.09
	GaussianDenseHMM	60.35%	−57771627.53	163.75

5   Conclusions

In response to the discussion in the paper proposing DenseHMMs, we came up with a generalization of this model to continuous emissions and adopted both proposed learning schemas. The idea of embedding hidden states gives an improvement by providing an additional, dense representation of the hidden states. Incorporating discretization and region-based co-occurrence allowed us to adopt the fast, unconstrained learning algorithm to a continuous problem.

We have evaluated our model to present its theoretical properties and capabilities to work in a real-world-related scenario. GaussianDenseHMM tended to be less time-consuming for big benchmarks. It also worked well with smaller amounts of data in terms of state decoding accuracy. In the scenario set in context or recommender systems, our model tended to work comparably well in a shorter time.

In future work, we suggest exploring the model's abilities for very large numbers of states as well as speeding it up for small data and studying other discretization techniques.

Acknowledgement. This work was supported by the Polish National Science Centre (NCN) under grant OPUS-18 no. 2019/35/B/ST6/04379.

References

1. Boeker, M., Hammer, H.L., Riegler, M.A., Halvorsen, P., Jakobsen, P.: Prediction of schizophrenia from activity data using hidden Markov model parameters. Neural Comput. Appl. **35**, 5619–5630 (2022)
2. Dempster, A.P., Laird, N.M., Rubin, D.B.: Maximum likelihood from incomplete data via the EM algorithm. J. Roy. Stat. Soc. Ser. B (Methodol.) **39**, 1–38 (1977)
3. Harper, F.M., Konstan, J.A.: The MovieLens datasets: history and context. ACM Trans. Interact. Intell. Syst. **5**, 19:1–19:19 (2016)
4. Hsiao, R., Tam, Y., Schultz, T.: Generalized Baum-Welch algorithm for discriminative training on large vocabulary continuous speech recognition system. In: IEEE International Conference on Acoustics, Speech, and Signal Processing, pp. 3769–3772 (2009)

5. Huang, K., Fu, X., Sidiropoulos, N.D.: Learning hidden Markov models from pairwise co-occurrences with application to topic modeling. In: International Conference on Machine Learning, vol. 80, pp. 2073–2082 (2018)

6. Kingma, D.P., Ba, J.: Adam: a method for stochastic optimization. In: International Conference for Learning Representations (2015)

7. Lecun, Y., Bottou, L., Bengio, Y., Haffner, P.: Gradient-based learning applied to document recognition. Proc. IEEE **86**, 2278–2324 (1998)

8. Lorek, P., Nowak, R., Trzcinski, T., Zieba, M.: FlowHMM: flow-based continuous hidden Markov models. In: Advances in Neural Information Processing Systems, vol. 35, pp. 8773–8784 (2022)

9. Sicking, J., Pintz, M., Akila, M., Wirtz, T.: DenseHMM: learning hidden markov models by learning dense representations. In: Advances in Neural Information Processing Systems (2020)

10. Zhang, F., Han, S., Gao, H., Wang, T.: A Gaussian mixture based hidden Markov model for motion recognition with 3D vision device. Comput. Electr. Eng. **83**, 106603 (2020)

11. Zhang, M., Jiang, X., Fang, Z., Zeng, Y., Xu, K.: High-order hidden Markov model for trend prediction in financial time series. Phys. A **517**, 1–12 (2019)

An Efficient Enhanced-YOLOv5 Algorithm for Multi-scale Ship Detection

Jun Li[ID], Guangyu Li[ID], Haobo Jiang[(✉)], Weili Guo[(✉)], and Chen Gong

Key Laboratory of Intelligent Perception and Systems for High-Dimensional
Information of Ministry of Education, School of Computer Science and Engineering,
Nanjing University of Science and Technology, Nanjing, China
{jun_li,guangyu.li2017,jiang.hao.bo,wlguo,chen.gong}@njust.edu.cn

Abstract. Ship detection has gained considerable attentions from
industry and academia. However, due to the diverse range of ship types
and complex marine environments, multi-scale ship detection suffers from
great challenges such as low detection accuracy and so on. To solve
the above issues, we propose an efficient enhanced-YOLOv5 algorithm
for multi-scale ship detection. Specifically, to dynamically extract two-
dimensional features, we design a MetaAconC-inspired adaptive spatial-
channel attention module for reducing the impact of complex marine
environments on large-scale ships. In addition, we construct a gradient-
refined bounding box regression module to enhance the sensitivity of loss
function gradient and strengthen the feature learning ability, which can
relieve the issue of uneven horizontal and vertical features in small-scale
ships. Finally, a Taylor expansion-based classification module is established which increases the feedback contribution of gradient by adjusting the first polynomial coefficient vertically, and improves the detection
performance of the model on few sample ship objects. Extensive experimental results confirm the effectiveness of the proposed method.

Keywords: Multi-scale Ship Detection · Improved YOLOv5
Network · Attention Module

1 Introduction

Ship detection is a critical aspect of maritime supervision and plays an essential role in intelligent maritime applications such as sea area monitoring, port
management, and safe navigation [24]. In recent years, various methods such
as foreground segmentation [4,16,25], background subtraction [3,22], and horizon detection [12,24] have been widely explored and have made considerable
progress. However, traditional ship detection methods often lack robustness and
may have limited applicability in the presence of complex noise interference.

Supported by the National Science Fund of China under Grant 62006119.
J. Li and G. Li — Equal contributions.

Meanwhile, owing to the development of deep learning in object detection, deep learning-based object detectors have achieved significant advancements. For example, Faster r-cnn [17] is a classic two-stage detection method that employs a region proposal network to generate detection boxes directly. SSD [10] enhances the detection accuracy of multi-scale objects by conducting object detection on multiple feature layers. CenterNet [5] is a detection method that detects the center point and size of an object without anchor. The YOLO [1,13–15,23] series are classic single-stage object detection methods that extract multi-scale features via a backbone network and a feature pyramid network, while introducing an anchor frame mechanism to enhance the model's robustness.

Inspired by these deep learning-based detection methods above, there is a growing research efforts towards deep learning-based ship detection. Region proposal network-based methods [7,9] and regression-based methods [2,19] have made certain progress. However, various issues, such as false detection and missed detection, persist in ship detection due to factors like the influence of background noise on the sea surface, the uneven distribution of horizontal and vertical features of ships, and the different sizes of ships.

To relieve the issues above, we propose a novel efficient enhanced-YOLOv5 algorithm for multi-scale ship detection. Specifically, in order to mitigate the issue that complex marine environments disrupting large-scale ships, we propose a MetaAconC-inspired dynamic spatial-channel attention module that extracts two-dimensional features, mitigating the environmental impact on large-scale ships. Aiming at the problem of uneven horizontal and vertical features of small-scale ships, we design a gradient-refined bounding box regression module to increase the gradient sensitivity, enhancing the learning ability of the algorithm on small-scale ship features. In order to relieve the challenge that sensitivity of the cross entropy function to class imbalance, we establish a Taylor expansion-based classification module, by adjusting the first polynomial coefficient vertically to increase the contribution feedback of the gradient, improving the detection performance of the model on few sample ship objects. To summarize, our main contributions are as follows:

- We propose a novel efficient enhanced-YOLOv5 algorithm for multi-scale ship detection, where a MetaAconC-inspired dynamic spatial-channel attention module is designed to mitigate the influence of complex marine environments on large-scale ships.
- To mitigate the problem of uneven horizontal and vertical features of small-scale ships, we design an effective gradient-refined bounding box regression module to enhance the learning ability of the algorithm on small-scale ship features.
- To further relieve the challenge that sensitivity to class imbalance, we also construct a Taylor expansion-based classification module to increase feedback contribution and improve the detection performanceon few sample ships.

2 Method

2.1 Overall Framework

The structure of an efficient enhanced-YOLOv5 algorithm is shown in Fig. 1. The algorithm comprises several components: a backbone network that extracts features from three different scales. The MetaAconC-inspired dynamic spatial-channel attention module which located in the three feature processing channels behind the backbone network to focus on the feature refinement of multi-scale ships, and a feature pyramid network for feature enhancement. Finally, the detection heads generate the final predictions, and our proposed modules, namely the gradient-refined bounding box regression module and the Taylor expansion-based classification module improve accuracy through gradient calculations and backpropagation during training.

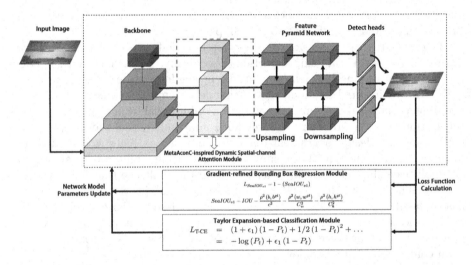

Fig. 1. The pipeline of an efficient enhanced-YOLOv5 algorithm framwork.

2.2 MetaAconC-Inspired Dynamic Spatial-Channel Attention Module

Due to the large span of large-scale ships in the image, its learned feature distribution tends to be largely split, which may potentially confuse the object semantics, thereby presenting limited detection accuracy. Especially in the complex marine environments, the semantic information of ships is easily polluted by background noise, which makes it difficult to learn. To mitigate the influence of complex marine environments on large-scale ships, we propose a MetaAconC-inspired dynamic spatial-channel attention module as shown in Fig. 2. In detail,

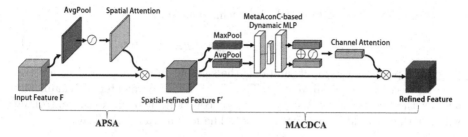

Fig. 2. The overview of the MetaAconC-inspired dynamic spatial-channel attention module. APSA denotes the average pooling-based spatial attention module. MACDCA denotes the MetaAconC-inspired dynamic channel attention module.

the average pooling-based spatial attention module obtains the intra-channel relationship of input features. Secondly, the MetaAconC-inspired dynamic channel attention module dynamically summarize the spatial relationships of features. As such, our module effectively learns the multi-dimensional features information of ships and the impact of complex marine environments on the noise of large ships is mitigated.

Average Pooling-Based Spatial Attention Module. The module integrates ship characteristic information between different channels, and further eliminates the negative impact of complex marine environment on large-scale ships through the similar semantic characteristics of background noise in channel dimensions. After obtaining the feature $F \in R^{H \times W \times C}$ through the CSPDarkNet53 backbone network from the input image, we input F into the average pooling-based spatial attention module to obtain global information by utilizing global averaging pooling of channel dimensions, followed by sigmoid function to produce spatial-refined attention weight $\in R^{H \times W \times 1}$, which is then multiplied with the input feature F to obtain spatial-refined feature F', which is fed into the next module.

MetaAconC-Inspired Dynamic Channel Attention Module. Since background noise is not invariable in spatial dimension, and a variety of unnecessary noise will be formed in complex marine environment, we designed the module to dynamically adjust attention mode, better learn ship characteristics, and effectively reduce the interference of dynamic background noise. This module conducts global average pooling and maximum pooling of spatial dimensions to F', and add the results through a two-layer neural network based on the MetaACON function [11] and sigmoid activation function to obtain channel-refined attention weight $\in R^{1 \times 1 \times C}$. Finally, we multiply this weight with feature F' to obtain refined feature. The smooth maximum function has been utilized to expand the Maxout function, resulting in the Acon series activation functions. The MetaAcon function allows the adaptive activation of neurons through the modification of a parameter, denoted by γ, which is defined as follows:

$$f_{(x)} = (p_1 - p_2)\, x \cdot \sigma\left(\gamma\,(p_1 - p_2)\, x\right) + p_2 x, \tag{1}$$

where x represents the input, and σ is the sigmoid function. p_1 and p_2 are two channel-wise learnable parameters. The channel-wise parameter γ dynamically adjusts the activation of neurons through convolution operations, controlling whether they should be activated or not. The formula for γ is given by:

$$\gamma = \sigma W_1 W_2 \sum_{h=1}^{H} \sum_{w=1}^{W} x_{c,h,w}, \tag{2}$$

where W_1 and W_2 represent two 1×1 convolution layers.

2.3 Gradient-Refined Bounding Box Regeression Module

The CIOU loss [27] is a widely used bounding box regression loss, which plays a crucial role in the YOLOv5 algorithm. However, CIOU loss has two main drawbacks in correspondence learning. (i) First, the current approach only takes into account the aspect ratio of the bounding box, without considering the actual height and width of the object. Ships are not all regular rectangles, and the aspect ratio of different ship types varies greatly. For example, the shape of the fishing boat is very slender, small in height but large in width. However, In order to better accommodate tourists, ships such as passenger ships and cruise ships are very tall compared to their width. As a consequence, the differences in aspect ratios of ships can hinder the accurate fitting of ships with varying shapes especially small-scale ships, leading to misidentification and missed detections. (ii) Second, the loss function gradient remains constant, which renders the model insensitive to fitting multi-scale objects, making small-scale ship detection more challenging.

To mitigate the issue (i), we divide the aspect ratio into height and width, and calculate them respectively [26]. In this way, the fitting direction of the regression module is closer to the shape of the ship. The width-height loss directly minimizes the width-height difference between the target box and the bounding box so that the model can better fit the ships with different shapes, which is defined as follows:

$$L_{SeaIOU_{v1}} = 1 - (SeaIOU_{v1}), \tag{3}$$

where $SeaIOU_{v1}$ is defined as:

$$SeaIOU_{v1} = IOU - \frac{\rho^2\,(b, b^{gt})}{c^2} - \frac{\rho^2\,(w, w^{gt})}{C_w^2} - \frac{\rho^2\,(h, h^{gt})}{C_h^2}, \tag{4}$$

where b and b^{gt} represent the center points of the bounding box and target box, respectively. $\rho(\cdot)$ represents the Euclidean distance. c represents the area of the smallest enclosing box that covers both boxes. C_w and C_h are the width and height of the minimum circumscribed frame that covers both boxes.

To mitigate the issue (ii), we establish a gradient-refined bounding box regression module that increases the gradient sensitivity of the loss function. Specifically, we modify the invariance of the gradient by applying a logarithmic function. The absolute gradient value decreases with the increase of the overlap, which is more favorable for bounding box regression. As such, when the distance between the boxes is far away, its gradient absolute value is larger, which is more conducive to the detection of small-scale ships. This approach enhances the contribution of small-scale ships to the feature learning ability of the model. The formula for the modified loss function is defined as:

$$L_{SeaIOU_{v2}} = \alpha \cdot \ln \alpha - \alpha \cdot \ln(\beta + (SeaIOU_{v1})), \tag{5}$$

where α and β represent parameters that control the gradient sensitivity of the loss function.

2.4 Taylor Expansion-Based Classification Module

The cross entropy loss is a popular classification loss, which plays a crucial role in the YOLOv5 algorithm, which is defined as:

$$L_{CE} = - \log(P_t) = \sum_{j=1}^{\infty} 1/j \, (1 - P_t)^j = (1 - P_t) + 1/2 \, (1 - P_t)^2 \ldots, \tag{6}$$

where P_t is the model's prediction probability of the ground-truth class.

However, it is sensitive to class imbalance. The cross-entropy loss assumes that the classes are balanced, which may result in the model becoming biased towards the majority class and failing to capture the features of the minority class. Specifically, In the training process, it back-propagates each type of ship according to the same contribution, making the model more inclined to learn the ship object with a large number of samples. However, the learning efficiency of the ship object with a few sample is very low, which greatly limits the detection performance of ships with few samples. In the application of ship detection, the sample number of ships is very uneven. Some ship types are very common, during training, more samples can be provided for the model to learn features and improve the detection performance. However, some ship types are not as common as the above ships, and the number of their samples is very small. It is difficult for the ship detection model to get enough learning samples in the training stage, so it is difficult to learn the characteristics of ships with few samples. Expanding datasets is a feasible approach, but it costs a lot. Therefore, it is necessary to optimize the training strategy.

To mitigate the issue, we establish a Taylor expansion-based classification module which presents the loss function as a linear combination of polynomial function. We get its gradient formula based on the cross entropy loss function, which is shown as:

$$- \frac{dL_{CE}}{dP_t} = \sum_{j=1}^{\infty} (1 - P_t)^{j-1} = 1 + (1 - P_t) + (1 - P_t)^2 \ldots \tag{7}$$

From the above formula, it can be seen that the first term of the cross entropy loss function is the largest, which is 1. The subsequent terms are smaller and smaller, which means that the first term contributes the most to the gradient gain. By adjusting the first polynomial coefficient vertically [8], we increase the feedback contribution of cross-entropy gradient. This module further strengthens the fitting ability and alleviates the sensitivity to class imbalance, which is defined as:

$$L_{T-CE} = (1 + \epsilon_1)(1 - P_t) + 1/2(1 - P_t)^2 + \ldots = -\log(P_t) + \epsilon_1(1 - P_t), \quad (8)$$

where ϵ_1 represents the parameter we adjusted in the first polynomial coefficient.

In this way, the sensitivity of the classification module to the number of samples is improved, the problem of low gradient gain of few sample ships is alleviated, and the detection performance of the model for few sample ship templates is enhanced

3   Experiments

3.1   Experimental Settings

Dataset. In this paper, we evaluate the performance of the proposed method on the SeaShips dataset [20], a well-known large-scale and precisely annotated maritime surveillance dataset released by Wuhan University in 2018. The dataset collected by the coastal land-based camera in Hengqin, Zhuhai, including 6 types of ships with different sizes, contains 31,455 images, 7,000 of which are publicly available. We divide the pictures according to the official scale. The training set and the validation set are 1750, and the remaining 3500 are used as the test set. The detection difficulties include ship size change, complex background interference and so on. In this dataset, the size of the fishing boat object is small, the sample size of the passenger ship is small, thus the detection accuracy of the algorithm for them is one of the main indicators to verify the performance of the model to small-scale ship object and few sample ship object.

Evaluation Indicators. We adopt evaluation indicators of COCO dataset, including $mAP_{0.5}$, $AP_{0.5}$, $mAP_{0.75}$, and $AP_{0.75}$. AP (Average Precision) is the area enclosed by the X-axis and Y-axis plots using *Recall* and *Precision* respectively. $AP_{0.5}$ and $AP_{0.75}$ are APs at IoU threshold of 0.5 and 0.75, respectively. For multi-object detection, each object would have an AP value first, and then take the weighted average to obtain mAP (Mean Average Precision).

Implementation Details. For our experiments, one GeForce RTX 2080ti GPU card is used, and the CUDA version is 10.0. The cuDNN version is 7.5.1, and the PyTorch version is 1.2.0. All models are trained for 300 epochs with batch size of 4, an initial learning rate of 1e-2, which is then reduced to a minimum of 1e-4 using a cosine annealing algorithm. We utilize the sgd optimizer with momentum 0.937 and weight decay 5e-4. All models are deployed according to the above Settings. YOLOv5 network is the original network of our method. We set $\alpha = 5$, $\beta = 4$ and $\epsilon_1 = 1$. In order to demonstrate the efficacy of the proposed method, we conduct an experimental comparison with the other conventional object detection methods on the Seaships dataset.

Table 1. Detection results on the Seaships dataset. It shows $mAP_{0.5}$ and $AP_{0.5}$ in each class. The bold number has the highest score in each column.

Model	$mAP_{0.5}$	Bulk cargo carrier	Container ship	Fishing boat	General cargo ship	Ore carrier	Passenger ship
Faster r-cnn[1]	0.949	0.958	**0.994**	0.906	0.966	0.950	0.917
Faster r-cnn[2]	0.946	0.927	0.990	0.917	0.969	0.938	0.933
SSD300[2]	0.935	0.949	0.987	0.888	0.962	0.930	0.893
SSD300[3]	0.891	0.918	0.967	0.809	0.925	0.898	0.831
YOLOv3	0.941	0.952	0.983	0.923	0.968	0.943	0.878
YOLOv4	0.921	0.901	0.975	0.901	0.937	0.918	0.894
Shao	0.874	0.876	0.903	0.783	0.917	0.881	0.886
YOLOv5	0.952	0.953	0.988	0.940	0.974	0.935	0.922
Ours	**0.966**	**0.961**	0.991	**0.956**	**0.982**	**0.951**	**0.952**

[1] ResNet50 [6] is selected as the backbone network.
[2] VGG16 [21] is selected as the backbone network.
[3] MobileNetv2 [18] is selected as the backbone network.

Table 2. Detection results on the Seaships dataset. It shows $mAP_{0.75}$ and $AP_{0.75}$ in each class. The bold number has the highest score in each column.

Model	$mAP_{0.75}$	Bulk cargo carrier	Container ship	Fishing boat	General cargo ship	Ore carrier	Passenger ship
Faster r-cnn[1]	0.658	0.608	0.806	0.553	0.767	0.576	0.636
Faster r-cnn[2]	0.650	0.537	0.852	0.582	0.731	0.569	0.629
SSD300[2]	0.673	0.704	0.903	0.509	0.789	0.609	0.525
SSD300[3]	0.491	0.499	0.745	0.286	0.594	0.445	0.373
YOLOv3	0.631	0.612	0.860	0.474	0.727	0.647	0.470
YOLOv4	0.506	0.487	0.673	0.360	0.579	0.472	0.467
YOLOv5	0.762	0.769	0.920	**0.669**	**0.850**	**0.721**	0.646
Ours	**0.785**	**0.788**	**0.940**	0.660	0.848	0.706	**0.767**

[1] ResNet50 [6] is selected as the backbone network.
[2] VGG16 [21] is selected as the backbone network.
[3] MobileNetv2 [18] is selected as the backbone network.

3.2 Quantitative Analysis

As shown in Table 1, we conduct an experimental comparison of $mAP_{0.5}$ and $AP_{0.5}$ with the other eight classical object detection methods on the Seaships dataset. The proposed method achieves a high $mAP_{0.5}$ of 96.6%, with the 3 ship classes having the highest AP values. In particular, for passenger ship with a smaller sample, $AP_{0.5}$ reaches 95.2%, an improvement of 3% over the original network. In addition, For small-scale fishing boat, $AP_{0.5}$ reaches 95.6%, an increase of 1.6% over the original network. Compared to Faster r-cnn [17] with various backbone networks, our proposed method alleviates the interference of complex environment by adding the proposed attention module, with $mAP_{0.5}$

increasing by 1.7% and 1.9%. Compared to SSD [10] with various backbone networks, our proposed method further enhance multi-scale features, with $mAP_{0.5}$ increasing by 3.1% and 7.5%. Specifically, for fishing boat, $AP_{0.5}$ increases by 6.8% and 14.7%. Compared to the YOLO series networks [1,15], our proposed method improves the feature description power of the model for multi-scale ships and achieves higher detection accuracy. Compared to Shao [19], our proposed method increases $mAP_{0.5}$ by 9.2% by reducing the complex environment interference and sample imbalance sensitivity with the proposed regeression and classification module. Particularly for fishing boat and container ship, $AP_{0.5}$ increases by 17.3% and 8.8%.

In order to further verify the performance of our proposed model more strictly, we experimentally compare $mAP_{0.75}$ and $AP_{0.75}$ with five other classical object detection methods on the Seaships dataset. Table 2 presents the performance of different methods on Seaships, our proposed method also achieves the highest detection performance of 78.5%, an improvement of 2.3% over the original network. It's worth noting that passenger ship with fewer samples, $AP_{0.75}$ reaches 76.7%, an improvement of 12.1% over the original network. In conclusion, our proposed method is more effective than other classical methods for improving the accuracy of multi-scale ship detection.

Table 3. Ablation experimental results of module on seaships Dataset.

Model	$mAP_{0.5}$	Bulk cargo carrier	Container ship	Fishing boat	General cargo ship	Ore carrier	Passenger ship
YOLOv5	0.952	0.953	0.988	0.940	0.974	0.935	0.922
+b	0.954	0.944	0.988	0.948	0.977	0.943	0.921
+b + c	0.963	0.958	**0.993**	0.956	0.976	0.948	0.950
+a + b + c (ours)	**0.966**	**0.961**	0.991	**0.956**	**0.982**	**0.951**	**0.952**

3.3   Ablation Studies

Table 3 displays the effect of the three proposed modules on the performance of the method. To ensure fair comparison, we use the same experimental setup for all the methods. a represents the MetaAconC-inspired dynamic spatial-channel attention module, b represents the gradient-refined bounding box regression module and c represents the Taylor expansion-based classification module.

The original network YOLOv5 achieves the $mAP_{0.5}$ of 95.2%. After b is added, The method improves small-scale ship detection performance by increasing gradient sensitivity, resulting in an $AP_{0.5}$ increase of 0.8% for fishing boat. Then the method enhances accuracy further by adding c to reduce class imbalance sensitivity, yielding an overall $mAP_{0.5}$ improvement of 0.9%, and $AP_{0.5}$ improved by 2.9% for passenger ships with fewer samples. After adding a, by focusing on the extraction of ship characteristics, the influence of complex Marine environment is weakened. Our method combined with the proposed attention module raises the $mAP_{0.5}$ to 96.6%, 1.4% higher than the original network YOLOv5. Experimental results show that our modules significantly improve ship detection performance across different sizes and ship types.

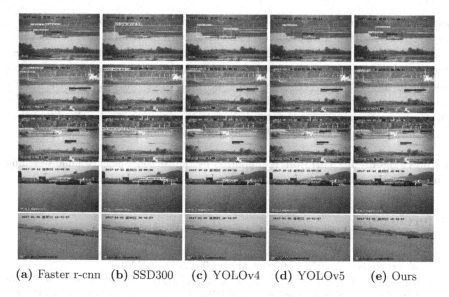

(a) Faster r-cnn   (b) SSD300   (c) YOLOv4   (d) YOLOv5   (e) Ours

Fig. 3. Qualitative comparison of different methods on Seaships.

3.4 Qualitative Analysis

Figure 3 illustrates the ship detection performance of our proposed method and the other classical methods under various complex conditions. From the first line, it can be seen that in the occlusion case, Faste r-cnn gets a duplicate bounding boxes due to the region proposal network. SSD300, YOLOv4 and YOLOv5 all miss the bulk cargo carrier that is hiding from each other. And from the fourth line, except our method, the other object detectors do not detect the obscured passenger ship. As can be seen from the second and third lines, when multi-scale ships exist at the same time, Faste r-cnn also produces redundant detection boxes. SSD and YOLOv4 fail to detect small fishing ships. Our original network YOLOv5 can not handle the detection of multi-scale ships well, resulting in the detection of small ships, while missing the detection of large ships across the whole map. By adding the proposed attention module, our proposed method alleviates the problem of semantic information fragmentation of large ships and detects these ships well. As can be seen from the fifth line, for the small ship object scenario, the position of the detection box of Faste r-cnn is offset and SSD failes to detect the small-scale fishing ship. It can be concluded that our proposed method effortlessly handles these situations with ease.

4   Conclusion

In this paper, we have proposed an efficient enhanced-YOLOv5 algorithm for multi-scale ship detection. Our approach consists of three components, specifically a metaAconC-inspired dynamic spatial-channel attention module has been designed to reduce the impact of complex marine environments on large-scale ships. Also, We have mitigated the issue of uneven horizontal and vertical features of small-scale ships by constructing a gradient-refined bounding box regression module. Moreover, we have proposed a Taylor expansion-based classification module to alleviate the sensitivity to class imbalance and improve the detection performance to few sample ships. The experimental results demonstrate the effectiveness of our proposed method. In future work, our model should further improve its ability to detect small-scale ships in complex marine environments.

Acknowledgement. This work is supported by the National Science Fund of China under Grant 62006119.

References

1. Bochkovskiy, A., Wang, C.Y., Liao, H.Y.M.: Yolov4: Optimal speed and accuracy of object detection. arXiv preprint arXiv:2004.10934 (2020)
2. Chen, J., Xie, F., Lu, Y., Jiang, Z.: Finding arbitrary-oriented ships from remote sensing images using corner detection. IEEE Geosci. Remote Sens. Lett. **17**(10), 1712–1716 (2019)
3. Chen, Z., Yang, J., Kang, Z.: Moving ship detection algorithm based on gaussian mixture model. In: 2018 3rd International Conference on Modelling, Simulation and Applied Mathematics (MSAM 2018), pp. 197–201. Atlantis Press (2018)
4. Dong, C., Feng, J., Tian, L., Zheng, B.: Rapid ship detection based on gradient texture features and multilayer perceptron. Infrared Laser Eng. **48**(10), 1026004–1026004 (2019)
5. Duan, K., Bai, S., Xie, L., Qi, H., Huang, Q., Tian, Q.: CenterNet: keypoint triplets for object detection. In: Proceedings of the IEEE/CVF International Conference on Computer Vision, pp. 6569–6578 (2019)
6. He, K., Zhang, X., Ren, S., Sun, J.: Deep residual learning for image recognition. In: Proceedings of the IEEE Conference on Computer Vision and Pattern Recognition, pp. 770–778 (2016)
7. Kim, K., Hong, S., Choi, B., Kim, E.: Probabilistic ship detection and classification using deep learning. Appl. Sci. **8**(6), 936 (2018)
8. Leng, Z., Tan, M., Liu, C., Cubuk, E.D., Shi, X., Cheng, S., Anguelov, D.: PolyLoss: a polynomial expansion perspective of classification loss functions. arXiv preprint arXiv:2204.12511 (2022)
9. Li, Q., Mou, L., Liu, Q., Wang, Y., Zhu, X.X.: HSF-NET: multiscale deep feature embedding for ship detection in optical remote sensing imagery. IEEE Trans. Geosci. Remote Sens. **56**(12), 7147–7161 (2018)
10. Liu, W., et al.: SSD: Single Shot MultiBox Detector. In: Leibe, B., Matas, J., Sebe, N., Welling, M. (eds.) ECCV 2016. LNCS, vol. 9905, pp. 21–37. Springer, Cham (2016). https://doi.org/10.1007/978-3-319-46448-0_2
11. Ma, N., Zhang, X., Liu, M., Sun, J.: Activate or not: Learning customized activation. In: Proceedings of the IEEE/CVF Conference on Computer Vision and Pattern Recognition, pp. 8032–8042 (2021)
12. Prasad, D.K., Prasath, C.K., Rajan, D., Rachmawati, L., Rajabaly, E., Quek, C.: Challenges n video based object detection in maritime scenario using computer vision. arXiv preprint arXiv:1608.01079 abs/1608.01079 (2016)

13. Redmon, J., Divvala, S., Girshick, R., Farhadi, A.: You only look once: unified, real-time object detection. In: Proceedings of the IEEE Conference on Computer Vision and Pattern Recognition (CVPR), pp. 779–788 (2016)

14. Redmon, J., Farhadi, A.: Yolo9000: better, faster, stronger. In: Proceedings of the IEEE Conference on Computer Vision and Pattern Recognition, pp. 6517–6525 (2017)

15. Redmon, J., Farhadi, A.: Yolov3: an incremental improvement. arXiv preprint arXiv:1804.02767 (2018)

16. Ren, L., Ran, X., Peng, J., Shi, C.: Saliency detection for small maritime target using singular value decomposition of amplitude spectrum. IETE Tech. Rev. **34**(6), 631–641 (2017)

17. Ren, S., He, K., Girshick, R., Sun, J.: Faster R-CNN: towards real-time object detection with region proposal networks. Adv. Neural. Inf. Process. Syst. **28**, 91–99 (2015)

18. Sandler, M., Howard, A., Zhu, M., Zhmoginov, A., Chen, L.C.: MobileNetV2: inverted residuals and linear bottlenecks. In: Proceedings of the IEEE Conference on Computer Vision and Pattern Recognition, pp. 4510–4520 (2018)

19. Shao, Z., Wang, L., Wang, Z., Du, W., Wu, W.: Saliency-aware convolution neural network for ship detection in surveillance video. IEEE Trans. Circuits Syst. Video Technol. **30**(3), 781–794 (2019)

20. Shao, Z., Wu, W., Wang, Z., Du, W., Li, C.: SeaShips: a large-scale precisely annotated dataset for ship detection. IEEE Trans. Multimedia **20**(10), 2593–2604 (2018)

21. Simonyan, K., Zisserman, A.: Very deep convolutional networks for large-scale image recognition. arXiv preprint arXiv:1409.1556 (2014)

22. Wang, B., Dong, L., Zhao, M., Xu, W.: Fast infrared maritime target detection: binarization via histogram curve transformation. Infrared Phys. Technol. **83**, 32–44 (2017)

23. Wang, C.Y., Bochkovskiy, A., Liao, H.Y.M.: YOLOv7: trainable bag-of-freebies sets new state-of-the-art for real-time object detectors. arXiv preprint arXiv:2207.02696 (2022)

24. Ye, C., Lu, T., Xiao, Y., Lu, H., Qunhui, Y.: Maritime surveillance videos based ships detection algorithms: a survey. J. Image Graphics **27**, 2078–2093 (2022)

25. Zhang, Y., Li, Q.Z., Zang, F.: Ship detection for visual maritime surveillance from non-stationary platforms. Ocean Eng. **141**, 53–63 (2017)

26. Zhang, Y.F., Ren, W., Zhang, Z., Jia, Z., Wang, L., Tan, T.: Focal and efficient IOU loss for accurate bounding box regression. Neurocomputing **506**, 146–157 (2022)

27. Zheng, Z., Wang, P., Liu, W., Li, J., Ye, R., Ren, D.: Distance-IOU loss: faster and better learning for bounding box regression. In: Proceedings of the AAAI Conference on Artificial Intelligence, vol. 34, pp. 12993–13000 (2020)

Double-Layer Blockchain-Based Decentralized Integrity Verification for Multi-chain Cross-Chain Data

Weiwei Wei, Yuqian Zhou$^{(\boxtimes)}$, Dan Li$^{(\boxtimes)}$, and Xina Hong

College of Computer Science and Technology, Nanjing University of Aeronautics and Astronautics, Nanjing, China
`zhouyuqian@nuaa.edu.cn`, `lidansusu007@163.com`

Abstract. With the development of blockchain technology, issues like storage, throughput, and latency emerge. Multi-chain solutions are devised to enable data sharing across blockchains, but in complex cross-chain scenarios, data integrity faces risks. Due to the decentralized nature of blockchain, centralized verification schemes are not feasible, making decentralized cross-chain data integrity verification a critical and challenging problem. In this paper, based on the ideas of *"governing the chain by chain"* and *"double layer blockchain"*, we propose a double-layer blockchain-based decentralized integrity verification scheme. We construct a supervision-chain by selecting representative nodes from multiple blockchains, which is responsible for cross-chain data integrity verification and recording results. Specifically, our scheme relies on two consensus phases: integrity consensus for verification and block consensus for result recording. We also integrate a reputation system and an election algorithm within the supervision-chain. Through security analysis and performance evaluation, we demonstrate the security and effectiveness of our proposed scheme.

Keywords: Data integrity · Double-Layer blockchain · Cross chain · Decentralized verification · Multi-chain architecture

1 Introduction

Blockchain technology has gained significant attention across industries in recent years. Initially introduced as the underlying technology for cryptocurrency, blockchain has evolved into a disruptive innovation with the potential to revolutionize numerous sectors, including finance, supply chain management, healthcare, and more [14,18]. However, the widespread adoption of blockchain faces the scalability issue, which arises from the inherent design of traditional blockchain networks. As the number of participants and transactions increases, the single-chain architecture encounters limitations in terms of throughput, latency, and storage requirements.

To solve these problems, some researchers have introduced innovative solutions such as multi-chain architecture and cross-chain technology [6]. For example, Kang et al. propose a multi-chain federated learning (FL) framework, in

© The Author(s), under exclusive license to Springer Nature Singapore Pte Ltd. 2024
B. Luo et al. (Eds.): ICONIP 2023, LNCS 14452, pp. 264–279, 2024.
https://doi.org/10.1007/978-981-99-8076-5_19

which multiple blockchains are customized for specific FL tasks and individually perform learning tasks for privacy protection [7]. Multiple blockchains interact and collaborate with each other through cross-chain techniques, enabling a scalable, flexible, and communication-efficient decentralized FL system.

These multi-chain architectures all require data sharing among multiple block-chains. Existing researches [4,16,17] focus only on how to achieve cross-chain interaction and collaboration, without solving the problem of data integrity in multi-chain data sharing. Only when the integrity of the cross-chain data is confirmed can cross-chain applications effectively engage in data exchange and collaboration, thus achieving the objectives and advantages of a multi-chain architecture. Therefore, research is needed to ensure cross-chain data integrity among multiple chains.

The data integrity problem in the cloud storage environment has been extensively studied. The provable data possession (PDP) scheme, along with its various iterations [1,10,19], is widely used to address the data integrity problem in the cloud. However, the integrity of data sharing in a multi-chain environment is fundamentally different as: 1) In a multi-chain architecture, it is necessary to verify the integrity of cross-chain data distributed among multiple receiving blockchains for a specific piece of data. In contrast, in cloud storage scenario, the integrity verification focuses on data within a single cloud. 2) A multi-chain architecture consists of multiple decentralized blockchains, lacking the centralized control found in cloud storage scenario.

Additionally, to alleviate the heavy computational burden on users, various research studies employ third-party auditors (TPA) to check the data integrity on the untrusted cloud [5,11]. However, the centralized TPA, which can never be fully trusted, will weaken the decentralized nature of the blockchain.

In this paper, we propose a decentralized scheme based on the idea of double-layer blockchain [2] to ensure cross-chain data integrity across multiple blockchains. We utilize representative nodes from each blockchain in the multi-chain architecture to construct a supervision-chain, which is responsible for cross-chain data integrity verification. Additionally, based on the lightweight sampling method and the Boneh-Lynn-Shacham (BLS) signature, we provide a probabilistic integrity guarantee. Furthermore, we also provide detailed descriptions of the reputation system, node election algorithm, and block consensus process for the proposed supervision-chain. The main contributions of this paper are summarized as follows:

- This paper is the first to apply the idea of a double-layer blockchain to the field of cross-chain data integrity verification. Representative nodes are extracted from a multi-chain architecture to construct a supervision-chain, which is responsible for verifying the integrity of cross-chain data and recording the verification results.
- A cross-chain data integrity verification process is designed within the supervision-chain, where nodes collaborate with each other for decentralized verification. Additionally, leveraging lightweight sampling algorithms and BLS signature, we achieve an efficient verification process.

– To improve the efficiency of the block consensus process, this paper introduces a block consensus committee. Moreover, the reputation system of the
supervision-chain and election algorithm for the block consensus committee
are meticulously designed.

The remainder of this paper is organized as follows. Section 2 gives a brief
introduction to the preliminaries covered in this paper. Section 3 defines the
system model, threat models, and design goals. Section 4 presents the proposed
scheme. Section 5 conducts the security analysis. Section 6 evaluates the performance of the proposed cross-chain data integrity verification scheme. Section 7
concludes this paper and points out the future work.

2   Preliminaries

2.1   Bilinear Pairing

Bilinear pairing [12] is based on cryptography, which relies on a difficult hypothesis similar to the elliptic curve discrete logarithm problem, which can often be
used to reduce the problem in one group to an easier problem in another group.
Let G and G_T be two multiplicative cyclic groups of large prime order q. A
map function $e : G \times G \to G_T$ is a bilinear pairing only when it satisfies three
properties below:

– Bilinear: For $u, v \in G$ and $a, b \in Z_q^*$, $e\left(u^a, v^b\right) = e(u, v)^{ab}$;
– Non-Degeneracy: $e(g, g)$ is a generator of G_T;
– Computability: For $\forall u, v \in G$, there exists efficient algorithms to compute
  $e(u, v)$.

2.2   BLS Signature

The Boneh-Lynn-Shacham (BLS) signature [9] is a cryptographic scheme widely
used to help senders certificate their messages. It works on top of elliptic curves
and employs bilinear pairing to perform verification. Assume that a sender is
equipped with a public/private key pair (pk, sk) $\left(sk \in \mathbb{Z}_q^* \text{ and } pk = g^{sk}\right)$. To
generate a signature sig for a given message mes, the sender maps mes to the
elliptic curve with a secure hash function $hash()$. Then, it generates signature
sig from its private key, i.e., $sig = hash(mes)^{sk}$. A receiver can verify mes with
the bilinear mapping function $e()$ mentioned in Sect. 2.1 based on the sender's
public key pk and message signature sig. If Eq. (1) holds, the received message
mes is correct.

$$e(pk, hash(mes)) \overset{?}{=} e(g, sig) \tag{1}$$

BLS signature's security is ensured by the hardness of solving the Computational Diffie-Hellman (CDH) problem. In our scheme, each node in the
supervision-chain has a randomly chosen unique sk as its private key. Then,
its corresponding public key pk is generated by g^{sk}.

2.3   Verifiable Random Function

Verifiable random function (VRF) is a cryptographic function that provides pseudo-random and publicly verifiable values, e.g., the one introduced in [3] based on bilinear pairing. Specifically, given a random seed x, a user u equipped with a public/private key pair (pk, sk) can generate a random value $f_{sk}(x)$ by Eq. (2) and a tag $\pi_{sk}(x)$ by Eq. (3).

$$f_{sk}(x) = e(g, g)^{1/(x+sk)} \tag{2}$$

$$\pi_{sk}(x) = g^{1/(x+sk)} \tag{3}$$

Tag $\pi_{sk}(x)$ is used to prove the correctness of $f_{sk}(x)$. With both $f_{sk}(x)$ and $\pi_{sk}(x)$, a receiver can verify the correctness of $f_{sk}(x)$ based on u's public key pk. If both Eq. (4) and Eq. (5) hold, the random value $f_{sk}(x)$ is correctly generated by u.

$$e\left(g^x \times pk, \pi_{sk}(x)\right) \overset{?}{=} e(g, g) \tag{4}$$

$$e\left(g, \pi_{sk}(x)\right) \overset{?}{=} f_{sk}(x) \tag{5}$$

3   Problem Statement

3.1   System Model

In a multi-chain architecture with n consortium blockchains, the problem is that the blockchain which possesses the original data d intends to verify the integrity of the cross-chain data stored in the receiving blockchains. These representative nodes are selected from each blockchain to form a supervision-chain, which also is a consortium blockchain. We refer to the original blockchains in the multi-chain architecture as the sub-layer and the supervision-chain as the main layer, forming a double-layer framework [2]. In the supervision-chain, each node can read data from the blockchain within the sub-layer it belongs to. Therefore, cross-chain data integrity verification from each blockchain in the sub-layer can be done within the supervision-chain. There are n nodes in supervision-chain, denoted as $Node = \{node_i | 1 \le i \le n\}$. Each node $node_i \in Node$ has a public/private key pair $(pk_{node_i}, sk_{node_i})$ and is identified by its public key pk_{node_i}. Here, $sk_{node_i} = x_i \in Z_p^*, pk_{node_i} = g^{x_i} \in G$.

Assume that a blockchain in the sub-layer has sent data d by cross-chain method to each of $k(k < n)$ received blockchains, and the corresponding nodes in the supervision-chain denoted as $Node_d \subseteq Node$. The system model consists of four parts, including the sending-chain, receiving-chains, supervision-chain and other blockchains in multi-chain architecture. The system model is shown in Fig. 1.

Sending-Chain: The original data owner, who sends the data d to the receiving-chains using cross-chain methods, intends to verify the integrity of the cross-chain data d.

Receiving-Chain: Receiving-chain is the recipient of the cross-chain data, and it is subject to data integrity verification conducted by the supervision-chain.
Other Blockchain: The other blockchain in the multi-chain architecture act as participant in the construction of the supervision-chain.
Supervision-Chain: The supervision-chain is constructed by representative nodes from the blockchains in the multi-chain architecture and is responsible for verifying the integrity of cross-chain data in receiving-chains.

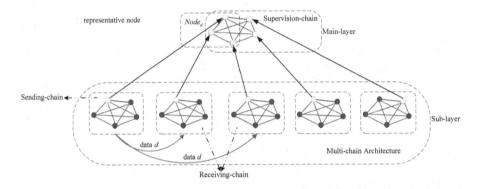

Fig. 1. System model

3.2   Threat Models

Assume that the representative node of the blockchain sending the data d is $node_s \in Node_d$, and the representative node of the blockchain receiving the data d is $node_r \in Node_d$. During the process of data integrity verification in the supervision-chain, the following threats exist:

- **Unexpected Failures.** Faults such as hardware failures, software exceptions and cyber attacks may cause cross-chain data to be corrupted.
- **Modification Attack.** The cross-chain data may be modified by the receiving-chain before being stored on the chain.
- **Freeriding Attack.** A $node_r$ may reuse an integrity proof message from another honest $node'_r$ to pass the integrity verification.
- **Prediction Attack.** If these nodes participating in the block consensus in the supervision-chain are easily predictable in advance, external adversaries can easily attack the block consensus process.

3.3   Design Goals

Under the above system model and threat model, our scheme should meet the following three goals.

- **Decentralized Verification.** In the context of cross-chain scenario, using centralized entities for data integrity verification will weaken the decentralization of blockchain systems. By distributing the verification process across multiple nodes, the overall system becomes more resilient and resistant to attacks or manipulations.
- **Correctness.** The proposed scheme should ensure that the supervision-chain can correctly verify the integrity of cross-chain data by integrity verification process.
- **Security.** The proposed scheme should prevent from modification attacks, freeriding attacks, and prediction attacks.

4   Double-Layer Blockchain-Based Decentralized Integrity Verification Scheme

4.1   Overview

In our scheme, we use the supervision-chain to verify cross-chain data integrity in blockchains of sub-layer and record the results. In order to achieve these goals, we employ two consensus protocols, one for integrity consensus and the other for block consensus [9]. The integrity consensus aims to achieve consensus on the verification result of a given cross-chain data d. The block consensus is utilized within the system to achieve consensus on the blocks that will be recorded on the blockchain. We assume that in the sub-layer, a blockchain possessing the original data d aims to verify the integrity of cross-chain data stored on the receiving blockchains. We refer to the representative node of the blockchain which sent the original data d as $node_s$ and refer to the representative nodes of the blockchains which received cross-chain data as $node_r \in Node_d$.

In summary, a complete scheme consists of two phases: integrity consensus and block consensus. In the first phase, the system reaches a consensus on the verification result. If enough results has generated in the first phase, the second phase starts. In the second phase, the system reaches a consensus on the transaction information to record it on the blockchain.

Next, we provide a detailed explanation of the integrity consensus process and the block consensus process. For simplicity, we give some notations in Table 1.

4.2   Integrity Consensus

Firstly, we describe the sampling algorithm used in the integrity verification process. Then we give the detailed integrity consensus process.

4.2.1   Sampling Algorithm Inspired by the sampling algorithm proposed in [15], $node_s$ generate sampling parameters sp_r for each node $node_r \in Node_d$ by the following steps. Assume that the number of $Node_d$ is $k + 1$. $Per(x, y)$ is a pseudo-random permutation function, where x is a random number and y is the total number of data blocks to be permuted.

Table 1. The notations in our scheme.

Notation	Description
d	The cross-chain data
n	The total number of nodes in the supervision-chain
$Node_d$	The corresponding nodes which possess data d
$node_s$	The representative node of the sending-chain in the supervision-chain
$node_r$	The representative node of the receiving-chain in the supervision-chain
$node_b$	The block consensus committee
k	The total number of representative nodes of the receiving-chains
m	The number of nodes in the block consensus committee
$Per(x, y)$	A pseudo-random permutation function
$hash()$	A hash function
f	The number of malicious nodes that the blockchain system can tolerate

Step 1. For data d, which is divided into n data blocks, $node_s$ uses $Per(x, y)$ to process the index array $\{1, 2, ..., n\}$ and get a randomly sorted index array $SIA = \{index_1, index_2, ..., index_n\}$.

Step 2. $node_s$ divide SIA into k subsets, i.e. $SIA = \{C_1, C_2, ..., C_k\}$, satisfying the intersection of k subsets is empty and the union is $\{1, 2, ..., n\}$. These k subsets are used as the challenged index sets for $Node_d$. Generally, each subset has about $\lfloor n/k \rfloor$ elements, where $\lfloor n/k \rfloor$ represents the integer part of n/k. In particular, the last subset has $n - (k-1)\lfloor n/k \rfloor$ elements.

4.2.2 Integrity Consensus Process Assume that $node_s$ would like to verify the integrity of cross-chain data d in $node_r \in Node_d$. The integrity consensus process goes through four steps, as shown in Fig. 2.

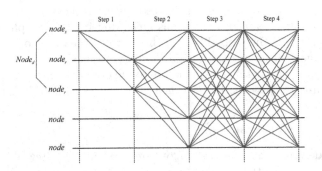

Fig. 2. Integrity consensus process

Step 1 Verification Request. $node_s$ use the sampling algorithm (see Sect. 4.2.1 Sampling Algorithm) to generate sampling parameters sp =

$\{C_1, C_2, ..., C_k\}$. Then it samples data blocks from d according to the sampling parameter $C_i \in sp$. Then it generates a Merkle Hash Tree(MHT) with a hash function $hash()$ based on the sampled data blocks. Next, $node_s$ calculate the BLS signature of the root node of MHT as the reference integrity proof for $node_r$, denoted as sig_r.

$$sig_r = hash(root)^{sk_{node_s}} \tag{6}$$

Finally, $node_s$ send verification request verification request vr $=<$ $pk_s, sp_r, d_{id}, sig_r >$ to each $node_r \in Node_d$.

Here, pk_s represents the unique identifier of $node_s$, sp_r denotes the sampling parameters for $node_r$, d_{id} is the the unique identifier of cross-chain data d and sig_r signifies the reference integrity proof for $node_r$.

Step 2 Integrity Proof Generation. Upon the receipt of a verification request, $node_r$ generates a MHT from its own cross-chain data d based on the specified sampling parameters sp_r and use the root of MHT to calculate its own integrity proof sig'_r.

$$sig'_r = hash(root')^{sk_{node_r}} \tag{7}$$

Then $node_r$ broadcasts the integrity proof message ipm $=<$ pk_s, pk_r, $sig_r, sig'_r >$ in the supervision-chain. Here, pk_r represents the unique identifier of $node_r$.

Step 3 Verification Response. Upon the receipt of an integrity proof message ipm from $node_r$, a node $node_i$ checks if Eq. (8) holds.

$$e(pk_s, sig'_r) \overset{?}{=} e(pk_r, sig_r) \tag{8}$$

If Eq. (8) holds, the ipm is valid and the cross-chain data in the blockchain of sub-layer is intact. Otherwise, the integrity of cross-chain data is corrupted. After validating all received ipm, $node_i$ broadcasts the verification response $vo = <d.id, result, list>$, where $result$ is the summary of the verification results and $list$ is a set of pk belonging to the nodes for which the integrity proof is invalid. Specially, if all equations holds, $result$ is true and $list$ is empty.

Step 4 Agreement. When a node $node_i$ receives $\lceil 2n/3 \rceil$ same verification response vo, the final verification conclusion is made. In detail, if $node_i$ receives $\lceil 2n/3 \rceil$ $result = true$, it thinks all nodes in $Node_d$ possess intact cross-chain data. Otherwise, $node_i$ thinks these nodes appearing $\lceil 2n/3 \rceil$ times in $list$ do not possess intact cross-chain data. After confirming the final result, $node_i$ broadcasts a integrity consensus commit in the supervision-chain. When a node receives $\lceil 2n/3 \rceil$ integrity consensus commits, it ends the integrity consensus process.

4.3   Block Consensus

In this subsection, we describe the whole block consensus process. Firstly, the reputation system and the election algorithm used in the block consensus process will be introduced. Then the detailed block consensus process will be given.

4.3.1   Reputation System In the supervision-chain, there are two types of rewards as incentives to motivate node to participate in integrity verification process.

Transaction Reward. A transaction reward is provided by the $node_s$ and allocated to those nodes that honestly verify integrity proof message ipm.

Block Reward. This reward is produced by the system to encourage nodes to participate in the maintenance of the supervision-chain, similar to most blockchain systems. For a node to gain block reward, it needs to satisfy the following two conditions: First, it is within the block consensus committee. Second, it actively and honestly participates in the block consensus process. The rewards gained by each node in the past are recorded on the blockchain, and the reputation of each node is calculated based on its historical rewards. In the election process of the consensus committee, nodes with a higher reputation score will be given prioritization.

In the reputation system, a node's reputation is determined by three factors: 1) The reputation score is higher when the node actively engages in a substantial number of integrity verification. Merely being honest without significant participation will not yield a high reputation score. 2) The reputation decreases when there are more occurrences of concealing inappropriate behavior. Each instance of such behavior results in a deduction of rewards, thereby lowering the reputation. 3) Recent behavior holds considerable weight in shaping the reputation score.

To meet the mentioned factors, we use an exponential moving average algorithm with bias correction to calculate the reputation of a node.

$$
r^t_{node_i} = \begin{cases} 0 & t = 0 \\ \dfrac{\rho \times p^t_{node_i} + (1-\rho) \times r^{t-1}_{node_i}}{1-\rho^t} & t > 0 \end{cases} \tag{9}
$$

Here, $p^t_{node_i}$ is the total amount of rewards from the recent transactions that have not been packed in block, $\rho \in (0,1)$ is a weighting factor, and a larger ρ indicates a higher weight on the impact of recent transactions on reputation. $1-\rho^t$ is a bias correction factor that ensures the stability of the reputation score over time.

4.3.2   Election Algorithm During the block consensus process, we first elect a block consensus committee. The block consensus is then carried out within this committee, and the consensus-reaching block is subsequently synchronized to other nodes. Using a block consensus committee to achieve block consensus has the advantage of improving the efficiency of block consensus. A drawback is that it reduces the number of malicious nodes (denoted as f) that the supervision-chain system can tolerate. Given that these nodes in the supervision-chain are diligently chosen as representatives from the sub-layer consortium blockchain, the incidence of malicious nodes within the supervision-chain is notably low. Thus, this drawback is acceptable. If the block consensus committee (m nodes) can be predicted based on information available on the blockchain, adversaries can disrupt the consensus process by attacking $\lceil 2m/3 \rceil$ nodes in the consensus committee. Therefore, it is unsafe to always select the top m nodes with the highest reputation scores to form the block consensus committee. Based on the reputation system(see Sect. 4.3.1 Reputation System), we use VRF (Verifiable Random Function) [9,13] to elect the consensus committee, ensuring that only nodes with reputation score surpassing a certain threshold have the chance to be selected for the block consensus committee. The block consensus committee election goes through two steps, as follows:

Step 1 Candidate preparation. Only nodes with reputation score exceeding a certain threshold are qualified to become candidates. By referring to their own reputation records stored on the blockchain, a node can easily determine whether it meets the criteria necessary to become a candidate. If eligible, $node_i$ generates a competition request $cr_i =< pk_{node_i}, f_{sk_{node_i}}(x), \pi_{sk_{node_i}}(x) >$ based on a random seed x and its private key sk_{node_i}. Here, $f_{sk_{node_i}}(x)$ and $\pi_{sk_{node_i}}(x)$ are calculated as follows:

$$f_{sk_{node_i}}(x) = e(g,g)^{1/(x+sk_{node_i})} \tag{10}$$

$$\pi_{sk_{node_i}}(x) = g^{1/(x+sk_{node_i})} \tag{11}$$

Then it broadcasts competition request cr_i in the supervision-chain.

Step 2 Leader and Member Determination. Upon receiving a competition request cr_i from node $node_i$, each node $node_j \in Node$ performs the necessary checks: 1) It checks that if node $node_i$ has sufficient reputation to qualify as a candidate. 2) It validates the correctness of the competition request cr by verifying the following two equations:

$$e(g^x \cdot pk, \pi_{sk}(x)) = e(g,g) \tag{12}$$

$$e(g, \pi_{sk}(x)) = f_{sk}(x) \tag{13}$$

If multiple candidates simultaneously possess valid competition requests, the node with the highest value of $f_{sk_{node}}(x)$ is selected as the leader. Next, select $3f$ nodes with higher values of $f_{sk_{node}}(x)$ from the candidates who were not successful in the election to become members of the block consensus committee.

4.3.3 Block Consensus Process

When multiple integrity consensus processes are completed and there is enough transaction information to be packaged into a block, the block consensus process starts. A block consensus process consists of three steps, as shown in Fig. 3.

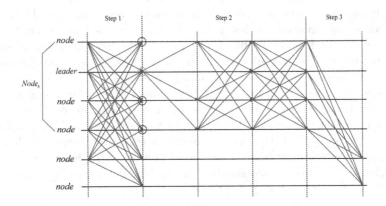

Fig. 3. Block consensus process

Step 1 Election. A block consensus committee $Node_b$ is elected (see Sect. 4.3.2 Election Algorithm).

Step 2 Consensus. The leader packs the transactions(several final consensus information reached during the integrity consensus phase) into a new block. Specifically, based on the messages received, it allocates the transaction rewards provided by $node_s$ to these nodes that verify the integrity proof message ipm honestly. Next, it allocates negative transaction rewards to these representative nodes of the sub-layer blockchain that do not honestly store cross-chain data d. Next, it updates the reputations of all nodes in the supervision-chain with Eq. (9). Specifically, the related rewards and reputations are also packed into the block. The leader then broadcasts the block in the block consensus committee $Node_b$ for validation. Upon receiving the block, the node $node_i \in Node_b$ checks correctness of current round information. If passed, $node_i$ broadcasts a prepare message to claim its ready state. Once $node_i$ obtains more than $\lceil 2m/3 \rceil$ prepares message, it begins to verify the content of the block based on the information obtained during the integrity consensus phase. If passed, $node_i$ broadcasts the commit message to other nodes in $Node_b$. Finally, if $node_i$ receives more than $\lceil 2m/3 \rceil$ commit messages, it will accept the new block and append it to the ledger.

Step 3 Synchronization. All nodes in the block consensus committee $Node_b$ respond to other nodes in the supervision-chain. A node will accept the new block if more than $f + 1$ same blocks are received, where f is the maximum number of malicious nodes that the blockchain system can tolerate. At the end, the leader receives the block reward, while the other nodes in $Node_b$ receive a block reward that is less than the leader's.

5 Security Analysis

In this section, we provide a brief evaluation of the correctness and security of the proposed scheme.

Theorem 1. *If a blockchain honestly stores cross-chain data d, its representative node can pass the integrity verification during the integrity consensus phase.*

Proof. The correctness of the Eq. (8) can be proved as follow:

$$e(pk_s, sig'_r) = e(g^{sk_s}, hash(root')^{sk_r}) = e(g^{sk_r}, hash(root)^{sk_s}) = e(pk_r, sig_r)$$

If a node $node_r$ have the intact data, it can generate a valid sig'_r. Therefore, it can pass other nodes' verification(Step 3 in 4.2.2 Integrity Consensus Process).

Theorem 2. *A node $node_i \in Node_d$ cannot reuse an integrity proof message from an honest node $node_r$ to attack the integrity consensus.*

Proof. For a integrity proof message $ipm =< pk_s, pk_r, sig_r, sig'_r >$, it contains only signatures sig_r and sig'_r but not the original hash tags. It is impossible for $node_i$ to forge a signature due to the hardness of solving the CDH problem [8]. Moreover, $node_i$ may reuse the existing signatures but change the identity of $node_r$ from pk_r to pk_i in ipm to forge an integrity proof message $ipm' =< pk_s, pk_i, sig_r, sig'_r >$. But this behavior can be easily detected by Eq. (8).

Theorem 3. *If the cross-chain data, denoted as d, is divided into n blocks, the probability that the supervision-chain successfully detects a dishonest blockchain that does not store d accurately is at least $Pr = 1 - (\frac{n-c}{n})^t$. In this equation, c represents the number of altered data blocks within d, and t stands for the number of challenged blocks.*

Proof. Based on the sampling algorithm, the challenged blocks on k nodes in $Node_d$ are completely unrepeatable and their union is exactly all the data blocks of d. For a node $node_r \in Node_d$, the probability of finding d modified is equal to the probability of at least challenging one modified data block. In other words, we can calculate the probability that at least one modified block is challenged on $node_r$ during the integrity consensus phase as follows:

$$Pr = 1 - \left(\frac{n-c}{n}\right)\left(\frac{n-c-1}{n-1}\right)\left(\frac{n-c-2}{n-2}\right)\cdots\left(\frac{n-c-t+1}{n-t+1}\right).$$

Given an arbitrary integer $i \leq n$, there is $\frac{n-c-i}{n-i} \geq \frac{n-c-i-1}{n-i-1}$. Hence, the following inequality holds:

$$Pr > 1 - \left(\frac{n-c}{n}\right)^t.$$

Theorem 4. *During the election process of the block consensus committee, any node can verify the correctness of a competition request cr.*

Proof. For a competition request $cr =< pk, f_{sk}(x), \pi_{sk}(x) >$, the correctness of it can be verified as follows:

$$e(g^x \cdot pk, \pi_{sk}(x)) = e(g^x \cdot g^{sk}, g^{1/(x+sk)}) = e(g, g)$$

$$e(g, \pi_{sk}(x)) = e(g, g^{1/(x+sk)}) = e(g, g)^{1/(x+sk)} = f_{sk}(x)$$

If both of the above two equations hold true, then cr is generated by the node with the public key pk.

6 Performance Evaluation

In this section, we conduct a series of experiments to evaluate the performance of our scheme.

6.1 Experimental Settings

We implement the integrity verification process based on the Java Pairing-Based Cryptography Library (JPBC) version 2.0.0, which performs the mathematical operations underlying pairing-based cryptosystems. In our experiments, we choose the type A pairing parameters in JPBC library, which the group order is 160 bits and the base field order is 512 bits. And our experiments are conducted on a PC laptop which runs Windows 10 on an Intel Core i5 CPU at 2.50 GHz and 8 GB DDR4 RAM. To get more precise results, each experiment is conducted 100 trials.

6.2 Experimental Results and Analysis

We focus on the evaluation related to computation cost, communication cost and detection precision in the integrity verification process. The detailed analysis is as follows.

Computation Cost. We set the number of nodes required for data integrity verification from 2 to 22. For the same block size of 8 KB, we set the data size of d to 4 MB and 8 MB, respectively, and measure the time required for the integrity verification process. As Fig. 4, the larger the file, the longer it takes to complete the integrity verification process. This is due to the increase in the number of data blocks, which leads to a higher number of samples being taken from each node. Consequently, a larger number of MHT nodes need to be computed during the verification process, resulting in increased time consumption. Furthermore, we can also observe from Fig. 4 that as the number of nodes to be verified increases, the time consumption also increases, which aligns with our expectations and falls within an acceptable range.

Communication Cost. We set the total number of nodes in the system from 5 to 25, with 4 nodes required for data integrity verification. For the same block size of 8 KB, we set the data size of d to 4 MB and 8 MB, respectively, and measure the communication cost for the integrity verification process. From Fig. 5, it can be observed that as the data size increases, the communication overhead remains relatively constant. This is because the increase in data size results in a larger number of data blocks, but it only leads to an increase in the number of blocks sampled within the data d of each node. Given that the number of integrity consensus nodes remains unchanged, the number of integrity proofs that need to be sent (i.e., the MHT root of BLS signatures) also remains constant. As a result, the communication overhead during the integrity verification process remains approximately unchanged. Furthermore, as shown in Fig. 5, for the same data size, the communication overhead increases with the number of nodes in the system. This is attributed to the fact that a larger number of nodes necessitates increased communication to achieve integrity consensus.

Detection Precision. We set the number of data blocks to 45,000 and investigate the relationship between detection accuracy and the number of sampled blocks under different corruption rates. As shown in Fig. 6, for each different data corruption rate, our scheme achieves close to 100% detection accuracy with only a small number of sampled data blocks.

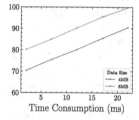
Time Consumption (ms)

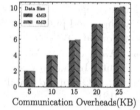
Communication Overheads(KB)

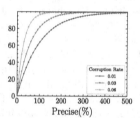
Precise(%)

Fig. 4. The computation cost in the integrity verification process.

Fig. 5. The communication cost in the integrity verification process.

Fig. 6. The detection rate for data corruption.

7 Conclusion

To solve the problem of cross-chain data integrity in the multi-chain architecture, we propose a double-layer blockchain-based decentralized integrity verification scheme based on the ideas of "*governing the chain by chain*" and "*double-layer blockchain*". In detail, We construct a supervision-chain by selecting representative nodes from multiple blockchains. And the supervision-chain is responsible for integrity verification and recording the corresponding results. To achieve decentralized data integrity verification, we propose an integrity consensus protocol.

To record the verification results, we propose the block consensus protocol. And we utilize a block consensus committee to enhance the efficiency of the block consensus process. The security analysis and performance evaluation demonstrate the security and effectiveness of our proposed scheme.

In the future, we aim to investigate the verification scheme for ensuring the integrity of cross-chain data processing results in multi-chain architectures.

Acknowledgement. This work is supported by the Fundamental Research Funds for the Central Universities (Grant No. NS2023047), the National Key Research and Development Program of China (Grant No. 2020YFB1005500) and Postgraduate Research & Practice Innovation Program of NUAA (No. xcxjh20221616).

References

1. Ateniese, G., et al.: Provable data possession at untrusted stores. In: Proceedings of the 14th ACM Conference on Computer and Communications Security, pp. 598–609 (2007)
2. Ding, Q., Gao, S., Zhu, J., Yuan, C.: Permissioned blockchain-based double-layer framework for product traceability system. IEEE Access **8**, 6209–6225 (2019)
3. Dodis, Y., Yampolskiy, A.: A Verifiable Random Function with Short Proofs and Keys. In: Vaudenay, S. (ed.) PKC 2005. LNCS, vol. 3386, pp. 416–431. Springer, Heidelberg (2005). https://doi.org/10.1007/978-3-540-30580-4_28
4. Jiang, Y., Wang, C., Wang, Y., Gao, L.: A cross-chain solution to integrating multiple blockchains for iot data management. Sensors **19**(9), 2042 (2019)
5. Jin, H., Jiang, H., Zhou, K.: Dynamic and public auditing with fair arbitration for cloud data. IEEE Trans. Cloud Comput. **6**(3), 680–693 (2016)
6. Kan, L., Wei, Y., Muhammad, A.H., Siyuan, W., Gao, L.C., Kai, H.: A multiple blockchains architecture on inter-blockchain communication. In: 2018 IEEE International Conference on Software Quality, Reliability and Security Companion (QRS-C), pp. 139–145. IEEE (2018)
7. Kang, J., et al.: Communication-efficient and cross-chain empowered federated learning for artificial intelligence of things. IEEE Trans. Netw. Sci. Eng. **9**(5), 2966–2977 (2022)
8. Li, B., He, Q., Chen, F., Jin, H., Xiang, Y., Yang, Y.: Inspecting edge data integrity with aggregate signature in distributed edge computing environment. IEEE Trans. Cloud Comput. **10**(4), 2691–2703 (2021)
9. Li, B., He, Q., Yuan, L., Chen, F., Lyu, L., Yang, Y.: EdgeWatch: collaborative investigation of data integrity at the edge based on blockchain. In: Proceedings of the 28th ACM SIGKDD Conference on Knowledge Discovery and Data Mining, pp. 3208–3218 (2022)
10. Li, J., Zhang, L., Liu, J.K., Qian, H., Dong, Z.: Privacy-preserving public auditing protocol for low-performance end devices in cloud. IEEE Trans. Inf. Forensics Secur. **11**(11), 2572–2583 (2016)
11. Liu, J., Huang, K., Rong, H., Wang, H., Xian, M.: Privacy-preserving public auditing for regenerating-code-based cloud storage. IEEE Trans. Inf. Forensics Secur. **10**(7), 1513–1528 (2015)
12. Menezes, A.: An introduction to pairing-based cryptography. Recent Trends Crypt. **477**, 47–65 (2009)

13. Micali, S., Rabin, M., Vadhan, S.: Verifiable random functions. In: 40th annual Symposium on Foundations of Computer Science (cat. No. 99CB37039), pp. 120–130. IEEE (1999)
14. Nakamoto, S.: Bitcoin: A peer-to-peer electronic cash system. Decentralized Bus. Rev. 21260 (2008)
15. Qiao, L., Li, Y., Wang, F., Yang, B.: Lightweight integrity auditing of edge data for distributed edge computing scenarios. Ad Hoc Netw. **133**, 102906 (2022)
16. Wood, G.: Polkadot: vision for a heterogeneous multi-chain framework. White paper **21**(2327), 4662 (2016)
17. Xiao, X., Yu, Z., Xie, K., Guo, S., Xiong, A., Yan, Y.: A Multi-blockchain Architecture Supporting Cross-Blockchain Communication. In: Sun, X., Wang, J., Bertino, E. (eds.) ICAIS 2020. CCIS, vol. 1253, pp. 592–603. Springer, Singapore (2020). https://doi.org/10.1007/978-981-15-8086-4_56
18. Yaga, D., Mell, P., Roby, N., Scarfone, K.: Blockchain technology overview. arXiv preprint arXiv:1906.11078 (2019)
19. Yu, Y., et al.: Identity-based remote data integrity checking with perfect data privacy preserving for cloud storage. IEEE Trans. Inf. Forensics Secur. **12**(4), 767–778 (2016)

Inter-modal Fusion Network with Graph Structure Preserving for Fake News Detection

Jing Liu[1], Fei Wu[1(✉)], Hao Jin[1], Xiaoke Zhu[2], and Xiao-Yuan Jing[1,3]

[1] Nanjing University of Posts and Telecommunications, Nanjing 210023, China
wufei_8888@126.com, 1222056329@njupt.edu.cn
[2] Henan University, Kaifeng 475001, China
whuzxk@whu.edu.cn
[3] Wuhan University, Wuhan 430072, China

Abstract. The continued ferment of fake news on the network threatens the stability and security of society, prompting researchers to focus on fake news detection. The development of social media has made it challenging to detect fake news by only using uni-modal information. Existing studies tend to integrate multi-modal information to pursue completeness for information mining. How to eliminate modality differences effectively while capturing structure information well from multimodal data remains a challenging issue. To solve this problem, we propose an **In**ter-modal **F**usion network with **G**raph **S**tructure **P**reserving (**IF-GSP**) approach for fake news detection. An inter-modal cross-layer fusion module is designed to bridge the modality differences by integrating features in different layers between modalities. Intra-modal and cross-modal contrastive losses are designed to enhance the inter-modal semantic similarity while focusing on modal-specific discriminative representation learning. A graph structure preserving module is designed to make the learned features fully perceive the graph structure information based on a graph convolutional network (GCN). A multi-modal fusion module utilizes an attention mechanism to adaptively integrate cross-modal feature representations. Experiments on two widely used datasets show that IF-GSP outperforms related multi-modal fake news detection methods.

Keywords: Contrastive Learning · Fake News Detection · Graph Convolutional Network · Multi-modal Fusion

1 Introduction

Although social media provides us easier access to information, we are inevitably caught in the dilemma of widespread dissemination of fake news. The spread of fake news may lead to innocent readers being deceived, exploited, and even result in incalculable consequences. It is extremely costly and time-consuming

B. Luo et al. (Eds.): ICONIP 2023, LNCS 14452, pp. 280–291, 2024.
https://doi.org/10.1007/978-981-99-8076-5_20

to identify fake news manually. Automated fake news detection technology is in urgent need of attention.

Earlier, fake news detection technology focused on using text messages to detect fake news. Ma *et al.* [14] employed recurrent neural networks (RNNs) to capture useful information for fake news detection. Yu *et al.* [24] adopted convolutional neural networks (CNNs) to extract features and model high-level interaction among features to realize fake news detection. With the rapid advancement of social media, researchers begin to leverage image information to identify fake news. Jin *et al.* [9] believed that images can be valuable in exposing fake news and explored visual and statistical features of images for fake news detection. Making up indistinguishable fake news tends to rely on using both textual and visual information. Although uni-modal methods can provide discriminative information about fake news, they still suffer from the issue of insufficient information. In contrast, multiple modalities can provide the correlation and complementarity information for fake news detection task more comprehensively. MVAE [11] introduces a multi-modal variational auto-encoder (VAE) to learn a unified representation of visual and textual information to conduct fake news detection. Moreover, increasing fake news detection methods [5] tend to take multi-modal data into account at the same time.

How to effectively reduce the inter-modal heterogeneity and preserve the modal-specific semantic integrity as much as possible is the key to the multi-modal fake news detection task. Although many multi-modal fake news detection models have been proposed, how to eliminate modality differences effectively with capturing structure information well has not been well studied. To address this issue, in this paper, we propose a novel fake news detection approach named inter-modal fusion network with graph structure preserving, which adequately explores and utilizes information in image and text modalities. To reduce the modality differences, inter-modal cross-layer fusion (ICF) module performs interaction between modalities and adopts contrastive learning technology to make cross-modal and intra-modal semantically similar features more compact. The graph structure preserving (GSP) module reconstructs multi-modal features and integrates structure information to ensure the integrity of information mining in order to facilitate the sufficiency of information utilization. The multi-modal fusion (MMF) module aims to adaptively assign weights to text and image modalities to obtain discriminative features.

The contributions of this paper are as follows:

- Focusing on jointly inter-modal differences reduction issue and structure information capture for fake news detection task, we propose an **Inter-modal Fusion** network with **Graph Structure Preserving (IF-GSP)** for fake news detection.
- IF-GSP eliminates the differences between modalities via inter-modal cross-layer fusion module. The graph structure preserving module makes the structure information of each modality embed the learned features to enhance the discriminative ability. Finally, the multi-modal feature representations

are aggregated by adaptively learning optimal weights for image and text modalities.

- The experimental results on Twitter [2] and Weibo [8] demonstrate that our model outperforms previous works and achieves state-of-the-art performance on both datasets.

2   Related Works

2.1   Fake News Detection

Uni-modal Fake News Detection. Based on deep attention, CAR [4] utilizes RNN to capture textual features to detect fake news. Qian *et al.* [16] used CNN to extract textual features at the sentence and word levels and analyzed responses of users to generate responses to new articles to assist fake news detection. MVNN [15] utilizes multi-domain visual information by dynamically merging the information of frequency and pixel domains for fake news detection.

Multi-modal Fake News Detection. Jin *et al.* [8] proposed att-RNN which utilizes long short term memory network (LSTM) to learn the joint representation of social contexts and texts to detect fake news. Wang *et al.* [21] argued that event-invariant features are beneficial for fake news detection and proposed a model that focuses on exploring transferable features with removing non-transferable event-specific features. MCAN [22] stacks multiple co-attention layers to learn inter-dependencies between textual and visual features to fuse multi-modal features. Chen *et al.* [5] proposed CAFE which aligns features between image and text modalities, and then learns the cross-modal ambiguity by calculating the Kullback-Leibler (KL) divergence between the distributional divergence of uni-modal features, and aggregates multi-modal features to conduct multi-modal fake news detection finally. MKEMN [25] focuses on multi-modal information and utilizes conceptual knowledge as supplement to enhance fake news detection effect. Zhou *et al.* [26] designed SAFE which transforms the image into text via a pre-trained model, and then measures the similarity between them to detect fake news. MCNN [23] concentrates on the consistency of multi-modal data and captures the similarity of multi-modal information for fake news detection.

However, this kind of methods fail to jointly take the issue of inter-modal differences reduction and structure information exploration into account adequately. In this paper, we fuse multi-modal features by modeling the inter-modal interaction with the constraint of structure information embedding.

2.2   Graph Neural Networks

With the popularity of graph neural networks (GNNs) [17], more and more GNN methods have been proposed. For example, the graph convolutional network (GCN) [13] automatically captures node features by aggregating information from neighboring nodes. The graph sample and aggregate (GraphSAGE) [6] predicts the labels of unlabeled nodes by sampling and aggregating information from neighboring nodes. The graph attention network (GAT) [19] specifies varying weights for different nodes in a neighborhood by stacking graph attentional layers, allowing nodes to take neighbor information into account. In recent years, GNN has been successfully introduced into fake news detection task. Based on GNN, Vaibhav et al. [18] modeled the relationship between sentences in the news to perform fake news detection. Bian et al. [1] leveraged bi-directional GCN to process propagation and dispersion of rumors.

In this paper, we focus on utilizing the strong representation ability of GCN to make features learn structure information of text and image modalities which is regarded as a supplement to the semantic information.

2.3   Contrastive Learning

Contrastive learning technology constraints feature learning process by maximizing the similarity between anchor and positive samples and minimizing the similarity between anchor and negative samples. Many contrastive learning methods have achieved significant success. MOCO [7] regards contrastive learning as a dictionary query task. SimCLR [3] uses contrastive loss to learn representation after a composition of data augmentation. Khosla et al. [12] designed supervised contrastive loss with effectively utilizing label information. Recently, contrastive learning has been used to the fake news detection task. Wang et al. [20] aimed to achieve more accurate image-text alignment by using contrastive learning technology for fake news detection.

In this paper, we jointly perform intra-modal and cross-modal contrastive losses to improve the semantically similarity for each modality and across modalities and disperse semantically dissimilar features within and between modalities, respectively.

3   Our Approach

In this paper, a news post contains information of text and image modalities. We utilize N news as the training set, where the image modality is denoted as $I = \{I_1, \ldots, I_N\}$, and the text modality is denoted as $T = \{T_1, \ldots, T_N\}$. $Y = \{y_1, \ldots, y_N\}$ is the label matrix corresponding to $D = \{I, T\}$, where $y_i \in (0, 1)$. Specifically, 0 represents real news and 1 represents fake news.

We propose an end-to-end learning approach IF-GSP, which aims to learn feature representations with strong discriminability for fake news detection. IF-GSP consists of three main components: (1) to mitigate modality differences, an inter-modal cross-layer fusion module is designed to integrate cross-modal features and perform contrastive learning within and between modalities to enhance discriminative ability of features; (2) to ensure the integrity of multi-modal information utilization, a graph structure preserving module is designed to make modal-specific features preserve the graph structure information that acts as a supplement information with respect to discriminative information of features;

(3) a multi-modal fusion module is designed to learn the optimal weight for each modality to integrate cross-modal features. The overall architecture of IF-GSP is shown in Fig. 1.

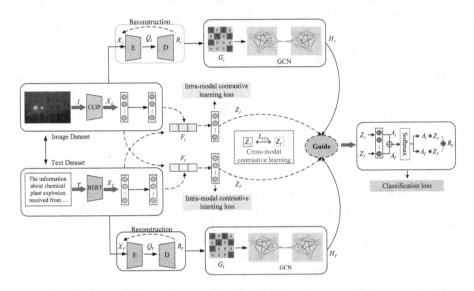

Fig. 1. The framework structure of IF-GSP.

For texts, we use BERT [10], which is a popular pre-trained language model built on Transformer that is trained using unsupervised learning on a large corpus, to encode text features T to obtain X_T. For images, we utilize ViT-B/32 to encode image features I into X_I.

3.1 Inter-modal Cross-Layer Fusion (ICF) Module

In order to deal with modality differences, we specially design an inter-modal cross-layer fusion module which consists of a two-layer image encoder and a two-layer text encoder to perform feature interaction. Inter-modal heterogeneity issue can be alleviated by promoting information exchange between different modalities across layers with interaction. Specifically, we fuse the output of the second layer of image encoder with the output of the first layer of text encoder to obtain F_I with the *cat* operation, and take the similar operation to obtain F_T. Then, we adopt fully-connected networks to further extract features to obtain more discriminative feature representations $Z_I = \{z_I^1, \ldots, z_I^N\}$ and $Z_T = \{z_T^1, \ldots, z_T^N\}$, respectively.

Although the inter-modal cross-layer fusion part is employed to ease the modality differences issue, the inter-modal differences problem still exist. For this reason, we further design cross-modal and intra-modal contrastive learning part to supervise the learning of features for image and text modalities. Cross-modal contrastive loss is implemented to focus on the consistency between modalities, which minimizes the distance of semantically similar features and maximizes the distance of those that are dissimilar across modalities. Intra-modal contrastive loss is designed to consider intra-modal similarity in each modality, so that the within-calss features have larger similarity, while the different-calss features have less similarity. Utilizing contrastive loss from both intra-modal and inter-modal perspectives can improve the discriminative ability of feature representations, and explore the consistency of cross-modal information, thereby helping to reduce the semantic gap between modalities.

The cross-modal contrastive loss is defined as follows:

$$L_{CCL} = \sum_{k \in K} \frac{-1}{|P(k)|} \sum_{p \in P(k)} \log \frac{\exp\left(z_C^k \cdot z_C^p / \tau\right)}{\sum_{a \in A(k)} \exp\left(z_C^k \cdot z_C^a / \tau\right)} \tag{1}$$

where $k \in K \equiv \{1, \ldots, 2N\}$ denotes the index of a feature representation within $Z_C = \{Z_I, Z_T\} = \left\{z_C^1, \ldots, z_C^{(2N)}\right\}$, $A(k) \equiv K/\{k\}$. $S = \{s_1, \ldots, s_{2N}\}$ is the label of Z_C, and $P(k) \equiv \{p \in A(k) : s_p = s_k\}$ is the set of indices of all positive feature representations for the anchor z_C^k.

Taking image modality as an example, the intra-modal contrastive loss is defined as follows.

$$L_{ICL_I} = \sum_{j \in J} \frac{-1}{|P(j)|} \sum_{p \in P(j)} \log \frac{\exp\left(z_I^j \cdot z_I^p / \tau\right)}{\sum_{a \in A(j)} \exp\left(z_I^j \cdot z_I^a / \tau\right)} \tag{2}$$

where $j \in J \equiv \{1, \ldots, N\}$ denotes the index of a feature representation within $Z_I = \{z_I^1, \ldots, z_I^N\}$, $A(j) \equiv J/\{j\}$. $Y = \{y_1, \ldots, y_N\}$ is the label of Z_I, and $P(j) \equiv \{p \in A(j) : y_p = y_j\}$ is the set of indices of all positive feature representations for the anchor z_I^j. The intra-modal contrastive loss of text modality L_{ICL_T} can be obtained similarly.

3.2   Graph Structure Preserving (GSP) Module

To preserve as much complete semantic information as possible, we utilize GNN with shared parameters to perform feature learning for each modality with considering multi-modal correlation. To achieve unified dimension of features across modalities, we firstly map X_I and X_T to a common latent space via auto-encoders. The corresponding output of the image and text encoder Q_I and Q_T will be utilized to perform further discriminative feature learning. We constrain the auto-encoders by reconstruction loss, which encourages the auto-encoder to preserve the semantic information of the feature vector as much as possible when performing latent space projection. The reconstruction loss based on mean squared error (MSE) loss is defined as follows:

$$L_R = MSE(X_I, R_I) + MSE(X_T, R_T) \tag{3}$$

where R_I and R_T are the outputs of corresponding decoders.

Taking image modality as an example, given a graph $\mathcal{G}_I(\nu_I, \varepsilon_I)$, it denotes the graph of size N with nodes $Q_I^i \in \nu_I$ and edges $(Q_I^i, Q_I^j) \in \varepsilon_I$. Based on label information, we define the adjacency matrix as follows:

$$G_I^{ij} = \begin{cases} 0, y_i \neq y_j \\ 1, y_i = y_j \end{cases} \tag{4}$$

The graph convolution process for each layer is defined as follows:

$$O_{(l)} = ReLu\left(D^{-1/2}G_I D^{-1/2}O_{(l-1)}W_{(l)}\right) \tag{5}$$

where $D_{ii} = \sum_j G_I^{ij}$ and $W_{(l)}$ is the convolution filter of the lth layer. $O_{(l-1)}$ and $O_{(l)}$ indicate the corresponding input and output of GCN, respectively. Q_I is the input of the first layer of GCN, and the output of the last layer is formulated as H_I. For text modality, we obtain H_T.

We design the graph structure preserving loss to help GCN guide the feature to remember structure information which can be viewed as a supplement to the semantic information. By compelling Z_I and H_I (Z_T and H_T) closer together, the graph structure information can be preserved for the level of Z_I and Z_T, which is helpful to represent and utilize features more comprehensively.

The graph structure preserving loss is defined as follows:

$$L_G = MSE(Z_I, H_I) + MSE(Z_T, H_T) \tag{6}$$

3.3   Multi-modal Fusion (MMF) Module

Considering different modalities play different roles in the decision-making process, we design attention-based fusion network to adaptively fuse Z_I and Z_T. To be more detailed, attention-based fusion network is designed to utilize an attention function $F^C\left(\cdot; \theta^C\right)$, i.e., a fully-connected layer activated by the Sigmoid function and parameterized by θ^C, to obtain the attention coefficients of Z_I and Z_T with Eq. 7, which are denoted as A_I and A_T.

$$\begin{cases} A_I = F^C\left(Z_I; \theta^C\right) \\ A_T = F^C\left(Z_T; \theta^C\right) \end{cases} \tag{7}$$

After that, we use the softmax function to further normalize these coefficients. And then, we obtain the fused cross-modal representation, which is denoted as $R_C = \{r_C^1, \ldots, r_C^N\}$.

$$R_C = A_I Z_I + A_T Z_T \tag{8}$$

We design a label classifier which consists of a fully-connected network, to detect fake news and design cross-entropy based classification loss to improve feature discrimination. The classification loss is defined as follows:

$$L_C = -\frac{1}{N}\sum_{i=1}^{N}\left(y_i \log P\left(r_C^i\right) + (1 - y_i)\log\left(1 - P\left(r_C^i\right)\right)\right) \tag{9}$$

where $P\left(r_C^i\right)$ is the predicted label probability of r_C^i.

According to the Eqs. (1), (2), (3), (6) and (9), the total loss is defined as follows:

$$L_{TOTAL} = L_{CCL} + L_{ICL_I} + L_{ICL_T} + L_G + L_C + L_R \qquad (10)$$

4 Experiment

4.1 Datasets

We use two widely used datasets which are described in detail as follows. For the Twitter dataset which was released for MediaEval Verifying Multimedia Use task, we follow [5] to process the Twitter dataset. For the Weibo dataset which has been widely used in prior multimodal fake news detection works, the real samples were collected from Xinhua News Agency, an authoritative news source of China. The fake samples were gathered by crawling the official fake news debunking system of Weibo over a time span from May 2012 to January 2016. The details of both datasets are presented in Table 1.

Table 1. The detailed information of dataset.

Statistics	Twitter	Weibo
Fake news in training set	5,007	3,749
Real news in training set	6,840	3,783
News in test set	1,406	1,996

4.2 Baseline

To evaluate the effectiveness of IF-GSP, we compare the performance of IF-GSP with following baselines, i.e., att-RNN [8], EANN [21], MVAE [11], MKEMN [25], SAFE [26], MVNN [23] and CAFE [5]. To perform a fair comparison, for att RNN, we remove social information, and for EANN, we remove the event discriminator component.

4.3 Implementation Details

The evaluation metrics include Accuracy (ACC), Precision, Recall, and F1-score (F1). We use the batch size of 128 and train the model using SGD with an initial learning rate of 0.001 for 100 epochs. Following [12], we set the temperature factor τ in Eqs. (1) and (2) as 0.07. All codes are implemented with PyTorch and run on a computer with NVIDIA GeForce GTX 1080Ti GPU card. Our IF-GSP requires training time about 3 s per epoch on Twitter, and around 2 s per epoch on Weibo.

Table 2. The result comparison between IF-GSP and other baseline methods.

	Method	ACC	Fake news			Real News		
			P	R	F1	P	R	F1
Twitter	att-RNN	0.664	0.749	0.615	0.676	0.589	0.728	0.651
	EANN	0.648	0.810	0.498	0.617	0.584	0.759	0.660
	MVAE	0.745	0.801	0.719	0.758	0.689	0.777	0.730
	MKEMN	0.715	0.814	0.756	0.708	0.634	0.774	0.660
	SAFE	0.762	0.831	0.724	0.774	0.695	0.811	0.748
	MCNN	0.784	0.778	0.781	0.779	0.790	0.787	0.788
	CAFE	0.806	0.807	0.799	0.803	0.805	0.813	0.809
	IF-GSP	**0.899**	**0.998**	**0.803**	**0.890**	**0.830**	**0.999**	**0.906**
Weibo	att-RNN	0.772	0.854	0.656	0.742	0.720	**0.889**	0.795
	EANN	0.795	0.806	0.795	0.800	0.752	0.793	0.804
	MVAE	0.824	0.854	0.769	0.809	0.802	0.875	0.837
	MKEMN	0.814	0.823	0.799	0.812	0.723	0.819	0.798
	SAFE	0.816	0.818	0.815	0.817	0.816	0.818	0.817
	MCNN	0.823	0.858	0.801	0.828	0.787	0.848	0.816
	CAFE	0.840	0.855	0.830	0.842	0.825	0.851	0.837
	IF-GSP	**0.905**	**0.895**	**0.929**	**0.912**	**0.918**	0.879	**0.898**

4.4   Results

Table 2 presents the experimental results of IF-GSP compared to the baselines. The best values are marked in bold. In terms of ACC and F1, IF-GSP outperforms all compared methods on each dataset. Specifically, IF-GSP achieves the highest accuracy of 0.899 and 0.905 on Twitter and Weibo, respectively. And for fake news, it achieves the highest F1 of 0.890 and 0.912 on Twitter and Weibo, respectively. IF-GSP outperforms compared state-of-the-art methods on Twitter and Weibo by a large margin. It is because we can fully utilize the complementarity and diversity among text and image modalities and design an interactive manner to bridge the inter-modal semantic gap. ICF module performs feature interaction by integrating different-layer features in different modalities and uses contrastive learning to enhance the similarity of information within modality and the correlation between modalities. Moreover, we believe the graph structure information can provide another perspective as supplement to explore information. The GSP module utilizes GCN to make features learn the structure information, which is conducive to ensure semantic integrity of multi-modal feature.

4.5   Ablation Experiments

To verify the effectiveness of important components of IF-GSP, we perform an ablation study and the results are summarized in Table 3. We divide IF-GSP into four variants to demonstrate the importance of each module.

- IF-GSP-1: The version of IF-GSP that removes the inter-layer cross-layer fusion part in the ICF module.
- IF-GSP-2: The version of IF-GSP that removes cross-modal and intra-modal contrastive learning part in the ICF module.
- IF-GSP-3: The version of IF-GSP that removes the GSP module.
- IF-GSP-4: The version of IF-GSP that replaces the MMF module by using direct fusion manner with the *cat* operation.

Table 3. The result of ablation experiments.

	Method	ACC	Fake news			Real News		
			P	R	F1	P	R	F1
Twitter	IF-GSP-1	0.858	0.867	0.852	0.859	0.849	0.864	0.856
	IF-GSP-2	0.895	1.00	0.794	0.885	0.823	1.00	0.903
	IF-GSP-3	0.890	0.981	0.799	0.881	0.825	0.984	0.897
	IF-GSP-4	0.888	0.983	0.794	0.878	0.821	0.985	0.896
	IF-GSP	**0.899**	0.998	0.803	**0.890**	0.830	0.999	**0.906**
Weibo	IF-GSP-1	0.883	0.890	0.887	0.888	0.875	0.879	0.877
	IF-GSP-2	0.850	0.841	0.882	0.861	0.862	0.814	0.837
	IF-GSP-3	0.897	0.894	0.911	0.903	0.900	0.880	0.890
	IF-GSP-4	0.895	0.919	0.878	0.898	0.871	0.914	0.892
	IF-GSP	**0.905**	0.895	0.929	**0.912**	0.918	0.879	**0.898**

From the results in Table 3 where the best values of evaluation metrics are marked in bold, we can obtain the following conclusions: (1) inter-modal cross-layer fusion module performs feature interaction between modalities by integrating different-layer features between text and image modalities to effectively reduce modality differences. (2) Cross-modal contrastive loss and intra-modal contrastive loss are beneficial to enhance discrimination ability of features. Intra-modal contrastive learning emphasizes the similarity within modality, while cross-modal contrastive learning models the correlation between image and text modalities, increasing the similarity of multi-modal within-class features and distinguishing dissimilar features from different modalities. (3) The graph structure preserving module leverages the representation learning capability of the graph network and makes neighbor information as an auxiliary information with respect to semantic information, which can fully explore and utilize the useful information in the multiple modalities. (4) The multi-modal fusion module takes the importance of different modalities into account with leveraging attention mechanism to assign optimal weights to each modality.

5   Conclusion

In this paper, we have proposed a novel fake news detection approach named inter-modal fusion network with graph structure preserving. IF-GSP reduces modality differences by performing inter-modal cross-layer fusion. With the graph structure information preserving being taken into consideration, it performs feature reconstruction and graph learning, such that the graph structure information is embedded to make features have strong discriminative ability. The multi-modal fusion module provides an effective scheme for feature fusion, which makes the complementarity between image and text modalities fully explored. Experimental studies on two widely used datasets, *i.e.,* Twitter and Weibo, show that IF-GSP is effective and outperforms the state-of-the-art models. Ablation experiments also show that each module of IF-GSP has its own significance.

Acknowledgement. This work was supported by the National Natural Science Foundation of China (No. 62076139), 1311 Talent Program of Nanjing University of Posts and Telecommunications, and Postgraduate Research & Practice Innovation Program of Jiangsu Province (No. SJCX22_0289).

References

1. Bian, T., et al.: Rumor detection on social media with bi-directional graph convolutional networks. In: AAAI Conference on Artificial Intelligence, pp. 549–556 (2020)
2. Boididou, C., Papadopoulos, S., Zampoglou, M., Apostolidis, L., Papadopoulou, O., Kompatsiaris, Y.: Detection and visualization of misleading content on Twitter. Int. J. Multimed. Inf. Retr. **7**(1), 71–86 (2018)
3. Chen, T., Kornblith, S., Norouzi, M., Hinton, G.: A simple framework for contrastive learning of visual representations. In: International Conference on Machine Learning, pp. 1597–1607 (2020)
4. Chen, T., Li, X., Yin, H., Zhang, J.: Call attention to rumors: deep attention based recurrent neural networks for early rumor detection. In: Trends and Applications in Knowledge Discovery and Data Mining, pp. 40–52 (2018)
5. Chen, Y., et al.: Cross-modal ambiguity learning for multimodal fake news detection. In: The World Wide Web Conference, pp. 2897–2905 (2022)
6. Hamilton, W., Ying, Z., Leskovec, J.: Inductive representation learning on large graphs. In: Advances in Neural Information Processing Systems, pp. 1024–1034 (2017)
7. He, K., Fan, H., Wu, Y., Xie, S., Girshick, R.: Momentum contrast for unsupervised visual representation learning. In: IEEE/CVF Conference on Computer Vision and Pattern Recognition, pp. 9729–9738 (2020)
8. Jin, Z., Cao, J., Guo, H., Zhang, Y., Luo, J.: Multimodal fusion with recurrent neural networks for rumor detection on microblogs. In: ACM International Conference on Multimedia, pp. 795–816 (2017)
9. Jin, Z., Cao, J., Zhang, Y., Zhou, J., Tian, Q.: Novel visual and statistical image features for microblogs news verification. IEEE Trans. Multimed. **19**(3), 598–608 (2016)

10. Devlin, J., Chang, M.-W., Lee, K., Toutanova, K.: BERT: pre-training of deep bidirectional transformers for language understanding. In: North American Chapter of the Association for Computational Linguistics: Human Language Technologies, pp. 4171–4186 (2019)
11. Khattar, D., Goud, J.S., Gupta, M., Varma, V.: MVAE: multimodal variational autoencoder for fake news detection. In: The World Wide Web Conference, pp. 2915–2921 (2019)
12. Khosla, P., et al.: Supervised contrastive learning. In: Advances in Neural Information Processing Systems, pp. 18661–18673 (2020)
13. Kipf, T.N., Welling, M.: Semi-supervised classification with graph convolutional networks. arXiv preprint arXiv:1609.02907 (2016)
14. Ma, J., et al.: Detecting rumors from microblogs with recurrent neural networks. In: International Joint Conference on Artificial Intelligence, pp. 3818–3824 (2016)
15. Qi, P., Cao, J., Yang, T., Guo, J., Li, J.: Exploiting multi-domain visual information for fake news detection. In: IEEE International Conference on Data Mining, pp. 518–527 (2019)
16. Qian, F., Gong, C., Sharma, K., Liu, Y.: Neural user response generator: fake news detection with collective user intelligence. In: International Joint Conference on Artificial Intelligence, pp. 3834–3840 (2018)
17. Scarselli, F., Gori, M., Tsoi, A.C., Hagenbuchner, M., Monfardini, G.: The graph neural network model. IEEE Trans. Neural Netw. 20(1), 61–80 (2008)
18. Vaibhav, V., Mandyam, R., Hovy, E.: Do sentence interactions matter? Leveraging sentence level representations for fake news classification. In: Graph-Based Methods for Natural Language Processing, pp. 134–139 (2019)
19. Veličković, P., Cucurull, G., Casanova, A., Romero, A., Lio, P., Bengio, Y.: Graph attention networks. arXiv preprint arXiv:1710.10903 (2017)
20. Wang, L., Zhang, C., Xu, H., Zhang, S., Xu, X., Wang, S.: Cross-modal contrastive learning for multimodal fake news detection. arXiv preprint arXiv:2302.14057 (2023)
21. Wang, Y., et al.: EANN: event adversarial neural networks for multi-modal fake news detection. In: ACM SIGKDD International Conference on Knowledge Discovery & Data Mining, pp. 849–857 (2018)
22. Wu, Y., Zhan, P., Zhang, Y., Wang, L., Xu, Z.: Multimodal fusion with co-attention networks for fake news detection. In: Findings of the Association for Computational Linguistics, ACL-IJCNLP, pp. 2560–2569 (2021)
23. Xue, J., Wang, Y., Tian, Y., Li, Y., Shi, L., Wei, L.: Detecting fake news by exploring the consistency of multimodal data. Inf. Process. Manage. 58(5), 102610 (2021)
24. Yu, F., Liu, Q., Wu, S., Wang, L., Tan, T., et al.: A convolutional approach for misinformation identification. In: International Joint Conference on Artificial Intelligence, pp. 3901–3907 (2017)
25. Zhang, H., Fang, Q., Qian, S., Xu, C.: Multi-modal knowledge-aware event memory network for social media rumor detection. In: ACM International Conference on Multimedia, pp. 1942–1951 (2019)
26. Zhou, X., Wu, J., Zafarani, R.: SAFE: similarity-aware multi-modal fake news detection. In: Lauw, H.W., Wong, R.C.-W., Ntoulas, A., Lim, E.-P., Ng, S.-K., Pan, S.J. (eds.) PAKDD 2020. LNCS (LNAI), vol. 12085, pp. 354–367. Springer, Cham (2020). https://doi.org/10.1007/978-3-030-47436-2_27

Learning to Match Features
with Geometry-Aware Pooling

Jiaxin Deng[1,2], Xu Yang[1,2], and Suiwu Zheng[1,3]([✉])

[1] State Key Laboratory of Multimodal Artificial Intelligence, Institute of
Automation, Chinese Academy of Sciences, Beijing 100190, China
{dengjiaxin2021,xu.yang}@ia.ac.cn

[2] University of Chinese Academy of Sciences, Beijing 100190, China

[3] Huizhou Zhongke advanced manufacturing Limited Company, Huizhou 516000,
China
suiwu.zheng@ia.ac.cn

Abstract. Finding reliable and robust correspondences across images
is a fundamental and crucial step for many computer vision tasks, such
as 3D-reconstruction and virtual reality. However, previous studies still
struggle in challenging cases, including large view changes, repetitive
pattern and textureless regions, due to the neglect of geometric constraint in the process of feature encoding. Accordingly, we propose a
novel GPMatcher, which is designed to introduce geometric constraints
and guidance in the feature encoding process. To achieve this goal, we
compute camera poses with the corresponding features in each attention
layer and adopt a geometry-aware pooling to reduce the redundant information in the next layer. By these means, an iterative geometry-aware
pooing and pose estimation pipeline is constructed, which avoids the
updating of redundant features and reduces the impact of noise. Experiments conducted on a range of evaluation benchmarks demonstrate that
our method improves the matching accuracy and achieves the state-of-
the-art performance.

Keywords: Image matching · Pose estimation · Transformer

1 Introduction

Accurate and robust correspondences for local features between image pairs are
essential and crucial for many geometric computer vision tasks, including structure from motion (SfM) [1,2], simultaneous localization and mapping (SLAM)
[3,4], and visual localization [5].

Traditional methods usually involve a classical pipeline, where correspondences are established by matching detected and described keypoints, and then
outlier estimators, e.g. RANSAC [6,7], are employed to find the final matches.
However, these methods tend to struggle in challenging cases such as large viewpoint variations, changing appearances and textureless regions, due to the lack
of enough repetitive keypoints and context information.

B. Luo et al. (Eds.): ICONIP 2023, LNCS 14452, pp. 292–302, 2024.
https://doi.org/10.1007/978-981-99-8076-5_21

To mitigate this problem, the advanced matchers such as SuperGlue [8] propose to use an attention-based graph neural network, which utilizes global information from all keypoints, to implement contextual descriptor augmentation. Yet, their excellent matching performance lies on the quadratic time complexity of the attention computation, which reduces the efficiency in real applications. While some following works introduce more efficient variations [9,10], they suffer from less accurate and robust performance.

Some other studies, including LoFTR [11], ASpanFormer [12], etc., focus on the detector-free image matching pipeline with transformers, as the performance of previous matching methods extremely depend on the stage of keypoint detection and description, which struggles in challenging cases. These methods have obtained remarkable performance, especially in the textureless regions. However, the operations of flatten in transformer destroy the inherent 2D-structure of the images. The consistence of corresponding features and geometric constraints are not considered, causing the failure in cases with large-scale viewpoint changes, as they sink into local similarities and fail to find the most crucial correspondences.

In this paper, we introduce a geometry-aware matching framework based on transformers, which aims to find the dominant features which are able to give accurate matches and exclude the uninformative noise based on the geometric constraints. Inspired by the previous method LoFTR [11], we use self and cross attention to collect global information and update the feature maps. Based on the observation that the attention scores in each layer represent the magnitude of correlation with other features, we can easily distinguish features which include key messages. The central challenge of our framework is to combine the geometric constraints with the attention-based module, while we solve this problem by computing the pose layer by layer, which aims to guide the pool operation.

We summarize our contributions in three aspects: 1) We propose a framework that combines the pose estimation and feature augmentation for feature matching, which enable geometric consistence in the process of feature encoding. 2) A novel geometry-aware pooling operation is proposed, which avoid unnecessary feature updation and reduce the noise interference. 3) State-of-the-arts results on extensive set of benchmarks are achieved. Our method outperforms the current matching pipeline in the benchmark of homography estimation, outdoor and indoor relative pose estimation. Evaluation on the challenging visual localization also prove the competitiveness of our method.

2   Related Works

Local Feature Matching. The traditional matching pipeline usually involves detecting [13–16], describing and matching a set of keypoints. The heuristic matching methods, such as NN matching and its variants [17,18], are widely used to find the correspondences for a long time. Despite their fast performance, these approaches usually suffer from large appearance changes. To address this issue, SuperGlue [8] introduces transformers with self and cross attention to utilize contextual information and achieve impressive performance. However, the quadratic complexity of attention operation reduces the efficiency. Some variants [9,10] attempt to improve the efficiency by optimizing the structure, but they sacrifice the accuracy to a certain extent. Another shortage of these detector-based methods is that the performance hardly depends on the features of the detector, which cannot guarantee robust and reliable performance in challenging cases.

To capture richer contexts in extreme conditions, such as low texture areas and repetitive patterns, the detector-free methods [11,12,19,20] reduce the detection stage and conduct the matching task end-to-end. Due to the strong ability of feature encoding and interaction with transformers, this series of methods hardly outperform the detector-based methods above. However, due to the lack of geometric constraints, they still fail to handle the cases with large viewpoint changes.

Efficient Attention. Numerous approaches have been proposed to address the quadratic time complexity of the attention mechanism [21]. These methods aim to reduce the complexity by incorporating various techniques such as learning a linear projection function [22,23], employing a token selection classifier [24, 25], or utilizing shared attention [26], among others. While these methods are usually designed for special downstream tasks, transferring them into the feature correspondence task directly is not feasible. In contrast, our approach is proposed to combine the geometric constrains and the efficient attention, and adjusts the pool operation adaptively.

3   The Proposed Method

The pipeline of our network structure is shown in Fig. 1. Our framework processes in a coarse-to-fine manner. Taking an image pair I_A, I_B as input, the network first uses a CNN backbone to extract initial coarse features $F_A^0, F_B^0 \in R^{\frac{H}{8} \times \frac{W}{8}}$ and fine feature maps $\hat{F}_A^0, \hat{F}_B^0 \in R^{\frac{H}{2} \times \frac{W}{2}}$. The coarse maps are passed through our geometry-aware attention model to establish coarse matches, while the fine features are used to refine the final result.

3.1   Feature Encoder

CNN is widely used for local features extraction due to the inductive bias of translation equivariance [13–16]. We use a convolutional neural network as the feature encoder, which extract both coarse features and fine features for each image separately. Specifically, we extract 1/8 of the original image dimension as coarse features and 1/2 of the original image dimension as fine features [11]. Compared to using the raw images directly, this design reduces the number of features to be processed.

3.2   Geometry-Aware Attention Module

Positional Encoding. As the previous methods, we first use a multi-layer perception (MLP) denoted as f_{enc}, where the features become position-dependent.

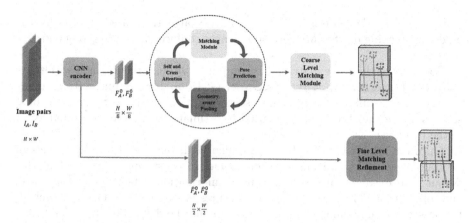

Fig. 1. Pipeline of our method. As features become more discriminative, more correct matches can be found, leading to more precise poses. The pose is utilized to provide geometric guidance to find more matches and discard redundant keypoints with geometry-aware pooling.

Specifically, for each feature $\mathbf{f_i}$ in the feature map, the output $\mathbf{f_i'} = f_{enc}(\mathbf{f_i}, \mathbf{p_i})$, where $\mathbf{p_i}$ denotes the position coordinate of $\mathbf{f_i}$. For simplicity, we denote the position-encoded coarse feature maps as F_A^0, F_B^0 backwards, corresponding to image pairs I_A, I_B respectively.

Feature Augmentation. Following the previous works [8, 11], the feature maps are fed into the self and cross attention module for feature augmentation. In the Vision Transformer, attention is adopted in the set of query (denoted as Q), key (denoted as K), and value (denoted as V). For the feature correspondence task, taken the flatten feature maps F_A^i, F_B^i as input, the output G_A^i, G_B^i can be calculated as:

$$G_A^i = \mathbf{FFN}(F_A^i, \; softmax(F_A^i {F_A^i}^T)F_A^i, \; softmax(F_A^i {F_B^i}^T)F_B^i) \tag{1}$$

$$G_B^i = \mathbf{FFN}(F_B^i, \; softmax(F_B^i {F_B^i}^T)F_B^i, \; softmax(F_B^i {F_A^i}^T)F_A^i) \tag{2}$$

where **FFN** denotes a feed forward network, which involves concatenation, layer normalization and linear layers.

Geometry-Aware Pooling. Based the observation that rebundant and uninformative features are ubiquitous in the attention layers, which not only increase computational costs, but also provide noise and interference information. Moreover, local similarities caused by repetitive patterns can easily lead to the inconsistence of the correspondences. To solve these problems, we use geometry-aware pooling, which delete the features with less potential of correspondences and avoid numerous invalid calculation. Different from the matching pipeline in LoFTR [11], the pose is estimated in every attention-based layer, which is used for screening out feature correspondences conforming to the geometric constraints.

Specifically, the score matrix S_i is computed through the augmented feature maps G_A^i, G_B^i, where $S_i(x, y) = \frac{1}{\theta}<G_A^i(x), G_B^i(y)>$. Then a dual-softmax operator is used to obtain the matching scores M_i. The matches whose matching score is above the threshold θ_m is used to estimate the fundamental matrix P_i in the i-th layer.

Once the fundamental matrix P_i is calculated, the operation of pooling can be conducted under the guidance of geometric constraints. The goal of this design is to extract features with crucial information, which give an accurate and robust pose. Therefore, we choose features in two aspects: 1) The matching features whose matching score is above the threshold θ_m, denoted as H_A^i, H_B^i respectively. 2) We use the self and cross attention score matrix (denoted as $R_S^i, R_C^i \in R^{m \times n \times h}$) in the i-th layer, to screen out the features with more complex relation. We choose the features whose average self and cross score surpass the threshold of θ_s and θ_c respectively, denoted as J_A^i, J_B^i, as well as their corresponding features in the other image under the guidance of fundamental matrix P_i, denoted as K_A^i, K_B^i.

Finally, we get the sets of features after the operation of geometry-aware pooling, as

$$F_A^{i+1} = \{H_A^i \cup J_A^i \cup K_A^i\} \subseteq G_A^i \qquad (3)$$

$$F_B^{i+1} = \{H_B^i \cup J_B^i \cup K_B^i\} \subseteq G_B^i \qquad (4)$$

3.3   Matching Determination

We use the same scheme as LoFTR [11] to obtain final correspondences, including a coarse matching stage and a refinement stage. Similar to the process of matching score calculation, the coarse matching stage conducts a dual-softmax operation on the score matrix S, where $S(x, y) = \frac{1}{\theta}<G_A(x), G_B(y)>$ (θ is a temperature parameter). Once the coarse matches M_c are generated, they would be fed into the correlation-based refinement block to get the final matching results, which is the same as LoFTR [11].

3.4   Loss Formulation

Our loss is composed of three aspects, including coarse-level loss L_c, fine-level loss L_f, as well as the pose-consistency loss L_p.

Following LoFTR [11], The coarse-level loss L_c is the cross entropy loss of the dual-softmax score matrix P

$$L_c = -\frac{1}{|M_{gt}|} \sum_{(i,j) \in M_{gt}} log(P(i, j)). \qquad (5)$$

where M_{gt} is the ground truth matches.

The fine-level loss L_f is supervised by the L2-distance between the final matches $M_f(i, j)$ and ground truth reprojection coordinates, which is formulated in detail in LoFTR [11].

In each layer, we calculate the pose-consistency loss L_p^i, as

$$L_p^i = \alpha||P^i - P_{gt}|| + \frac{1}{N_{gt}} \sum_k f_{sampson}(P^i, a_k^{gt}, b_k^{gt}). \qquad (6)$$

where P^i, P_{gt} are predict fundamental matrix in the i-th layer and the ground truth fundamental matrix, (a_k^{gt}, b_k^{gt}) is a ground truth match and N_{gt} is the total numbers of matching pairs, $f_{sampson}$ is the Sampson distance which is formulated in detail in [26].

The final loss is calculated as

$$L = L_c + L_f + \frac{1}{T} \sum_i^T L_p^i \qquad (7)$$

where T is the number of attention-based layers.

4   Experiments

In this section, we demonstrated the performance of our method on several donwnstream tasks, including homography estimation, pose estimation and visual localization.

4.1   Homography Estimation

Dataset. We evaluate our method on the HPatchses dataset [28], which is widely used for the evaluation of local descriptors and contains totally 108 sequences under large illumination and viewpoint changes.

Baselines. We compare our methods with current state-of-the-art methods: 1) detector-based methods, including D2-Net [16], R2D2 [13], Patch2Pix [29], SuperGlue [8] and SGMNet [9] which are on top of SuperPoints [14]. 2) detector-free methods, including DRC-Net [30], LoFTR [11], QuadTree [19] and Aspanformer [12].

Evaluation Metrics. Following [11], We compute the corner error between the images warped with estimated homograpy matrix and the ground truth homograpy matrix. Then we set the threshold of $3/5/10$ to identify the correctness, and the area under the cumulative curve (AUC) of the corner error indicates the matching accuracy.

Result. The results of homography estimation is shown in Table 1. The AUC of the corner error is demonstrated, and the best performance is in bold. As can be seen, the proposed method outperformed the existing stat-of-the-art methods on the threshold of $3/5$ px, and achieves competitive performance on the threshold of 10 px.

Table 1. Results of homography estimation on the HPathes dataset

Category	Method	Homography est. AUC(%)		
		@3 px	@5 px	@10 px
Detector-based	D2-Net [16]+NN	23.2	35.9	53.6
	R2D2 [13]+NN	50.6	63.9	76.8
	Patch2Pix [29]	59.2	69.6	80.9
	SP [14]+SuperGlue [8]	53.9	67.4	81.4
	SP [14]+SGMNet [9]	51.8	65.4	80.3
Detector-free	DRC-Net [30]	50.6	56.1	68.4
	LoFTR [11]	65.9	75.6	84.6
	QuadTree [19]	64.8	76.7	84.4
	Aspanformer [12]	66.1	77.1	**85.3**
	GPMatcher (ours)	**66.8**	**77.3**	85.1

4.2 Relative Pose Estimation

Dataset. We evaluate pose estimation on indoor dataset ScanNet [31] and outdoor dataset Megadepth [32] respectively.

The ScanNet dataset [31]is an indoor 2D-3D dataset that collects 1513 indoor scenes, including rgb, depth, and three-dimensional point cloud data. We choose 1500 RGB image pairs for testing, which contains large viewpoint changes, extensive repetitive patterns, and abundant textureless regions.

The Megdepth dataset [31] contains 1M Internet images of 196 outdoor scenes, each of which has been reconstructed by COLMAP. Similarly, we choose 1500 test pairs for the following evaluation.

Evaluation Metrics. Following [11], We first compute the pose error, which is defined as the maximum of angular error in rotation and translation between the predict and ground truth. Then we calculate the AUC of the above error up to the threshold of $5°/10°/20°$, which indicates the accuracy of pose estimation.

Result. The results of indoor and outdoor relative pose estimation is shown in Tables 2 and 3, respectively. The AUC of the pose error is demonstrated, and the best performance is in bold. It is demonstrated that our method achieves overall best performance on both indoor and outdoor pose estimation among all the relative methods.

4.3 Visual Localization

Dataset. We choose the Aachen Day-Night v1.1 datasets [33] for our visual localization evaluation, which consists of a scene model built upon 6697 images with 824 day-time images and 197 night-time images as queries.

Table 2. Evaluation on ScanNet dataset [31] for indoor pose estimation

Method	Indoor pose est. AUC(%)		
	@5°	@10°	@20°
D2-Net [16] + NN	5.2	14.5	27.9
SP [14] + SuperGlue [8]	16.2	33.8	51.8
SP [14] + SGMNet [9]	15.4	32.1	48.3
DRC-Net [30]	7.7	17.9	30.5
LoFTR [11]	16.9	33.6	50.6
QuadTree [19]	24.9	44.7	61.8
Aspanformer [12]	25.6	46.0	63.3
GPMatcher (ours)	**26.1**	**47.1**	**63.7**

Table 3. Evaluation on Megadepth [32] dataset for outdoor pose estimation

Method	Outdoor pose est. AUC (%)		
	@5°	@10°	@20°
SP [14] + SuperGlue [8]	42.2	61.2	75.9
SP [14] + SGMNet [9]	40.5	59.0	73.6
DRC-Net	27.0	42.9	58.3
LoFTR [11]	52.8	69.2	81.2
QuadTree [19]	54.6	70.5	82.2
Aspanformer [12]	55.3	71.5	83.1
GPMatcher (ours)	**55.6**	**71.9**	**83.3**

Table 4. Evaluation on Aachen Day-Night v1.1 datasets [33] for outdoor visual localiztion

Method	Day	Night
	$(0.25\,\mathrm{m}, 2°)/(0.5\,\mathrm{m}, 5°)/(1\,\mathrm{m}, 10°)$	
SP [14]+SuperGlue [8]	**89.8**/**96.1**/**99.4**	77.0/90.6/**100.0**
LoFTR [11]	88.7/95.6/99.0	78.5/90.6/99.0
Aspanformer [12]	89.4/95.6/99.0	77.5/**91.6**/99.5
GPMatcher (ours)	89.6/**96.2**/98.9	**79.0**/91.3/99.5

Evaluation Metrics. Following the Long-Term Visual Localization Benchmark [34], we report the accuracy of localization with matching pairs generated by HLoc.

Result. The evaluation of outdoor visual localization on Aachen Day-Night v1.1 datasets [33] is shown in Table 4. It is demonstrated that our method achieves best performance on some cases and competitive performance on the other cases.

4.4 Ablation Study

To validate the effectiveness of different design components of our method, we conduct ablation experiments on the Megadepth dataset [32]. The experiment is conducted as Sect. 4.2 above. Specifically, we compare three designs of pooling structure: 1) None pooling operation: A design which hold all of the features. 2) Semi-pooling: Pooling according to the attention scores, which discard 1/4 features in each layer. 3) Geometry-aware pooling: Full design of our proposed method.

Table 5. Ablation Study on Megadepth dataset [32]

Method	Outdoor pose est. AUC(%)		
	@5°	@10°	@20°
None pooling	52.4	69.2	81.1
Semi-pooling	53.5	69.3	81.6
Full design	**55.6**	**71.9**	**83.3**

As in presented Table 5, our method improve overall performance by a considerable margin, validating the essentiality of our network designs.

5 Conclusion

In this paper, we have proposed a novel geometric-aware feature matching framework GPMatcher based on transformers, which is capable of combining the geometric consistence into the attention-based module. State-of-the-art results validates the effectiveness of our method. We believe that this paper would potentially provide new insights on learning feature matching. With more engineering optimizations, we are looking forward to wider application of our method in real use.

Acknowledgement. This work was supported by the National Key Research and Development Program of China under Grant 2020AAA0105900, and partly by National Natural Science Foundation (NSFC) of China (grants 61973301, 61972020), and partly by Youth Innovation Promotion Association CAS.

References

1. Schonberger, J.L., Frahm, J.M.: Structure-from-motion revisited. In: CVPR (2016)
2. Resindra, A., Torii, A., Okutomi, M.: Structure from motion using dense CNN features with keypoint relocalization. IPSJ Trans. Comput. Vis. Appl. **10**, 6 (2018)
3. Mur-Artal, R., Montiel, J.M.M., Tardos, J.D.: ORB-SLAM: a versatile and accurate monocular slam system. IEEE Trans. Robot. **31**, 1147–1163 (2015)

4. Mur-Artal, R., Tardos, J.: ORB-SLAM2: an open-source slam system for monocular, stereo and RGB-D cameras. IEEE Trans. Robot. **33**, 1255–1262 (2016)
5. Sattler, T., Weyand, T., Leibe, B., Kobbelt, L.: Image retrieval for image-based localization revisited. In: BMVC (2012)
6. Fischler, M.A., Bolles, R.C.: Random sample consensus: a paradigm for model fitting with applications to image analysis and automated cartography. Commun. ACM **24**, 381–395 (1981)
7. Barath, D., Matas, J., Noskova, J.: MAGSAC: marginalizing sample consensus. In: CVPR (2019)
8. Sarlin, P.-E., DeTone, D., Malisiewicz, T., Rabinovich, A.: SuperGlue: learning feature matching with graph neural networks. In: CVPR (2020)
9. Chen, H., et al.: Learning to match features with seeded graph matching network. In: ICCV (2021)
10. Shi, Y., Cai, J.-X., Shavit, Y., Mu, T.-J., Feng, W., Zhang, K.: ClusterGNN: cluster-based coarse-to-fine graph neural network for efficient feature matching. In: CVPR (2022)
11. Sun, J., Shen, Z., Wang, Y., Bao, H., Zhou, X.: LoFTR: detector-free local feature matching with transformers. In: CVPR (2021)
12. Chen, H., et al.: ASpanFormer: detector-free image matching with adaptive span transformer. In: Avidan, S., Brostow, G., Cissé, M., Farinella, G.M., Hassner, T. (eds) Computer Vision, ECCV 2022. LNCS, vol. 13692, pp. 20–36. Springer, Cham (2022). https://doi.org/10.1007/978-3-031-19824-3_2
13. Revaud, J., et al.: R2D2: repeatable and reliable detector and descriptor. In: NeurIPS (2019)
14. DeTone, D., Malisiewicz, T., Rabinovich, A.: SuperPoint: self-supervised interest point detection and description. In: CVPRW (2018)
15. Luo, Z., et al.: ASLfeat: learning local features of accurate shape and localization. In: CVPR (2020)
16. Dusmanu, M., et al.: D2-Net: a trainable CNN for joint description and detection of local features. In: CVPR (2019)
17. Lowe, D.G.: Distinctive image features from scale-invariant keypoints. IJCV **60**, 91–110 (2004). https://doi.org/10.1023/B:VISI.0000029664.99615.94
18. Rublee, E., Rabaud, V., Konolige, K., Bradski, G.: ORB: an efficient alternative to SIFT or SURF. In: ICCV (2011)
19. Tang, S., Zhang, J., Zhu, S., Tan, P.: Quadtree attention for vision transformers. In: ICLR (2021)
20. Huang D, Chen Y, Xu S, et al.: Adaptive assignment for geometry aware local feature matching. In: CVPR (2023)
21. Vaswani, A., et al.: Attention is all you need. In: NeurIPS (2017)
22. Katharopoulos, A., Vyas, A., Pappas, N., Fleuret, F.: Transformers are RNNs: fast autoregressive transformers with linear attention. In: ICML (2020)
23. Wang, S., Li, B.Z., Khabsa, M., Fang, H., Ma, H.: LinFormer: self-attention with linear complexity. arXiv preprint (2020)
24. Fayyaz, M., et al.: Adaptive token sampling for efficient vision transformers. In: Avidan, S., Brostow, G., Cissé, M., Farinella, G.M., Hassner, T. (eds.) Computer Vision, ECCV 2022. LNCS, vol. 13671, pp. 396–414. Springer, Cham (2022). https://doi.org/10.1007/978-3-031-20083-0_24
25. Lee, J., Lee, Y., Kim, J., Kosiorek, A., Choi, S., Teh, Y.W.: Set transformer: a framework for attention-based permutation-invariant neural networks. In: ICML (2019)

26. Chen, B., et al.: PSViT: better vision transformer via token pooling and attention sharing. In: arXiv preprint (2021)
27. Hartley, R., Zisserman, A.: Multiple view geometry in computer vision. Cambridge University Press (2003)
28. Balntas, V., Lenc, K., Vedaldi, A., Mikolajczyk, K.: HPatches: a benchmark and evaluation of handcrafted and learned local descriptors. In: CVPR (2017)
29. Zhou, Q., Sattler, T., Leal-Taixe, L.: Patch2Pix: epipolar-guided pixel-level correspondences. In: CVPR (2021)
30. Li, X., Han, K., Li, S., Prisacariu, V.: Dual-resolution correspondence networks. In: NeurIPS (2020)
31. Dai, A., Chang, A.X., Savva, M., Halber, M., Funkhouser, T., Nießner, M.: ScanNet: richly-annotated 3D reconstructions of indoor scenes. In: CVPR (2017)
32. Li, Z., Snavely, N.: MegaDepth: learning single-view depth prediction from internet photos. In: CVPR (2018)
33. Zhang, Z., Sattler, T., Scaramuzza, D.: Reference pose generation for long-term visual localization via learned features and view synthesis. IJCV **129**, 821–844 (2020)
34. Toft, C., et al.: Long-term visual localization revisited. IEEE T-PAMI **44**, 2074–2088 (2020)

PnP: Integrated Prediction and Planning for Interactive Lane Change in Dense Traffic

Xueyi Liu[1,2], Qichao Zhang[1,2(✉)], Yinfeng Gao[2,3], and Zhongpu Xia[4]

[1] State Key Laboratory of Multimodal Artificial Intelligence Systems, Institute of Automation, Chinese Academy of Sciences, Beijing 100190, China
zhangqichao2014@ia.ac.cn
[2] School of Artificial Intelligence, University of Chinese Academy of Sciences, Beijing 100049, China
[3] School of Automation and Electrical Engineering, University of Science and Technology Beijing, Beijing 100083, China
[4] Beijing, China

Abstract. Making human-like decisions for autonomous driving in interactive scenarios is crucial and difficult, requiring the self-driving vehicle to reason about the reactions of interactive vehicles to its behavior. To handle this challenge, we provide an integrated prediction and planning (PnP) decision-making approach. A reactive trajectory prediction model is developed to predict the future states of other actors in order to account for the interactive nature of the behaviors. Then, n-step temporal-difference search is used to make a tactical decision and plan the tracking trajectory for the self-driving vehicle by combining the value estimation network with the reactive prediction model. The proposed PnP method is evaluated using the CARLA simulator, and the results demonstrate that PnP obtains superior performance compared to popular model-free and model-based reinforcement learning baselines.

Keywords: Lane change · Decision-making · Reinforcement learning

1 Introduction

Recently, self-driving vehicles (SDVs) have received a great deal of attention from academic and industry communities. With the advancement of Artificial Intelligence technology, data-driven algorithms are being applied progressively to automatic driving decision-making systems, demonstrating enormous potential [5,9,18]. Due to the strong interactivity and uncertainty in complex interaction scenarios, SDVs still face numerous challenges to make human-like behaviors such as lane changes in dense traffic [16], unprotected left turn at intersections [11]. Generally, humans are naturally able to address those social interaction scenarios, since we own an inherent ability to anticipate how other drivers will

---

Z. Xia—Independent Researcher.

B. Luo et al. (Eds.): ICONIP 2023, LNCS 14452, pp. 303–316, 2024.
https://doi.org/10.1007/978-981-99-8076-5_22

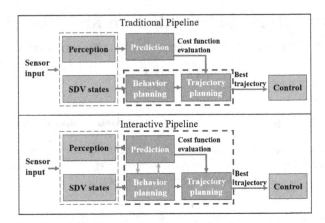

Fig. 1. The traditional and interactive pipeline in autonomous driving.

react to our actions. This is one of the primary reasons why SDVs can not handle complex interaction scenarios efficiently.

As shown in Fig. 1, in the conventional pipeline, the prediction and planning modules are tackled sequentially, where planning is the downstream task of prediction. After obtaining the predicted trajectories of other vehicles, the SDV selects a tactical decision behavior (behavior planning) and plans the optimal trajectory (motion planning) for the control module to execute. In other words, the planning results have no effect on the trajectory predictions of other system participants, leading to the passive role of the SDV in the system [10].

Recent research identifies similar challenges in the conventional pipeline and attempts to incorporate how the ego vehicle influences interactive vehicles by bridging prediction and planning [8]. Model-based DRL algorithms commonly construct a dynamic model that the agent uses to make decisions. [20] introduces an uncertainty-aware model-based reinforcement learning for end-to-end autonomous driving, resulting in improved learning efficiency and performance. In [17], dynamic horizon value estimation is designed based on the world model from Dreamer [6] for lane changes on the highway. Inspired by AlphaGo [14], the integration of tree search with learning techniques can also be effectively employed in autonomous driving systems. [2] applies the n-step temporal-difference (TD) search [15] for robot navigation, where the navigation planning process involves the utilization of a non-reactive prediction model. [19] proposes an environment model based on self-attention and develops model-based RL algorithms upon the model. In other words, the interaction information only contains the future states of the ego vehicle and other participants, while failing to account for how the behavior of the ego affects interactive vehicles.

Motivated by this, we propose PnP, a novel integrated prediction and planning learning method for interactive scenarios that demonstrates the reasoning ability of interactive vehicles' responses to the tactical behavior of the ego vehicle. The contributions of our research can be summarized as follows:

- We develop a reactive trajectory prediction model based on Long Short-Term Memory (LSTM) and Graph Convolution Network (GCN), which incorporates deduction capacity to predict how interactive vehicles react to the behavior of the ego vehicle.
- We introduce PnP, an integrated prediction and planning learning method for interactive lane changes, which enables the ego vehicle to be an active actor with interactive reasoning ability by integrating the reactive trajectory prediction with the TD search.
- To enhance the interactivity of the vehicles, we construct a challenging lane-change scenario in dense traffic using the high-fidelity simulator CARLA, and experimental results prove that PnP is superior to popular model-free and model-based reinforcement learning baselines.

2   Background

2.1   Markov Decision Process (MDP)

The decision-making process in dense traffic can be modeled as an MDP. Generally, MDP is described as a 5-tuple (S, A, T, R, γ), where S denotes the state space and A denotes the action space. $T(s_{t+1}|s_t, a_t)$ is the state transition probability that the system transitions to the next state s_{t+1} after taking action $a_t \in A$ at state $s_t \in S$, and $r_t = R(s_t, a_t)$ yields the reward for executing action a_t at state s_t. $\gamma \in (0, 1)$ is the discount factor. For the autonomous driving task, it is difficult to represent the probability distributions T and P explicitly. The policy $\pi(a|s)$ maps states to actions. At each time step t, the agent selects a feasible action a_t at current state s_t, which causes the system transitions to next state s_{t+1} with probability $T(s_{t+1}|s_t, a_t)$. Meanwhile, the agent receives a numerical reward $r_t = R(s_t, a_t)$. The goal of the agent is to learn an optimal policy by maximizing the cumulative reward:

$$\pi^* = \arg\max_{\pi} E\Big[\sum_{i=0}^{\infty} \gamma^t r_t\Big] \tag{1}$$

2.2   TD Search

TD search incorporates the value function approximation and bootstrapping into simulation-based tree search, which is much more efficient than Monte-Carlo Tree Search (MCTS) with an equal number of simulations. Using the current state of the system as the root node, the agent simulates forward search based on the transition model and reward model. And the value function is estimated by sampling data from the current policy. After every n-step of simulation, the value function can be updated employing TD learning:

$$V(s_t) \leftarrow V(s_t) + \alpha \left(\sum_{m=0}^{n-1} \gamma^m r_{t+m} + \gamma^n V(s_{t+n}) - V(s_t)\right) \tag{2}$$

where $V(s_t)$ denotes the value estimation for s_t, and α is the learning rate.

2.3 Trajectory Prediction

As a model-based learning method, the environment model is required for TD search. Note that the reward function can be designed explicitly, and the unknown transition model should be reconstructed. Different from the world model in [17], we utilize the reactive trajectory prediction to traverse the tree to the next state for interactive vehicles over multiple time steps. The trajectory prediction problem can be formulated as utilizing the past states of traffic participants to estimate their future states, i.e., to obtain the future waypoints of N vehicles. Let $X_t^{ego} = \{x_t^0\}$ represents the state of the ego vehicle at time t and $X_t^{nei} = \{x_t^1, x_t^2, ..., x_t^N\}$ represents other vehicles', where $x_t^i \in \mathbb{R}^d$ is the d dimensional state of the i-th actor. Given the decision action of the ego vehicle at time t, the future path $X_{t+1:t+T-1}^{ego}$ can be obtained according to the trajectory planning module. Using past states sequence $X_{t-\tau:t}^{nei}$, prediction module should forecast the future waypoints $\hat{X}_{t+1:t+T}^{nei}$, where T represents the prediction length.

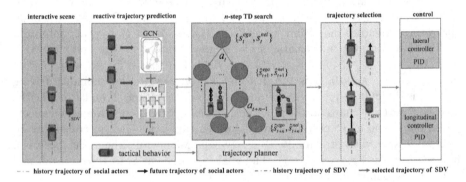

Fig. 2. Schematic diagram of the proposed method PnP.

3   Method

To address interactive lane-change scenarios, we propose the PnP framework illustrated in Fig. 2. It consists of reactive trajectory prediction and n-step TD search. With the tactical behavior query, the reactive trajectory prediction and the trajectory planner of the ego vehicle generate the next states over multiple time steps for social actors and the ego vehicle, respectively. With the future states and the designed reward function, the value estimation network is trained based on n-step TD learning. Then, the expected future return for each action of the ego vehicle in the current state can be estimated more accurately. Finally, the action with the highest value is executed as the tactical behavior. The corresponding trajectory is selected for the control module.

3.1 Reactivate Trajectory Prediction

Different from previous trajectory prediction work [4,21], this section builds a reactive trajectory prediction model considering the reactions of social actors to the tactical behavior of the ego vehicle. We present a novel architecture to capture interactive features. The model performs the following operations to complete prediction tasks: (1) obtaining individual temporal features from the history trajectory of each social actor based on LSTM; (2) utilizing GCN to extract global spatial features; (3) predicting the future trajectory with the spatiotemporal features and the interaction flag in a step-wise rollout manner. The reactive trajectory prediction network is shown in Fig. 3.

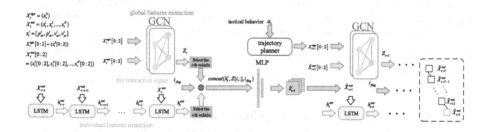

Fig. 3. Overall structure of the reactive trajectory prediction model.

LSTM for Individual Features. LSTM has been used to learn the patterns of sequential data [1]. The historical trajectories of vehicles are fed into the standard LSTM network, which can obtain the individual features of each vehicle:

$$h_t^i = f_{LSTM}(h_{t-1}^i, x_t^i), i \in \{1, 2, ..., N\} \tag{3}$$

where $x_t^i = \left[p_{xt}^i, p_{yt}^i, v_{xt}^i, v_{yt}^i\right]$ means the input of LSTM cell. p_x^i, p_y^i and v_x^i, v_y^i represent the longitudinal and lateral position and velocity of i-th agent in two coordinate directions, respectively. Finally, the individual temporal features of the environmental vehicles are obtained from the LSTM network:

$$h_t^{nei} = f_{LSTM}(X_{t-\tau:t}^{nei}) = \{h_t^1, h_t^2, ..., h_t^N\} \tag{4}$$

GCN for Global Features. GCN is a deep learning model for processing graph-structured data, capable of extracting global information from a spatial perspective [13]. In the task of vehicle trajectory prediction, we employ a two-layer GCN with a symmetric adjacency matrix. Taking the positions $X_t^{ego}[0:2], X_t^{nei}[0:2]$ as input, the information goes through a feature extraction layer, then extracted features are concatenated together to form the node feature matrix $H^{(0)}$:

$$H^{(0)} = f_{concat}(f_r(X_t^{ego}[0:2]), f_h(X_t^{nei}[0:2])) \tag{5}$$

where f_{concat} is the connecting function, f_r, f_h are the extract layers.

The adjacency matrix adjusts the connection relationships between nodes by transforming and integrating features. After applying the softmax operation, the weights of connections between nodes are converted into a probability distribution along the row dimension. By multiplying and normalizing with node features, effective information propagation and feature aggregation can be achieved on the graph.

$$H^{(1)} = \text{Softmax}(H^{(0)}W_a H^{(0)^T})H^{(0)}W_s^{(0)}$$
$$Z_t = H^{(2)} = \text{Softmax}(H^{(1)}W_a H^{(1)^T})H^{(1)}W_s^{(1)} \tag{6}$$

where the weights $W_a, W_s^{(0)}, W_s^{(1)}$ are optimized through gradient descent. H represents the node feature matrix after dimension transformation, and the output Z_t represents the global spatial features. The entire graph convolutional network computation process can be represented by the following formula:

$$Z_t = f_{GCN}(X_t^{ego}[0:2], X_t^{nei}[0:2]) \tag{7}$$

Reactive Trajectory Prediction Combining the Interaction Signal: In the reactive trajectory prediction model, we concatenate the corresponding spatiotemporal features with the interaction signal i_{flag} between the ego vehicle and social vehicles. As the interactive signal, when there is an interaction between the environment vehicle and the ego vehicle, i_{flag} is set to 1; in other cases, it is set to 0. After the connected features are fed into the fully connected layer, the future state for each social vehicle at the next time step can be predicted.

$$\hat{X}_{t+1}^{nei} = f_{MLP}(h_t^{nei}, Z_t, i_{flag}) \tag{8}$$

According to the tactical behavior a_t of the ego vehicle, we can obtain the future positions of the ego vehicle by the trajectory planner. With the future states of the ego vehicle and social vehicles, the individual and global features from LSTM and GCN can be updated sequentially. Finally, we can obtain the predicted future trajectories of all social vehicles:

$$\hat{X}_{t+1:t+T}^{nei} = f_P(X_{t-\tau:t}^{nei}, X_{t+1:t+T-1}^{ego}, i_{flag}) \tag{9}$$

For the training of the reactive trajectory prediction model, the root mean squared error (RMSE) is chosen as the loss function:

$$J = \sqrt{\frac{1}{T}\sum_{j=1}^{T}\left(x_{t+j}^i - \hat{x}_{t+j}^i\right)^2} \tag{10}$$

where $\hat{x}_{t+j}^i$ is the predicted future state of i-th social vehicle at time step $t+j$, and x_{t+j}^i is the corresponding ground truth.

3.2   Integrated Prediction and Planning

In this section, we introduce the integrated PnP based on the TD search method for handling the interactive lane-change task in dense traffic. As a model-based algorithm, TD search utilizes tree structures to simulate and update the value function. In the driving task, the reactive trajectory prediction model is learned as the transition function of interactive vehicles, considering the reciprocal influences among these vehicles. Firstly, introduce the decision-making problem setup for lane changes.

State Space. The system state includes the position information of the ego vehicle and other vehicles. The state at time t is represented as the following formula:

$$s_t = \{s_t^{ego}, s_t^{nei}\} = \left(s_t^0, s_t^1, ..., s_t^N\right) \tag{11}$$

and $s_t^i = (p_{xt}^i, p_{yt}^i), i \in \{0, 1, ..., N\}$, is also the first two dimensional characteristics of x_t^i. For the TD search algorithm, when using the value function to estimate the state value, the future state is required. The estimation of future states for environment vehicles can be derived from the trajectories inferred by the reactive prediction model. The future state of the ego vehicle is provided by the trajectory planning module with the decision action. The state at time $t+1$ can be represents as $\hat{s}_{t+1} = \{\hat{s}_{t+1}^{ego}, \hat{s}_{t+1}^{nei}\}$.

Action Space. In this work, we use a discrete action space for lane changes. The discrete actions are set as {change lane left, change lane right, and stay in the current lane}. Then, the tactical behavior is forwarded to the trajectory planner[1], which generates the future driving trajectories of the ego vehicle.

Reward Function. Safety and efficiency are crucial in the context of intense traffic. In addition, the reward function is designed for the following purposes:

- The SDV is expected to be closely oriented with the lane centerline.
- Once encountering a sluggish vehicle, the SDV is encouraged to make the change decision.
- The collision should be avoided while driving.

In general, the reward function is as follows:

$$r_t = \begin{cases} 10.0, & \text{if success} \\ -10.0, & \text{if collision} \\ 0.2 \times r_m, & \text{else} \end{cases} \tag{12}$$

where r_m is calculated by the relative distance. When the SDV stays in the current lane, the r_m term is calculated as $r_m = -(1 - |p_x^0 - p_{lc}|/(0.5 \times w_l))$, where p_x^0 is the lateral position of the ego vehicle, p_{lc} is the lateral position of the closest waypoint in current lane centerline, and w_l is the width of the current

---

[1] https://github.com/enginBozkurt/MotionPlanner.

lane. This reward encourages the ego vehicle to complete lane-change actions in the designed experimental scenarios.

Transition Model. Transition model is required for n-step TD search. As previously mentioned, the transition model is divided into two parts for PnP: the future states of social vehicles are predicted by reactive trajectory prediction network $f_P(X_{t-\tau:t}^{nei}, X_{t+1:t+T-1}^{ego}, i_{flag})$; the future state of the ego vehicle is obtained according to the trajectory planner.

After constructing the decision model, our temporal-difference search method can be employed. TD search is a simulation-based search algorithm, where the agent simulates from a root state and samples with the transition and reward model. Due to the low dimensional action space, we can simulate the experience with a limited rollout depth.

A neural network f_V, with parameters θ_V, is used to approximate the value function $V(s_t)$. The value of current state s_t is estimated using the following formula, with a rollout depth of d:

$$f_V(s_t) = r(s_t, a_t) + \sum_{m=1}^{d-1} \gamma^m r(\hat{s}_{t+m}, a_{t+m}) + \gamma^d f_V(\hat{s}_{t+d}) \tag{13}$$

To balance the exploration and exploitation during the learning process, the ϵ-greedy policy is adopted to choose the action:

$$a_t = \begin{cases} \arg\max_{a_t \in A} r(s_t, a_t) + \gamma f_V(\hat{s}_{t+1}), \textit{with } 1 - \epsilon_t \textit{ probability} \\ random \ \ a_t \in A, \ \ with \ \epsilon_t \ probability \end{cases} \tag{14}$$

To make the training process more stable, a target value network is adopted. The update frequency of the target value network is slower compared to the value network, which benefits the convergence of the network. We use $f_{V'}$ represents the target network with parameters $\theta_{V'}$. The network parameters are updated through gradient descent. After adding the target network, the TD error is represented by the following formula.

$$\delta = r(s_t, a_t) + \sum_{m=1}^{d-1} \gamma^m r(\hat{s}_{t+m}, a_{t+m}) + \gamma^d f_{V'}(\hat{s}_{t+d}) - f_V(s_t) \tag{15}$$

4   Experiments

4.1   Experiment Setup

To verify the performance of the proposed algorithm, we select the high-fidelity simulator CARLA [3], which can customize the test scenario and generate personalized traffic flow. Based on the CARLA simulator, we construct a passive lane-change scenario in dense traffic including social vehicles with different driving styles. It is a standard straight road with three lanes. The ego vehicle is initialized in the right lane. Considering accidents or temporal traffic control, the preceding vehicle gradually slows down. The ego vehicle has to make a lane change maneuver to the middle lane. This situation is frequently observed on urban roads during peak hours and is also applicable to merging scenarios.

We set two types of social vehicles in the target lane. Aggressive vehicles won't decelerate when the preceding vehicle executes a lane change. The social vehicle type is sampled to be cooperative or aggressive from a binary uniform distribution. The lane-change test scenario terminates as the ego completes a lane change, surpasses the specified position, or encounters a collision.

To prove the validity of the prediction algorithm, we chose the LSTM predictor and the same structure of the PnP predictor method without i_{flag} as the comparative approach. We use average displacement error(ADE) and final displacement error(FDE) to evaluate results. The metric ADE computes average L2 norm between the ground-truth trajectory and the predicted trajectory over all future time steps, while FDE measures the L2 norm at the final time step.

To evaluate the performance of the decision-making model, we train the value network function using the TD search method in the built scenario and compare PnP with the following methods:

- DQN [12] uses experience replay to address the issue of sample correlation, and target network to enhance the stability.
- DreamerV2 [7] samples latent states from a discrete distribution and utilizes KL balancing techniques to ensure the accuracy of model reconstruction.

In addition, the trained models are tested in the complex scenario with 100 episodes. And the evaluation metrics encompass the success rate of lane-changing and collision rate, taking into account efficiency and safety. In the testing, we check if the ego vehicle collides with other agents at every time step during the simulation. The success rate metric is calculated when the ego changes to the target lane without any collisions. The collision rate metric is calculated when the vehicle collides with other vehicles during the driving process.

4.2   Reactive Trajectory Prediction Model Evaluation

A well-balanced dataset can significantly improve the accuracy and robustness of trajectory prediction models. An interactive dataset is collected by rule-based driving vehicles in the aforementioned CARLA scenarios. The training set contains 1000 trajectories, and the testing set includes 400 trajectories. The selection of interactive vehicles is based on relative distance and the lane in which the ego vehicle is located. The slow-moving vehicle ahead of the ego vehicle is always the interactive vehicle, while the interactive vehicles in the target lane are the closest two vehicles to the ego vehicle at the current time step. In Fig. 4, the interactive vehicles are highlighted with red boxes.

After training 100 epochs in the training set, the experiment results are shown in the Table 1. The table compares the performance of different prediction algorithms, showing that the prediction model achieves the highest accuracy compared to other network models in both overall and interactive scenarios.

Table 1. Testing displacement error of different prediction models.

Scenario	Metric	LSTM	LSTM+GCN	LSTM+GCN+i_{flag}
All scenario	ADE	0.42	0.39	**0.30**
	FDE	0.89	0.72	**0.58**
Interactive scenario	ADE	0.87	0.75	**0.57**
	FDE	1.88	1.51	**1.17**

The results demonstrate that the LSTM predictor cannot accurately predict the future trajectory of vehicles with different styles. Because the model lacks consideration for the differences in how the two types of vehicles react to the behavior of the ego vehicle. Benefiting from the well-designed network structure, the reactive prediction model without interaction signals has improved the performance of the LSTM network, but still cannot perfectly distinguish the two types of social vehicles. Considering the interactive signals helps the reactive prediction model more accurately predict future states. The proposed reactive trajectory prediction model reduces the ADE and FDE in the entire scene to **0.30** and **0.58**. Additionally, the model reduces ADE and FDE to **0.57** and **1.17** in the interactive scene.

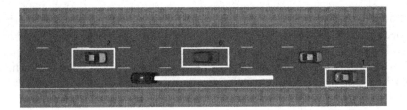

Fig. 4. Diagram of the interactive vehicles selection.

4.3   Lane Change Decision Performance

After training 250,000 steps in the scenario, the curves of each algorithm are shown in Fig. 5. The solid lines depict the mean scores and shaded areas indicate the standard deviation of five trials over different random seeds. The model-based algorithm, DreamerV2, demonstrates the highest return, followed by PnP. And PnP exhibits the least variance, indicating its high stability.

The performance of the above algorithms in the test scenario is as shown in Table 2. PnP achieves the highest success rate of **85%** in the lane-changing task, outperforming other baseline algorithms. This result demonstrates the advantage of model-based algorithms and showcases the effectiveness of integrated prediction and decision-making methods.

According to the training curve, although DreamerV2 achieves higher returns, its performance in lane-changing tasks is not as good as PnP. Through analysis, it can be determined that negative rewards are given to the vehicle when it stays in the current lane. To prevent collisions with vehicles in the target lane, the vehicle driven by PnP temporarily stays in the current lane, receiving continuous negative rewards, resulting in a lower return. Figure 6 shows the decision comparison between PnP and DreamerV2 in this scenario.

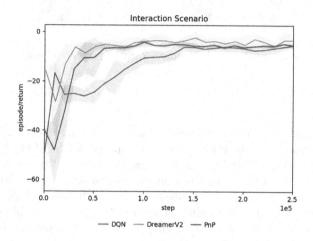

Fig. 5. Training curves of model-based and model-free methods.

Table 2. Performances of different algorithms on lane changing in dense traffic.

Model	Success rate (%)	Collision rate (%)
DQN [12]	51	49
DreamerV2 [7]	68	32
PnP (Ours)	**85**	**15**

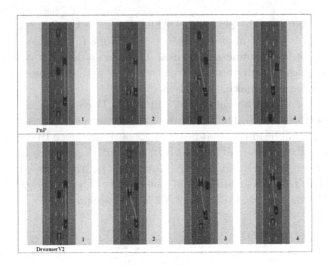

Fig. 6. The behaviors of PnP and DreamerV2 in the presence of aggressive vehicles.

5   Conclusion

This paper proposes an integrated prediction and planning learning method. And the reactive trajectory prediction model is composed of LSTM and GCN, which can extract the temporal and spatial features. Considering the interactive signal, our model can predict the diverse responses of social actors to the lane-changing behavior of the ego vehicle. By connecting the prediction and planning, PnP improves the application of prediction models to downstream planning tasks, whose success rate in the overtaking lane-change task is superior to other baselines.

Acknowledgements. This work was supported by the National Key Research and Development Program of China under Grants 2022YFA1004000, National Natural Science Foundation of China (NSFC) under Grants No. 62173325 and CCF Baidu Open Fund.

References

1. Chauhan, N.S., Kumar, N.: Traffic flow forecasting using attention enabled Bi-LSTM and GRU hybrid model. In: Tanveer, M., Agarwal, S., Ozawa, S., Ekbal, A., Jatowt, A. (eds.) Neural Information Processing, ICONIP 2022. CCIS, vol. 1794, pp. 505–517. Springer, Singapore (2023). https://doi.org/10.1007/978-981-99-1648-1_42
2. Chen, C., Hu, S., Nikdel, P., Mori, G., Savva, M.: Relational graph learning for crowd navigation. In: 2020 IEEE/RSJ International Conference on Intelligent Robots and Systems (IROS), pp. 10007–10013. IEEE (2020)

3. Dosovitskiy, A., Ros, G., Codevilla, F., Lopez, A., Koltun, V.: CARLA: an open urban driving simulator. In: Conference on Robot Learning, pp. 1–16. PMLR (2017)
4. Gu, J., Sun, C., Zhao, H.: DenseTNT: end-to-end trajectory prediction from dense goal sets. In: Proceedings of the IEEE/CVF International Conference on Computer Vision, pp. 15303–15312 (2021)
5. Guo, Y., Zhang, Q., Wang, J., Liu, S.: Hierarchical reinforcement learning-based policy switching towards multi-scenarios autonomous driving. In: 2021 International Joint Conference on Neural Networks (IJCNN), pp. 1–8. IEEE (2021)
6. Hafner, D., Lillicrap, T., Ba, J., Norouzi, M.: Dream to control: Learning behaviors by latent imagination. In: International Conference on Learning Representations (2019)
7. Hafner, D., Lillicrap, T.P., Norouzi, M., Ba, J.: Mastering Atari with discrete world models. In: International Conference on Learning Representations (2020)
8. Hagedorn, S., Hallgarten, M., Stoll, M., Condurache, A.: Rethinking integration of prediction and planning in deep learning-based automated driving systems: a review. arXiv preprint arXiv:2308.05731 (2023)
9. Li, D., Zhao, D., Zhang, Q., Chen, Y.: Reinforcement learning and deep learning based lateral control for autonomous driving [application notes]. IEEE Comput. Intell. Mag. **14**(2), 83–98 (2019)
10. Liu, J., Zeng, W., Urtasun, R., Yumer, E.: Deep structured reactive planning. In: 2021 IEEE International Conference on Robotics and Automation (ICRA), pp. 4897–4904. IEEE (2021)
11. Liu, Y., Gao, Y., Zhang, Q., Ding, D., Zhao, D.: Multi-task safe reinforcement learning for navigating intersections in dense traffic. J. Franklin Inst. (2022)
12. Mnih, V., et al.: Human-level control through deep reinforcement learning. Nature **518**(7540), 529–533 (2015)
13. Palmal, S., Arya, N., Saha, S., Tripathy, S.: A multi-modal graph convolutional network for predicting human breast cancer prognosis. In: Tanveer, M., Agarwal, S., Ozawa, S., Ekbal, A., Jatowt, A. (eds.) Neural Information Processing, ICONIP 2022. Communications in Computer and Information Science, vol. 1794, pp. 187–198. Springer, Singapore (2023). https://doi.org/10.1007/978-981-99-1648-1_16
14. Silver, D., et al.: Mastering the game of go with deep neural networks and tree search. Nature **529**(7587), 484–489 (2016)
15. Silver, D., Sutton, R.S., Müller, M.: Temporal-difference search in computer go. Mach. Learn. **87**, 183–219 (2012)
16. Wang, J., Zhang, Q., Zhao, D.: Highway lane change decision-making via attention-based deep reinforcement learning. IEEE/CAA J. Automatica Sinica **9**(3), 567–569 (2021)
17. Wang, J., Zhang, Q., Zhao, D.: Dynamic-horizon model-based value estimation with latent imagination. IEEE Trans. Neural Netw. Learn. Syst. (2022)
18. Wang, J., Zhang, Q., Zhao, D., Chen, Y.: Lane change decision-making through deep reinforcement learning with rule-based constraints. In: 2019 International Joint Conference on Neural Networks (IJCNN), pp. 1–6. IEEE (2019)
19. Wen, J., Zhao, Z., Cui, J., Chen, B.M.: Model-based reinforcement learning with self-attention mechanism for autonomous driving in dense traffic. In: Tanveer, M., Agarwal, S., Ozawa, S., Ekbal, A., Jatowt, A. (eds.) Neural Information Processing, ICONIP 2022. LNCS, vol. 13624, pp. 317–330. Springer, Cham (2023). https://doi.org/10.1007/978-3-031-30108-7_27

20. Wu, J., Huang, Z., Lv, C.: Uncertainty-aware model-based reinforcement learning: methodology and application in autonomous driving. IEEE Trans. Intell. Veh. **8**, 194–203 (2022)
21. Zhao, X., Chen, Y., Guo, J., Zhao, D.: A spatial-temporal attention model for human trajectory prediction. IEEE CAA J. Autom. Sinica **7**(4), 965–974 (2020)

Towards Analyzing the Efficacy of Multi-task Learning in Hate Speech Detection

Krishanu Maity[1]([✉]), Gokulapriyan Balaji[2], and Sriparna Saha[1]

[1] Department of Computer Science and Engineering, Indian Institute of Technology Patna, Patna 801103, India
{krishanu_2021cs19,sriparna}@iitp.ac.in
[2] Indian Institute of Information and Technology, Design and Manufacturing, Kancheepuram, Chennai, India

Abstract. Secretary-General António Guterres launched the United Nations Strategy and Plan of Action on Hate Speech in 2019, recognizing the alarming trend of increasing hate speech worldwide. Despite extensive research, benchmark datasets for hate speech detection remain limited in volume and vary in domain and annotation. In this paper, the following research objectives are deliberated (a) performance comparisons between multi-task models against single-task models; (b) performance study of different multi-task models (fully shared, shared-private) for hate speech detection, considering individual dataset as a separate task; (c) what is the effect of using different combinations of available existing datasets in the performance of multi-task settings? A total of six datasets that contain offensive and hate speech on the accounts of race, sex, and religion are considered for the above study. Our analysis suggests that a proper combination of datasets in a multi-task setting can overcome data scarcity and develop a unified framework.

Keywords: Hate Speech · Data scarcity · Single Task · Multi-Task

1 Introduction

Our world's communication patterns have changed dramatically due to the rise of social media platforms, and one of those changes is an increase in improper behaviors like the usage of hateful and offensive language in social media posts. On 15 March 2021, an independent United Nations human right expert said that social media has too often been used with "relative impunity" to spread hate, prejudice and violence against minorities[1]. Hate speech [15] is any communication that disparages a person or group on the basis of a characteristic such as color, gender, race, sexual orientation, ethnicity, nationality, religion, or other features. Hate speech detection is crucial in social media because it helps in ensuring a safe and inclusive online environment for all users. Even though social media platforms provide space for people to connect, share, and engage

---

· [1] https://news.un.org/en/story/2021/03/1087412.

B. Luo et al. (Eds.): ICONIP 2023, LNCS 14452, pp. 317–328, 2024.
https://doi.org/10.1007/978-981-99-8076-5_23

with each other, the anonymity and ease of access to these platforms also make them attractive platforms for those who engage in hate speech.

Hate speech has serious consequences and can cause significant harm to its targets. It can lead to increased discrimination, bullying, and even physical violence. Moreover, it can contribute to the spread of misinformation, stoke fear and division, and undermine the fabric of society. The harm that hate speech causes is amplified in online spaces, where the reach and impact of messages can be much greater than in the real world. According to the Pew Research Center, 40% of social media users have experienced some sort of online harassment[2]. According to the FBI, there were 8,263 reported hate crime incidents in 2020, which represents an increase of almost 13% from the 7,314 incidents reported in 2019[3]. Between July and September 2021, Facebook detected and acted upon 22.3 million instances of hate speech content[4]. A study found that from December 2019 to March 2020, there was a substantial 900% surge in the number of tweets containing hate speech directed towards Chinese people and China[5]. These hate posts that are supposedly safe on social media create real-world violence and riots. This warrants the requirement for the detection and control of hate speech.

That is why social media companies have taken steps to detect and remove hate speech from their platforms. This is a challenging task, as hate speech often takes many different forms and is difficult to define. In addition, there is often a fine line between free speech and hate speech, and companies must balance these competing interests while still protecting users from harm. It is important to note that hate speech detection is not just a technical challenge, it is also a societal challenge. Companies must understand the cultural and historical context of hate speech to develop policies and algorithms that are fair and effective. It is also important to ensure that hate speech detection does not undermine freedom of expression, or discriminate against marginalized groups.

Over the last decade, plenty of research has been conducted to develop datasets and models for automatic online hate speech detection on social media [17,25]. The efficacy of hate speech detection systems is paramount because labeling a non-offensive post as hate speech denies a free citizen's right to express himself. Furthermore, most existing hate speech detection models capture only single type of hate speech, such as sexism or racism, or single demographics, such as people living in India, as they trained on a single dataset. Such types of learning negatively affect recall when classifying data that are not captured in the training examples. To build an effective machine learning or deep learning-based hate speech detection system, a considerable amount of labeled data is required. Although there are a few benchmark data sets, their sizes are often limited and they lack a standardized annotation methodology.

---

[2] https://www.pewresearch.org/internet/2017/07/11/online-harassment-2017/.

[3] https://www.fbi.gov/news/press-releases/fbi-releases-2019-hate-crime-statistics.

[4] https://transparency.fb.com/data/community-standards-enforcement/hate-speech/facebook/.

[5] https://l1ght.com/Toxicity_during_coronavirus_Report-L1ght.pdf.

In this work, we address three open research questions related to building a more generic model for textual hate speech detection.

(i) **RQ1:** *Does multi-task learning outperform single-task learning and single classification model trained using merged datasets?* This research question pertains to the advantage of multi-task learning for various datasets over other training strategies. When multiple datasets are available, the most intuitive method of training is to merge the datasets and train the model in a single-task learning setting. Different datasets are considered individual tasks in multi-task settings.

(ii) **RQ2:** *Which type of multi-task model performs the best across a wide range of benchmark datasets?* Two widely used multi-task frameworks, Fully shared (FS) and Shared private (SP) with adversarial training (Adv), have been explored to investigate which one is preferable for handling multiple datasets.

(iii) **RQ3:** *What combination of datasets improve or degrade the performance of the multi-task learning model?* This question addressed the effect of different dataset combinations on model performance. Different dataset combinations bring knowledge from various domains. For n datasets ($n >= 2$), there are ($2^n - n - 1$) possible combinations, each containing at least two datasets. The study on the improvement of performance on the grounds of complementary or contrasting properties of datasets plays an important role in the selection of datasets for multi-task learning.

This current paper addresses the above-mentioned questions by developing three multi-task learning models: fully shared, shared-private, and adversarial, as well as presenting insights about dataset combinations and investigating the performance improvement of multi-task learning over single-task learning and a single model trained using a merged dataset.

2 Related Work

Text mining and NLP paradigms have previously been used to examine a variety of topics related to hate speech detection, such as identifying online sexual predators, detecting internet abuse, and detecting cyberterrorism [22].

Detecting hateful and offensive speech presents challenges in understanding contextual nuances, addressing data bias, handling multilingual and code-switching text, adapting to the evolving nature of hate speech, dealing with subjectivity and ambiguity, countering evasion techniques, and considering ethical considerations [6]. These challenges necessitate robust and adaptable methodologies, including deep learning and user-centric approaches, to enhance hate speech detection systems. A common approach for hate speech detection involves combining feature extraction with classical machine learning algorithms. For instance, Dinakar et al. [3] utilized the Bag-of-Words (BoW) approach in conjunction with a Naïve Bayes and Support Vector Machines (SVMs) classifier. Deep Learning, which has demonstrated success in computer vision, pattern recognition, and speech processing, has also gained significant momentum in natural language processing (NLP). One significant advancement in this direction was the introduction of embeddings [14], which have proven to be useful when combined with classical machine learning algorithms for hate speech detection [13], surpassing the performance of the BoW approach. Furthermore, other Deep Learning methods have been explored, such as the utilization of Convolutional Neural Networks (CNNs) [27], Recurrent Neural Networks (RNNs) [4], and hybrid models combining the two [9]. Another significant development was the introduction of transformers, particularly BERT, which exhibited exceptional performance in a recent hate speech detection competition, with seven out of the top ten performing models in a subtask being based on BERT [26].

2.1 Works on Single Dataset

The work by Watanabe et al. [25] introduced an approach that utilized unigrams and patterns extracted from the training set to detect hate expressions on Twitter, achieving an accuracy of 87.4% in differentiating between hate and non-hate tweets. Similarly, Davidson et al. [2] collected tweets based on specific keywords and crowdsourced the labeling of hate, offensive, and non-hate tweets, developing a multi-class classifier for hate and offensive tweet detection. In a separate study, a dataset of 4500 YouTube comments was used by authors in [3] to investigate cyberbullying detection, with SVM and Naive Bayes classifiers achieving overall accuracies of 66.70% and 63% respectively. A Cyberbullying dataset was created from Formspring.me in a study by authors in [20], and a C4.5 decision tree algorithm with the Weka toolkit achieved an accuracy of 78.5%. CyberBERT, a BERT-based framework created by [17], exhibited cutting-edge performance on Twitter (16k posts), Wikipedia (100k posts) and Formspring (12k posts) datasets. On a hate speech dataset of 16K annotated tweets, Badjatiya et al [1] conducted extensive tests with deep learning architectures for learning semantic word embeddings, demonstrating that deep learning techniques beat char/word n-gram algorithms by 18% in terms of F1 score.

2.2 Works on Multiple Datasets

Talat et al. [23] experimented on three hate speech datasets with different annotation strategies to examine how multi-task learning mitigated the annotation bias problem. Authors in [21] employed a transfer learning technique to build a single representation of hate speech based on two independent hate speech datasets. Fortuna et al. [5] merged two hate speech datasets from different social media (one from Facebook and another from Twitter) and examined that adding data from a different social network allowed to enhance the results.

Although there are some attempts in building a generalized hate speech detection model based on multiple datasets, none of them has addressed the insight on (i) how to combine datasets; (ii) is multi-tasking better than single task setup and a single model trained using merged dataset, (iii) which type of multitasking is better: FS or SP.

Table 1. Source, statistics and domain of six hate speech datasets used in our experiments

Dataset	# Samples	# Classes and #Samples in each class	Source	Domain
D1 [2]	24783	3: Hate speech (1430), Offensive (19190), Neither (4163)	Twitter	Hate, Offensive
D2 [7]	10703	2: Non-hate (9507), Hate (1196)	Stormfront forum	Race, Religion
D3 [24]	10141	3: Racism (12), Sexism (2656), None (7473)	Twitter	Race, Sexism
D4 [12]	7005	2: Non Hate-Offensive (4456), Hate and Offensive (4456)	Twitter	Hate, Offensive
D5 [16]	10000	2: Non-hateful (5790), Hateful (4210)	Twitter	Immigrants, Sexism
D6 [11]	31962	2: Non-hate (29720), Hate (2242)	Twitter	Race, Sexism

3   Dataset Description

Six datasets (Table 1) are selected in an attempt to understand the effect of using multiple datasets and to conduct experiments. These datasets include examples of hate, offensiveness, racism, sexism, religion, and prejudice against immigrants. Even though the samples differ in terms of annotation style, domain, demography, and geography, there is common ground in terms of hate speech.

4   Methodology

To investigate how multiple hate speech datasets can help in building a more generalized hate speech detection model, we have experimented with two widely used multi-task frameworks (Fig. 1), i.e., Fully shared and Shared Private, developed by [10]. In the feature extraction module (Fig. 2), we employed Glove [18] and FastText [8] embedding to encode the noisy social media data efficiently. The joint embedding is passed through a convolution layer followed by max pooling to generate the local key phrase-based convoluted features. In the FS model, the final output from the CNN module is shared over n task-specific channels, one for each dataset (task). For the SP model, individual CNN representation from each of the tasks is passed through the corresponding task-specific output layer. In addition to task-specific layers, there is a shared layer (Fully Connected layer) to learn task invariant features for the SP model. The adversarial loss is added in model training to make shared and task-specific layers' feature spaces mutually exclusive [19].

5   Experimental Results and Analysis

This section describes the results of single task setting, multi-task setting of three models for different combinations of 6 benchmark datasets. The experiments are intended towards addressing the following research questions:

- **RQ1:** How does multi-task learning enhance the performance of hate speech detection compared to single task learning and single task based on a merged dataset?

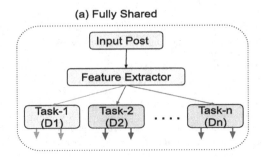

(a) Fully Shared

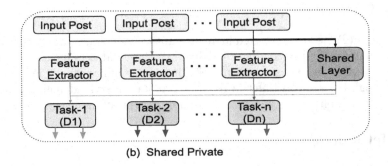

(b) Shared Private

Fig. 1. (a) Fully shared and (b) Shared private multi-task frameworks.

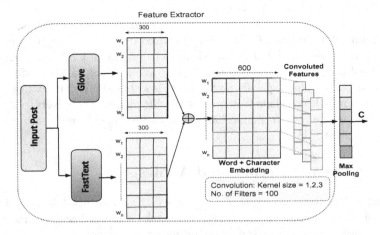

Fig. 2. Feature extraction module based on Glove and FastText joint embedding followed by CNN

- **RQ2**: Which type of multi-task learning model provides the best results among the three models?
- **RQ3**: Which combination of the benchmark datasets should be used for obtaining the best results from multi-task learning?

The experiments were performed on 5-fold cross-validation on the datasets and the results are evaluated in terms of accuracy value. The values mentioned inside the brackets are the improvements or decrements in accuracy compared to single-task learning. Keeping the size of the datasets in mind, a batch size of 8 was found optimal and configurations such as the ReLU activation function, and 5e−4 learning rate were chosen and the models were trained for 20 epochs.

Table 2. Single-task learning performance with individual datasets and merged datasets

Dataset Combination	Single Task		
	STL	Merged (All)	Merged (-D1)
D1	91.28	20.33	-
D2	87.6	84.96	88.97
D3	82.89	71.71	73.63
D4	63.81	63.74	64.88
D5	70.05	58.5	59.9
D6	94.75	87.63	92.87

Table 3. Multi-task Learning Performance

Dataset Combination	Multi Task			
	FS	FS - adv	SP	SP - adv
D1	92.68 (+1.40)	93.63 (+2.35)	95.04 (+3.76)	95.59 (+4.31)
D2	90.20 (+2.60)	89.02 (+1.42)	88.70 (+1.10)	89.53 (+1.93)
D3	83.81 (+1.12)	83.62 (+0.73)	86.79 (+3.90)	86.95 (+4.06)
D4	67.88 (+4.07)	66.25 (+2.44)	66.10 (+2.29)	65.53 (+1.72)
D5	71.45 (+1.40)	71.67 (+1.62)	74.80 (+4.75)	75.00 (+4.95)
D6	96.16 (+1.41)	95.72 (+0.97)	96.70 (+1.95)	96.78 (+2.03)

5.1 RQ1: Single Task vs Merging All vs Multi-task

In Table 2, the accuracy of single task learning is compared with a model trained after merging all datasets and with a multitasking framework. It is evident from this table that the performance of single-task learning is better than that of the model trained using a merged version of all the datasets. However, when dataset 1 which performed very poorly was removed from the merged set and experiments are again conducted, the accuracy values for datasets 2 and 4 are improved over the single-task learning accuracies. The selection of datasets that

Table 4. Experimental results of Fully Shared, Shared Private models under multi-task settings with 2 datasets combinations; Like, in (D3-D5) combination, 1st and 2nd represent the performance of D3 and D5, respectively

Dataset Combination	Fully Shared		Shared Private	
	1st	2nd	1st	2nd
D1-D2	93.33 (+2.05)	90.19 (+2.59)	94.05 (+2.77)	88.00 (+0.4)
D1-D3	93.55 (+2.27)	83.34 (+0.45)	94.01 (+2.73)	84.07 (+1.18)
D1-D4	93.54 (+2.26)	68.88 (+5.07)	93.93 (+2.65)	64.48 (+0.67)
D1-D5	93.35 (+2.07)	72.55 (+2.50)	93.40 (+2.12)	74.60 (+4.55)
D1-D6	92.39 (+1.11)	95.22 (+0.47)	94.61 (+3.33)	95.51 (+0.76)
D2-D3	89.86 (+2.26)	83.39 (+0.50)	89.37 (+1.77)	84.96 (+2.07)
D2-D4	90.55 (+2.95)	67.74 (+3.93)	88.27 (+0.67)	64.45 (+0.64)
D2-D5	90.00 (+2.4)	73.20 (+3.15)	89.25 (+1.65)	74.05 (+4.00)
D2-D6	90.43 (+2.83)	95.52 (+0.77)	88.46 (+0.86)	95.77 (+1.02)
D3-D4	83.88 (+0.99)	67.38 (+3.57)	84.22 (+1.33)	65.24 (+1.43)
D3-D5	83.00 (+0.11)	71.90 (+1.85)	84.57 (+1.68)	74.75 (+4.70)
D3-D6	83.44 (+0.55)	95.18 (+0.43)	84.17 (+1.28)	95.86 (+1.11)
D4-D5	68.09 (+4.28)	71.59 (+1.54)	65.31 (+1.50)	73.25 (+3.20)
D4-D6	67.09 (+3.28)	96.04 (+1.29)	65.42 (+1.61)	96.20 (+1.45)
D5-D6	72.05 (+2.00)	95.95 (+1.20)	73.80 (+3.75)	96.30 (+1.55)

are used to form the merged dataset for developing a unified model plays a significant role in the performance of the system. When the combination of datasets is selected after analyzing the domain, supplementary and complementary information available with the dataset, the unified model becomes more generalized. But blindly combining all the datasets leads to decreased performance of the unified model trained on the merged dataset. In multi-task settings (see Table 3), the performances on all the datasets are improved significantly over both single-task learning and single-task training on a merged dataset. In a multi-task setting, hate speech detection from a single dataset is considered an individual task. This concept proves to provide an edge to the model for its ability to generalize and perform better compared to the other training settings.

5.2 RQ2: Fully Shared vs. Shared Private (+/− Adversarial Training)

Among the models trained over multiple datasets as shown in Tables 4 and 5, there is no clear winner that can be selected. However, with the benchmark datasets used in our experiments, the shared private model proves to be the better model among its alternatives. This could be due to the training of shared and task-specific layers on the datasets which provide in-depth knowledge and prioritize the information from both these layers. But, the absence of such an ability to prioritize shared knowledge inhibits the performance of the fully shared network. As proof of this, the accuracies for datasets 1, 3, 5, and 6 among all the combinations are higher in the shared private model compared to the fully

Table 5. Experimental results of Fully Shared - Adversarial, Shared Private - Adversarial models under multi-task settings with 2 datasets combinations; Like, in (D3-D5) combination, 1st and 2nd represent the performance of D3 and D5, respectively

Dataset Combination	Fully Shared - Adversarial		Shared Private - Adversarial	
	1st	2nd	1st	2nd
D1-D2	93.51 (+2.23)	88.89 (+1.29)	94.69 (+3.41)	87.80 (+0.20)
D1-D3	93.67 (+2.39)	83.30 (+0.41)	94.96 (+3.68)	85.50 (+2.61)
D1-D4	93.60 (+2.32)	66.94 (+3.13)	94.67 (+3.39)	64.74 (+0.93)
D1-D5	93.28 (+2.00)	73.01 (+2.96)	94.71 (+3.43)	75.00 (+4.95)
D1-D6	92.30 (+1.02)	94.98 (+0.23)	94.39 (+3.11)	95.93 (+1.18)
D2-D3	89.95 (+2.35)	83.28 (+0.39)	88.51 (+0.91)	84.17 (+1.28)
D2-D4	90.03 (+2.43)	66.87 (+3.06)	87.85 (+0.25)	64.54 (+0.73)
D2-D5	89.74 (+2.14)	73.24 (+3.19)	88.01 (+0.44)	72.85 (+2.80)
D2-D6	90.47 (+2.87)	95.47 (+0.72)	87.98 (+0.38)	95.91 (+1.16)
D3-D4	84.05 (+1.16)	66.83 (+3.02)	84.78 (+1.89)	64.77 (+0.96)
D3-D5	83.96 (+1.07)	72.11 (+2.06)	84.65 (+1.76)	74.98 (+4.93)
D3-D6	84.02 (+1.13)	95.50 (+0.75)	84.71 (+1.82)	95.95 (+1.20)
D4-D5	68.36 (+4.55)	71.52 (+1.47)	64.71 (+0.90)	73.92 (+3.87)
D4-D6	66.91 (+3.10)	95.83 (+1.08)	64.47 (+0.66)	96.66 (+1.91)
D5-D6	72.13 (+2.08)	95.98 (+1.23)	74.00 (+3.95)	96.45 (+1.70)

shared. However, interestingly the accuracy values of dataset 2 (D2) are better in a fully shared model. A possible explanation for this pattern could be in the source of the datasets. Unlike other datasets which were tweets, D2 belongs to a different source of social media posts.

When adversarial training is incorporated, the performance improves in datasets that have common ground/features. However, when the combination includes datasets of different sources, then the performance of the shared private adversarial model worsens compared to the shared private model. The adversarial layer alters the knowledge attained by the shared layer in such a way as to make the feature space of shared and specific layers to be mutually exclusive. This creates a more generalization causing deterioration in the performance. Fully shared adversarial is also similar in nature but the accuracy is hampered more compared to the shared private adversarial making this pattern difficult to predict or understand.

5.3   RQ3: Datasets Combination

From Table 6 and 7, it can be observed that the improvement in individual dataset compared to single task learning is limited as the number of datasets have increased (most of the time, the combination of two datasets performs better than the combination of three datasets). This could be due to the difficulty in generalizing the model on various datasets. The best performance is observed when using datasets of similar sizes and sources. An interesting insight was observed when datasets having information on different domains boost the

Table 6. Fully Shared Model Performance with 3 datasets combination

Dataset Combination	Fully Shared		
	1st	2nd	3rd
D1-D2-D3	92.27 (+0.99)	89.72 (+2.12)	83.44 (+0.55)
D1-D2-D4	92.25 (+0.97)	89.86 (+2.26)	68.31 (+4.50)
D1-D2-D6	92.21 (+0.93)	89.82 (+2.22)	95.06 (+0.31)
D1-D3-D4	92.35 (+1.07)	82.95 (+0.06)	68.74 (+4.93)
D1-D3-D5	91.97 (+0.69)	83.05 (+0.16)	71.15 (+1.10)
D1-D4-D5	91.83 (+0.55)	69.20 (+5.39)	70.95 (+0.90)
D2-D3-D5	90.05 (+2.45)	83.41 (+0.52)	71.60 (+1.55)
D2-D4-D6	90.01 (+2.41)	66.88 (+3.07)	95.17 (+0.42)
D3-D4-D5	83.40 (+0.51)	67.52 (+3.71)	71.15 (+1.10)
D4-D5-D6	67.38 (+3.57)	71.20 (+1.15)	94.90 (+0.15)

Table 7. Shared Private Model Performance with 3 datasets combination

Dataset Combination	Shared Private		
	1st	2nd	3rd
D1-D2-D3	94.67 (+3.39)	88.70 (+1.10)	84.33 (+1.44)
D1-D2-D4	94.57 (+3.29)	88.45 (+0.85)	65.02 (+1.21)
D1-D2-D6	94.59 (+3.31)	88.53 (+0.93)	95.02 (+0.27)
D1-D3-D4	94.45 (+3.17)	83.80 (+0.91)	64.64 (+0.83)
D1-D3-D5	95.05 (+3.77)	83.64 (+0.75)	72.24 (+2.19)
D1-D4-D5	94.49 (+3.21)	63.94 (+0.13)	72.67 (+2.62)
D2-D3-D5	88.78 (+1.18)	83.49 (+1.20)	72.22 (+2.17)
D2-D4-D6	88.51 (+0.91)	64.55 (+0.74)	95.77 (+1.02)
D3-D4-D5	84.05 (+1.16)	64.42 (+0.61)	73.43 (+3.38)
D4-D5-D6	64.67 (+0.86)	73.31 (+3.26)	95.88 (+1.13)

performance of each other significantly. For example, datasets 1 and 6 belonging to the same source have samples emphasizing different domains. Dataset 1 having samples that are majorly offensive gains shared knowledge on the attack of women and immigrants from dataset 6. Dataset 6 too learns knowledge of contrasting domains from dataset 1 that help generalize the model to tackle new samples.

6   Conclusion and Future Work

In this paper, an attempt was made to create a hate speech detection model that was trained on different datasets. To improve the performance and generality of the model, multi-task learning was leveraged. With the help of this methodology and careful examination of the datasets, a robust model that identifies and prevents various domains of hate attacks can be built, thus creating a safe and trustworthy space for users in social media. The contributions of the current work are twofold: (a) Experiments conducted across different types of settings and models help us develop a multi-task system that can be trained on datasets from different domains and detect hate speech in a generalized manner. (b) Studies were conducted on the effect of combinations and increase in datasets in a multi-task setting to improve the decision-making process of setting up new hate speech detection systems.

In the future, we would like to work on multi-modal hate speech detection systems that can help us monitor a plethora of social media.

Acknowledgements. The Authors would like to acknowledge the support of Ministry of Home Affairs (MHA), India, for conducting this research.

References

1. Badjatiya, P., Gupta, S., Gupta, M., Varma, V.: Deep learning for hate speech detection in tweets. In: Proceedings of the 26th International Conference on World Wide Web Companion, pp. 759–760 (2017)
2. Davidson, T., Warmsley, D., Macy, M., Weber, I.: Automated hate speech detection and the problem of offensive language. In: Proceedings of the International AAAI Conference on Web and Social Media, vol. 11, pp. 512–515 (2017)
3. Dinakar, K., Reichart, R., Lieberman, H.: Modeling the detection of textual cyber-bullying. In: 2011 Proceedings of the International Conference on Weblog and Social Media. Citeseer (2011)
4. Do, H.T.T., Huynh, H.D., Van Nguyen, K., Nguyen, N.L.T., Nguyen, A.G.T.: Hate speech detection on Vietnamese social media text using the bidirectional-LSTM model. arXiv preprint arXiv:1911.03648 (2019)
5. Fortuna, P., Bonavita, I., Nunes, S.: Merging datasets for hate speech classification in Italian. In: EVALITA@ CLiC-it (2018)
6. Fortuna, P., Nunes, S.: A survey on automatic detection of hate speech in text. ACM Comput. Surv. (CSUR) **51**(4), 1–30 (2018)
7. de Gibert, O., Perez, N., García-Pablos, A., Cuadros, M.: Hate speech dataset from a white supremacy forum. In: Proceedings of the 2nd Workshop on Abusive Language Online (ALW2), Brussels, Belgium, October 2018, pp. 11–20. Association for Computational Linguistics (2018). https://doi.org/10.18653/v1/W18-5102. https://www.aclweb.org/anthology/W18-5102
8. Grave, E., Bojanowski, P., Gupta, P., Joulin, A., Mikolov, T.: Learning word vectors for 157 languages. arXiv preprint arXiv:1802.06893 (2018)
9. Maity, K., Saha, S.: BERT-capsule model for cyberbullying detection in code-mixed Indian languages. In: Métais, E., Meziane, F., Horacek, H., Kapetanios, E. (eds.) NLDB 2021. LNCS, vol. 12801, pp. 147–155. Springer, Cham (2021). https://doi.org/10.1007/978-3-030-80599-9_13
10. Maity, K., Saha, S., Bhattacharyya, P.: Emoji, sentiment and emotion aided cyber-bullying detection in Hinglish. IEEE Trans. Comput. Soc. Syst. **10**, 2411–2420 (2022)
11. Malik, J.S., Pang, G., van den Hengel, A.: Deep learning for hate speech detection: a comparative study. arXiv preprint arXiv:2202.09517 (2022)
12. Mandl, T., et al.: Overview of the HASOC track at FIRE 2019: hate speech and offensive content identification in Indo-European languages. In: Proceedings of the 11th Forum for Information Retrieval Evaluation, pp. 14–17 (2019)

13. Mehdad, Y., Tetreault, J.: Do characters abuse more than words? In: Proceedings of the 17th Annual Meeting of the Special Interest Group on Discourse and Dialogue, pp. 299–303 (2016)

14. Mikolov, T., Sutskever, I., Chen, K., Corrado, G.S., Dean, J.: Distributed representations of words and phrases and their compositionality. In: Advances in Neural Information Processing Systems, vol. 26 (2013)

15. Nockleby, J.T.: Hate speech in context: the case of verbal threats. Buff. L. Rev. **42**, 653 (1994)

16. i Orts, Ò.G.: Multilingual detection of hate speech against immigrants and women in Twitter at SemEval-2019 task 5: frequency analysis interpolation for hate in speech detection. In: Proceedings of the 13th International Workshop on Semantic Evaluation, pp. 460–463 (2019)

17. Paul, S., Saha, S.: CyberBERT: BERT for cyberbullying identification. Multimed. Syst. **28**, 1897–1904 (2020)

18. Pennington, J., Socher, R., Manning, C.D.: GloVe: global vectors for word representation. In: Proceedings of the 2014 Conference on Empirical Methods in Natural Language Processing (EMNLP), pp. 1532–1543 (2014)

19. Preoţiuc-Pietro, D., Liu, Y., Hopkins, D., Ungar, L.: Beyond binary labels: political ideology prediction of Twitter users. In: Proceedings of the 55th Annual Meeting of the Association for Computational Linguistics (Volume 1: Long Papers), pp. 729–740 (2017)

20. Reynolds, K., Kontostathis, A., Edwards, L.: Using machine learning to detect cyberbullying. In: 2011 10th International Conference on Machine Learning and Applications and Workshops, vol. 2, pp. 241–244. IEEE (2011)

21. Rizoiu, M.A., Wang, T., Ferraro, G., Suominen, H.: Transfer learning for hate speech detection in social media. arXiv preprint arXiv:1906.03829 (2019)

22. Simanjuntak, D.A., Ipung, H.P., Nugroho, A.S., et al.: Text classification techniques used to faciliate cyber terrorism investigation. In: 2010 Second International Conference on Advances in Computing, Control, and Telecommunication Technologies, pp. 198–200. IEEE (2010)

23. Talat, Z., Thorne, J., Bingel, J.: Bridging the gaps: multi task learning for domain transfer of hate speech detection. In: Golbeck, J. (ed.) Online Harassment. HIS, pp. 29–55. Springer, Cham (2018). https://doi.org/10.1007/978-3-319-78583-7_3

24. Waseem, Z., Hovy, D.: Hateful symbols or hateful people? Predictive features for hate speech detection on Twitter. In: Proceedings of the NAACL Student Research Workshop, San Diego, California, June 2016, pp. 88–93. Association for Computational Linguistics (2016). http://www.aclweb.org/anthology/N16-2013

25. Watanabe, H., Bouazizi, M., Ohtsuki, T.: Hate speech on Twitter: a pragmatic approach to collect hateful and offensive expressions and perform hate speech detection. IEEE Access **6**, 13825–13835 (2018)

26. Zampieri, M., Malmasi, S., Nakov, P., Rosenthal, S., Farra, N., Kumar, R.: SemEval-2019 task 6: identifying and categorizing offensive language in social media (OffensEval). arXiv preprint arXiv:1903.08983 (2019)

27. Zimmerman, S., Kruschwitz, U., Fox, C.: Improving hate speech detection with deep learning ensembles. In: Proceedings of the Eleventh International Conference on Language Resources and Evaluation, LREC 2018 (2018)

Exploring Non-isometric Alignment Inference for Representation Learning of Irregular Sequences

Fang Yu[1], Shijun Li[1]([✉]), and Wei Yu[2]([✉])

[1] School of Computer Science, Wuhan University, Wuhan 430072, Hubei, China
shjli@whu.edu.cn
[2] Institute of Artificial Intelligence, School of Computer Science, Wuhan University, Wuhan, China
yuwei@whu.edu.cn

Abstract. The development of Internet of Things (IoT) technology has led to increasingly diverse and complex data collection methods. This unstable sampling environment has resulted in the generation of a large number of irregular monitoring data streams, posing significant challenges for related data analysis tasks. We have observed that irregular sequence sampling densities are uneven, containing randomly occurring dense and sparse intervals. This data imbalance tendency often leads to overfitting in the dense regions and underfitting in the sparse regions, ultimately impeding the representation performance of models. Conversely, the irregularity at the data level has limited impact on the deep semantics of sequences. Based on this observation, we propose a novel Non-isometric Alignment Inference Architecture (NAIA), which utilizes a multi-level semantic continuous representation structure based on inter-interval segmentation to learn representations of irregular sequences. This architecture efficiently extracts the latent features of irregular sequences. We evaluate the performance of NAIA on multiple datasets for downstream tasks and compare it with recent benchmark methods, demonstrating NAIA's state-of-the-art performance results.

Keywords: Representation learning · Non-isometric Alignment Inference · Irregular sequences · Continuous latent representation

1 Introduction

This paper presents a Non-isometric Alignment Inference Architecture (NAIA) for self-supervised representation of irregular sequences. Irregular sequences exist in various scientific and engineering fields in the form of non-equidistant time series, including healthcare [1, 2], meteorology [3], financial markets [4], and industrial production [5]. For example, in mobile health monitoring [6], improper device wearing by users can lead to temporary interruptions in recording. Additionally, in smart device monitoring [7], sensors only sample when predefined triggering events occur. This random and irregular sampling results in variations in observation intervals and densities, generating

© The Author(s), under exclusive license to Springer Nature Singapore Pte Ltd. 2024
B. Luo et al. (Eds.): ICONIP 2023, LNCS 14452, pp. 329–340, 2024.
https://doi.org/10.1007/978-981-99-8076-5_24

irregular sequences. Irregularly sampled sequences pose challenges to deep learning models based on fully-observed, fixed-size feature representations [9]. On one hand, irregular sequences introduce potential structural complexity to the corresponding neural networks, and the lack of temporal alignment in the data hinders the self-supervised training process. On the other hand, the advancement in hardware has led to a gradual increase in the sampling rates of various sensors, making models more prone to over-rely on patterns in dense regions while tending to neglect sparse data intervals. This can result in local overfitting or underfitting, consequently reducing the model's representational performance [8].

Existing research primarily focuses on constructing specialized network structures for such irregular sequences [10, 11]. The recently proposed non-isometric time series imputation autoencoder (mTAN) [12] achieves more advanced sequence classification and imputation performance by utilizing attention mechanisms. In fact, Gaussian process regression, as a classical machine learning algorithm, can represent the variance of input samples through posterior probability inference [13]. However, its computational complexity is $O(n3)$, and it lacks a deep neural network structure, making it difficult to extract deep-level semantics from input sequences. The main contributions of the NAIA model are as follows:

- It maps the input data to a continuous latent function in the latent space, enabling any query time related to the input to generate feedback information in the latent space. This achieves continuous representation and temporal alignment for irregular sequences.
- By designing the JS module embedded in the latent space, NAIA allows irregular intervals to generate boundaries based on their own latent distribution stability. The model only needs to train on randomly samples from each interval, facilitating lightweight training and mitigating the issue of local overfitting caused by uneven data sampling.
- Two paths have been specifically designed for the propagation of time information. They inject query information into NAIA from both the front-end and back-end of the latent space, enabling the model to be trained under query guidance and enhancing the performance of representation learning.

2 Related Work

The most classical model in time series tasks is the Autoregressive Integrated Moving Average (ARIMA). Its simple and effective statistical properties and theoretical architecture based on Box-Jenkins methodology [14] make it widely applicable to various forecasting or reconstruction tasks. However, coping with today's big data tasks, its performance in extracting deep semantics of time series could be much better. Moreover, ARIMA does not adapt well to non-isometric time series. On the other hand, filter-based methods infer the sequence values at new time points given from historical data, including Han-Gyu Kim et al. [15] use recurrent neural networks to infer and impute the missing data on the sequence. However, this method can only impute the data before the observation point.

Among probability-based methods, Gaussian process regression (GPR) [13] is an important architecture for dealing with irregular sequences. GPR can input an analyzable joint posterior distribution based on the non-isometric characteristics of an irregular sequence. One of the covariance matrices converts the non-isometric structure of the observed samples into an interpolated uncertainty measure. One problem with GPR is that the positive definite constraint on its covariance matrix can hinder performance in a multivariate setting. One standard solution is constructing GPR in multiple sequence dimensions and using separable temporal kernel functions [17]. However, the construction process requires each dimension to share the same kernel function, thus hindering its interpolation performance, the efficiency of the operation is significantly reduced [18].

Among the self-attention mechanisms, various variants of Transformer [19] have achieved significant performance gains in various deep learning tasks. It has sparked the attention and interest in multi-headed self-attention mechanisms [22]. The attention mechanism is effective for capturing long-term dependencies in data. Research has focused on using it for long-term time series forecasting and anomalous pattern recognition. Notable advancements have been achieved in these areas. However, it still has some limitations, including squared time/memory complexity and susceptibility to error accumulation caused by various decoder variants. And then, various new variants, represented by Informer [16], were proposed. They improved the Transformer architecture and reduced the complexity, and the performance of capturing semantic patterns of time series became enhanced [23–25]. In 2022, a new multi-head attention network was proposed. It Makes significant improvements in the performance of irregular sequence representations based on the Transformer [12]. This method uses a multi-headed temporal cross-attention encoder to embed secular values.

3 Non-isometric Alignment Inference

NAIA is a mechanism capable of aligning inference and providing continuous representation for non-uniform sequences and their intervals. This continuous latent representation enables NAIA to accommodate irregular inputs and generate corresponding reconstructions or predictions at any given query time.

3.1 Preliminaries

The irregular sequence $X(T)$ can be decomposed into N subintervals along the time dimension, and the i-th sub-interval is:

$$x(t_{i,1:j_i}) = [x(t_{i,1}), x(t_{i,2}), \ldots, x(t_{i,j_i})] \subset X(T) \tag{1}$$

where $x(t_{i,1:j_i})$ represents the i-th sub-interval consisting of j_i samples arranged in chronological order. Assuming z_i is a latent distribution of $x(t_{i,1:j_i})$, we aim to approximate z_i with a posterior probability that follows a relatively stable latent probability distribution:

$$q[z_i | t_{i,1:j_i}, x(t_{i,1:j_i})] \sim d[x(t_{i,1:j_i})] \tag{2}$$

The symbol q represents the posterior probability, z_i represents the latent representation corresponding to $x(t_{i,1:j_i})$, and $d[x(t_{i,1:j_i})]$ represents the latent distribution of data $x(t_{i,1:j_i})$ in the sub-interval. When $d[x(t_{i,1:j_i})]$ remains stable, the corresponding time interval $t_{i,1:j_i}$ is a stable sub-interval. The representation of irregular sequences using data within stable sub-intervals faces three challenges. Firstly, the latent representation in neural networks is discrete, and only when z_i is in continuous form can it provide feedback output for any given query time. Secondly, irregular sequences need to be automatically partitioned into stable intervals as the input data changes. Lastly, when the samples within sub-intervals are input to the neural network, it needs to be based on the assumption of fully-observed, fixed-size, which is not satisfied by irregular sequences.

3.2 Theoretical Analysis

We train the model by setting queries to achieve representation learning. For irregular sequences, the only label available is the value corresponding to the query time, making the entire process self-supervised. The prior association between the query value and the observed data can be expressed as follows:

$$\max p\left[x\left(t_{i,1:j_i} + \lambda_i\right)|t_{1,1:j_1}, \ldots, t_{i:1:j_i}, t_{i,1:j_i} + \lambda_i, x\left(t_{1,1:j_1}\right), \ldots, x\left(t_{i,1:j_i}\right)\right] \quad (3)$$

Here, p represents a probability symbol, λ_i represents the time increment required to reach the query time from the current time. By integrating and taking the logarithm of Eq. (3), we obtain:

$$\max \log \int_z \left\{ q\left[z|t_{1,1:j_1}, \ldots, t_{i:1:j_i}, t_{i,1:j_i} + \lambda_i, x\left(t_{1,1:j_1}\right), \ldots, x\left(t_{i,1:j_i} + \lambda_i\right)\right] \cdot \right.$$
$$\left. \log p\left[x\left(t_{i,1:j_i} + \lambda_i\right)|t_{1,1:j_1}, \ldots, t_{i,1:j_i} + \lambda_i, x\left(t_{1,1:j_1}\right), \ldots, x\left(t_{i,1:j_i}\right)\right] \right\} dz \quad (4)$$

For the formal simplicity of the inference process, let $t_i = (t_{i,1}, t_{i,2}, \ldots, t_{i,j_i})$, which gives:

$$\log p[x(t_i + \lambda_i)|t_{1:i}, t_i + \lambda_i, x(t_{1:i})]$$
$$= \underbrace{\mathbb{E}_{q[z|t_{1:i}, t_i+\lambda_i, x(t_{1:i}), x(t_i+\lambda_i)]} \left\{ \log \frac{p[z, x(t_i + \lambda_i)|t_{1:i}, t_i + \lambda_i, x(t_{1:i})]}{q[z|t_{1:i}, t_i + \lambda_i, x(t_{1:i}), x(t_i + \lambda_i)]} \right\}}_{ELBO} +$$
$$D_{KL}\{q[z|t_{1:i}, t_i + \lambda_i, x(t_{1:i}), x(t_i + \lambda_i)]||p[z|t_{1:i}, t_i + \lambda_i, x(t_{1:i}), x(t_i + \lambda_i)]\} \quad (5)$$

where D_{KL} stands for KL-divergence and ELBO stands for evidence lower bound, representing the expectation part of Eq. (5). Since DKL > 0, maximizing Eq. (4) is equivalent to maximizing this ELBO. The query time may break through the current stable interval into the next adjacent interval, and we use the JS-divergence constructed on the time increment to measure the consistency of the interval distribution:

$$JS\left[d\left(x\left(t_{i,1:j_i}\right)\right)||d\left(x\left(t_{i,1:j_i}+\lambda_i\right)\right)\right] = \frac{1}{2} D_{KL}\left(d\left(x\left(t_{i,1:j_i}\right)\right)\left\|\frac{d\left(x\left(t_{i,1:j_i}\right)\right) + d\left(x\left(t_{i,1:j_i}+\lambda_i\right)\right)}{2}\right.\right) +$$
$$\frac{1}{2} D_{KL}\left(d\left(x\left(t_{i,1:j_i}+\lambda_i\right)\right)\left\|\frac{d\left(x\left(t_{i,1:j_i}\right)\right) + d\left(x\left(t_{i,1:j_i}+\lambda_i\right)\right)}{2}\right.\right) > \theta \geq 1 \quad (6)$$

where JS denotes the JS-divergence, $d(*)$ denotes the prior distribution of the data and satisfies Eq. (2), and θ denotes the interval stability parameter, which is used to specify the degree of impact of additional data on the consistency of the interval distribution. Equation (6) utilizes the JS-divergence to measure the consistency of the distribution of the queried data $x(t_{i,1:j_i}+\lambda_i)$ relative to the current data $x(t_{i,1:j_i})$ under the effect of time increments. NAIA uses the method with latent distribution for interval partitioning to avoid local feature traps and distinguish stable intervals in terms of high-level semantics. In addition, JS-divergence has symmetry and provides a uniform metric for distribution differences. When the difference between two intervals satisfies Eq. (6), it indicates a significant deviation of the query time from the original distribution, which does not satisfy the consistency. A new stable interval needs to be created to shelter the $x(t_{i,1:j_i}+\lambda_i)$:

$$x(t_{i,1:j_i}+\lambda_i) \xrightarrow{set\ up} x(t_{i+1,1:j_{i+1}}), \quad JS > \theta. \tag{7}$$

where θ denotes the distribution difference threshold. When the JS-divergence in Eq. (6) is less than or equal to θ, the distribution difference between the new data and the original data is considered minor, and the original interval absorbs the incremental time on the data, i.e.

$$x(t_{i,1:j_i}+\lambda_i) \xrightarrow{absorb} x(t_{i,1:j_i+\lambda_i}), \quad JS \leq \theta. \tag{8}$$

Once the interval partition is complete, we factorize the ELBO into an easily optimized form:

$$\max ELBO$$

$$= \max \mathbb{E}_{q[z|t_{1:i},t_i+\lambda_i,x(t_{1:i}),x(t_i+\lambda_i)]} \left\{ \sum_{j=j_i}^{j_i+\lambda_i} \log p[x(t_{i,j}+\lambda_i)|t_i,z] \right\} - \tag{9}$$

$$\min D_{KL}\{q[z|t_{1:i},t_i+\lambda_i,x(t_{1:i}),x(t_i+\lambda_i)]||p[z|t_{1:i},x(t_{1:i})]\}$$

In Eq. (9), the ELBO is decomposed into two terms; the former is the expectation of the prior query value concerning the variational posterior of z. It is not analyzable but can be approximated by neural networks. The z in the latent space is a state-continuous functional representation, meaning it can receive λ_i at the arbitrary size. The latter part of the ELBO is a KL-divergence to the probabilities q and p. Equation (9) aligns the non-isometric loss function. Maximizing the ELBO reduces the difference between the incremental time reconstruction and the query value. Minimizing the KL- divergence makes the reconstruction probability of incremental time obey the density function di, making Eq. (9) regularization.

3.3 Non-isometric Alignment Inference Architecture Network

We design a complete inference architecture based on Non-isometric Alignment Inference (NAIA), which approximates the expectation part of ELBO and outputs query values based on the query times in the interval. In the training phase, the self-supervised signal is the query time, which comes from the next samples of the sequence. The latent space preserves a continuous functional representation z. In the forecasting or reconstruction phase, any query time can be chosen since z can output the corresponding density function based on any query. The computational diagram of NAIA is constructed according to Eq. (9).

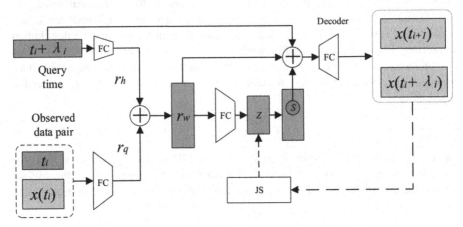

Fig. 1. Computational diagram of NAIA

as shown in Fig. 1, the time $t_{i,1:j_i}$ and the corresponding value constitute the observed data pair for the i-th set of inputs of NAIA, while $t_{i,1:j_i} + \lambda_i$ is used as query information input. They are compressed by their respective fully connected neural network into abstract representations r_h and r_q, respectively. The information of these two representations is fused by additive operations into weak query representations r_w for the semantic association. r_w is encoded into latent functional representation z by a neural network. z then outputs a probability density function di according to the weak query semantics. A sampling of the density function produces a sampling representation s aligned with the query time. Finally, r_w and s and the query time $t_{i,1:j_i} + \lambda_i$ are concatenated and fed to the decoder to reconstruct the sequence corresponding to the query time. The time operator $\oplus$ represents the concatenate operation. The query data $x(t_{i,1:j_i} + \lambda_i)$ is involved in the training of NAIA as a self-supervised signal. And whether the output data break the T_i interval or not is decided by the JS module on the dashed path according to Eq. (7). When the input sample does not break the i-th interval, the self-supervised signal is kept as $x(t_{i,1:j_i} + \lambda_i)$; otherwise, the self-supervised signal is changed to $x(t_{i+1,1:j_{i+1}})$.

The query time undergoes neural network encoding and additive fusion on the first path. Its semantics is weakened, but the hierarchy becomes deeper, favoring a more restrained alignment of the observed temporal semantics toward the self-supervised signal. r_w is again compressed through FCs into a latent functional representation z. The above process corresponds to the expected probability q of Eq. (9). The second path of the query time injects it directly into the decoder. The query time is not processed by the neural network, preserving shallow temporal semantics. This query serves as a constraint on the supervised signal, forcing the output density function of the functional representation to be close to the latent distribution of query values.

When the model is trained, a portion of the sequence of the current interval is sampled. Samples that break the distribution consistency measure are assigned to the next interval by the JS module so that the samples retained in each interval represent the distribution of that interval. The input of the i-th interval corresponds to the conditional part of the prior of Eq. (4). The queries are incremental in time, and the sequence values corresponding to the query times are set as self-supervised signals. In this way, each current input is supervised by the value of the next query time induced by the query. To achieve temporal alignment in deep semantics, we use $r_w = r_h \oplus r_q$, to realize the semantic association of observed data representations with query representations, where r_w is called a weak query representation.

The functional representation z outputs a density function d_i in the latent space based on the weak representation r_w of the query, and samples di to obtain the representation sample s. Since z inherits the deep weak semantics of the query, s is weakly aligned with the self-supervised signal. NAIA fuses the query, sampling representation s, and weak query representation into r by concatenation operation $r = (t + \lambda) \oplus s \oplus r_w$ where $\oplus$ denotes the concatenation operation, and finally, the decoder reconstructs query value. The decoder also serves as a framework interface for NAIA and can choose different types of neural networks for replacement. This work focuses on inference architecture, so the most common fully connected neural network is used here.

4 Experiments

The performance of deep neural networks in representation learning cannot be directly tested. As a general-purpose inference architecture, NAIA can be applied to various downstream tasks. We explore the representation performance of NAIA by using five real-world datasets for sequence forecasting and anomaly detection tasks. The Electricity dataset [20, 21] contains the electricity consumption of 321 customers. SMD [27] is from a large technology company, and the data contains 38 dimensions with a true anomaly rate of 0.042. PSM [28] is a time series of 26 application server nodes recorded by eBay Inc. MSL [29] is 55 dimensions from NASA's Mars Science Laboratory rover, with a true anomaly ratio of 0.105. SWaT [30] is 51 sensors from a continuously operating water infrastructure system, with a true anomaly ratio of 0.121. The optimizer uses ADAM [31], where the exponential decay factors for the first- and second-order moments are set to: 0.87 and 0.995, respectively, and the learning rate is set to 0.001. Each data set is randomly divided, with 80% of the training set and 20% of the test set. The interval stability parameter θ was set to 0.05. The deviation parameter h was set to 1 by default.

We set up three sets of neural network configuration schemes given to NAIA. NAIA: all neural network interfaces are docked to fully connected neural networks. NAIA-R: Sequences were embedded in RNN instead of a fully connected one. NAIA-D/2: The number of neural network layers in the decoder part of the default configuration is reduced by half, and the effect of the interference decoder on the performance of the functional representation is examined.

The experiment contains four baselines: the MTGP based on a Gaussian process fitting (GPR) [17], which builds a temporal model for each dimension by creating a temporal representation kernel with a task representation kernel and then by their Hadamard product. GRU-D [32] is a recurrent neural network redesigned based on gating units, which decomposes the input part into three variables based on GRU, i.e., variable, mask, and time interval, to make the inference more adequate. VAE-RNN is based on variational auto-encoder architecture [26], based on neural differential equation approach ODE-RNN-ODE [30], which uses ODE to construct an encoder and decoder for continuous representation of samples. The informer is based on a self-attention mechanism [16] to improve the model's predictive power for long-term dependence by modifying the attention mechanism.

4.1 Forecasting Performance

In this section, the sequence forecasting task is used to validate the representation performance and timeliness of the model, and MSE is used as the experiment metric. The relative training elapsed time rate at optimal performance is used as the timeliness metric of the training representation, and the elapsed time of NAIA is set at 100 units. During the training phase, part of the data set is randomly masked, and an overall masking rate of m (the proportion of the masked data to the total data) is to generate irregular sequences and examine the model's ability to adapt to irregular sequences.

As shown in Table 1, m denotes the overall masking ratio, and it can be found that when the masking ratio is 0 (the input sequence is equally spaced and complete). In this case, NAIA does not show a significant advantage. In addition, NAIA maintains the lead among all other masking ratios. In addition, NAIA-R achieves the best performance in all masking ratios. The targeted design of the encoders can improve the forecasting performance of NAIA, and the neural network interface of NAIA gives some extensibility to this architecture. NAIA-D/2 does not show any significant performance degradation compared to NAIA, which indicates that NAIA's performance comes mainly from the embedding and representation modules rather than the decoder.

4.2 Anomaly Detection Performance

Anomaly detection is another downstream task that indirectly demonstrates the representational performance of the model. We perform anomaly detection using four real datasets with existing anomalous data and mask sequences in the same manner as described in Sect. 4.1. The model is required to reconstruct the removed data based on contextual patterns, and anomaly patterns are detected by examining the reconstruction MSE error (set to 0.03). The performance of anomaly detection is measured using the F1 score. Experimental results are shown in Table 2.

Table 2 shows that the informer model with self-attention performs best on all four datasets at $m = 0$. However, NAIA beats the other baselines when irregular sequences exist ($m > 0$), especially with the NAIA-R structure. NAIA is less affected by increasing masking ratios than other methods. Its F1 score drops slightly as m grows. This is because NAIA uses representative samples to handle non-equally spaced input and determines

Table 1. Forecasting performance on Electricity (MSE as %)

Strategy	$m = 0$	$m = 0.2$	$m = 0.4$	Elapsed time
MTGP	52.65 ± 0.37	57.79 ± 0.16	63.36 ± 0.02	2.9*
GRU-D	41.32 ± 0.86	44.73 ± 0.60	46.52 ± 0.69	428.3
VAE-RNN	42.09 ± 0.27	46.91 ± 0.18	49.58 ± 0.71	470.5
ODE-RNN-ODE	39.26 ± 0.30	41.02 ± 0.49	47.67 ± 0.35	245.3
Informer	$\mathbf{28.92 \pm 0.38}$	39.22 ± 0.52	45.09 ± 0.63	735.9
NAIA	33.25 ± 0.28	$\mathbf{34.18 \pm 0.58}$	$\mathbf{39.83 \pm 0.82}$	100.0
NAIA-R	31.04 ± 0.47	$34.56 \pm 0.39*$	$36.45 \pm 0.98*$	131.64
NAIA-D/2	36.52 ± 0.56	35.51 ± 0.47	40.74 ± 0.20	98.27

Table 2. Anomaly detection performance (F1 as %)

Dataset	SMD			PSM			MSL			SWaT		
Metric	m = 0	m = 0.2	m = 0.4	m = 0	m = 0.2	m = 0.4	m = 0	m = 0.2	m = 0.4	m = 0	m = 0.2	m = 0.4
MTGP	57.25	53.21	47.42	70.20	69.74	62.53	68.97	65.32	52.28	45.73	43.17	38.42
GRU-D	62.47	59.23	55.01	68.65	64.11	57.28	69.18	68.12	62.52	57.95	56.06	50.21
VAE-RNN	59.53	56.39	51.22	69.31	65.09	61.40	72.57	71.34	61.21	54.24	51.79	47.63
ODE-RNN-ODE	75.59	72.08	63.70	85.88	81.64	76.57	86.33	82.10	75.35	81.29	78.23	70.05
Informer	**88.35**	82.83	72.47	**92.32**	83.28	79.94	**89.02**	82.42	77.16	**85.59**	80.87	76.49
NAIA	84.19	83.98	80.72	87.79	85.57	84.95	85.07	84.71	84.35	83.62	82.65	81.87
NAIA-R	85.93	**85.62**	**81.06**	90.13	**87.25**	**85.11**	88.28	**86.17**	**85.92**	85.20	**84.20**	**83.81**
NAIA-D/2	83.76	83.55	80.34	85.08	83.78	81.04	82.42	83.26	82.75	81.49	79.56	79.30

the stable interval by the latent distribution of the output from the latent function representation. Therefore, the masking at the original sequence level has little impact on the latent distribution of the stable interval AS the data lies. The latent distribution of the stable interval fills in the local data gaps within the interval at a higher semantic level.

4.3 Ablation Experiments

We explore the contribution of key components to model performance through ablation experiments. The metric of performance test metric remains consistent with Sect. 4.2. The functional representation (FP) is responsible for receiving sample functions and transforming them continuously into density functions. We then remove the first path (P1) and the second path (P2) of the query times separately to examine the effect of missing deep/shallow semantic paths on the model. Finally, we remove the stability interval constraints (SIC) based on JS-divergence discrimination, allowing the sequence to mark query times in an uninhibited manner. The experimental results are shown in Table 3.

Table 3. Ablation results (F1 as %)

Strategy	SMD		PSM		MSL		SWaT	
	m = 0.2	m = 0.4	m = 0.2	m = 0.4	m = 0.2	m = 0.4	m = 0.2	m = 0.4
FP -	64.29	56.51	73.98	57.19	63.35	59.41	68.22	52.05
P1 -	70.49	68.78	71.75	69.43	74.28	65.20	70.12	67.32
P2 -	73.67	70.21	75.92	71.95	77.47	68.09	74.28	66.50
SIC-	79.76	71.65	82.90	77.92	78.38	72.37	79.55	70.97
NAIA	83.98	80.72	85.57	84.95	82.42	77.16	82.65	81.87

as shown in Table 3 (minus signs in the first column represent the removal of the component), the F1 score decreases the most when the FP is removed. Furthermore, it is sensitive to the change in masking ratio, and increasing the masking ratio degrades the F1 scores. This phenomenon indicates that the continuous property of functional representation plays a vital role in the semantic alignment of irregular sequences. The absence of path 1 has a more significant impact on performance than path 2, indicating that the deep weak semantics are more important for the model representation than the shallow semantics. Even so, the absence of only path 2 still significantly reduces the F1 score, implying the effectiveness of the interplay mechanism between deep and shallow semantics.

5   Conclusion

This paper introduces NAIA, a method for approximating the distribution among intervals of irregular sequences. By constructing continuous latent function representations, we capture the temporal semantic information of sequences. NAIA approximates the semantic associations between observed sequences and queries as density functions and aligns the sampling points of irregular sequences with query times, enabling self-supervised learning of the model. The computational architecture of NAIA extracts deep weak semantics and shallow strong semantics of query times. These semantics are injected into the pre- and post-stages of the latent representation, replacing the self-attention mechanism and reducing the representation complexity of irregular sequences to O(n). Experimental evaluations on five datasets demonstrate that NAIA achieves state-of-the-art performance compared to four recent baselines. Future research directions include further exploring the representation mechanisms of latent functions in the latent space and finding more suitable output methods for sequence distributions.

References

1. Marlin, B.M., Kale, D.C., Khemani, R.G., Wetzel, R.C.: Unsupervised pattern discovery in electronic health care data using probabilistic clustering models. In: Proceedings of the 2nd ACM SIGHIT International Health Informatics Symposium, pp. 389–398 (2012)
2. Yadav, P., Steinbach, M., Kumar, V., Simon, G.: Mining electronic health records (EHRs) a survey. ACM Comput. Surv. (CSUR). **50**, 1–40 (2018)
3. Schulz, M., Stattegger, K.: SPECTRUM: spectral analysis of unevenly spaced paleoclimatic time series. Comput. Geosci. **23**, 929–945 (1997)
4. Dogariu, M., Ştefan, L.-D., Boteanu, B.A., Lamba, C., Kim, B., Ionescu, B.: Generation of realistic synthetic financial time-series. ACM Trans. Multimedia Comput. Commun. Appl. **18**, 1–27 (2022)
5. Cheng, P., et al.: Asynchronous fault detection observer for 2-D markov jump systems. IEEE Trans Cybern. **52**, 13623–13634 (2021)
6. Cheng, L.-F., Stück, D., Quisel, T., Foschini, L.: The Impact of Missing Data in User-Generated mHealth Time Series
7. Song, X., Sun, P., Song, S., Stojanovic, V.: Event-driven NN adaptive fixed-time control for nonlinear systems with guaranteed performance. J. Franklin Inst. **359**, 4138–4159 (2022)
8. Fang, G., et al.: Up to 100x faster data-free knowledge distillation. In: Proceedings of the AAAI Conference on Artificial Intelligence, pp. 6597–6604 (2022)
9. Jiang, Y., Yin, S., Kaynak, O.: Performance supervised plant-wide process monitoring in industry 4.0: a roadmap. IEEE Open J. Ind. Electron. Soc. **2**, 21–35 (2021)
10. Horn, M., Moor, M., Bock, C., Rieck, B., Borgwardt, K.: Set functions for time series. In: International Conference on Machine Learning, pp. 4353–4363. PMLR (2020)
11. Tan, Q., et al.: DATA-GRU: dual-attention time-aware gated recurrent unit for irregular multivariate time series. In: Proceedings of the AAAI Conference on Artificial Intelligence, pp. 930–937 (2020)
12. Narayan Shukla, S., Marlin, B.M.: Multi-time attention networks for irregularly sampled time series. arXiv e-prints. arXiv-2101 (2021)
13. Rasmussen, C.E., Williams, C.K.I.: Gaussian Processes for Machine Learning. The MIT Press (2005). https://doi.org/10.7551/mitpress/3206.001.0001
14. Box, G.E.P., Jenkins, G.M., Reinsel, G.C., Ljung, G.M.: Time Series Analysis: Forecasting and Control. John Wiley & Sons (2015)
15. Kim, H.-G., Jang, G.-J., Choi, H.-J., Kim, M., Kim, Y.-W., Choi, J.: Recurrent neural networks with missing information imputation for medical examination data prediction. In: 2017 IEEE International Conference on Big Data and Smart Computing (BigComp), pp. 317–323. IEEE (2017)
16. Zhou, H., et al.: Informer: beyond efficient transformer for long sequence time-series forecasting. In: Proceedings of the AAAI conference on artificial intelligence, pp. 11106–11115 (2021)
17. Bonilla, E. V, Chai, K., Williams, C.: Multi-task Gaussian process prediction. Adv Neural Inf Process Syst. **20** (2007)
18. Shukla, S.N., Marlin, B.: Interpolation-prediction networks for irregularly sampled time series. In: International Conference on Learning Representations (2018)
19. Vaswani, A., et al.: Attention is all you need. In: Advances in Neural Information Processing Systems, vol. 30, (2017)
20. Trindade, A.: ElectricityLoadDiagrams20112014. Data Set, UCI Machine Learning Repository (2015)
21. Wang, K., et al.: Multiple convolutional neural networks for multivariate time series prediction. Neurocomputing **360**, 107–119 (2019)

22. Zhang, J., Li, X., Tian, J., Luo, H., Yin, S.: An integrated multi-head dual sparse self-attention network for remaining useful life prediction. Reliab. Eng. Syst. Saf. **233**, 109096 (2023)

23. Liu, S., et al.: Pyraformer: low-complexity pyramidal attention for long-range time series modeling and forecasting. In: International Conference on Learning Representations (2021)

24. Wu, H., Xu, J., Wang, J., Long, M.: Autoformer: decomposition transformers with auto-correlation for long-term series forecasting. Adv. Neural. Inf. Process. Syst. **34**, 22419–22430 (2021)

25. Zhou, T., Ma, Z., Wen, Q., Wang, X., Sun, L., Jin, R.: Fedformer: frequency enhanced decomposed transformer for long-term series forecasting. In: International Conference on Machine Learning, pp. 27268–27286. PMLR (2022)

26. Fabius, O., van Amersfoort, J.R.: Variational recurrent auto-encoders. arXiv e-prints. arXiv-1412 (2014)

27. Su, Y., Zhao, Y., Niu, C., Liu, R., Sun, W., Pei, D.: Robust anomaly detection for multivariate time series through stochastic recurrent neural network. In: Proceedings of the 25th ACM SIGKDD International Conference on Knowledge Discovery & Data Mining, pp. 2828–2837 (2019)

28. Abdulaal, A., Liu, Z., Lancewicki, T.: Practical approach to asynchronous multivariate time series anomaly detection and localization. In: Proceedings of the 27th ACM SIGKDD Conference on Knowledge Discovery & Data Mining, pp. 2485–2494 (2021)

29. Hundman, K., Constantinou, V., Laporte, C., Colwell, I., Soderstrom, T.: Detecting spacecraft anomalies using LSTMs and nonparametric dynamic thresholding. In: Proceedings of the 24th ACM SIGKDD International Conference on Knowledge Discovery & Data Mining, pp. 387–395 (2018)

30. Shen, L., Li, Z., Kwok, J.: Timeseries anomaly detection using temporal hierarchical one-class network. Adv. Neural. Inf. Process. Syst. **33**, 13016–13026 (2020)

31. Zhang, Z.: Improved adam optimizer for deep neural networks. In: 2018 IEEE/ACM 26th International Symposium on Quality of Service (IWQoS), pp. 1–2. IEEE (2018)

32. Che, Z., Purushotham, S., Cho, K., Sontag, D., Liu, Y.: Recurrent neural networks for multivariate time series with missing values. Sci. Rep. **8**, 6085 (2018)

Retrieval-Augmented GPT-3.5-Based Text-to-SQL Framework with Sample-Aware Prompting and Dynamic Revision Chain

Chunxi Guo, Zhiliang Tian[✉], Jintao Tang, Shasha Li, Zhihua Wen, Kaixuan Wang, and Ting Wang[✉]

College of Computer, National University of Defense Technology, Changsha, China
{chunxi,tianzhiliang,tangjintao,shashali,zhwen,wangkaixuan18,
tingwang}@nudt.edu.cn

Abstract. Text-to-SQL aims at generating SQL queries for the given natural language questions and thus helping users to query databases. Prompt learning with large language models (LLMs) has emerged as a recent approach, which designs prompts to lead LLMs to understand the input question and generate the corresponding SQL. However, it faces challenges with strict SQL syntax requirements. Existing work prompts the LLMs with a list of demonstration examples (i.e. question-SQL pairs) to generate SQL, but the fixed prompts can hardly handle the scenario where the semantic gap between the retrieved demonstration and the input question is large. In this paper, we propose a retrieval-augmented prompting method for an LLM-based Text-to-SQL framework, involving sample-aware prompting and a dynamic revision chain. Our approach incorporates sample-aware demonstrations, which include the composition of SQL operators and fine-grained information related to the given question. To retrieve questions sharing similar intents with input questions, we propose two strategies for assisting retrieval. Firstly, we leverage LLMs to simplify the original questions, unifying the syntax and thereby clarifying the users' intentions. To generate executable and accurate SQLs without human intervention, we design a dynamic revision chain that iteratively adapts fine-grained feedback from the previously generated SQL. Experimental results on three Text-to-SQL benchmarks demonstrate the superiority of our method over strong baseline models.

Keywords: Large language model · Text-to-SQL · Prompt learning

1 Introduction

Text-to-SQL task aims to convert natural language question (NLQ) to structured query language (SQL), allowing non-expert users to obtain desired information from databases [1,2]. As databases are popular in various scenarios involving different domains (e.g., education and financial systems, etc.), it is desirable to train a model that generalizes well across multiple domains. To facilitate cross-domain generalization [3,4], researchers adapt encoder-decoder architecture [5,6], reducing the requirement for specific domain knowledge via end-to-end training. These approaches require diverse and extensive training data to train the model, which is prohibitively expensive [7].

Recent progress focuses on large language models (LLMs) (e.g., GPT-3 [8], Codex [9] and GPT-4 [10]) with prompt learning [11], which refers to using specific prompts or instructions to generate desired responses. Rajkumar et al. [12] and Liu et al. [13] evaluate several prompt learning baselines for Text-to-SQL tasks. Their findings show that though it is natural for LLMs to generate text sequences, generating SQL is still a challenge due to the SQL's strict syntax requirements. To address these issues, inspired by few-shot learning [11], existing work employs prompting the LLMs with a list of demonstration examples (i.e. question-SQL pairs) to generate SQL queries. However, they typically rely on manual labour to create static demonstration examples tailored to specific tasks. DIN-SQL [14] selects pre-defined samples from each category, SELF-DEBUGGING [15] explains the code to LLMs but without explanation demonstration. These methods employ a static demonstration, meaning that the demonstration examples provided to LLMs are fixed and do not adapt or change across different examples. These static demonstration examples hardly adapt to the scenarios where the semantic gap between retrieved demonstrations and the input question is large, which is called retrieval bias [16], commonly appearing in the retrieval-augmented generation.

Inspired by [17], we argue that providing dynamic demonstrations can be adaptive to specific samples and schema for SQL generation. Dynamic examples enable the SQL generation to accommodate various scenarios. By adjusting to specific instances, demonstrations can be customized to incorporate the necessary query structure, logical operations, and question semantics. This adaptability makes it easier to generate SQL queries that are relevant and appropriate for various situations.

In this paper, we propose retrieval-augmented prompts for an LLM-based Text-to-SQL model, which contains sample-aware prompting and a dynamic revision chain. Specifically, we propose to retrieve similar SQL queries to construct prompts with sample-aware demonstration examples. Notice that users often ask questions in different expressions, even if they have the same intention and SQL query. It makes the model hard to retrieve helpful examples. To solve this issue, we propose to extract the question's real intention via two strategies: Firstly, we simplify original questions through LLMs to clarify the user's intentions and unify the syntax for retrieval. Secondly, we extract question skeletons for retrieving items with similar question intents. To produce executable and accurate SQL, we design a dynamic revision chain, generating SQL queries by iteratively adapting to fine-grained feedback according to the previous version of the generated SQL. The feedback includes SQL execution results, SQL explanations, and related database contents. This dynamic chain manages to generate executable and accurate SQL through automatic interaction between the language model and the database without human intervention.

Our contributions are as follows: (1) We develop a retrieval-augmented framework for Text-to-SQL tasks by prompting LLMs with sample-aware demonstrations. (2) We propose a dynamic revision chain, which adapts to the previously generated SQL with fine-grained feedback. (3) Our method outperforms strong baseline models on three Text-to-SQL benchmarks.

2   Related Work

2.1   Encoder-Decoder SQL Generation

SQL generation tasks have achieved significant advancements through the utilization of encoder-decoder architectures [2].

On the encoder side, Guo et al. [18] proposed IRNET, using attention-based Bi-LSTM for encoding and an intermediate representation-based decoder for SQL prediction. Later, [19,20] introduced graph-based encoders to construct schema graphs and improve input representations. Works such as RATSQL [1], SDSQL [5], LGESQL [21], S^2SQL [22], R^2SQL [23], SCORE [24], and STAR [25] further improved structural reasoning by modelling relations between schemas and questions. GRAPHIX-T5 [6] overcomes the limitations of previous methods by incorporating graph representation learning in the encoder. Concurrently, RASAT [26] also provided T5 with structural information by adding edge embedding into multi-head self-attention.

On the decoder side, we divide the methods into four categories: sequence-based methods (BRIDGE [27], PICARD [3]) directly translate NLQ into SQL query token by token, template-based methods (X-SQL [28], HydraNet [29]) employ predefined templates to regulate SQL generation and ensure structural coherence, stage-based methods (GAZP [30], RYANSQL [31]) first establish a coarse-grained SQL framework and then fills in the missing details in the frame which calls slot-filling methodologies, and hierarchical-based methods (IRNet [32], RAT-SQL [1]) generate SQL according to grammar rules in a top-down manner, resulting in a tree-like structure.

2.2   LLM-Based SQL Generation

LLM-based models recently emerge as a viable option for this task [12,13]. For effectively utilizing, it is important to design appropriate in-context demonstration [33] and chain-of-thought (CoT) [34] strategies that can elicit its ability [7].

In terms of searching for demonstrations, DIN [14] selects a fair number of demonstration examples from each category (e.g. simple classes, non-nested complex classes and nested complex classes), but they are fixed. Moreover, Guo et al. [35] adaptively retrieve intention-similar SQL demonstration examples through de-semanticization of the questions. However, none of these methods can solve the ambiguous and varied questioning of realistic scenarios.

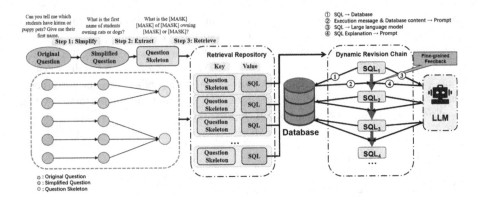

Fig. 1. Framework overview: The left half shows retrieval repository construction in three steps. The top three sentences are three specific instances each. The Green dashed box presents the training set. The right half is a dynamic revision chain with SQL queries generated by LLM iterations as nodes (green boxes). The output of steps 2 and 4 are collectively referred to as fine-grained feedback. (Color figure online)

As for the CoT prompting strategy, DIN-SQL [14] follows a least-to-most [36] prompting method, decomposing Text-to-SQL task into subtasks and solves them one by one. Pourreza and Chen et al. explore self-correction [14,15], where the LLM explain the question and SQL, providing valuable feedback for improvement. Tian et al. [37] propose interactive generation with editable step-by-step explanations, combining human intervention with LLM generation to refine the final SQL output. Additionally, Sun et al. [38] explore execution-based self-consistent prompting methods.

Nonetheless, creating task-specific demonstration examples [14,15,37,38] demands manual labour. Inspired by some retrieval-related research [16,39,40], we develop a retrieval-augmented framework for Text-to-SQL tasks. Our method works through automatic interaction between the LLMs and the databases without human intervention. Moreover, explaining to itself and simple feedback alone [14,15] are weak for digging out errors for correction. Our approach takes into account all three aspects of fine-grained feedback, which interact with each other to create effective feedback.

3 Methodology

Our framework consists of two modules as shown in Fig. 1: (1) **Retrieval Repository:** (see Sect. 3.1) We construct a retrieval repository with simplified questions added and then use question skeletons to retrieve sample-aware SQL demonstration examples. (2) **Dynamic Revision Chain:** (see Sect. 3.2) We further revise the generated SQL queries by adding fine-grained feedback.

3.1   Retrieval Repository

We construct a retrieval repository consisting of multiple key-value retrieval items, where the keys represent the question skeletons and the values are k sample-aware SQL queries. These processes enable us to generate demonstration examples that showcase the desired behaviours of the LLM. Our method involves: (1) Simplifying original questions to unify various questioning styles (see Sect. 3.1.1). (2) Extracting question skeletons to construct a retrieval repository (see Sect. 3.1.2). (3) Retrieving SQL queries according to skeleton similarities (see Sect. 3.1.3).

3.1.1   Question Simplification

We simplify natural language questions by prompting the LLM with instructions. In this way, we can avoid the frustration of unusual questioning styles and enhance the syntax and wording variety in the repository.

Specifically, we construct a prompt template $prompt(.)$: *"Replace the words as far as possible to simplify the question, making it syntactically clear, common and easy to understand: [QUESTION]"*, where *"[QUESTION]"* represents the original natural language question. We then obtain the simplified question by feeding $prompt(Q)$ into the LLM. We maintain a consistent temperature setting in the language model to ensure that all simplified sentences exhibit the same probability distribution.

3.1.2   Question Skeleton Extraction

We then extract question skeletons, including both original questions and simplified questions. We follow the method proposed by Guo et al. [35] to obtain question skeletons. This process removes specific schema-related tokens from the questions, focusing solely on the structure and intent. Finally, we take the (question skeleton, SQL) pairs from the training set and store them in the retrieval repository. Note that the number of samples in the retrieval repository is twice as large as the training set, due to the addition of the simplified samples.

Let $\mathcal{D}_{\text{train}}$ represent the training set, and R denotes the retrieval repository. The original natural language question is denoted as Q_o, while Q_r represents the simplified question. The question skeletons are denoted as S_o and S_r for the original and simplified questions, respectively. We formalize the composition of the retrieval repository as follows:

$$R = \{(S_o, \text{SQL}), (S_r, \text{SQL}) \mid (Q_o, \text{SQL}) \in \mathcal{D}_{\text{train}}\}.$$

3.1.3   Sample Retrieval

The retrieval process searches for the most similar question skeletons and returns their corresponding SQL queries from the retrieval repository. This search is based on the semantic similarity between the skeleton of the new question and the items' keys in R.

Specifically, given a new question $\widetilde{Q_o}$, we first obtain its simplified sentence $\widetilde{Q_r}$, and their corresponding question skeletons $\widetilde{S_o}$ and $\widetilde{S_r}$, following the same method used in previous two subsections (see Sects. 3.1.1 and 3.1.2). Then we calculate the cosine similarity scores s_o between the semantic vector of question skeleton $\widetilde{S_o}$ and all question skeletons S in R. Similarly, we also compute the cosine similarity scores s_r for simplified question skeleton $\widetilde{S_r}$ using the formula: $s = \cos\left(\mathbf{f}(S) \cdot \mathbf{f}(\widetilde{S})\right)$, where $\mathbf{f}(.)$ represents an off-the-shelf semantic encoder[1]. Here, $\widetilde{S}$ will be instantiated as $\widetilde{S_o}$ and $\widetilde{S_r}$, and s will be instantiated as s_o and s_r, correspondingly.

From these scores, we select the top-k retrieval samples with the highest rankings. Let k_1 and k_2 denote the number of samples retrieved from the original question skeleton $\widetilde{S_o}$ and the simplified question skeleton $\widetilde{S_r}$ respectively, such that $k = k_1 + k_2$. We then concatenate the k samples to form a demonstration example as input to the LLM. Our retrieval repository offers LLMs with sample-aware SQL examples, which display a more practical answer space.

3.2 Dynamic Revision Chain

We employ LLMs to generate an initial SQL query, and then we iteratively revise the generated SQL queries based on fine-grained feedback, forming a dynamic revision chain. The dynamic revision chain consists of the SQL queries generated by the LLM iteration as nodes and the prompts provided to the LLM as edges. With minimal human intervention, LLMs interact with databases to generate accurate and executable SQL queries in two stages of the dynamic revision chain: (1) assembling prompt based on the fine-grained feedback (see Sect. 3.2.1), and (2) generating SQL via iterative prompting (see Sect. 3.2.2).

3.2.1 Fine-Grained Feedback

We collect three fine-grained pieces of information based on the SQL generated in the previous iteration. The intuition is that various information hampers LLMs' focus, so they struggle to extract necessary data from extensive and complex databases. Thus, we should progressively narrow down the scope and prioritize the most likely information. The fine-grained feedback in our approach consists of three aspects of information:

(1) **Execution Error Feedback**: We feed the SQL query generated by LLM into the database engine (i.e. SQLite) for execution. We then obtain the error messages reported during the execution and add them to the prompt. It checks whether the predicted SQL can be executed correctly, and reports the specifics of the error (e.g. *"no such table: [TABLE]"*, *"no such function: YEAR"*, *"misuse of aggregate: COUNT()"*). By incorporating the execution error messages into the prompt, LLM can learn from its errors. This helps to generate queries that follow the SQL syntax rules.

---

[1] We utilize SBERT [41] in our experiment.

(2) **Natural Language Explanation**: We prompt the LLM with instructions, converting the SQL predicted in the previous iteration back into its corresponding natural language expression. Specifically, we construct an instruction: *"What does this SQL query mean? What are the differences between the predicted meaning and the question meanings above?"*. The LLM identifies semantic gaps and fills them by explaining the meaning of its own generated SQL and comparing it to the meaning of the original question.

(3) **Related Database Contents:** We provide the LLM with content details about the database tables and columns involved in the SQL queries predicted in the previous iteration, including the possible values involved in the question. It aims to allow LLMs to simulate execution and thus generate more contextually relevant and accurate SQL queries.

Overall, the fine-grained feedback approach aims to enable LLMs to learn from their mistakes, understand the meaning of the SQL queries generated and use contextual information in the database to generate more accurate and relevant SQL queries. By addressing challenges and focusing on important aspects, the methodology aims to help the LLm better extract the necessary data from complex databases and improve the performance of its query generation.

3.2.2   Iterative SQL Generation

Based on prompts with fine-grained feedback, the LLM iteratively generates SQL queries. The intuition for iterative generation is that one iteration of fine-grained feedback might not check for all mistakes, whereas multiple iterations of feedback generation are more likely to get progressively closer to the gold answer.

Specifically, we concatenate three fine-grained feedback components with the previously generated SQL in each iteration, feeding them into the LLM. We then obtain a new SQL and collect new fine-grained feedback based on it, proceeding so to iterative generation. Let's denote the previous SQL query generated by the LLM as SQL_{prev} and the current SQL query as SQL_{curr}. The fine-grained feedback components are represented as F_{error} for execution error feedback, F_{NL} for natural language explanation, and F_{DB} for related database contents. At each iteration i, the LLM generates a new SQL query $SQL_{curr}^{(i)}$ by incorporating the fine-grained feedback components:

$$SQL_{curr}^{(i)} = \text{LLM}(SQL_{prev}, F_{error}^{(i)}, F_{NL}^{(i)}, F_{DB}^{(i)}).$$

After executing $SQL_{curr}^{(i)}$ using the database engine, we obtain the result $R_{prev}^{(i)}$ from the previous iteration and $R_{curr}^{(i)}$ from the current iteration. To avoid infinite loops, we set a maximum number of iterations N_{max}. The termination condition is defined as: $R_{prev}^{(i)} = R_{curr}^{(i)}$ or $i = N_{max}$. This control mechanism ensures that the generated SQL queries converge to an optimal and executable solution within a reasonable timeframe.

In this iterative feedback loop, we enable a dynamic interaction between the LLM and the database engine, maximizing the generation of executable SQL without extensive human intervention.

4   Experiments

4.1   Experimental Setup

4.1.1   Setting

We evaluate our method on text-davinci-003, which offers a balance between capability and availability. We apply FAISS [42] for storing the question skeletons and efficient retrieval followed by Guo et al. [35]. For the initial simplification of questions, we set temperature $\tau = 1.0$. When generating SQL samples, we set temperature $\tau = 0.5$. For the number of retrieval samples, we assign $k_1 = 4$ and $k_2 = 4$. The maximum number of iterations N is 8.

4.1.2   Datasets

(1) **Spider** [43] is a large-scale benchmark of cross-domain Text-to-SQL across 138 different domain databases. (2) **Spider-Syn** [44] is a challenging variant based on Spider that eliminates explicit alignment between questions and database schema by synonym substitutions. (3) **Spider-DK** [45] is also a variant dataset based on Spider with artificially added domain knowledge.

4.1.3   Evaluation

We consider two key metrics: execution accuracy (EX) and test-suite accuracy (TS) [46]. EX measures the accuracy of the execution results by comparing them with the standard SQL queries, while TS measures whether the SQL passes all EX evaluations for multiple tests, generated by database augmentation. Note that EX is the most direct indication of the model performance in Text-to-SQL, although it contains false positives. Exact match evaluation is not performed, as multiple correct SQLs exist for one query. We use the official TS evaluation procedure, while for EX, we slightly modify the evaluation procedure due to the need to decouple the fine-tuning-based models for independent evaluation.

4.1.4   Baselines

We compare to two groups of methods:
Fine-Tuning T5-3B Baselines: PICARD [3] is a technique that constrains auto-regressive decoders in language models through incremental parsing; **RASAT** [26], which incorporates relation-aware self-attention into transformer models while also utilizing constrained auto-regressive decoders; and **RESD-SQL** [5], which introduces a ranking-enhanced encoding and skeleton-aware decoding framework to effectively separate schema linking and skeleton parsing.

Prompting LLMs Baselines: As for the large language models, we use two variants of the Codex family [9,12] (**Davinci** and **Cushman**), **PaLM-2** [38,47], the **GPT-4** model [10,14] and the **ChatGPT** model [13]. In addition to a simple baseline assessment model, we choose several recent LLM-based works. **DIN** [14] decompose the Text-to-SQL tasks into sub-tasks: schema linking, query classification and decomposition, SQL generation, and self-correction; then performing few-shot prompting with GPT-4 [10]. **SELF-DEBUGGING** [15] adds error messages to the prompt and conducts multiple rounds of few-shot prompting for self-correction. **Few-shot SQL-PaLM** [38] adopts an execution-based self-consistency prompting approach.

4.2   Main Results

4.2.1   Performance on Spider Dataset

Table 1 shows how ours performed on Spider compared to baseline methods. Across all three datasets, our methods achieve the highest of execution accuracy (EX) and test suite accuracy (TS).

Table 1. Performance comparison on Spider with various methods. "-" indicates that the results are not available. Schema indicates that the prompt contains the SQL for creating the database tables (i.e. tables, columns, value and its type)[4].

Model	Method	EX	TS
T5-3B	T5-3B + PICARD [3]	79.3	69.4
	RASAT + PICARD [26]	80.5	70.3
	RESDSQL-3B + NatSQL [5]	84.1	73.5
Codex-cushman	Few-shot [12]	61.5	50.4
	Few-shot + Schema [12]	63.7	53.0
Codex-davinci	Few-shot [12]	60.8	51.2
	Few-shot + Schema [12]	67.0	55.1
	Few-shot [14]	71.0	61.5
	DIN-SQL [14]	75.6	69.9
	SELF-DEBUGGING [15]	84.1	-
PaLM2	Few-shot SQL-PaLM [38]	82.7	77.3
	Fine-tuned SQL-PaLM [38]	82.8	78.2
GPT-4	Zero-shot [14]	72.9	64.9
	Few-shot [14]	76.8	67.4
	DIN-SQL [14]	82.8	74.2
ChatGPT	Zero-shot [13]	70.1	60.1
Text-davinci	Zero-shot	73.1	71.6
	Ours	**85.0 (0.9↑)**	**83.2 (5.9↑)**

Ours exhibits strong performance on test-suite accuracy, which exceeds the next-best method results in fine-tuning and prompting by 9.7% and 5.9% respectively. In terms of EX, ours outperforms the next best method in both fine-tuning and prompting by 0.9%.

Comparison with Zero-Shot Prompting Models: Across all three metrics, ours surpasses Codex, ChatGPT and even GPT-4 models utilizing zero-shot prompting, despite they employ the prescribed format as outlined in the official guidelines[5]. This indicates that although LLMs are trained using a specific format, their acquired competencies become internalized and subsequently expanded for application within more flexible formats.

Comparison with Few-Shot Prompting Models: Ours also outperforms all models in a few-shot setting. The closest EX performance compared to ours is SELF-DEBUGGING, with an iterative prompting strategy as well, but we still outperform it by 0.9%. Notice that the two methods with similar few-shot prompting in the Codex-davinci model, the latter performs 10% better than the former in both EX and TS. It indicates that the selection of demonstration examples (easy, non-nested complex, and nested complex classes) [14] plays a significant role. While ours uses an adaptive sample-aware method brings 8.2% more effective than this static demonstration, which suggests that incorporating more effective prompts is crucial for LLMs to understand new specific tasks.

4.2.2   Performance on Spider-SYN and DK Datasets

Table 2 shows that ours is significantly more robust than baseline methods for Spider variants. As compared to Spider-SYN, ours improved 4.5% on EX and 12.6% on TS. Surprisingly, ours improved by 13.6% over the previous SOTA on Spider-DK.

4.3   Various Difficulty Levels Analysis

As shown in Table 3, we evaluate our effectiveness at various difficulty levels, which are determined by the number of SQL keywords used, the presence of nested sub-queries, and the utilization of column selections or aggregations. The results show that ours outperforms the other models at all levels except for the easy level, where it is worse than SQL-PaLM. The improvement in performance with increasing difficulty levels indicates that our model's strengths become more pronounced as the queries become more challenging. This suggests that our model excels in handling complex SQL queries.

4.4   Ablation Study

Figure 2 demonstrates with and without each of the two modules at four complexity levels. It shows that the exclusion of any of the modules leads to a

---

[5] https://platform.openai.com/examples/default-sqltranslate.

Table 2. Evaluation of our method on Spider-SYN and Spider-DK datasets[10]. "-" indicates that the results are not available[11].

SPIDER-SYN			
	Method	EX	TS
Fine-tuning	T5-3B + PICARD [3]	69.8	61.8
	RASAT + PICARD [26]	70.7	62.4
	RESDSQL-3B + NatSQL [5]	76.9	66.8
Prompting	ChatGPT [13]	58.6	48.5
	Text-davinci	60.7	60.3
	Few-shot SQL-Palm [38]	74.6	67.4
	Fine-tuned SQL-Palm [38]	70.9	66.4
	Ours	**81.4(4.5↑)**	**80.0(12.6↑)**
SPIDER-DK			
	Method	EX	TS
Fine-tuning	T5-3B + PICARD [3]	62.5	-
	RASAT + PICARD [26]	63.9	-
	RESDSQL-3B + NatSQL [5]	66.0	-
Prompting	ChatGPT [12]	62.6	-
	Text-davinci	66.2	-
	Few-shot SQL-Palm [38]	66.5	-
	Fine-tuned SQL-Palm [38]	67.5	-
	Ours	**81.1 (13.6↑)**	-

Table 3. Test-suite accuracy at various complexity levels on Spider. The first four rows are from [38], which is described as execution accuracy data in the original paper [14], but actually, it is test-suite accuracy data.

Prompting	Easy	Medium	Hard	Extra	All
Few-shot (CodeX-davinci) [14]	84.7	67.3	47.1	26.5	61.5
Few-shot (GPT-4) [14]	86.7	73.1	59.2	31.9	67.4
DIN-SQL (CodeX-davinci) [14]	89.1	75.6	58.0	38.6	69.9
DIN-SQL (GPT-4) [14]	91.1	79.8	64.9	43.4	74.2
Few-shot SQL-PaLM (PaLM2) [38]	**93.5**	84.8	62.6	48.2	77.3
Fine-tuned SQL-PaLM (PaLM2) [38]	**93.5**	85.2	68.4	47.0	78.2
Ours (Text-davinci)	91.9	**88.6**	**75.3**	**63.9**	**83.2**

decrease in performance at all levels of difficulty, in terms of hard and extra levels. Decreases in model performance are similar for the w/o revise and w/o simplify settings. Note that both modules of our method are most effective in improving the Spider-DK's easy level by 13.6% each, which requires additional domain knowledge. This suggests that the simplification strategy and dynamic revision chain strategy contribute to a variety of generalisation issues.

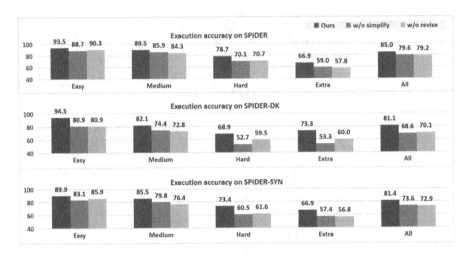

Fig. 2. Ablation study of our model components at various complexity levels across three datasets. w/o simplify refers to using a direct retrieval of the question skeletons rather than a strategy of simplifying questions. w/o revise refers to removing the dynamic revision chain module.

We found that removing the simplification module resulted in a significant drop in model performance, particularly in the DK dataset where the overall drop was 12.5%. The impact at different difficulty levels is in descending order of extra, hard, easy, and medium. This is possibly due to the fact that the model can incorporate more external knowledge as a supplementary description when simplifying, especially in the case of more SQL components. Note that w/o simplify is rather more effective for solving easy-level problems than medium-level ones, probably because the execution accuracy of easy-level problems is already high and short sentences are more likely to cause ambiguity.

Without the revision module, model performance suffers more as the difficulty level increases. On Spider-DK the model performance decreases by 11.0%, especially on easy-level and extra-level by 13.6% and 13.3% respectively. As higher difficulty levels require more knowledge, this suggests that the fine-grained feedback in the revision module effectively complements the domain knowledge required for SQL generation.

4.5   Iterative Round Analysis

From Fig. 3, we observe that the major improvement comes from the first two iteration turns. We note that in addition to the 4.6% improvement in the first iteration of Spider, the other two datasets studied for generalisability, Spider-DK and Spider-SYN, also showed a slight improvement in accuracy in the second iteration over the first. This indicates that iterative feedback of fine-grained information from a dynamic revision chain helps to deal with more complex generalisation problems, comparable to multiple reasoning needs to progressively derive the target answer.

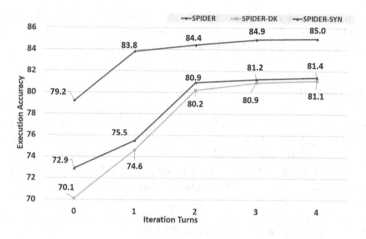

Fig. 3. Analysis of dynamic SQL revision chain with different numbers of iteration rounds across three datasets: Spider, Spider-SYN, and Spider-DK.

4.6   Case Study

To demonstrate our model, we show a comparison of predicted SQL queries in Fig. 4 using ChatGPT [12], DIN-SQL [14], SQL-PaLM [38] and Ours.

In the first example, since the question obviously mentions *"French"*, the general models will be confused about the exact value of the column *"citizenship"* even if they pick it out. Noting that a SQL query must match the exact word mentioned to find the correct answer. Our approach provides the exact value of the database content involved in the first fine-grained iteration, which leads to a golden answer.

The second example requires only the selection of one item, whereas DIN-SQL and SQL-PaLM both select two. ChatGPT incorrectly uses the aggregate function *COUNT()*, which in this case is required in conjunction with *GROUP BY*. Our approach self-corrects the error in the second fine-grained iteration by interpreting the SQL interpretation in natural language.

Question:	What are the names of the singers who are not French citizens?	
ChatGPT :	SELECT Name FROM singer WHERE Citizenship <> 'French'	✗
DIN-SQL:	SELECT Name FROM singer WHERE Citizenship != 'French'	✗
SQL-PaLM:	SELECT name FROM singer WHERE citizenship ! = "French"	✗
Ours:	SELECT singer.Name FROM singer WHERE singer.Citizenship != "France"	√
Question:	Please show the record formats of orchestras in ascending order of count.	
ChatGPT :	SELECT Orchestra, Major_Record_Format FROM orchestra ORDER BY COUNT(∗) ASC	✗
	(Error feedback: misuse of aggregate: COUNT())	
DIN-SQL:	SELECT Major_Record_Format, COUNT(∗) as count FROM orchestra GROUP BY Major_Record_Format ORDER BY count ASC	✗
SQL-PaLM:	SELECT major_record_format , count(∗) FROM orchestra GROUP BY major_record_format ORDER BY count(∗) Asc	✗
Ours:	SELECT Orchestra.Major_Record_Format FROM Orchestra GROUP BY Orchestra.Major_Record_Format ORDER BY COUNT(∗) ASC	√

Fig. 4. Two illustrative cases from Spider [43]. Blue-coloured text is the correct generation, while the red-coloured text indicates the wrong generation. On the right hand side, √ means correct SQL while × means wrong. (Color figure online)

5   Conclusion

We propose retrieval-augmented prompts for an LLM-based Text-to-SQL model. By utilizing sample-aware prompting and a dynamic revision chain, we address the challenge of retrieving helpful examples and adapting the generated SQL based on fine-grained feedback. Experimental results on three Text-to-SQL benchmarks demonstrate the effectiveness of our method.

Acknowledgements. Research on this paper was supported by National Natural Science Foundation of China (Grant No. 62306330).

References

1. Wang, B., Shin, R., Liu, X., Polozov, O., Richardson, M.: RAT-SQL: relation-aware schema encoding and linking for text-to-SQL parsers. ACL (2020)
2. Cai, R., Xu, B., Zhang, Z., Yang, X., Li, Z., Liang, Z.: An encoder-decoder framework translating natural language to database queries. In: IJCAI, July 2018 (2018)
3. Scholak, T., Schucher, N., Bahdanau, D.: PICARD: parsing incrementally for constrained auto-regressive decoding from language models. In: EMNLP (2021)
4. Cai, R., Yuan, J., Xu, B., Hao, Z.: SADGA: structure-aware dual graph aggregation network for text-to-SQL. In: NIPS, December 2021 (2021)
5. Li, H., Zhang, J., Li, C., Chen, H.: Decoupling the skeleton parsing and schema linking for text-to-SQL. arXiv arXiv:2302.05965 (2023)
6. Li, J., Hui, B., et al.: Graphix-T5: mixing pre-trained transformers with graph-aware layers for text-to-SQL parsing. arXiv arXiv:2301.07507 (2023)
7. Zhao, W.X., Zhou, K., Li, J., et al.: A survey of large language models. arXiv preprint arXiv:2303.18223 (2023)
8. Brown, T., Mann, B., Ryder, N., et al.: Language models are few-shot learners. In: NIPS, vol. 33, pp. 1877–1901 (2020)
9. Chen, M., Tworek, J., Jun, H., Yuan, Q., de Oliveira Pinto, H.P., et al.: Evaluating large language models trained on code. arXiv arXiv:2107.03374 (2021)

10. OpenAI: GPT-4 technical report (2023)
11. Liu, P., Yuan, W., Fu, J., Jiang, Z., Hayashi, H., Neubig, G.: Pre-train, prompt, and predict: a systematic survey of prompting methods in natural language processing. ACM Comput. Surv. **55**(9), 1–35 (2023)
12. Rajkumar, N., Li, R., Bahdanau, D.: Evaluating the text-to-SQL capabilities of large language models. arXiv arXiv:2204.00498 (2022)
13. Liu, A., Hu, X., Wen, L., Yu, P.S.: A comprehensive evaluation of ChatGPT's zero-shot text-to-SQL capability. arXiv arXiv:2303.13547 (2023)
14. Pourreza, M., Rafiei, D.: DIN-SQL: decomposed in-context learning of text-to-SQL with self-correction. arXiv preprint arXiv:2304.11015 (2023)
15. Chen, X., Lin, M., Schärli, N., Zhou, D.: Teaching large language models to self-debug. arXiv preprint arXiv:2304.05128 (2023)
16. Song, Y., et al.: Retrieval bias aware ensemble model for conditional sentence generation. In: ICASSP, pp. 6602–6606. IEEE (2022)
17. Min, S., et al.: Rethinking the role of demonstrations: what makes in-context learning work? arXiv preprint arXiv:2202.12837 (2022)
18. Guo, J., et al.: Towards complex text-to-SQL in cross-domain database with intermediate representation. arXiv preprint arXiv:1905.08205 (2019)
19. Bogin, B., Berant, J., Gardner, M.: Representing schema structure with graph neural networks for text-to-SQL parsing. In: ACL, September 2019 (2019)
20. Chen, Z., et al.: ShadowGNN: graph projection neural network for text-to-SQL parser. In: NAACL, June 2021 (2021)
21. Cao, R., Chen, L., et al.: LGESQL: line graph enhanced text-to-SQL model with mixed local and non-local relations. In: ACL, July 2021 (2021)
22. Hui, B., Geng, R., Wang, L., et al.: S²SQL: injecting syntax to question-schema interaction graph encoder for text-to-SQL parsers. In: ACL, pp. 1254–1262 (2022)
23. Hui, B., Geng, R., Ren, Q., et al.: Dynamic hybrid relation exploration network for cross-domain context-dependent semantic parsing. In: AAAI, May 2021 (2021)
24. Yu, T., Zhang, R., Polozov, A., Meek, C., Awadallah, A.H.: SCore: pre-training for context representation in conversational semantic parsing. In: ICLP (2021)
25. Cai, Z., Li, X., Hui, B., Yang, M., Li, B., et al.: STAR: SQL guided pre-training for context-dependent text-to-SQL parsing. In: EMNLP, October 2022 (2022)
26. Qi, J., Tang, J., He, Z., et al.: RASAT: integrating relational structures into pre-trained Seq2Seq model for text-to-SQL. In: EMNLP, pp. 3215–3229 (2022)
27. Lin, X.V., Socher, R., Xiong, C.: Bridging textual and tabular data for cross-domain text-to-SQL semantic parsing. In: EMNLP, pp. 4870–4888 (2020)
28. He, P., Mao, Y., Chakrabarti, K., Chen, W.: X-SQL: reinforce schema representation with context. arXiv arXiv:1908.08113 (2019)
29. Lyu, Q., Chakrabarti, K., Hathi, S., Kundu, S., Zhang, J., Chen, Z.: Hybrid ranking network for text-to-SQL. arXiv preprint arXiv:2008.04759 (2020)
30. Zhong, V., Lewis, M., Wang, S.I., Zettlemoyer, L.: Grounded adaptation for zero-shot executable semantic parsing. In: EMNLP, pp. 6869–6882 (2020)
31. Choi, D., Shin, M.C., Kim, E., Shin, D.R.: RYANSQL: recursively applying sketch-based slot fillings for complex text-to-SQL in cross-domain databases. In: CL (2021)
32. Guo, J., et al.: Towards complex text-to-SQL in cross-domain database with intermediate representation. In: ACL, pp. 4524–4535 (2019)
33. Brown, T., Mann, B., et al.: Language models are few-shot learners. In: NIPS, vol. 33, pp. 1877–1901 (2020)
34. Wei, J., et al.: Chain-of-thought prompting elicits reasoning in large language models. In: Advances in Neural Information Processing Systems, vol. 35, pp. 24824–24837 (2022)

35. Guo, C., et al.: A case-based reasoning framework for adaptive prompting in cross-domain text-to-SQL. arXiv preprint arXiv:2304.13301 (2023)
36. Zhou, D., et al.: Least-to-most prompting enables complex reasoning in large language models. arXiv preprint arXiv:2205.10625 (2022)
37. Tian, Y., Li, T.J.J., Kummerfeld, J.K., Zhang, T.: Interactive text-to-SQL generation via editable step-by-step explanations. arXiv preprint arXiv:2305.07372 (2023)
38. Sun, R., et al.: SQL-PaLM: improved large language model adaptation for text-to-SQL. arXiv arXiv:2306.00739 (2023)
39. Tian, Z., Bi, W., Li, X., Zhang, N.L.: Learning to abstract for memory-augmented conversational response generation. In: ACL, pp. 3816–3825 (2019)
40. Wen, Z., et al.: GRACE: gradient-guided controllable retrieval for augmenting attribute-based text generation. In: Findings of ACL 2023, pp. 8377–8398 (2023)
41. Reimers, N., Gurevych, I.: Sentence-BERT: sentence embeddings using Siamese BERT-networks. In: EMNLP-IJCNLP, pp. 3982–3992. ACL (2019). https://aclanthology.org/D19-1410
42. Johnson, J., Douze, M., Jegou, H.: Billion-scale similarity search with GPUs. IEEE Trans. Big Data **7**, 535–547 (2019)
43. Yu, T., Zhang, R., et al.: Spider: a large-scale human-labeled dataset for complex and cross-domain semantic parsing and text-to-SQL task. In: EMNLP, June 2019 (2019)
44. Gan, Y., Chen, X., Huang, Q., Purver, M., et al.: Towards robustness of text-to-SQL models against synonym substitution. In: ACL, July 2021 (2021)
45. Gan, Y., Chen, X., Purver, M.: Exploring underexplored limitations of cross-domain text-to-SQL generalization. In: EMNLP, December 2021 (2021)
46. Zhong, R., Yu, T., Klein, D.: Semantic evaluation for text-to-SQL with distilled test suites. In: EMNLP, pp. 396–411 (2020)
47. Anil, R., et al.: PaLM 2 technical report. arXiv preprint arXiv:2305.10403 (2023)

Improving GNSS-R Sea Surface Wind Speed Retrieval from FY-3E Satellite Using Multi-task Learning and Physical Information

Zhenxiong Zhou[1], Boheng Duan[2(✉)], and Kaijun Ren[2(✉)]

[1] School of Computer Science and Technology, National University of Defense Technology, Changsha, China
zhouzhenxiong@nudt.edu.cn
[2] School of Meteorology and Oceanography, National University of Defense Technology, Changsha, China
{bhduan,renkaijun}@nudt.edu.cn

Abstract. Global Navigation Satellite System Reflectometry (GNSS-R) technology has great advantages over traditional satellite remote sensing detection of sea surface wind field in terms of cost and timeliness. It has attracted increasing attention and research from scholars around the world. This paper focuses on the Fengyun-3E (FY-3E) satellite, which carries the GNOS II sensor that can receive GNSS-R signals. We analyze the limitations of the conventional sea surface wind speed retrieval method and the existing deep learning model for this task, and propose a new sea surface wind speed retrieval model for FY-3E satellite based on a multi-task learning (MTL) network framework. The model uses the forecast product of Hurricane Weather Research and Forecasting (HWRF) model as the label, and inputs all the relevant information of Delay-Doppler Map (DDM) in the first-level product into the network for comprehensive learning. We also add wind direction, U wind and V wind physical information as constraints for the model. The model achieves good results in multiple evaluation metrics for retrieving sea surface wind speed. On the test set, the model achieves a Root Mean Square Error (RMSE) of 2.5 and a Mean Absolute Error (MAE) of 1.85. Compared with the second-level wind speed product data released by Fengyun Satellite official website in the same period, which has an RMSE of 3.37 and an MAE of 1.9, our model improves the performance by 52.74% and 8.65% respectively, and obtains a better distribution.

Keywords: FY-3E · HWRF · MTL · Wind speed retrieval

1 Introduction

GNSS-R technology is a method of detecting information about the Earth's surface by receiving the signals reflected from the GNSS satellites to the Earth's

B. Luo et al. (Eds.): ICONIP 2023, LNCS 14452, pp. 357–369, 2024.
https://doi.org/10.1007/978-981-99-8076-5_26

surface [1]. It does not require a signal transmitter, only a signal receiver. With the increasing number of GNSS satellites, GNSS-R technology has the advantages of global coverage, low cost and fast acquisition, and has been widely applied in various fields [4]. One of them is using GNSS-R technology to retrieve sea surface wind fields, which is an emerging technology [5]. In the United States, NASA launched the CYGNSS series of satellites in 2016 to monitor and warn of tropical cyclones, and released a series of sea surface wind speed retrieval products [16]; in China, the FY-3E satellite launched in 2021 also carried a GNSS-R receiver [20], and released the first and second level sea surface wind speed retrieval products in the following year [10]. Their methods of retrieving sea surface wind speed are based on a large amount of historical data, constructing empirical physical models, and obtaining the retrieved sea surface wind speed [19]. However, under normal observation, most of the weather conditions are stable, and the observed wind speeds are low. There is less data on high wind speeds under extreme weather conditions. Therefore, the results obtained by retrieving empirical physical models perform better at low wind speeds, but have larger errors at high wind speeds [15]. The accuracy of wind speed retrieval has an important impact on the observation and prediction of extreme weather, which causes huge economic losses every year. Therefore, improving the accuracy of GNSS-R sea surface wind speed retrieval has extremely important value.

To tackle the above problem, this paper first uses the wind speed forecasted by the HWRF model as the label, which is also mentioned in the release of the CYGNSS second-level wind speed V3.0 version. We collected the HWRF model data for the past month, and matched it with the FY-3E data at the same time, and obtained a high wind data set. Based on the multi-task learning neural network model [7], we designed a sea surface wind field retrieval model for FY-3E. The model inputs all the information of DDM in all FY-3E first-level products as input data. In the purely data-driven neural network model, we added wind direction and other information as physical constraints. Besides the main task of wind speed, we set up three subtasks of U wind, V wind and wind direction. Through experiments and validation, compared with the FY-3E second-level products, our model effectively improved the overall sea surface wind speed retrieval accuracy of FY-3E.

This thesis makes three contributions:

- Using U wind, V wind, wind direction and other physical information to constrain the neural network model and improve its performance.
- Fully exploiting all the information of DDM in FY-3E first-level products for wind speed retrieval.
- Improving the accuracy of sea surface wind speed retrieval compared with the traditional empirical physical function method, and using the multi-task learning method to retrieve the FY-3E first-level products.

The rest of the paper is organized as follows: Sect. 2 reviews related work and limitations. Section 3 describes the data used as labels and input, introduces the model and defines the evaluation metrics. Section 4 provides detailed information on the experimental procedure and its results. Section 5 concludes with some remarks.

2   Related Work

Besides the empirical physical model function method used by the institutions to release products to retrieve sea surface wind speed, in recent years, many scholars have also started to use machine learning methods to retrieve GNSS-R sea surface wind speed. Jennifer et al. [14] used the ANN method to input several parameters in the empirical physical model function into the neural network for learning; Balasubramaniam et al. [2] added satellite latitude and longitude, geophysical information and other variables, and input them into the ANN neural network; Chu et al. [3] selected 33 variables from the product information to participate in the network model; Hammond et al. [7] used the first-level product DDM image obtained by satellite scanning as the input, and used the CNN neural network model; in addition, there are Munoz-Martin et al. [12,13,17,21], who introduced third-party data sources such as wave height, SMAP precipitation data, sea salt density, etc., as corrections added to the neural network.

All these methods of using machine learning to retrieve high sea surface wind speed, although they have improved the retrieval accuracy of sea surface wind field to some extent, they all have some problems. The first problem: no matter what network model is used in all these papers, the data labels are either the ECMWF Reanalysis v5 (ERA5) [8] or the second Modern-Era Retrospective analysis for Research and Applications(MERR-2) [6] global historical reanalysis data. Due to the characteristics of the historical reanalysis data itself, it shows good accuracy when the sea surface wind speed is low. But in the case of higher wind speed, the reanalysis data will produce serious underestimation [11]. Therefore, all of the above papers are focused on studying the low wind speed sea surface wind field, so that the overall effect of the model has been greatly improved at low wind speed, but there is no specific research on higher wind speed. The second problem: due to the traditional physical empirical function method of retrieving sea surface wind speed, most of the designed neural networks output reference physical empirical function method to extract features from DDM image part features such as Leading Edge Slope (LES), Normalised Bistatic Radar Cross Section (NBRCS), DDM Average(DDMA) and results for fitting. These models also selectively screened the information in the first-level products when selecting and adding variables for input. For example, 33 pieces of information were selected from 119 pieces of information; only processed DDM was used and Raw DDM was directly ignored; only 5×3 DDMA area was extracted from 17×11 area; all the information of DDM in the first-level products was not fully utilized. The third problem: whether it is the traditional empirical physical formula method or the neural network method, only the final wind speed information is used. And the label and forecast products give both U wind and V wind, and can also calculate the wind direction. The above methods ignore the physical information implied therein.

3   Methodology

3.1   FY-3E and HWRF Data

The subject of this paper is the FY-3E satellite. Since the GNOS II sensor carried by FY-3E can receive GNSS-R signals, compared with the CYGNSS satellite that can only receive GPS signals [18], it can also receive signals from Galileo and Beidou positioning satellites [9], so the data source is more abundant. The training data used in this paper is the first-level product of FY-3E. We obtained all data in May 2023, about 1.1 million data, through the website. The first-level product contains six elements: Channel, DDM, Receiver, Specular, Time and Transmitter. In addition to selecting all 31 variables of the DDM element as shown in Table 1, we also added two spatial position-related elements SP_lon and Sp_lat in Specular to participate in the network. Among these variables, except for Ddm_raw_data which is a two-dimensional variable with dimension 122×20 and $Ddm\_effective\_area$ which is a two-dimensional variable with dimension 9×20, all other variables are one-dimensional variables.

Table 1. DDM variables

No	1	2	3	4
Name	Ddm_brcs_factor	$Ddm\_doppler\_refer$	$Ddm\_effective\_area$	$Ddm\_kurtosis$
No	5	6	7	8
Name	Ddm_noise_m	Ddm_noise_raw	Ddm_noise_source	Ddm_peak_column
No	9	10	11	12
Name	Ddm_peak_delay	$Ddm\_peak\_doppler$	$Ddm\_peak\_power\_ratio$	Ddm_peak_raw
No	13	14	15	16
Name	Ddm_peak_row	Ddm_peak_snr	Ddm_power_factor	$Ddm\_quality\_flag$
No	17	18	19	20
Name	Ddm_range_refer	Ddm_raw_data	$Ddm\_skewness$	Ddm_sp_column
No	21	22	23	24
Name	Ddm_sp_delay	Ddm_sp_dles	$Ddm\_sp\_doppler$	Ddm_sp_les
No	25	26	27	28
Name	Ddm_sp_nbrcs	$Ddm\_sp\_normalized\_snr$	Ddm_sp_raw	$Ddm\_sp\_reflectivity$
No	29	30	31	
Name	Ddm_sp_row	Ddm_sp_snr	$Sp\_delay\_doppler\_flag$	

The training label uses the forecast data of the HWRF model [18]. HWRF (Hurricane Weather Research and Forecasting) model is a numerical model for predicting the track and intensity of tropical cyclones (hurricanes, typhoons). It was jointly developed by the NOAA and the NRL of the United States, aiming to provide high-precision hurricane forecast information. We collected the data forecasted by HWRF in the whole month of May. Since the longer the forecast time, the larger the deviation from reality, considering that the forecast time interval of the HWRF model is 3H and the release product time interval is 6 h, we chose every six hours of 0 o'clock and the first forecast time, i.e., 3 h time data to ensure the accuracy and richness of the data to the greatest extent, and finally obtained about 700k valid data.

Since FY-3E and HWRF data are inconsistent in terms of spatiotemporal continuity and resolution, data matching is required for FY-3E and HWRF data. When processing the FY-3E first-level product, we found that its spatial and temporal distribution was uneven, so we gridded it. In terms of spatial gridding, we used the method of taking the average, and averaged the grid points within 0.25 range of latitude and longitude as the value of that grid point. In terms of temporal resolution, because the time resolution of the label data HWRF is 1 h, and the time resolution of the FY-3E first-level product is 1 s, we needed to reduce the scale of the time resolution. We used the method of time window averaging, and took the average value of the points within half an hour before and after as the value of the point for the whole hour. The specific processing process is shown in Algorithm 1. The first part is to grid the FY-3E DDM variables. Referring to the spatiotemporal interval of HWRF, we rounded the latitude and longitude to 0.25°, and took the integer point time within half an hour interval as the time. Then, we took the average value of all the same type of variables at the same location and time as the value of this grid. The second part is to match the DDM variables and HWRF wind speed values at the same spatiotemporal location. Since the HWRF wind speed value is represented by U wind and V wind, we converted it to obtain the normal 10-meter height wind speed. Through matching, we successfully matched about 90k data, which was used as the training data set for this time.

3.2   Model Structure

The model is based on the 31 variable data of the FY-3E satellite first-level product DDM and the latitude and longitude position data. The model design is based on data-driven combined with physical information constraints, adding wind direction and horizontal and vertical wind components information. Considering the characteristics of the task, the model design is based on the multi-task learning neural network architecture, and obtains a new GNSS-R sea surface wind speed retrieval model with physical information constraints.

The model consists of three processes as shown in Fig. 1. The first process is the feature extraction of the two-dimensional image data DDM effective area and DDM raw data. The features of DDM effective area are extracted through the shared Conv convolutional network. DDM raw data needs to be convolved again to extract features separately because its size is much larger than DDM effective area. Then, all the features are concatenated with DDM 1D features and position to form training features. The second process is the preliminary training of the data, which is the shared network layer in the multi-task learning network. The training features are input into the ANN network for training. The third process sets different subtask learning network structures according to different learning tasks. Different network learning layers are designed for wind speed, wind direction, U wind and V wind respectively. And through the setting of different learning task weights, the final main task result is obtained to retrieve the sea surface wind speed.

Algorithm 1   FY-3E DDM variables with HWRF wind speed matching

---

Input:

longitude of HWRF wind speed $[hwrf\_long]$, latituede of HWRF wind speed$[\ hwrf\_lat]$,
the index of HWRF time $[hwrf\_time\_index]$,
longitude of DDM$[\ ddm\_long]$, latituede of DDM wind $[ddm\_lat]$,
time of DDM$[\ ddm\_time]$,variables of DDM $[ddm\_varaibles]$

1: **DDM Variable Gridding**($ddm\_long,\ ddm\_lat,\ ddm\_time,\ ddm\_varaibles$)
2:   $ddm\_long = ddm\_long\ /\ 0.25$
3:   $ddm\_lat = ddm\_lat\ /\ 0.25$
4:   **if** ($ddm\_time\ \%\ 3600 > 1800$)
5:     $ddm\_time = ddm\_time\ /\ 3600 + 1$
6:   **else:**
7:     $ddm\_time = ddm\_time\ /\ 3600$
8:   $unique\_data = [\ ]$
9:   **for** $i, (lat\_val, lon\_val, time\_val)$in enumerate(zip($ddm\_long,\ ddm\_lat,\ ddm\_time$)):
10:     $key = (lat\_val, lon\_val, time\_val)$
11:     **if** key not in $unique\_data$:
12:       $unique\_data[key] = [i]$
13:     **else:**
14:       $unique\_data[key].append(i)$
15:   **for** $indices$ in $unique\_data.values()$:
16:     $avg\_variables = mean(Ddm\_variables[indices])$
17:   **return** $avg\_variables$

18: **DDM_Variables_Wind_Match**($ddm\_long,\ ddm\_lat,\ ddm\_time,\ avg\_varaibles,$
                                            $hwrf\_long,\ hwrf\_lat,\ hwrf\_time$)
19:   $hwrf\_long = hwrf\_long\ /\ 0.25$
20:   $hwrf\_lat = hwrf\_lat\ /\ 0.25$
21:   **if** ($hwrf\_long == ddm\_long$ **and** $hwrf\_lat == ddm\_lat$ **and** $hwrf\_time == ddm\_time$)
12:     $u = HWRF\_wind["u10"][time\_index][lat\_index][long\_index]$
23:     $v = HWRF\_wind["v10"][time\_index][lat\_index][long\_index]$
24:     $match\_wind = \sqrt{v^2 + u^2}$
25:   **return** $match\_wind$

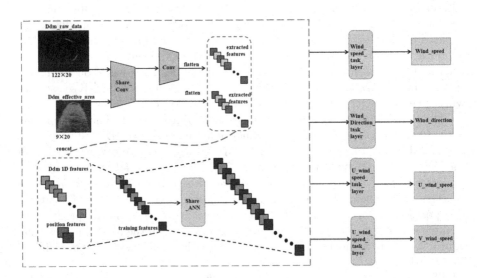

Fig. 1. Model structure

3.3   Metrics

In order to evaluate the training effect of the model, we chose Root Mean Square Error (RMSE), Mean Absolute Error (MAE), Mean Absolute Percentage Error (MAPE), Pearson's Correlation Coefficient and Coefficient(ρ) of Determination (r^2) as the evaluation indicators of the retrieval accuracy results. The specific formulas are as follows:

$$RMSE = \sqrt{\frac{1}{n}\sum_{i=1}^{n}(y_{hwrf(i)} - \hat{y}_{fy(i)})^2} \tag{1}$$

$$MAE = \frac{1}{n}\sum_{i=1}^{n}|y_{hwrf(i)} - \hat{y}_{fy(i)}| \tag{2}$$

$$MAPE = \frac{1}{n}\sum_{i=1}^{n}\left|\frac{y_{hwrf(i)} - \hat{y}_{fy(i)}}{y_{hwrf(i)}}\right| \times 100\% \tag{3}$$

$$\rho = \frac{\sum_{i=1}^{n}(y_{hwrf(i)} - y_{\bar{hwrf}})(y_{fy(i)} - y_{\bar{fy}})}{\sqrt{\sum_{i=1}^{n}(y_{hwrf(i)} - y_{\bar{hwrf}})^2}\sqrt{\sum_{i=1}^{n}(y_{fy(i)} - y_{\bar{fy}})^2}} \tag{4}$$

$$r^2 = 1 - \frac{\sum_{i=1}^{n}(y_{hwrf(i)} - \hat{y}_{fy(i)})^2}{\sum_{i=1}^{n}(y_{hwrf(i)} - \bar{y}_{fy(i)})^2} \tag{5}$$

where n is the number of samples, $y_{hwrf(i)}$ is the label value, $\hat{y}_{fy(i)}$ is the predicted value, $\bar{y}_{hwrf(i)}$ is the mean value of the label, and $\bar{y}_{fy(i)}$ is the mean value of the prediction. The smaller RMSE, MAE and MAPE are, the closer the model results are to the label, and the better the training effect. The larger Pearson's Correlation Coefficient(ρ) and Coefficient of Determination (r^2) are, the more the model fits the label data, and the better the training effect.

When comparing with FY-3E second-level products, we also chose P_{RMSE}, P_{MAE}, P_{MAPE} and P_{ρ} as evaluation indicators. It can be seen that compared with the traditional method, the new model has improved in the sea surface wind speed retrieval problem. The specific formulas are as follows:

$$P_{RMSE} = \frac{RMSE_1 - RMSE_2}{RMSE_1} \times 100\% \tag{6}$$

$$P_{MAE} = \frac{MAE_1 - MAE_2}{MAE_1} \times 100\% \tag{7}$$

$$P_{MAPE} = MAPE_1 - MAPE_2 \tag{8}$$

$$P_{\rho} = \frac{\rho_2 - \rho_1}{\rho_1} \times 100\% \tag{9}$$

4   Experiment and Result

4.1   Data Preprocessing

In the experiment, the data needs to be cleaned first. Because there are many missing places in the satellite observation data, these missing NaN values need to be removed. Some unreasonable data, such as -9999 wind speed and inf values, are also deleted. And remove the corresponding label of the training data set. When calculating the loss function, because the wind direction size is 0–360. If the loss is calculated directly, the wind direction loss greater than 180 will not conform to the actual situation. Therefore, we design the wind direction function loss to judge whether the difference between the two wind directions is greater than 180°. We keep it within 180°, if it is greater than 180°, then we use 360 minus the difference to get the angle of the wind direction.

4.2   Experiment

We split the processed data into training set, validation set and test set according to the ratio of 7:2:1. Before training, we set the seed to ensure the repeatability of the experiment. We choose a suitable learning rate, and set it to 0.001 after testing. We input the test set into the network model for training. When setting the loss for training, we notice that the wind direction loss is much larger than the wind speed loss in numerical size. If we simply add them, the model's training improvement effect on wind direction is not obvious. Therefore, we set the wind direction loss to a loss weight of 0.01 according to the numerical value. After adding U wind and V wind to the model, considering that U wind and V wind tasks are subtasks, we also set corresponding weights for U wind and V wind. After multiple debugging, we set the weights of U wind and V wind to 0.1.

4.3   Ablation Experiment

To evaluate the performance of the model and the influence of each subtask on the main task wind speed accuracy, we designed seven groups of control experiments. The results of the experiments are shown in Fig. 2. The data used in the figure is from the validation set, which is used to observe the training process of the model. The green line in the figure represents the change of the total loss of the model; the blue line is the change of the loss of the training set; the red line is the change of the loss of the main task wind speed. It can be seen from the figure that when the number of training reaches about 70, the model tends to converge. Comparing the training process of each model, we can find that whether adding any one of wind direction, U wind, V wind tasks, or adding several or all tasks, compared with single wind speed retrieval task, they all have a positive improvement on training network model. Among them, the wind_direction_u_v model with three subtasks of wind direction, U wind and V wind has the best training effect. Its model training loss is the lowest among all models at 4.05.

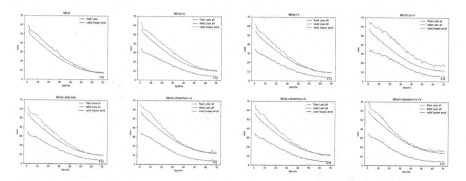

Fig. 2. The training process of each model

In terms of effect evaluation, we keep each model and input the test set to obtain the results as shown in Table 2. We can see that the wind_direction_u_v model performs excellently in five metrics: RMSE, MAE, MAPE, ρ and r2.

Table 2. Different models' result

Model	RMSE	MAE	MAPE	ρ	r2
wind	2.81	2.08	52.60	0.76	0.45
wind_direction	2.80	2.09	54.33	0.74	0.46
wind_u	2.82	2.15	53.76	0.75	0.43
wind_v	2.77	2.08	55.38	0.73	0.47
wind_u_v	3.00	2.32	60.71	0.72	0.35
wind_direction_u	2.57	1.89	52.14	0.75	0.54
wind_direction_v	2.66	1.99	51.54	0.77	0.51
wind_direction_u_v	**2.50**	**1.85**	**50.20**	**0.78**	**0.57**

4.4   Result

Comparing the output of the wind_direction_u_v model with the second-level sea surface wind speed retrieval product of FY-3E in the same time period, we can see that the accuracy of the wind_direction_u_v model is greatly improved as shown in Table 3. Among them, RMSE is reduced from 3.37 to 2.50, a decrease of 52.74%; MAE is reduced from 1.9 to 1.85, a decrease of 8.65%. These two metrics show that the accuracy of the wind_direction_u_v model has been greatly improved compared with the traditional method. MAPE is increased from 35.97 to 50.20, an increase of 19.41%; cov is increased from 0.64 to 0.78, an increase of 17.95%; this indicates that the result of the wind_direction_u_v model retrieval is more concentrated in distribution while improving the accuracy.

Table 3. Retrieval result copmared to FY-3E L2 product

Model	RMSE	MAE	MAPE	ρ
FY-3E_L2	3.37	1.9	35.97	0.64
wind_direction_u_v	2.50	1.85	50.20	0.78
Model	P_{RMSE}	P_{MAE}	P_{MAPE}	P_ρ
wind_direction_u_v	**52.74%**	**8.65%**	**19.41%**	**17.95%**

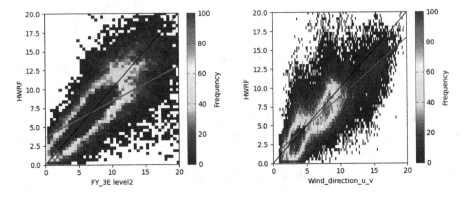

Fig. 3. Scatter density plot of winds under 20 m/s

We plot the output of the wind_direction_u_v model and the second-level sea surface wind speed retrieval product of FY-3E with HWRF wind speed as scatter density plots respectively. Through the scatter density plot, we can more intuitively see the advantages of the wind_direction_u_v model compared with the traditional method. From Fig. 3, we can see that when the wind speed is concentrated within 20 m/s, the red fitting line of the wind_direction_u_v model retrieval is more biased towards the middle; from Fig. 4, we can see that when the wind speed is higher than 20 m/s, the second-level product sea surface wind speed of FY-3E has a large deviation. The maximum wind speed value is close to 60m/s and deviates greatly from HWRF wind speed. The distribution and accuracy performance are poor. While the maximum sea surface wind speed retrieved by the wind_direction_u_v model is less than 30 m/s, and the distribution is concentrated, and it fits well with HWRF wind speed. This shows that the wind_direction_u_v model has a certain improvement effect on the second-level sea surface wind speed retrieval product of FY-3E in terms of accuracy results and distribution, whether in high or low wind speed.

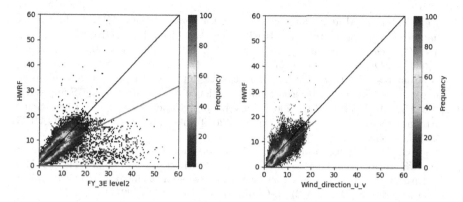

Fig. 4. Scatter density plot of all winds

5 Conclusion

This paper focuses on the GNSS-R sea surface wind speed retrieval research, and conducts it with the FY-3E satellite data as the subject. By exploring the traditional GNSS-R sea surface wind speed retrieval method and the current machine learning method, we find that there are problems such as insufficient DDM information mining and ignoring relevant physical information. We collect and match the data of HWRF for a whole month as the label. We take all the relevant information of DDM in FY-3E first-level product as the input. We use the multi-task learning model to fit with this research, and combine the wind direction, U wind and V wind physical information to constrain. By comparing with the FY-3E second-level wind speed product, we obtain a better GNSS-R sea surface wind speed retrieval model.

Besides achieving the expected effect of the model, we also have some additional findings. As one of the subtasks, the wind direction retrieval also achieved an accuracy within 20°. While in the traditional GNSS-R wind direction retrieval research, the accuracy is generally above 20°. We speculate that wind speed may also have some help for wind direction retrieval. In the future, we can further study this field.

Acknowlegement. The authors are grateful for the public FY-3E L1 data from the China National Satellite Meteorological Center. The authors are also grateful for the NOAA HWRF forecast products.

(FY_data available at: http://satellite.nsmc.org.cn/PortalSite/Data/Satellite.aspx)

(HWRF_data available at: https://ftpprd.ncep.noaa.gov/data/nccf/com/hwrf)

References

1. Asgarimehr, M., Zavorotny, V., Wickert, J., Reich, S.: Can GNSS reflectometry detect precipitation over oceans? Geophys. Res. Lett. **45**(22), 12–585 (2018)
2. Balasubramaniam, R., Ruf, C.: Neural network based quality control of CYGNSS wind retrieval. Remote. Sens. **12**(17), 2859 (2020). https://doi.org/10.3390/rs12172859

3. Chu, X., et al.: Multimodal deep learning for heterogeneous GNSS-R data fusion and ocean wind speed retrieval. IEEE J. Sel. Top. Appl. Earth Obs. Remote. Sens. **13**, 5971–5981 (2020). https://doi.org/10.1109/JSTARS.2020.3010879

4. Egido, A., Delas, M., Garcia, M., Caparrini, M.: Non-space applications of GNSS-R: from research to operational services. examples of water and land monitoring systems. In: IEEE International Geoscience & Remote Sensing Symposium, IGARSS 2009, 12–17 July 2009, University of Cape Town, Cape Town, South Africa, Proceedings, pp. 170–173. IEEE (2009). https://doi.org/10.1109/IGARSS.2009.5418033

5. Foti, G., et al.: Spaceborne GNSS reflectometry for ocean winds: first results from the UK TechDemosat-1 mission. Geophys. Res. Lett. **42**(13), 5435–5441 (2015)

6. Gelaro, R., et al.: The modern-era retrospective analysis for research and applications, version 2 (MERRA-2). J. Clim. **30**(14), 5419–5454 (2017)

7. Hammond, M.L., Foti, G., Gommenginger, C., Srokosz, M.: An assessment of CyGNSS v3.0 level 1 observables over the ocean. Remote. Sens. **13**(17), 3500 (2021). https://doi.org/10.3390/rs13173500

8. Hersbach, H., et al.: The ERA5 global reanalysis. Q. J. R. Meteorol. Soc. **146**(730), 1999–2049 (2020)

9. Huang, F., et al.: Characterization and calibration of spaceborne GNSS-R observations over the ocean from different beidou satellite types. IEEE Trans. Geosci. Remote. Sens. **60**, 1–11 (2022). https://doi.org/10.1109/TGRS.2022.3224844

10. Huang, F., et al.: Assessment of FY-3E GNOS-II GNSS-R global wind product. IEEE J. Sel. Top. Appl. Earth Obs. Remote. Sens. **15**, 7899–7912 (2022). https://doi.org/10.1109/JSTARS.2022.3205331

11. Li, X., Qin, X., Yang, J., Zhang, Y.: Evaluation of ERA5, ERA-interim, JRA55 and MERRA2 reanalysis precipitation datasets over the Poyang lake basin in China. Int. J. Climatol. **42**(16), 10435–10450 (2022)

12. Muñoz-Martín, J.F., Camps, A.: Sea surface salinity and wind speed retrievals using GNSS-R and l-band microwave radiometry data from FMPL-2 onboard the FSSCat mission. Remote. Sens. **13**(16), 3224 (2021). https://doi.org/10.3390/rs13163224

13. Pascual, D., Clarizia, M.P., Ruf, C.S.: Improved CYGNSS wind speed retrieval using significant wave height correction. Remote. Sens. **13**(21), 4313 (2021). https://doi.org/10.3390/rs13214313

14. Reynolds, J., Clarizia, M.P., Santi, E.: Wind speed estimation from CYGNSS using artificial neural networks. IEEE J. Sel. Top. Appl. Earth Obs. Remote. Sens. **13**, 708–716 (2020). https://doi.org/10.1109/JSTARS.2020.2968156

15. Ricciardulli, L., Mears, C.A., Manaster, A., Meissner, T.: Assessment of CYGNSS wind speed retrievals in tropical cyclones. Remote. Sens. **13**(24), 5110 (2021). https://doi.org/10.3390/rs13245110

16. Ruf, C., et al.: CYGNSS Handbook (2022)

17. Wang, C., Yu, K., Qu, F., Bu, J., Han, S., Zhang, K.: Spaceborne GNSS-R wind speed retrieval using machine learning methods. Remote. Sens. **14**(14), 3507 (2022). https://doi.org/10.3390/rs14143507

18. Warnock, A.M., Ruf, C.S., Morris, M.: Storm surge prediction with CYGNSS winds. In: 2017 IEEE International Geoscience and Remote Sensing Symposium, IGARSS 2017, Fort Worth, TX, USA, 23–28 July 2017, pp. 2975–2978. IEEE (2017). https://doi.org/10.1109/IGARSS.2017.8127624

19. Wu, J., et al.: Sea surface wind speed retrieval based on empirical orthogonal function analysis using 2019-2020 CYGNSS data. IEEE Trans. Geosci. Remote. Sens. **60**, 1–13 (2022). https://doi.org/10.1109/TGRS.2022.3169832

20. Yang, G., et al.: FY3E GNOS II GNSS reflectometry: mission review and first results. Remote. Sens. **14**(4), 988 (2022). https://doi.org/10.3390/rs14040988
21. Yueh, S., Chaubell, J.: Sea surface salinity and wind retrieval using combined passive and active l-band microwave observations. IEEE Trans. Geosci. Remote. Sens. **50**(4), 1022–1032 (2012). https://doi.org/10.1109/TGRS.2011.2165075

Incorporating Syntactic Cognitive in Multi-granularity Data Augmentation for Chinese Grammatical Error Correction

Jingbo Sun[1], Weiming Peng[2,3], Zhiping Xu[4], Shaodong Wang[1],
Tianbao Song[5(✉)], and Jihua Song[1]

[1] School of Artificial Intelligence, Beijing Normal University, Beijing 100875, China
{sunjingbo,202021210043}@mail.bnu.edu.cn, songjh@bnu.edu.cn
[2] Chinese Character Research and Application Laboratory, Beijing Normal
University, Beijing 100875, China
pengweiming@bnu.edu.cn
[3] Linguistic Data Consortium, University of Pennsylvania, Philadelphia 19104, USA
[4] China Mobile (Hang Zhou) Information Technology Co., Ltd., Hangzhou 310000,
China
[5] School of Computer Science and Engineering, Beijing Technology and Business
University, Beijing 100048, China
songtianbao@btbu.edu.cn

Abstract. Chinese grammatical error correction (CGEC) has recently attracted a lot of attention due to its real-world value. The current mainstream approaches are all data-driven, but the following flaws still exist. First, there is less high-quality training data with complexity and a variety of errors, and data-driven approaches frequently fail to significantly increase performance due to the lack of data. Second, the existing data augmentation methods for CGEC mainly focus on word-level augmentation and ignore syntactic-level information. Third, the current data augmentation methods are strongly randomized, and fewer can fit the cognition pattern of students on syntactic errors. In this paper, we propose a novel multi-granularity data augmentation method for CGEC, and construct a syntactic error knowledge base for error type *Missing and Redundant Components*, and syntactic conversion rules for error type *Improper Word Order* based on a finely labeled syntactic structure treebank. Additionally, we compile a knowledge base of character and word errors from actual student essays. Then, a data augmentation algorithm incorporating character, word, and syntactic noise is designed to build the training set. Extensive experiments show that the $F_{0.5}$ in the test set is 36.77%, which is a 6.2% improvement compared to the best model in the NLPCC Shared Task, proving the validity of our method.

Keywords: Grammatical error correction · Data augmentation ·
Multi-granularity knowledge base · Sentence structure grammar

B. Luo et al. (Eds.): ICONIP 2023, LNCS 14452, pp. 370–382, 2024.
https://doi.org/10.1007/978-981-99-8076-5_27

1   Introduction

Chinese Grammatical Error Correction (CGEC) is a research hotspot in the field of natural language processing (NLP) [6,7,12,16,17], with the goal of identifying and correcting all grammatical errors present in the text, such as *Spelling Errors, Missing and Redundant Components, Grammatical Structural Confusion, Improper Word Order* etc. [7], providing a practical solution for improving proofreading efficiency, reducing content risk, and aiding language teaching.

Recently, many studies are trying to continuously optimize the performance of CGEC through neural networks, and the best solution is currently based primarily on data-driven neural network models, such as Transformer [11]. However, the existing models still have the following problems: (1) Text containing complexity and a variety of errors in real-world scenarios is difficult to obtain and insufficient, and data-driven methods frequently produce poor results due to a lack of data. (2) To address the issue of insufficient real error samples, researchers have attempted to increase the scale of the training set through data augmentation. However, existing methods primarily focus on word-level augmentation, ignoring syntactic-level information. Figure 1 shows that the types of syntactic errors are diverse, making syntactic data augmentation difficult. (3) Current data augmentation methods are strongly randomized, and CGEC-oriented data augmentation samples are less reasonable and significantly different from the real errors, few of them can fit the pattern of students cognition of grammatical errors. In the literature [12], for example, the original sentence is "列车为前往机场的乘客提供了行李架(The train provided luggage racks for passengers going to the airport)", and the sentence with noise is "列车本应前往机场的朝兴乘客猫爪草了行李架(The train should have gone to the airport for the Chaoxing passengers catclawed the luggage racks)", with a large error gap between the faked sample and the real scenario.

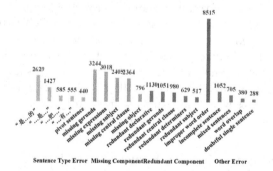

Fig. 1. Top five statistical results of syntactic errors in the HSK dynamic composition corpus. The most frequent error type in red is "*Improper Word Order*" (Color figure online).

To alleviate the above problems and integrate students cognition in CGEC training, a multi-granularity data augmentation algorithm based on the pattern of real error cognition is proposed, incorporating character, word, and syntactic noise. First, to address the issue that the existing data augmentation samples are not sufficiently realistic, we summarize the error information based on the realistic corpus and construct a knowledge base of lexical and syntactic errors. Meanwhile, to address the problem of existing data augmentation algorithms for fused syntactic noise, a linguistic rule-based method for *Improper word order* is developed, which accounts for the majority of errors in the corpus, to convert correct sentences into sentences containing errors based on the annotated syntactic pattern structure treebank, which can also improve the diversity and authenticity of syntactic errors in training data. To address the issue of insufficient data, we design a data augmentation method that incorporates multi-granularity noise. Because language cognition cannot leave the real context [4], to tackle the problem of cognitive pattern fusion, we argue that the knowledge base used is derived from real student essays, while the faked samples retain readability and conform to the error-generating habits of students, and the training data constructed on this basis can fit students cognition and provide interpretable of CGEC models. Finally, the Transformer-based CGEC method is used to generate samples.

2   Related Work

2.1   Chinese Grammatical Error Correction

The main frameworks of the established CGEC models include seq2seq [7,10,12] (similar to neural machine translation), seq2edit [3], seq2action [5], and non-autoregressive [6] etc. Transformer [12,17] is now the primary model utilized in the seq2seq framework for CGEC, in addition to CNN [9] and RNN [18]. Numerous studies have recently been carried out to enhance the performance of CGEC by improving the quality of training samples through data augmentation. Wang et al. [12] propose the MaskGEC model that improves CGEC by dynamically adding random masked tokens to the source sentences during training, increasing the diversity of training samples. To improve the training dataset, Zhao et al. [17] propose a data augmentation technique that corrupts a monolingual corpus to produce parallel data. Tang et al. [10] provide a novel data augmentation technique that includes word and character granularity errors. However, they all ignore syntactic information in the data augmentation.

2.2   Chinese Sentence Pattern Structure Treebank

Treebank is a meticulously annotated corpora that provides details of each sentence about the part-of-speech, syntactic structure, etc. The main types of treebanks that are widely used in NLP are phrase-structured treebanks [14] and dependency-structured treebanks [2]. In this paper, we employ a Sentence Pattern Structure treebank (SPST) based on sentence-based grammar, which uses the hierarchical diagrammatic syntactic analysis method to parse the structure of complex sentences suitable for Chinese grammar teaching and students cognition [8,15]. SPST contains diagrammatic scheme, XML storage structure, and sentence pattern structure expression (SPSE), the example is shown in Fig. 2. The application of SPST in the area of teaching Chinese as a second language demonstrates its applicability and effectiveness [8] and syntactic error sentence generation is implemented in Sect. 3.2 based on SPSE.

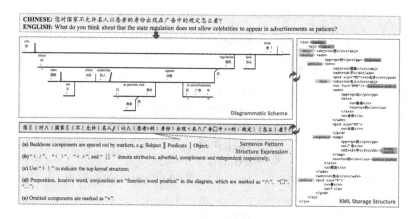

Fig. 2. An example and explanation of chinese SPST.

3 Proposed Method

3.1 Materials

In this paper, there are three granularities of noise: character, word, and syntactic, and an error knowledge base for each granularity separately is also designed.

Lexical Error Knowledge Base. In real-life situations, character and word errors are more frequent. The errors are caused by the fact that many Chinese characters and words have identical sound, morphology, and meaning. Based on the analysis of existing error instances, the high-frequency error samples from the HSK dynamic composition corpus are collected, and the extraction results are displayed in Table 1. Among them, the frequencies of char and word are 22,116 and 7,612, respectively.

Syntactic Error Knowledge Base. For *Missing and Redundant Components*, we extract classical sentences according to the Chinese second language syllabus and design regular expressions to identify the raw texts, e.g., the regular expression for the concessive complex sentence "虽然(although)...但是(but)..." is "虽然[\u4e00 − \u9fa5] + [,;] 但是". Our syntactic knowledge base contains 410 records.

Table 1. Examples of high-frequency errors from HSK dynamic composition corpus.

Wrong char	Correct char	Frequency	Wrong word	Correct word	Frequency
作	做	442	而	而且	178
象	像	334	发生	产生	105
咽	烟	254	决解	解决	101
建	健	248	经验	经历	96
份	分	205	权力	权利	89

Table 2. Examples of syntactic conversion rules.

Conversion description	Examples
①Subject exchange with object of prepositional structure	**Input:**我对历史很感兴趣(I am very interested in history)
	SPSE:我‖[对∧历史][很]感 ｜ 兴趣
	①**Output:**历史对我很感兴趣
②Subject + tense verb + epithet: subject postposition	**Input:**我跟你不一样(I am not like you)
	SPSE:我‖[跟∧你][不]一样
	②**Output:**跟你不一样我
③Subject + auxiliary verb + verb → auxiliary verb + subject + verb	**Input:**我想当翻译(I want to be a translator)
	SPSE:我‖想：当 ｜ 翻译
	③**Output:**想我当翻译
④Converting the locative to a complement; ⑤Prepositional-object structure as adverbial: put at the end; ⑥Complement place before predicate	**Input:**我在外交部工作十五年(I have worked in the Ministry of Foreign Affairs for 15 years)
	SPSE:我‖[在∧外交部]工作〈十五年〉
	④**Output:**我工作在外交部十五年
	⑤**Output:**我工作十五年在外交部
	⑥**Output:**我在外交部十五年工作

3.2   Syntactic Error Sentence Generation

For *Improper Word Order*, a syntactic error sentence generation method that generates error sentences directly by regular expressions based on SPSE is designed. For example, input sentence "我对历史很感兴趣(I am very interested in history)", the SPSE is "我‖[对∧历史][很]感 ｜ 兴趣". By swapping the positions of the two adverbials, the sentence in the wrong order is "我‖[很][对∧历史]感 ｜ 兴趣(I very am interested in history)". By deleting "我(I)", the sentence with the missing subject is "‖[对∧历史][很]感 ｜ 兴趣(am very interested in history)". This paper has designed 14 various types of error conversion rules, which generate 21,627 error sentences. This generated data is directly used for training without additional noise addition. Table 2 shows some of the conversion rules and examples.

3.3   Multi-granularity Data Augmentation

In the following, the materials from Sect. 3.1 are used to add noise to raw text for data augmentation, the multi-granularity data augmentation method is described in Algorithm 1. Syntactic-level noise is added to the current character and word noise studies [10,12]; additionally, its source is multi-granularity knowledge base rather than random noise. Syntactic data augmentation, unlike character and word augmentation, requires analyzing syntactic knowledge in a sentence and then performing operations for specified targets. Considering the specificity of the syntactic noise addition process, the input sentences are subjected to syntactic noise addition, word noise addition and character noise addition in order according to different ratios, resulting in the final output noise-added sentences. The noise addition ratio are controlled by parameters α_{syn}, α_{word}, α_{char}. Each noise addition process is implemented by insertion, deletion, and replacement. The proposed method is illustrated in Fig. 3.

Algorithm 1. Syntactic data augmentation algorithm

Input: The original correct sentence S_{clean}; syntactic knowledge base $\mathcal{K}_{syn}$; the length N of $\mathcal{K}_{syn}$; the regular expressions $\mathcal{REG}_{syn}$, insertion operations $\mathcal{I}$, deletion operations $\mathcal{D}$, and replacement operations $\mathcal{R}$ in $\mathcal{K}_{syn}$.
Output: The sentence S_{noise} after adding noise.
1: **function** SYNDA(S_{clean}, $\mathcal{K}_{syn}$)
2:    **for** $i = 1 \rightarrow i = N$ **do**
3:        **if** Match $(S_{clean}, \mathcal{REG}_{syn}^i)$ **then**
4:            $S_{noise} \leftarrow$ execute $\mathcal{R}^i$ or $\mathcal{D}^i$;
5:        **end if**
6:    **end for**
7:    **if** S_{noise} is null **then**
8:        Select $\mathcal{K}_{syn}^j$ and position $p < len(S_{clean})$ at random;
9:        $S_{noise} \leftarrow$ insert $\mathcal{I}^j$;
10:   **end if**
11: **end function**

3.4   Transformer for CGEC

Transformer [11] framework is adopted in previous works [10,17]. The proposed model is depicted in Fig. 3, where the encoder transforms the input text into a semantic vector, the decoder transforms the output vector of the current step based on the output encoded in the previous step, each vector corresponds to a Chinese character, and all the output characters are combined to produce the corrected sentence. The encoder consists of $N = 6$ identical blocks, each with a *Multi-Head Attention* (MHA) layer and a *Feed-Forward Network* (FFN) layer. The MHA mechanism employs a scaled dot-product in each attention layer, where Q, K and V denote the query matrix, key matrix, and value matrix of

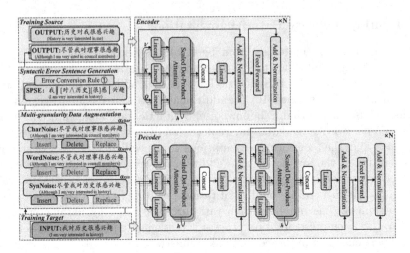

Fig. 3. An illustration of our method. In WordNoise sentence, "历史" (history) and "理事" (council member) are pronounced "lishi" in Chinese. In CharNoise sentence, the word "兴趣" (interested) is deleted to a single character "趣".

the attention layer, respectively, which are obtained by passing the input vector through different linear layers. The following is the MHA calculation procedure:

$$\text{Attention}(Q, K, V) = \text{Softmax}\left(\frac{QK^T}{\sqrt{d_k}}\right)V, \tag{1}$$

$$\text{MHA}(Q, K, V) = (h_1 \oplus h_2 \oplus \ldots \oplus h_n)W^O, h_i = \text{Attention}(Q_i, K_i, V_i), \tag{2}$$

where d_k is the embedding dimension, $\oplus$ means concatenation, h_i denotes the ith head of MHA and n is the number of heads.

4 Experiments and Results

4.1 Datasets

This section describes the data that is used in the experiments. **(a) NLPCC dataset**[1] which is presented in NLPCC 2018 Shared Task 2 is the most commonly used for CGEC in recent years, and it contains 1,220,734 sentence pairs of training set and 2,000 of test set. **(b) HSK dynamic composition corpus**[2] which collects the essay exam answer sheets of Hanyu Shuiping Kaoshi (HSK), which is Chinese pinyin for the Chinese Proficiency Test, which contains various grammatical errors. This dataset has 156,870 publicly available sentences for training. **(c) Chinese SPST**[3] which is a high-quality finely labeled corpus

---

[1] http://tcci.ccf.org.cn/conference/2018/dldoc/trainingdata02.tar.gz.
[2] http://hsk.blcu.edu.cn/.
[3] http://www.jubenwei.com/.

with syntactic information and the error correction sentence pairs for training are obtained using the sentence conversion rules proposed in Sect. 3.2. The sources of the corpus are Chinese second language teaching materials, totaling 12,103 sentences. The 21,627 sentence pairs are obtained through Sect. 3.2 for training. **(d) AI Challenger**[4] which is a large-scale corpus, contains 10,051,898 sentences. To obtain the sentence pairings utilized for training, these data are noised using the data augmentation method suggested in Sect. 3.3 for all sentences. The following experiments will select one million, two million, four million, and six million pairs to validate the influence of adding noise scale on model performance.

4.2 Parameters and Metrics

The implementation of our model is based on fairseq[5], using a Tesla V100-SXM2 32G GPU. The model hyperparameters are set as follows: Word embedding dimensions of 512 and shared weights for both source and target. Optimizer using Adam with β_1 and β_2 set to $(0.9, 0.98)$. The initial learning rate is 1×10^{-7}, which increases linearly to 1×10^{-3} over the first 4,000 training steps, followed by a gradual dropout until training is completed. The dropout is set to 0.3 and the batch size is set to 128. The beam search size is set to 12 during decoding.

As evaluation metrics, *Precision* (P), *Recall* (R), and $F_{0.5}$, and *Max Match* (M^2) are employed to measure the largest overlap of words between input and output. The formula for computation is as follows:

$$P = \frac{\sum_{i=1}^{N} |e_i \cap g_i|}{\sum_{i=1}^{N} |e_i|}, R = \frac{\sum_{i=1}^{N} |e_i \cap g_i|}{\sum_{i=1}^{N} |g_i|}, F_{0.5} = \frac{(1 + 0.5^2) \times R \times P}{0.5^2 \times P + R}, \quad (3)$$

where e is the edit set of the text to be modified by the model, g is the standard edit set of that text, and $|e_i \cap g_i|$ denotes the number of matches between the edit set of the model and the standard edit set for sentence i, which is calculated as shown in Eq. 4. The m2scorer toolkit[6] is used for testing.

$$|e_i \cap g_i| = \{e \in e_i \mid \exists g \in g_i, \text{ match } (e, g)\}. \quad (4)$$

4.3 Baselines

The first three are the CGEC models that performed well in NLPCC 2018 Shared Task 2, and the next four are the CGEC models that used data augmentation, the last one is a ensemble model for CGEC. **(a) YouDao** [1]. This method modifies the sentences separately using five different hybrid models. Finally, the language model reorders the output. **(b) AliGM** [18]. This method combines NMT-based GEC, SMT-based GEC, and rule-based GEC. **(c) BLCU** [9]. This

---

[4] https://challenger.ai/datasets/translation.
[5] https://github.com/pytorch/fairseq.
[6] https://github.com/nusnlp/m2scorer.

method employs a multi-layer convolutional seq2seq model. **(d) MaskGEC** [17]. This method improves CGEC by dynamically adding random masked tokens to source sentences during training, increasing the diversity of training samples. **(e) Word&Char-CGEC** [10]. This method proposes a data augmentation method that integrates word granularity noise for CGEC. **(f) C-Transformer** [13]. This method introduces a copy mechanism and constructs error samples to expand the training data based on a vocabulary of homomorphic and homophonic words. **(g) E-Transformer** [12]. This method adopts data augmentation by corrupting a monolingual corpus to produce parallel data. **(h) HRG** [3]. This method consists of a Seq2Seq model, a Seq2edit model, and an error detector.

4.4   Main Results

The experimental results are shown in Table 3, the $F_{0.5}$ of our model achieves 36.77%, which is a considerable improvement when compared to the top three baselines in the NLPCC 2018 Task 2. Compared to the C-Transformer and E-Transformer, the improvement is 2.72% and 2.36%, respectively. Our approach produces comparable results when compared to the Word&Char-CGEC and MaskGEC. Compared with HRG, our method improves 9.47% in precision and 2.21% in $F_{0.5}$. This demonstrates that the CGEC model is competitive when compared to the CGEC model that incorporates data augmentation. The experimental findings provided above show that applying data augmentation can increase the performance of the CGEC model, and our method is also an effective data augmentation scheme. The discrepancy between the MaskGEC and the Word&Char-CGEC is mostly due to the simple reason that the corpus applied for data augmentation in experiments differs from each of these two models, and the change in data sources will have some impact during training.

Table 3. Experimental result. The results are in percentage.

Line	Model	$P \uparrow$	$R \uparrow$	$F_{0.5} \uparrow$
1	AliGM	41.00	13.75	29.36
2	YouDao	35.24	18.64	29.91
3	BLCU	47.63	12.56	30.57
4	MaskGEC	44.36	22.18	36.97
5	Word&Char-CGEC	47.29	23.89	39.49
6	C-Transformer	38.22	23.72	34.05
7	E-Transformer	39.43	22.80	34.41
8	HRG	36.79	27.82	34.56
9	ours	46.26	20.20	36.77

4.5   Effect of Data Augmentation at Different Scales

To investigate the effect of the scale of noised data on model performance, five groups of tests with no noise data and noise data of one million, two million, four million, and six million are set. The experimental results displayed in Fig. 4 reveal that, when compared to the model without noise addition, the performance of the model with noise-added data improves by about 3%, with the best result obtained when the noise-added data reaches 2 million, and the improvement of the model performance decreases as the data size grows further. We also attempt to train more epochs on four and six million, and the experimental findings indicate that the $F_{0.5}$ of the model does not change much at 40 and 50 epochs compared to 30 epochs, showing that the model stabilized at 30 epochs. There are two reasons for the ineffectiveness of the four million and six million data. The first is that the monolingual corpus chosen in this paper contains spoken language, which does not conform to the characteristics of written language and deviates from the textual error correction field; the second is that, despite manual proofreading, the quality and distribution of the data require further improvement. Other comparative tests are carried out in this work using the 2 million noised data.

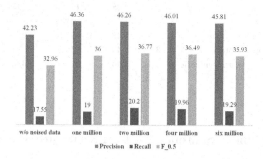

Fig. 4. Effect of data augmentation at different scales.

4.6   Effect of Data Augmentation at Different Granularities

We mainly test the usefulness of the proposed data augmentation strategy. The noise addition ratios of different granularities for the experiments are determined using statistics and analysis of the HSK dynamic composition corpus, as stated in the second column of Table 4. The Char&Word&Syntactic model has the best result, demonstrating that the data augmentation strategy of including multi-granularity noise is beneficial. When compared to models with only single noise-added, the model with character noise performs best, due to the fact that character errors are the most common in writing, and models with only word noise and syntactic noise cannot fit this error distribution. When comparing the models of char&word, char&syntactic, and word&syntactic, we discover that the model with char-granularity noise is the best for the same reason as stated above. When comparing the models of char&word and char&syntactic, the model with syntactic noise is more effective because the syntax contains not only information on syntactic structure but also information about words, which implies that the addition of noise with syntax proposed in this paper is superior to the previous data augmentation methods for CGEC. Furthermore, the process that generates word noise requires the splitting of sentences, and the correctness of the splitting affects the model effort. The syntactic knowledge base built in this paper currently comprises more than 400 rules, which is insufficient to match the model training requirements, indicating that the syntactic noise addition method suggested in this paper has room for further improvement.

Table 4. Effect of data augmentation at different granularities. The results are in percentage.

Model	Noise ratio	$P \uparrow$	$R \uparrow$	$F_{0.5} \uparrow$	$\Delta F_{0.5}$
w/o noised data	–	42.23	17.55	32.96	–
char	–	45.34	20.29	36.36	+3.4
word	–	42.2	16.38	32.08	–0.88
syntactic	–	38.77	14.45	29.01	–3.95
char&word	7 : 3	44.61	18.34	34.67	+1.71
char&syntactic	8 : 2	48.1	17.55	35.68	+2.72
word&syntactic	7 : 3	40.85	16.02	31.19	–1.77
char&word&syntactic	6 : 2.5 : 1.5	46.26	20.2	36.77	+3.81

4.7 Effect of Chinese Sentence Pattern Structure Treebank

In this section, two sets of comparison experiments are carried out, as indicated in Table 5, in order to test the enhanced effect of the Chinese SPST on our method. The $F_{0.5}$ is enhanced by 0.21% when the training samples created from the Chinese SPST are added on top of the NLPCC and HSK datasets. The $F_{0.5}$ value improves by 2.62% compared with the model in line 1, but is slightly lower than that of model 3 when the samples generated from the Chinese SPST are added to the NLPCC and HSK datasets and the 2 million noised data. The experimental results show that the model with the samples generated by the Chinese SPST is effective and syntactic cognitive information fusion can improve the ability of CGEC, but the improvement is not obvious, owing to the fact that there are only 21,627 training pairs in this corpus, which is insufficient in comparison to the existing data set for the model to learn enough syntactic knowledge from it.

Table 5. Effect of chinese SPST. The results are in percentage.

Line	Training data	$P \uparrow$	$R \uparrow$	$F_{0.5} \uparrow$	$\Delta F_{0.5}$
1	NLPCC+HSK	42.23	17.55	32.96	–
2	NLPCC+HSK+SPST	42.63	17.57	33.17	+0.21
3	NLPCC+HSK+2 million noised data	46.26	20.2	36.77	+3.81
4	NLPCC+HSK+SPST+2 million noised data	45.24	19.15	35.55	+2.62

5   Conclusion

In this paper, we propose a novel multi-granularity data augmentation method for CGEC. We build a multi-granularity knowledge base to fit the cognition pattern of students on syntactic errors, with 7,612 pieces of character error knowledge, 23,816 pieces of word error knowledge, and 426 pieces of syntactic error knowledge. And 21,627 samples are generated based on Chinese SPST. The experimental results prove effectiveness and show that syntactic cognitive information fusion and the ability to enhance the model. In the future, we will investigate the influence of different noise addition ratios on the model further. In addition, to further the use of Chinese SPST, we will target new error kinds for sample generation and explore expanding the finely annotated corpus.

Acknowledgments. This work was supported by the Beijing Natural Science Foundation (Grant No.4234081), the National Natural Science Foundation of China (Grant No.62007004), the Major Program of Key Research Base of Humanities and Social Sciences of the Ministry of Education of China (Grant No.22JJD740017) and the Scientific and Technological Project of Henan Province of China (Grant No.232102210077).

References

1. Fu, K., Huang, J., Duan, Y.: Youdao's winning solution to the NLPCC-2018 task 2 challenge: a neural machine translation approach to Chinese grammatical error correction. In: Zhang, M., Ng, V., Zhao, D., Li, S., Zan, H. (eds.) NLPCC 2018. LNCS (LNAI), vol. 11108, pp. 341–350. Springer, Cham (2018). https://doi.org/10.1007/978-3-319-99495-6_29
2. He, W., Wang, H., Guo, Y., Liu, T.: Dependency based Chinese sentence realization. In: Proceedings of ACL-AFNLP, pp. 809–816 (2009)
3. Hinson, C., Huang, H.H., Chen, H.H.: Heterogeneous recycle generation for Chinese grammatical error correction. In: Proceedings of COLING, pp. 2191–2201 (2020)
4. Kasper, G., Roever, C.: Pragmatics in Second Language Learning. Handbook of Research in Second Language Teaching and Learning, pp. 317–334 (2005)
5. Li, J., et al.: Sequence-to-action: grammatical error correction with action guided sequence generation. In: Proceedings of AAAI. vol. 36, pp. 10974–10982 (2022)
6. Li, P., Shi, S.: Tail-to-tail non-autoregressive sequence prediction for Chinese grammatical error correction. In: Proceedings of ACL, pp. 4973–4984 (2021)
7. Ma, S., et al.: Linguistic rules-based corpus generation for native Chinese grammatical error correction. In: Findings of EMNLP, pp. 576–589 (2022)

8. Peng, W., Wei, Z., Song, J., Yu, S., Sui, Z.: Formalized Chinese sentence pattern structure and its hierarchical analysis. In: Proceedings of CLSW, pp. 286–298 (2022)
9. Ren, H., Yang, L., Xun, E.: A sequence to sequence learning for Chinese grammatical error correction. In: Proceedings of NLPCC, pp. 401–410 (2018)
10. Tang, Z., Ji, Y., Zhao, Y., Li, J.: Chinese grammatical error correction enhanced by data augmentation from word and character levels. In: Proceedings of CCL, pp. 13–15 (2021)
11. Vaswani, A., et al.: Attention is all you need. In: Proceedings of NeurIPS. vol. 30, pp. 5998–6008 (2017)
12. Wang, C., Yang, L., Wang, y., Du, y., Yang, E.: Chinese grammatical error correction method based on transformer enhanced architecture. J. Chin. Inf. Process. **34**(6), 106–114 (2020)
13. Wang, Q., Tan, Y.: Chinese grammatical error correction method based on data augmentation and copy mechanism. CAAI Trans. Intell. Syst. **15**(1), 99–106 (2020)
14. Xue, N., Xia, F., Chiou, F.D., Palmer, M.: The Penn Chinese TreeBank: phrase structure annotation of a large corpus. Nat. Lang. Eng. **11**(2), 207–238 (2005)
15. Zhang, Y., Song, J., Peng, W., Zhao, Y., Song, T.: Automatic conversion of phrase structure TreeBank to sentence structure treebank. J. Chin. Inf. Process. **5**, 31–41 (2018)
16. Zhang, Y., et al.: MuCGEC: a multi-reference multi-source evaluation dataset for Chinese grammatical error correction. In: Proceedings of NAACL, pp. 3118–3130 (2022)
17. Zhao, Z., Wang, H.: MaskGEC: improving neural grammatical error correction via dynamic masking. In: Proceedings of AAAI. vol. 34, pp. 1226–1233 (2020)
18. Zhou, J., Li, C., Liu, H., Bao, Z., Xu, G., Li, L.: Chinese grammatical error correction using statistical and neural models. In: Proceedings of NLPCC, pp. 117–128 (2018)

Long Short-Term Planning for Conversational Recommendation Systems

Xian Li, Hongguang Shi, Yunfei Wang, Yeqin Zhang, Xubin Li, and Cam-Tu Nguyen[✉]

State Key Laboratory for Novel Software Technology, Nanjing University, Nanjing, China
{a81257,dream,woilfwang,zhangyeqin,lixubin}@smail.nju.edu.cn, ncamtu@nju.edu.cn

Abstract. In Conversational Recommendation Systems (CRS), the central question is how the conversational agent can naturally ask for user preferences and provide suitable recommendations. Existing works mainly follow the hierarchical architecture, where a higher policy decides whether to invoke the conversation module (to ask questions) or the recommendation module (to make recommendations). This architecture prevents these two components from fully interacting with each other. In contrast, this paper proposes a novel architecture, the long short-term feedback architecture, to connect these two essential components in CRS. Specifically, the recommendation predicts the long-term recommendation target based on the conversational context and the user history. Driven by the targeted recommendation, the conversational model predicts the next topic or attribute to verify if the user preference matches the target. The balance feedback loop continues until the short-term planner output matches the long-term planner output, that is when the system should make the recommendation.

Keywords: Conversational Recommendation Systems · Planning

1 Introduction

Traditional recommendation systems rely on user behavior history, such as ratings, clicks, and purchases, to understand user preferences. However, these systems often encounter challenges due to the data sparseness issue. Specifically, users typically rate or buy only a small number of items, and new users may not have any records at all. As a result, achieving satisfactory recommendation performance becomes difficult. Furthermore, traditional models struggle to address two critical questions without clear user guidance and proactive feedback: (a) What are users interested in? and (b) What are the reasons behind each system recommendation?

B. Luo et al. (Eds.): ICONIP 2023, LNCS 14452, pp. 383–395, 2024.
https://doi.org/10.1007/978-981-99-8076-5_28

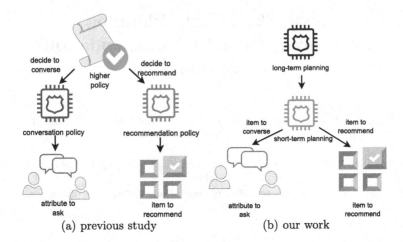

(a) previous study   (b) our work

Fig. 1. The architecture of previous studies (a) and our work (b)

The emergence of conversational recommender systems (CRSs) has fundamentally transformed traditional recommendation methods. CRSs facilitate the recommendation of products through multi-turn dialogues, enabling users and the system to dynamically interact using natural language. In these dialogues, the system not only elicits a user's preferences but also provides explanations for its recommended actions. Such capabilities are often absent in traditional recommendation approaches. Moreover, the conversational setting of CRSs presents a natural solution to the cold-start problem by allowing the system to proactively inquire about the preferences of new customers.

Most existing works for CRS use a hierarchical structure, where a higher policy determines whether to use the conversation module (to ask questions) or the recommendation module. This architecture prevents the conversation and the recommendation modules from fully interacting with each other. In contrast, this paper proposes a new approach where we plan the conversation and the recommendation by the same module, the short-term planner. Here, the short-term planner is influenced by a long-term planner that aims to model user long-term preference. Figure 1 demonstrates the main difference between our framework and the previous ones.

Our main constribution is three-fold: Firstly, it proposes a solution to combine user's past interactions and ongoing interactions (in conversations) into a long-term planner that takes into account the timestamps of these actions. Secondly, it presents a short-term planner that smoothly drives the conversations to the targeted item from the long-term planner. Lastly, it introduces a new dataset that captures practical aspects of both the recommendation module and the conversation module in the context of conversational recommender systems.

2   Related Work

Previous studies focus on the conversation (Light Conversation, Heavy Recommendation) or recommendation (Heavy Recommendation, Light Conversation).

2.1   Heavy Recommendation, Light Conversation

In this type of CRS, the aim is to understand user preferences efficiently and make relevant suggestions quickly. To achieve this, the system needs to focus on selecting the right attributes and asking the right questions. Sun et al. [16] and Lei et al. [7] proposed CRM and EAR, which train a policy of when and what attributes to ask. In such studies, the recommendation is made by an external recommendation model. Lei et al. [8] proposed SCPR that exploits a hierarchical policy to decide between asking and recommendation then invokes the corresponding components to decide what to ask and which to recommend. Deng et al. [3] proposed UNICORN, a unified model to predict an item to recommend or an attribute to ask the question. The model allows rich interactions between the recommendation and the conversation model.

These methods rely on a simple conversation module that is limited to templated yes/no responses. Our approach differs from these studies as we focus on real conversations where users can actively change the dialog flow and agents should smoothly change the topic towards the targeted recommendation.

2.2   Light Recommendation, Heavy Conversation

This kind of CRS puts a greater emphasis on understanding conversations and generating reasonable responses. These methods [2,10,11,14,19,20] also adopt a hierarchical policy that decides between recommendation and conversation, but additional strategies are used to bridge the semantic gap between the word space of the conversation module and the item space of the recommendation module. Li et al. [10] proposed REDIAL that exploits a switching decoder to decide between recommendation and conversation. Chen et al. [2] proposed KBRD, that exploited a switching network like REDIAL [10], but improved the interactions between the recommendation and conversation modules with entity linking and a semantic alignment between a recommendation item to the word space for response generation. Zhou et al. [19] proposed KGSF that uses word-oriented and entity-oriented knowledge graphs (KGs) to enrich data representations in the CRS. They aligned these two semantic spaces for the recommendation and the conversation modules using Mutual Information Maximization. These methods (REDIAL, KGSF, KBRD) do not plan what to ask or discuss, but generate responses directly based on the conversational history. Zhou et al. [20] introduced TG-Redial where topic prediction is used as a planner for conversations. However, there is still a higher policy to decide to invoke the conversation or the recommendation module. Similarly, Zhang et al. [18] predicted a sequence of sub-goals (social chat, question answering, recommendation) to guide the dialog model. Here, the goal prediction plays the role of higher policy.

Unlike these methods, we do not have two distinct modules for deciding what to ask and what to recommend. Instead, we assume a knowledge graph (KG) that connects attributes, topics, and items, and aim to plan the next entity node on the graph for grounding the dialog. Specifically, a long-term policy module, which has access to user historical interactions, predicts a targeted item in the KG. The short-term policy then exploits the targeted item and predicts either an attribute for conversation or an item for the recommendation. The objective of the short-term policy is to select a node in the KG so that the agent can smoothly drive the conversation to the long-term (targeted) item (Table 1).

3   Preliminaries

Notations for CRS. Formally, we are given an external knowledge graph $\mathcal{G}$ that consists of an entity set $\mathcal{V}$ and a relation set $\mathcal{E}$ (edges). The entity set $\mathcal{V}$ contains all entities, including items and non-items (e.g., item attributes). Alternatively, the knowledge graph can be denoted as a set of triples (edges) $\{(e^h, r, e^t)\}$ where $e^h, e^t \in \mathcal{V}$ are the head and tail entities and r indicates the relationship between them.

A CRS aids users in purchase decisions via conversation. During training, utterances from a user-agent conversation are labeled with entities from the knowledge graph $\mathcal{K}$. The agent's responses contain references to item entities for

Table 1. The description of the key symbols

	Description
$\mathcal{P}^u$	The user profile
$\mathcal{C}^u$	The current conversation
$\mathcal{S}^u$	The entity sequence with timestamp derived from $\mathcal{C}^u$ and the user profile $\mathcal{P}^u$
e^l	The targeted recommendation to be made in the upcoming turns (output of long-term plan)
e^s	The entity that will be grounded in upcomming turns (output of short-term plan)
w	The latest user utterance in $\mathcal{C}^u$
$\mathbf{z}^w$	The representation of the current utterance
$\mathbf{z}^e$	The representation of the current dialog entity sequence
$\mathcal{K}$	Knowledge graph entities

recommendations or non-item entities for clarification or chitchat. On the other hand, users refer to entities such as items or attributes to express their desired request.

In addition to the conversation history C, we also have access to the user profile for each user u. The user profile is a list of pairs (s_i, t_i), where s_i is an item entity that the user u has interacted with in the past, and t_i is the time of the interaction. Here, the interactions can be purchases, browsing, or other historical activities besides conversations.

Task Definition. Based on these notations, the CRS is defined as given a multi-type context data (i.e., conversation, knowledge graph, user profile), we aim to

(1) select entities from the knowledge graph to ground the next system response; (2) generate a response based on the selected entities. The selected entities may contain items (for the recommendation) or information about item attributes. The selected entities should be relevant to a dialog context (short-term user preference) and the user profile (long-term user preference).

4   Our Proposed Model

The LSTP's overall structure is depicted in Fig. 2. The process begins with the extraction of entities from the conversation, of which the timestamps are set to zero. Then, the user profile is combined with this sequence to create a unified entity sequence. A targeted item entity is then selected by the long-term planner from the knowledge graph based on the sequence $\mathcal{S}^u$. Lastly, the short-term planner picks a grounding entity (item/non-item), which is then utilized to produce the next system response.

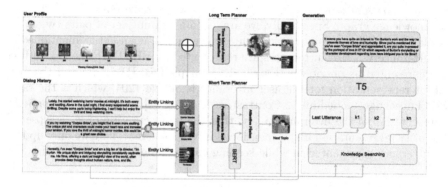

Fig. 2. The architecture of our LSTP framework

The long-term planner uses user profiles to more accurately target long-term user preferences, while the short-term planner considers recent utterances when planning the next entity in the conversation. The short-term planner is guided by the long-term planner to potentially lead to the long-term plan's target, while also ensuring the next entity is relevant to the dialogue history for a natural conversation. When a recommendation is necessary, the short-term planner should provide output consistent with the long-term planner.

4.1   Knowledge Representation and Grounding

Entities in the knowledge graph are represented by vectors in the same semantic space using Knowledge Graph Embeddings (KGE). This step is essential for both the long-term planner and the short-term one.

Knowledge Graph Embeddings. In this paper, we utilize TransE [1], which is available in the toolkit OpenKE [5], for knowledge graph embeddings. The main idea of TransE is that we represent entities and relations in the same semantic space so that if e^h should be connected to e^t via the relation r then $e^h + r \approx e^t$. Here, we use the same notation for entities (relations) and their vector representations. Formally, TransE learns a scoring function f as follows:

$$f(e^h, r, e^t) = -||e^h + r - e^t||_{1/2}$$

where $||_{1/2}$ is either L_1 or L_2 norm, and $e^h, e^t \in R^d$ with $d = 1024$ being the embedding dimension. The scoring function is larger if (e^h, r, e^t) is more likely to be a fact, i.e., e^h is connected to e^t via r in KG. Contrastive learning [5] is used to learn embeddings for all the entities and relations by enforcing the scores of true triples higher than those of negative (distorted) triples.

Entity Linking (or Knowledge Grounding). The objective of entity linking is to find entities previously mentioned in the dialog context. This is done by learning sentence representation so that it can be used to retrieve related entities from the knowledge graph. Specifically, we are given a training set of pairs $(\mathbf{w}_i, e_i)$ in which $\mathbf{w}_i$ indicates a conversational utterance and e_i is an entity mentioned in the utterance $\mathbf{w}_i$. User utterances are represented by BERT [4] whereas entities are represented as previously described. We exploit BERT large, and thus the output representation for the user utterance is of size 1204, which is the same with knowledge embeddings. Contrastive learning [5] is used to finetune the representation of the utterances $\mathbf{w}_i$ so that the representation of $\mathbf{w}_i$ is closer to the entity representation if e_i is mentioned in $\mathbf{w}_i$. Note that, here we only update the utterance representation while keeping entity embeddings unchanged.

4.2 Long-Term Planning

The goal of the Long-term Planner (LTP) is to anticipate the upcoming recommendation that can be made based on a series of entities from the user profile and ongoing conversation context. To train LTP, we randomly select conversational contexts and their respective recommendations to create a dataset. It's important to note that not every conversation turn results in a recommendation, so the recommended item may come several turns after the latest turn in the dialog context. LTP is designed to consider a user's past interactions to make recommendations further in the future.

The training set for LTP consists of triples $(\mathcal{P}^u, C^u, e^l)$. Here, $\mathcal{P}^u$ and C^u respectively represent the user profile and the context of the dialogue with the user u, and e^l indicates the targeted recommendation to be made in the upcoming turns. The entity sequence with timestamp derived from C^u and the user profile $\mathcal{P}^u$ is denoted as $\mathcal{S}^u = \{(e_1^s, t_1), \ldots, (e_l^s, t_l), (e_{l+1}^s, 0), \ldots, (e_{l+m}^s, 0)\}$, where l and m respectively represent the count of entities in the user profile and dialog history. To ensure uniformity, we set the length of the sequence $\mathcal{S}^u$ to be N and truncate the old entities in the sequence. The entity sequence can be then represented as $\mathcal{S}^u = \{(e_1^s, t_1), \ldots, (e_N^s, t_N)\}$, where the timestamp t_i is zero if the corresponding entity is mentioned in the current dialog context instead of the user profile. Padding is applied to the sequences $\mathcal{S}^u$ with length less than N.

We represent the sequence $\mathcal{S}^u$ by Multi-head Time-Aware Self-Attention (MH-TaSelfAttn) [9]. Unlike standard self-attention in Transformer [17], time-aware Self-attention (Ta-SelfAttn) takes into account the personalized interval between two user interactions (entities in $\mathcal{S}^u$) to calculate the attention score between them. By personalization, the user-specific minimum and maximum interval values are considered for modeling temporal information between two user interactions. Specifically, we initially represent each entity in $\mathcal{S}^u$ by knowledge embeddings, and then obtain the entity sequence representation as follows:

$$Z = \text{MH-TaSelfAttn}(\mathcal{S}^u) \tag{1}$$

Here $Z = \{\mathbf{z}_1, \ldots, \mathbf{z}_N\}$ is the sequence of output representations and $\mathbf{z}_i \in R^d$. To predict the upcoming recommendation, we obtain the last vector from Z as the sequence representation $\mathbf{z}^l$. We then measure the relevance between $\mathbf{z}^l$ and a candidate item using a dot product score. We then finetune the MH-TaSelfAttn layers to optimize the output representation so that the upcoming recommendation e^l is higher compared to other (negative) items. During inference, the item with the highest score $\tilde{e}^l$ is used as the predicted recommendation target.

4.3   Short-Term Planning

The purpose of the Short-term planner (STP) is to choose an entity that is related to the current dialog context and helps guide the conversation toward the LTP target. Intuitively, if the selected entity matches the LTP output, the next system response should provide a recommendation. The STP pays more attention to the current dialog when making decisions, unlike the LTP, which makes use of the user's historical actions. During STP training, the actual target (e^l) is used instead of the predicted (long-term) target $\tilde{e}^l$ for the upcoming recommendation. In addition, STP accepts as input the lastest user utterance $\mathbf{w}$ and the entity sequence $\mathcal{S}_c^u$, which is the part of $\mathcal{S}^u$ containing entities in the dialog context C^u. The representation for multi-type input of STP is obtained by:

$$\mathbf{z}^w = BERT(\mathbf{w}) \tag{2}$$
$$\mathbf{z}^e = \text{Pooling}[\text{MH-SelfAttn}(\mathcal{S}_c^u)] \tag{3}$$
$$\mathbf{z}^s = \text{Mean}[\text{SelfAttn}(\mathbf{z}^w, \mathbf{z}^e, e^l)] \tag{4}$$

where the last equation shows how the short-term representation is obtained by fusing the current utterance representation $\mathbf{z}^w$, the current dialog entity sequence $\mathbf{z}^e$ and the long-term target e^l. Here, Pooling indicates that we get the last item representation from MH-SelfAttn similarly to LTP, and SelfAttn indicates the standard self-attention operation in Transformer [17]. The STP is trained so that the next entity associated with the current context is higher compared to other entities. Like in LTP, only the additional layers in STP are finetuned, not entity embeddings. During inference, the item with the highest score $\tilde{e}^s$ is used as the prediction for the next grounding entity.

4.4 Plan-Based Response Generation

Given the next grounding entity $\tilde{e}^s$ from the STP, and let $\tilde{e}^s_{-1}$ be the grounding entity of the previous agent turn, knowledge search (Algorithm 1) aims to select a set of surrounding K entities to(one or two hops away) improve the context for smooth response generation. This returned list is then flattened and combined with the latest utterance $\mathbf{w}$ as input to the T5 model [13] for generating a response. During the training process, T5 is fine-tuned by optimizing the model to generate the correct response given the latest user utterance and correct grounding knowledge.

Algorithm 1. Knowledge Search

Require:
  Grounding entities for the previous and next agent turns $\tilde{e}^s_{-1}$, $\tilde{e}^s$;
  Knowledge Graph $\mathcal{K} = \{(e^h_k, r_k, e^t_k)\}^{N_{tri}}_{k=1}$ where e^h_k and e^t_k indicate the head and tail entities in the k-th triple;
Ensure:
  Extended grounding knowledge list K
1: Initialize $K = \emptyset$;
2: **for** $k = 1$ to N_{tri} **do**
3:    **if** $e^h_k = \tilde{e}^s_{-1}$ and $e^t_k = \tilde{e}^s$ **then**
4:       $K = K \cup (e^h_k, r_k, e^t_k)$
5:    **else if** $e^h_k = \tilde{e}^s_{-1}$ **and** $\exists r$ so that $(e^t_k, r, \tilde{e}^s) \in \mathcal{K}$ **then**
6:       $K = K \cup (e^h_k, r_k, e^t_k)$
7:    **else if** $e^h_k = \tilde{e}^s$ or $e^t_k = \tilde{e}^s_{-1}$ **then**
8:       $K = K \cup (e^h_k, r_k, e^t_k)$
9:    **end if**
10: **end for**
11: **return** $K[: 20]$;

---

5 Data Collection

Our assumption is that a user's current preference should be influenced by their ongoing dialogue and long-term interests reflected by their historical actions. However, current datasets do not provide sufficient information for our evaluation. For instance, the ReDial dataset lacks user profiles. On the other hand, although the TG-Redial dataset offers user profiles, they are not accompanied by timestamps essential for LSTP modeling. Therefore, we created our dataset, TAP-Dial (Time-Aware Preference-Guided Dial). Data gathering for TAP-Dial is similar to TG-Redial but with some differences. Firstly, every user profile comes with a timestamp, which is not present in TG-Redial. Secondly, although the conversation grounding task in TAP-Dial resembles the next topic prediction in TG-Redial, we propose that there is a unified knowledge graph that links item attributes and topics. More details on our data collection are provided below.

Table 2. Statistics of entities and relations in the knowledge graph

	Name	Number	Name	Number
Entity	Movie	5733	Date	8887
	Star	2920	Number	1223
	Types of Movies	31	Key words	2063
	Location	175	Constellation	12
	Profession	21	Awards	15816
Relation	The Constellation of	2691	The director of	2766
	The type of	14566	The release date of	7862
	The relative of	470	The country of	842
	The award records of	35245	The birth date of	2607
	The popularity of	5733	The profession of	8606
	The key words of	18369	The birthplace of	2852
	The representative works of	7668	The score of	5719
	The screenwriter of	2997	Collaborate with	1094
	The main actors of	14364	Star	14364

Data Collection and Knowledge Graph Construction. We focused on movies as our domain of research and obtained raw data from the Douban website. Our selection process involved filtering users with inaccurate or irrelevant information to create a user set of 2,693. In addition, we gathered a total of 5,433 movies that were the most popular at the time as our item set. We also gathered supplementary data including information about directors, actors, tags, and reviews. A knowledge graph is then constructed with entities and relations as shown in Table 2.

Dialog Flow Construction. To generate a list of recommendations for user conversations, we begin by selecting a set of targeted items. This is done by clustering the items in the user's history to determine a mixture of their preferences. We then choose the cluster centers as potential targets for recommendations, taking into account the most significant clusters and the timestamp.

Inspired by TG-Redial, we assume that the conversation should smoothly and naturally lead to the recommended items. Unlike TG-Redial, which relies on a separate topic set to ground non-recommendation turns, we use the knowledge graph's set of entities as potential grounding knowledge. In order to ensure smooth transitions between turns, we construct dialog flows consisting of lists of entities in the knowledge graph that gradually lead to the targeted items. Note that the first grounding entity can be randomly chosen.

Dialog Annotation. In the final stage, we recruit crowd-workers for writing the dialogs given dialog flows. We then received a total of 4416 dialogs for training, 552 dialogs for validation, and 552 for testing. Note that, all the dialogs are accompanied with grounding entities and targeted recommendations.

6   Experiments

6.1   Baselines and Metrics

In our experiments, we used several baselines, including REDIAL [10], KBRD [2], KGSF [19], TG-REDIAL [20]. All these baselines rely on sequence to sequence models as the bases for the conversation modules.

For evaluation, previous methods assume that there is an oracle policy that predefines recommendation turns, and evaluate the recommendation task and the conversation task separately. To ensure fairness, we compared our LSTP method with other methods on the recommendation and conversation tasks separately using a similar procedure. We used MRR [15], NGCG [6], HIT for the recommendation task, and BLUE [12], Distinct, and F1 for the generation task.

6.2   Main Results

Recommendation. The recommended task results are shown in Table 3. It is observable that TG-REDIAL outperforms the other baselines. This is because it incorporates both contextual and historical sequence information. LSTP achieves the best performance as it includes not only sequence information but also temporal interval information. In the long-term planning model, the accuracy can be improved by increasing the number of stacked attention modules, but it comes with time overhead. Therefore, a model with four stacked attention modules was chosen to balance between time and accuracy.

Table 3. Results of recommendation task

Model	NDCG			HIT	
	@1	@10	@50	@10	@50
REDIAL	0.002	0.010	0.048	0.005	0.013
KBRD	0.132	0.284	0.327	0.228	0.237
KGSF	0.103	0.222	0.263	0.177	0.186
TG-REDIAL	0.267	0.466	0479	0.399	0.404
LSTP	**0.301**	**0.474**	**0.481**	**0.417**	**0.418**

Dialog Generation. The generated task results in Table 4 indicate that LSTP performs the best. In comparison to the dialogue modules of other models, the advantage of LSTP lies in its ability to select relevant knowledge from the knowledge graph based on correct prediction results. Without the knowledge search module (LSTP w/o KS), the model's advantage would not be apparent, which also demonstrates the role of the Long-Short Term Planner (LSTP) module in predicting the next topic and making recommendations. Additionally, the distinctiveness of LSTP (w/o KS) is lower than the baseline, but the distinctiveness value of LSTP is significantly higher than the baseline. This confirms the significant impact of introducing external knowledge for diverse responses.

Table 4. The results of dialog generation task

Model	BLEU@1	BLEU@2	BLEU@3	BLEU@4	Dist@1	Dist@2	Dist@3	Dist@4	F1
REDIAL	0.168	0.020	0.003	0.001	0.017	0.242	0.500	0.601	0.21
KBRD	0.269	0.070	0.027	0.011	0.014	0.134	0.310	0.464	0.28
KGSF	0.262	0.058	0.021	0.007	0.012	0.114	0.240	0.348	0.26
TG-REDIAL	0.183	0.040	0.013	0.005	0.013	0.153	0.352	0.532	0.22
LSTP (w/o KS)	0.297	0.106	0.054	0.029	0.019	0.182	0.332	0.450	0.32
LSTP	**0.333**	**0.136**	**0.081**	**0.054**	**0.022**	**0.263**	**0.519**	**0.607**	**0.38**

6.3 Additional Analysis

Conversational Grounding Task. Following TG-Redial [20], we compare LSTP to several baselines including *MGCG* [11], Connv-Bert, Topic-Bert [20] on predicting entities to ground the conversation at non-recommendation turns. The results of the entity prediction for non-recommendation turns are shown in Table 5, demonstrating that the LSTP model achieves the best performance. This is partially because LSTP incorporates not only sequence information but also temporal interval information and sentence information.

Ablation Study. The results of our ablation study are presented in Table 6. Here, Over., Rec., Conv. respectively refer to the grounding prediction at all the turns, recommendation turns, or conversation turns. We implemented the Long Short-Term Planning (LSTP) with different variants of the short-term planner, where we include history to the short-term planner (w/ history), exclude the long-term planner (w/o long-term) or the latest user utterance (w/o latest utt). When integrating the user's historical interaction into the short-term planner, we observed a substantial enhancement in the recommendation outcomes but a significant deterioration in the conversation results. On the other hand, without the guidance of long-term planning (w/o long-term), the performance of recommendations suffered, partially demonstrating the importance role of the long-term planner. In contrast, the consideration of the latest user utterance did not seem to have a significant impact on the results, partially showing that entity linking might provide sufficient information for planning.

Table 5. Grounding entity prediction at non-recommendation turns

Model	HIT@1	HIT@3	HIT@5
Conv-Bert	0.169	0.245	0.285
Topic-Bert	0.251	0.348	0.394
MGCG	0.174	0.281	0.335
TG-REDIAL	0.219	0.327	0.382
LSTP	**0.312**	**0.447**	**0.482**

Table 6. HIT@1 of LSTP with variants of the Short-term planner

Model	Over.	Rec.	Conv.
LSTP	**0.308**	0.301	**0.312**
w history	0.279	**0.33**	0.254
w/o long-term	0.304	0.286	**0.312**
w/o lastest utt.	**0.308**	0.305	0.309

7 Conclusion

In this paper, we investigated the issue of the insufficient interaction between the dialogue and recommendation modules in previous CRS studies. We introduced LSTP model, which consists of a long-term model and a short-term module. The long-term model predicts a targeted recommendation based on long-term human interactions (historical interactions). The short-term model is able to predict the subsequent topic or attribute, thereby ascertaining if the user's preference aligns with the designated target. This harmonious feedback loop is continuously cycled until the output from the short-term planner matches the long-term planner's output. The equilibrium state indicates the system's optimal readiness for recommendation. We crafted a novel conversation dataset reflecting this dynamic. Experimental results on this dataset verify the effectiveness of our method.

Acknowledgements. We thank the data annotators for their meticulous work on dialogue annotation, which was pivotal for this research.

References

1. Bordes, A., Usunier, N., et al.: Translating embeddings for modeling multi-relational data. In: NIPS (2013)
2. Chen, Q., Lin, J., et al.: Towards knowledge-based recommender dialog system. In: EMNLP (2019)
3. Deng, Y., Li, Y., Sun, F., et al.: Unified conversational recommendation policy learning via graph-based reinforcement learning. In: SIGIR (2021)
4. Devlin, J., Chang, M.-W., et al.: BERT: pre-training of deep bidirectional transformers for language understanding. In: NAACL (2018)
5. Han, X., Cao, S., et al.: OpenKE: an open toolkit for knowledge embedding. In: EMNLP (2018)
6. Järvelin, K., Kekäläinen, J.: Cumulated gain-based evaluation of IR techniques. In: TOIS (2002)
7. Lei, W., He, X., et al.: Estimation-action-reflection: towards deep interaction between conversational and recommender systems. In: WSDM (2020)
8. Lei, W., Zhang, G., et al.: Interactive path reasoning on graph for conversational recommendation. In: KDD (2020)

9. Li, J., Wang, Y., McAuley, J.: Time interval aware self-attention for sequential recommendation. In: WSDM (2020)
10. Li, R., Kahou, S.E., et al.: Towards deep conversational recommendations. In: NIPS (2018)
11. Liu, Z., Wang, H., et al.: Towards conversational recommendation over multi-type dialogs. In: ACL (2020)
12. Papineni, K., Roukos, S., et al.: BLEU: a method for automatic evaluation of machine translation. In: ACL (2002)
13. Raffel, C., Shazeer, N., et al.: Exploring the limits of transfer learning with a unified text-to-text transformer. J. Mach. Learn. Res. **21**, 5485–5551 (2020)
14. Ren, X., Yin, H., et al.: Learning to ask appropriate questions in conversational recommendation. In: SIGIR (2021)
15. Salton, G., McGill, M.: Introduction to Modern Information Retrieval. McGraw-Hill Inc, New York (1984)
16. Sun, Y., Zhang, Y.: Conversational recommender system. In: SIGIR (2018)
17. Vaswani, A., Shazeer, N., et al.: Attention is all you need. In: NIPS (2017)
18. Zhang, J., Yang, Y., et al.: KERS: a knowledge-enhanced framework for recommendation dialog systems with multiple subgoals. In: Findings of EMNLP (2021)
19. Zhou, K., Zhao, W.X., et al.: Improving conversational recommender systems via knowledge graph based semantic fusion. In: KDD (2020)
20. Zhou, K., Zhou, Y., et al.: Towards topic-guided conversational recommender system. In: COLING (2020)

Gated Bi-View Graph Structure Learning

Xinyi Wang and Hui Yan$^{(\boxtimes)}$

School of Computer Science and Engineering, Nanjing University of Science and
Technology, Nanjing, China
{wangxinyi,yanhui}@njust.edu.cn

Abstract. Graph structure learning (GSL), which aims to optimize
graph structure and learn suitable parameters of graph neural networks
(GNNs) simultaneously, has shown great potential in boosting the performance of GNNs. As a branch of GSL, multi-view methods mainly
learn an optimal graph structure (final view) from multiple information sources (basic views). However, basic views' structural information
is insufficient, existing methods ignore the fact that different views can
complement each other. Moreover, existing methods obtain the final view
through simple combination, fail to constrain the noise, which inevitably
brings irrelevant information. To tackle these problems, we propose a
Gated Bi-View GSL architecture, named GBV-GSL, which interacts
two basic views through a selection gating mechanism, so as to "turn off"
noise as well as supplement insufficient structures. Specifically, two basic
views that focus on different knowledge are extracted from original graph
as two inputs of the model. Furthermore, we propose a novel view interaction technique based on selection gating mechanism to remove redundant
structural information and supplement insufficient topology while retaining their focused knowledge. Finally, we design a view attention fusion
mechanism to adaptively fuse two interacted views to generate the final
view. In numerical experiments involving both clean and attacked conditions, GBV-GSL shows significant improvements in the effectiveness and
robustness of structure learning and node representation learning. Code
is available at https://github.com/Simba9257/GBV-GSL.

Keywords: Graph neural networks · Graph structure learning ·
Gating mechanism

1 Introduction

Graph is capable of modeling real systems in diverse domains varying from
natural language and images to network analysis. Nowadays, as an emerging
technique, Graph Neural Networks (GNNs) [7,12,18] have achieved great success with their characteristic message passing scheme [5] that aims to aggregate
information from neighbors iteratively. So far, GNNs have shown superior performance in a wide range of applications, such as node classification [21,22] and
link prediction [23].

B. Luo et al. (Eds.): ICONIP 2023, LNCS 14452, pp. 396–407, 2024.
https://doi.org/10.1007/978-981-99-8076-5_29

GNNs are extremely sensitive to the quality of given graphs and thus require resilient and high-quality graph structures. However, we are not always provided with graph structures, such as in natural language processing [2,13] or computer vision [17]. Even if given the graph structures, due to the complexity of real information sources, the quality of graphs is often unreliable. On one hand, graph structures in real-world tend to be noisy, incomplete, adversarial, and heterophily (i.e., the edges with a higher tendency to connect nodes of different types). On the other hand, graphs sometimes suffer from malicious attacks, such that original structures are fatally destroyed. With attacked graphs, GNNs would be very vulnerable. As a result, there are various drawbacks prevailing in real graphs, which prohibits original structure from being the optimal one for downstream tasks.

Recently, graph structure learning (GSL) has aroused considerable attentions, which aims to learn optimal graph structure and parameters of GNNs simultaneously [25]. Current GSL methods can be roughly divided into two categories, single-view [4,9,10,24] based and multi-view based [1,14,16,19]. The single-view based GSL method is first proposed to estimate the optimal structure from one view, i.e., the given adjacency matrix, by forcing the learned structure to accord with some properties. For instance, LDS [4] samples graph structure from a Bernoulli distribution of adjacency matrix and learns them together with GNN parameters in a bi-level way. Pro-GNN [10] learns the graph structure with low rank, sparsity and feature smoothness constraints. However, considering that learning graph structure from one information source inevitably leads to bias and uncertainty [14], the multi-view based GSL method aims to extract multiple basic views from original structure, and then comprehensively estimate the final optimal graph structure based on these views. As an example, IDGL [1] constructs the structure by two type of views: normalized adjacency matrix and similarity matrix calculated with node embeddings. GEN [19] presents an iterative framework based on Bayesian inference. CoGSL [14] propose a compact GSL architecture by mutual information compression. Multi-view based methods are able to utilize multifaceted knowledge to make the final decision on GSL.

While promising, multi-view based GSL methods still have the following issues. (1) The complementarity of different views is ignored. Different views have their own focused knowledge. For example, the adjacency matrix focuses on the actual connection between nodes in reality, while the similarity matrix focuses on the similarity of node embeddings. These views may be insufficient, but they can complement each other to supplement insufficient knowledge. Existing multi-view methods augment the basic view separately, which ignores the complementarity between different views and may not be able to deeply explore the optimal graph structure. (2) Unrestricted information irrelevant to downstream tasks. An optimal graph structure should only contain the most concise information about downstream tasks (e.g., node labels), so that it can conduct the most precise prediction on labels. If the learned structure absorbs the information of labels as well as additional irrelevance from basic views, this structure is more prone to adversarial attacks when small perturbations are deployed on these irrelevant parts. Existing multi-view methods obtain the final view through simple combination, fail to constrain the irrelevant information from basic views to final view. Hence, the final structure inevitably involves additional noise and disassortative connections and are also vulnerable to perturbations.

To address these issues, considering that basic views are the information source of final view, it is vital to guarantee the quality of basic views. On one hand, basic views are also needed to contain the information about labels, which can fundamentally guarantee the performance of final view. On the other hand, these views also should be independent of each other, so that they can eliminate the redundancy and provide diverse knowledge about labels for final view. In addition, considering that the final view extracts information from basic views, we need to constrain the information flow from basic views to final view, which avoids irrelevant information and contributes to the robustness of the model. In this paper, we present GBV-GSL, an effective and robust graph structure learning framework that can adaptively optimize the topological graph structure and can achieve superior node representations. Specifically, we first carefully extract two basic views that focus on different knowledge from original structure as inputs, and utilize a view estimator to properly adjust basic views. With the estimated basic views, we propose a view interaction technique based on selection gating mechanism to remove redundant structural information and supplement insufficient topology. In this mechanism, the model first calculates the gating signal based on two estimated views, and the gating signal controls the proportion of their mixed information to their respective topological structures. The objective of gating is twofold: on the complement side, to control the importance given to mixed information, and on the denoise side, how much of the redundant information every view should "forget". Then, we further design a view attention fusion mechanism to automatically learn the importance weights of the interacted views, so as to adaptively fuse them. In the end, the label information is not only used for the final view, but also for supervising the classification results of two interacted views. Our contributions are summarized as follows:

- To our best knowledge, we are the first to utilize gating mechanism to study the optimal structure in GSL. We propose a view interaction module, which aims to insufficient structures as well as suppress irrelevant noise.
- We propose GBV-GSL, an effective and robust gated bi-view graph structure learning framework, which can achieve superior node representations.
- We validate the effectiveness of GBV-GSL compared with state-of-the-art methods on seven datasets. Additionally, GBV-GSL also outperforms other GSL methods on attacked condition, which further demonstrates the robustness of GBV-GSL.

2   Problem Definition

Let $\mathcal{G} = (\mathcal{V}, \xi)$ represent a graph, where $\mathcal{V}$ is the set of n nodes and ξ is the set of edges. All edges formulate an original adjacency matrix $A \in \mathbb{R}^{n \times n}$, where A_{ij}

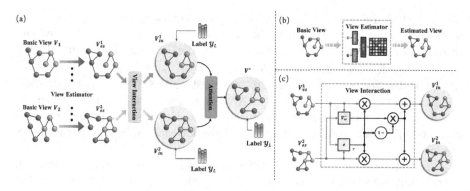

Fig. 1. The overview of our proposed GBV-GSL. (a) Model framework. (b) View estimator. (c) View interaction.

denotes the relation between nodes v_i and v_j. Graph $\mathcal{G}$ is often assigned with node feature matrix $X = [x_1, x_2, \ldots, x_N] \in \mathbb{R}^{n \times d}$, where x_i means the d dimensional feature vector of node i. In semi-supervised classification, we only have a small part of nodes with labels $\mathcal{Y}_L$. The traditional goal of graph structure learning for GNNs is to simultaneously learn an optimal structure and GNN parameters to boost downstream tasks.

As one typical architecture, GCN [12] is usually chosen as the backbone, which iteratively aggregates neighbors' information. Formally, the kth GCN layer can be written as:

$$GCN\left(A, H^{(k)}\right) = D^{-1/2} A D^{-1/2} H^{(k-1)} W^{(k)}, \tag{1}$$

where D is the degree matrix of A, and $W^{(k)}$ is weight matrix. $H^{(k)}$ represents node embeddings in the kth layer, and $H^{(0)} = X$. In this paper, we simply utilize $GCN(V, H)$ to represent this formula, where V is the view and H is the node features or embeddings.

3   The Proposed Model

In this section, we elaborate the proposed model GBV-GSL. The overall architecture is shown in Fig. 1(a). Our model begins with two basic views. Then, we utilize a view estimator to optimize two basic views separately. With two estimated views, we propose a view interaction technique based on selection gating mechanism to optimize them again. Next, we design a view attention fusion mechanism to adaptively fuse two interactive views to generate the final view. Finally, we provide the optimization objective of this model.

3.1   The Selection of Basic Views

Given a graph $\mathcal{G}$, GBV-GSL starts from extracting different structures. In this paper, we mainly investigate three widely-studied structures: (1) Adjacency matrix, which reflects the local structure; (2) Diffusion matrix, which represents the stationary transition probability from one node to other nodes and provides a global view of graph. Here, we choose Personal PageRank (PPR), whose closed-form solution [8] is $S = \alpha \left(I - (1 - \alpha)D^{-1/2}AD^{-1/2}\right)^{-1}$, where $\alpha \in (0, 1]$ denotes teleport probability in a random walk, I is a identity matrix, and D is the degree matrix of A; (3) KNN graph, which reflects the similarity in feature space. We utilize original features to calculate cosine similarity between each node pair, and retain top-k similar nodes for each node to construct KNN graph.

These three views contain the different properties from various angles, and we carefully select two of them as two basic views V_1 and V_2, which are the inputs of GBV-GSL.

3.2   View Estimator

Given two basic views V_1 and V_2, we need to further polish them so that they are more flexible to generate the final view. Here, we devise a view estimator for each basic view, shown in Fig. 1(b). Specifically, for basic view V_1, we first conduct a GCN [12] layer to get embeddings $Z^1 \in \mathbb{R}^{n \times d_{es}}$:

$$Z^1 = \sigma \left(GCN \left(V_1, X\right)\right), \tag{2}$$

where σ is non-linear activation. With embedding Z^1, probability of an edge between each node pair in V_1 can be reappraised. For target node i, we concatenate its embedding z_i^1 with embedding z_j^1 of another node j, which is followed by a MLP layer:

$$w_{ij}^1 = W_1 \cdot \left[z_i^1 \| z_j^1\right] + b_1, \tag{3}$$

where w_{ij}^1 denotes the weight between i and j, $W_1 \in \mathbb{R}^{2d_{es} \times 1}$ is mapping vector, and $b_1 \in \mathbb{R}^{2d_{es} \times 1}$ is the bias vector. Then, we normalize the weights for node i to get the probability p_{ij}^1 between node i and other node j. Moreover, to alleviate space and time expenditure, we only estimate limited scope S^1. For example, for adjacency matrix or KNN, we only inspect their k-hop neighbors, and for diffusion matrix, we only reestimate top-h neighbors for each node according to PPR values. Here, h and k are hyper-parameters. So, p_{ij}^1 is calculated as:

$$p_{ij}^1 = \frac{exp \left(w_{ij}^1\right)}{\sum_{k \in S^1} exp \left(w_{ik}^1\right)}. \tag{4}$$

In this way, we construct a probability matrix P^1, where each entry is calculated by Eq. (4). Combined with original structure, the estimated view is as follows:

$$V_{es}^1 = V_1 + \mu^1 \cdot P^1, \tag{5}$$

where $\mu^1 \in (0, 1)$ is a combination coefficient, and the i-th row of V_{es}^1, denoted as $V_{es\_i}^1$, shows new neighbors of node i in the estimated view. Estimating V_2 is similar to V_1 but with a different set of parameters, and we can get the estimated view V_{es}^2 finally.

3.3   View Interaction

The two estimated views V_{es}^1 and V_{es}^2 only have topological information contained in their respective basic structural views V_1 and V_2, which may be insufficient or noisy. To address this issue, we introduce a portion of their mixed topology information for V_{es}^1 and V_{es}^2, and referred to this process as view interaction, shown in Fig. 1(c). Specifically, we use a selection gating mechanism to make V_{es}^1 and V_{es}^2 interact with each other. First, the gating signal r is computed by

$$r = \sigma \left(V_{es}^1 \cdot U_r + b_r^1 + V_{es}^2 \cdot W_r + b_r^2 \right), \tag{6}$$

where $U_r, W_r \in \mathbb{R}^{n \times n}$ are learnable parameters, $b_r^1, b_r^2 \in \mathbb{R}^{n \times 1}$ are bias vector, σ is activation function $sigmoid(x) = \frac{1}{1+e^{-x}}$, and $r \in (0,1)$ represents the importance of the data passed by the gate. After having the gating signal r, taking V_{es}^1 as an example, we have it pass through the gate with $\overline{V_{es}} = \frac{V_{es}^1 + V_{es}^2}{2}$, to obtain the interactive view V_{in}^1:

$$V_{in}^1 = (1 - r) \cdot \overline{V_{es}} + r \cdot V_{es}^1, \tag{7}$$

here we use $\overline{V_{es}}$ to represent mixed topology information because it is simple and effective. Similarly, another interactive view V_{in}^2 is obtained from the same gating system:

$$V_{in}^2 = (1 - r) \cdot \overline{V_{es}} + r \cdot V_{es}^2, \tag{8}$$

3.4   View Attention Fusion

Now we have two interactive views V_{in}^1 and V_{in}^2. Considering the node label can be correlated with one of them or even their combinations, we use the attention mechanism $att \left(V_{in}^1, V_{in}^2 \right)$ to learn their corresponding importance (β^1, β^2) as follows:

$$\left(\beta^1, \beta^2 \right) = att \left(V_{in}^1, V_{in}^2 \right), \tag{9}$$

here $\beta^1, \beta^2 \in \mathbb{R}^{n \times 1}$ indicate the attention values of n nodes with embeddings V_{in}^1, V_{in}^2, respectively.

Here we focus on node i, where its embedding in V_{in}^1 is $V_{in\_i}^1 \in \mathbb{R}^{1 \times n}$ (i.e., the i-th row of V_{in}^1). We firstly transform the embedding through a nonlinear transformation, and then use one shared attention vector $q \in \mathbb{R}^{h' \times 1}$ to get the attention value ω_i^1 as follows:

$$\omega_i^1 = q^T \cdot tanh \left(W \cdot \left(V_{in\_i}^1 \right)^T + b \right). \tag{10}$$

Here $W \in \mathbb{R}^{h' \times h}$ is the weight matrix and $b \in \mathbb{R}^{h' \times 1}$ is the bias vector. Similarly, we can get the attention values ω_i^2 for node i in view V_{in}^2. We then normalize the attention values ω_i^1, ω_i^2 with $softmax$ function to get the final weight:

$$\beta_i^1 = softmax \left(\omega_i^1 \right) = \frac{exp \left(\omega_i^1 \right)}{exp \left(\omega_i^1 \right) + exp \left(\omega_i^2 \right)}. \tag{11}$$

Larger β_i^1 implies the corresponding view is more important. Similarly, $\beta_i^2 = softmax\left(\omega_i^2\right)$. For all the n nodes, we have the learned weights $\beta^1 = \left[\beta_i^1\right]$, $\beta^2 = \left[\beta_i^2\right] \in \mathbb{R}^{n \times 1}$, and denote $\beta_{in}^1 = diag\left(\beta^1\right)$, $\beta_{in}^2 = diag\left(\beta^2\right)$. Then we combine these two views to obtain the final view V^*:

$$V^* = \beta_{in}^1 \cdot V_{in}^1 + \beta_{in}^2 \cdot V_{in}^2. \tag{12}$$

3.5   Optimization Objective

V^* is a fusion of V_{in}^1 and V_{in}^2, indicating that better V_{in}^1 and V_{in}^2 can result in better V^*. Therefore, we optimize parameters Θ of classifiers for each view to improve the accuracy on given labels $\mathcal{Y}_L$. Specifically, we first utilize two-layer GCNs to obtain predictions of V_{in}^1 and V_{in}^2 :

$$\begin{aligned} O^1 &= softmax\left(GCN\left(V_{in}^1, \sigma\left(GCN\left(V_{in}^1, X\right)\right)\right)\right), \\ O^2 &= softmax\left(GCN\left(V_{in}^2, \sigma\left(GCN\left(V_{in}^2, X\right)\right)\right)\right), \end{aligned} \tag{13}$$

where σ is activation function. Similarly, we also can get the predictions of V^*:

$$O^* = softmax(GCN\left(V^*, \sigma\left(GCN\left(V^*, X\right)\right)\right)), \tag{14}$$

The parameters of GCNs involved in Eq. (13) and Eq. (14) are regarded as the parameters Θ of classifiers together. Θ can be optimized by evaluating the cross-entropy error over $\mathcal{Y}_L$:

$$\min_{\Theta} \mathcal{L}_{cls} = - \sum_{O \in \{O^1, O^2, O^*\}} \sum_{v_i \in y_L} y_i \, ln o_i, \tag{15}$$

where y_i is the label of node v_i, and o_i is its prediction. With the guide of labeled data, we can optimize the proposed GBV-GSL via back propagation and learn the embedding of nodes for classification.

4   Experiments

4.1   Experimental Setup

Datasets. We employ seven open datasets, including three non-graph datasets (i.e., Wine, Breast Cancer (Cancer) and Digits) available in scikit-learn [15], a blog graph Polblogs [10] and three heterophily graph datasets (i.e. Texas, Cornell, Wisconsin). Notice that for non-graph datasets, we construct a KNN graph as an initial adjacency matrix as in [1].

Baselines. We compare the proposed GBV-GSL with two categories of baselines: classical GNN models GCN [12], GAT [18], GraphSAGE [7] and four graph structure learning based methods Pro-GNN [10], IDGL [1], GEN [19], CoGSL [14].

Implementation Details. For three classical GNN models (i.e. GCN, GAT, GraphSAGE), we adopt the implementations from PyTorch Geometric library [3]. For Pro-GNN, IDGL, GEN and CoGSL, we use the source codes provided by authors, and follow the settings in their original papers with carefully tune. For the proposed GBV-GSL, we use Glorot initialization [6] and Adam [11] optimizer. We carefully select two basic views for different datasets as two inputs. We tune the learning rate for Adam optimizer from {0.1, 0.01, 0.001}. For combination coefficient μ, we test ranging from {0.1, 0.5, 1.0}. Finally, we carefully select total iterations T from {100, 150, 200}, and tune training epochs for each module from {1, 5, 10}. For fair comparisons, we set the hidden dimension as 16 and randomly run 10 times and report the average results for all methods.

For Pro-GNN, IDGL, GEN, CoGSL and our GBV-GSL, we uniformly choose two-layer GCN as backbone to valuate the learnt structure.

4.2 Node Classification

In this section, we evaluate the proposed GBV-GSL on semi-supervised node classification. For different datasets, we follow the original splits on training set, validation set and test set. To more comprehensively evaluate our model, we use two common evaluation metrics, including F1-macro and F1-micro. The results are reported in Table 1, where we randomly run 10 times and report the average results. As can be seen, the proposed GBV-GSL generally outperforms all the other baselines on all datasets, which demonstrates that GBV-GSL can boost node classification in an effective way. The huge performance superiority of GBV-GSL over the backbone GCN implies that the view interaction and classifiers of our model are collaboratively optimized and mutually reinforcing. In comparison with other GSL frameworks, our performance improvement demonstrates that the proposed framework is effective, and the learned structure with more effective information, which can provide better solutions.

4.3 Defense Performance

Here, we aim to evaluate the robustness of various methods. Cancer, Polblogs and Texas are adopted. We focus on comparing GSL models, because these models can adjust original structure, which makes them more robust than other GNNs. Specifically, we choose Pro-GNN from single-view based methods. And for multi-view based methods, IDGL and CoGSL are both selected.

To attack edges, we adopt random edge deletions or additions following [1,4]. Specifically, for edge deletions, we randomly remove 5%, 10%, 15% of original edges, which retains the connectivity of attacked graph. For edge addition, we randomly inject fake edges into the graph by a small percentages of the number of original edges, i.e. 25%, 50%, 75%. In view of that our GBV-GSL needs two inputs while other methods need one input, for a fair comparison, we deploy attacks on each of two inputs separately and on both of them together with the same percentages. We choose poisoning attack [20], where we firstly generate attacked graphs and then use them to train models. All the experiments are

Table 1. Quantitative results (% ± σ) on node classification. (bold: best)

datasets	metric	GCN	GAT	GraphSAGE	LDS	Pro-GNN	IDGL	GEN	CoGSL	**GBV-GSL**
Wine	F1-macro	94.1±0.6	93.6±0.4	96.3±0.8	93.4±1.0	97.3±0.3	96.3±1.1	96.4±1.0	97.5±0.6	**97.9±0.3**
	F1-micro	93.9±0.6	93.7±0.3	96.2±0.8	93.4±0.9	97.2±0.3	96.2±1.1	96.3±1.0	97.4±0.7	**97.8±0.3**
Cancer	F1-macro	93.0±0.6	92.2±0.2	92.0±0.5	83.1±1.5	93.3±0.5	93.1±0.9	94.1±0.8	93.5±1.2	**94.5±0.3**
	F1-micro	93.3±0.5	92.9±0.1	92.5±0.5	84.8±0.8	93.8±0.5	93.6±0.9	94.3±1.0	94.0±1.0	**94.9±0.3**
Digits	F1-macro	89.0±1.3	89.9±0.2	87.5±0.2	79.7±1.0	89.7±0.3	92.5±0.5	91.3±1.3	92.5±1.3	**92.7±1.0**
	F1-micro	89.1±1.3	90.0±0.2	87.7±0.2	80.2±0.9	89.8±0.3	92.6±0.5	91.4±1.2	92.5±1.2	**92.7±1.0**
Polblogs	F1-macro	95.1±0.4	94.1±0.1	93.3±2.5	94.9±0.3	94.6±0.6	94.6±0.7	95.2±0.6	95.5±0.2	**95.9±0.2**
	F1-micro	95.1±0.4	94.1±0.1	93.4±2.5	94.9±0.3	94.6±0.6	94.6±0.7	95.2±0.6	95.5±0.2	**95.9±0.2**
Texas	F1-macro	42.1±2.6	32.2±6.8	76.8±9.3	33.9±9.1	35.5±7.2	51.0±4.8	51.6±7.2	70.0±4.8	**78.5±4.8**
	F1-micro	61.1±1.3	59.2±4.6	84.9±5.4	59.7±7.0	60.8±6.1	64.9±3.0	73.4±6.7	80.8±2.6	**86.0±2.1**
Cornell	F1-macro	52.9±1.0	31.8±7.5	58.9±5.0	36.3±9.3	32.1±9.8	49.6±4.3	36.3±9.1	61.4±7.9	**75.6±4.5**
	F1-micro	66.8±1.2	58.1±2.8	73.2±3.9	63.8±7.8	63.0±7.9	64.9±2.2	65.6±6.7	76.5±2.1	**84.9±1.9**
Wisconsin	F1-macro	44.2±1.7	33.3±3.1	43.3±5.2	31.2±9.8	34.6±8.6	38.1±3.2	31.3±5.1	55.1±7.2	**77.1±7.3**
	F1-micro	65.1±0.8	58.4±2.7	77.8±3.8	53.1±6.7	55.7±7.0	58.8±3.9	55.1±8.1	83.3±3.4	**89.6±2.3**

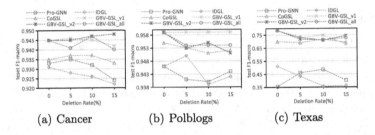

Fig. 2. Results of different models under random edge deletion.

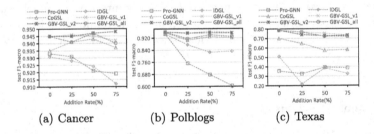

Fig. 3. Results of different models under random edge addition.

conducted 10 times and we report the average accuracy. The results are plotted in Figs. 2 and 3. Besides, the curves of "GBV-GSL_v1", "GBV-GSL_v2" and "GBV-GSL_all" mean the results that one of inputs of GBV-GSL is attacked and both of them are attacked, respectively.

From the figures, GBV-GSL consistently outperforms all other baselines under different perturbation rates by a margin for three cases. We also find that as the perturbation rate increases, the margin becomes larger, which indicates that our model is more effective with violent attack. Besides, "GBV-GSL_all" also performs competitive. Although both of its two inputs are attacked, "GBV-GSL_all" still outperforms other baselines.

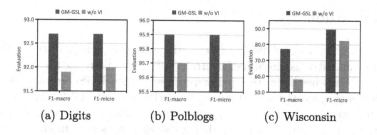

(a) Digits   (b) Polblogs   (c) Wisconsin

Fig. 4. Results with or without view interaction module.

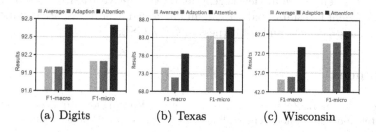

(a) Digits   (b) Texas   (c) Wisconsin

Fig. 5. Results on different fusions.

4.4 Ablation Studies

Analysis of View Interaction. With two estimated views, we propose a view interaction technique based on selection gating mechanism to optimize them again. To evaluate the effectiveness of view interaction, we compare the evaluation result of the final view trained with and without the view interaction on Digits, Polblogs and Wisconsin in Fig. 4. We can see a significant performance drop consistently for GBV-GSL on all datasets by turning off the view interaction component (i.e., using two estimated views for fusion directly). The reason is that, without our view interaction, the two estimated views only have topological information contained in their respective basic structural views, which may be insufficient or noisy. View interaction allows estimated views to retain a portion of their own structure while introducing their mixed structure, which removes noise and solves the problem of insufficient structure.

Analysis of View Attention Fusion. We use an attention fusion mechanism, which assigns weights to two interactive views based on the corresponding importance for each node as Eqs. (9)–(12) in Sect. 3.5. To verify the validation of this part, we design two more baselines. One is to simply average two interactive views as the final view. The other is to use adaptive fusion mechanism [14] to fuse them. We test on Digits, Texas and Wisconsin and show the results in Fig. 5, where "Attention" refers to attention fusion we introduce. We can see that the attention fusion we designed is the best behaved of three ways, which fully proves its effectiveness.

5   Conclusion

In order to solve the problems of neglecting the complementarity of different views and unrestricted information irrelevant to downstream tasks in existing multi-view GSL methods, this paper proposes a novel GSL framework GBV-GSL that introduce the selection gating mechanism into GSL. We designed a view interaction module that incorporates mixed information controlled by the same gating signal into two estimated views, so as to remove redundant structural information and supplement insufficient topology while retaining their focused knowledge. Then, the final view is generated by adaptive attention fusion. Extensive experimental results, under clean and attacked conditions, are conducted to verify the effectiveness and robustness of GBV-GSL.

References

1. Chen, Y., Wu, L., Zaki, M.: Iterative deep graph learning for graph neural networks: better and robust node embeddings. Adv. Neural Inf. Process. Syst. **33**, 19314–19326 (2020)
2. Chen, Y., Wu, L., Zaki, M.J.: Reinforcement learning based graph-to-sequence model for natural question generation. arXiv preprint arXiv:1908.04942 (2019)
3. Fey, M., Lenssen, J.E.: Fast graph representation learning with PyTorch geometric. arXiv preprint arXiv:1903.02428 (2019)
4. Franceschi, L., Niepert, M., Pontil, M., He, X.: Learning discrete structures for graph neural networks. In: International Conference on Machine Learning, pp. 1972–1982. PMLR (2019)
5. Gilmer, J., Schoenholz, S.S., Riley, P.F., Vinyals, O., Dahl, G.E.: Neural message passing for quantum chemistry. In: International Conference on Machine Learning, pp. 1263–1272. PMLR (2017)
6. Glorot, X., Bengio, Y.: Understanding the difficulty of training deep feedforward neural networks. In: Proceedings of the Thirteenth International Conference on Artificial Intelligence and Statistics, pp. 249–256. JMLR Workshop and Conference Proceedings (2010)
7. Hamilton, W., Ying, Z., Leskovec, J.: Inductive representation learning on large graphs. In: Advances in Neural Information Processing Systems, vol. 30 (2017)
8. Hassani, K., Khasahmadi, A.H.: Contrastive multi-view representation learning on graphs. In: International Conference on Machine Learning, pp. 4116–4126. PMLR (2020)
9. Jiang, B., Zhang, Z., Lin, D., Tang, J., Luo, B.: Semi-supervised learning with graph learning-convolutional networks. In: Proceedings of the IEEE/CVF Conference on Computer Vision and Pattern Recognition, pp. 11313–11320 (2019)
10. Jin, W., Ma, Y., Liu, X., Tang, X., Wang, S., Tang, J.: Graph structure learning for robust graph neural networks. In: Proceedings of the 26th ACM SIGKDD International Conference on Knowledge Discovery & Data Mining, pp. 66–74 (2020)
11. Kingma, D.P., Ba, J.: Adam: a method for stochastic optimization. arXiv preprint arXiv:1412.6980 (2014)
12. Kipf, T.N., Welling, M.: Semi-supervised classification with graph convolutional networks. arXiv preprint arXiv:1609.02907 (2016)

13. Linmei, H., Yang, T., Shi, C., Ji, H., Li, X.: Heterogeneous graph attention networks for semi-supervised short text classification. In: Proceedings of the 2019 Conference on Empirical Methods in Natural Language Processing and the 9th International Joint Conference on Natural Language Processing (EMNLP-IJCNLP), pp. 4821–4830 (2019)
14. Liu, N., Wang, X., Wu, L., Chen, Y., Guo, X., Shi, C.: Compact graph structure learning via mutual information compression. In: Proceedings of the ACM Web Conference 2022, pp. 1601–1610 (2022)
15. Pedregosa, F., et al.: Scikit-learn: machine learning in Python. J. Mach. Learn. Res. **12**, 2825–2830 (2011)
16. Pei, H., Wei, B., Chang, K.C.C., Lei, Y., Yang, B.: Geom-GCN: geometric graph convolutional networks. arXiv preprint arXiv:2002.05287 (2020)
17. Qi, X., Liao, R., Jia, J., Fidler, S., Urtasun, R.: 3D graph neural networks for RGBD semantic segmentation. In: Proceedings of the IEEE International Conference on Computer Vision, pp. 5199–5208 (2017)
18. Velickovic, P., Cucurull, G., Casanova, A., Romero, A., Lio, P., Bengio, Y., et al.: Graph attention networks. Stat **1050**(20), 10–48550 (2017)
19. Wang, R., et al.: Graph structure estimation neural networks. In: Proceedings of the Web Conference 2021, pp. 342–353 (2021)
20. Wu, T., Ren, H., Li, P., Leskovec, J.: Graph information bottleneck. Adv. Neural Inf. Process. Syst. **33**, 20437–20448 (2020)
21. Xu, K., Hu, W., Leskovec, J., Jegelka, S.: How powerful are graph neural networks? arXiv preprint arXiv:1810.00826 (2018)
22. Xu, K., Li, C., Tian, Y., Sonobe, T., Kawarabayashi, K.I., Jegelka, S.: Representation learning on graphs with jumping knowledge networks. In: International Conference on Machine Learning, pp. 5453–5462. PMLR (2018)
23. You, J., Ying, R., Leskovec, J.: Position-aware graph neural networks. In: International Conference on Machine Learning, pp. 7134–7143. PMLR (2019)
24. Zheng, C., et al.: Robust graph representation learning via neural sparsification. In: International Conference on Machine Learning, pp. 11458–11468. PMLR (2020)
25. Zhu, Y., Xu, W., Zhang, J., Liu, Q., Wu, S., Wang, L.: Deep graph structure learning for robust representations: a survey. arXiv preprint arXiv:2103.03036, vol. 14 (2021)

How Legal Knowledge Graph Can Help Predict Charges for Legal Text

Shang Gao[1,2], Rina Sa[3], Yanling Li[1,2(✉)], Fengpei Ge[4], Haiqing Yu[1], Sukun Wang[1], and Zhongyi Miao[1]

[1] College of Computer Science and Technology, Inner Mongolia Normal University, Hohhot 011517, People's Republic of China
wdmxsyf_gs@163.com, {cieclyl,ciecyhq,ciecwsk, ciecmzy}@imnu.edu.cn
[2] Key Laboratory of Infinite-dimensional Hamiltonian System and Its Algorithm Application, Ministry of Education, Hohhot 010000, People's Republic of China
[3] Department of Mathematics and Computer Engineering, Ordos Institute of Technology, Ordos 017000, People's Republic of China
sarina216@163.com
[4] Library, Beijing University of Posts and Telecommunications, Beijing 100876, People's Republic of China
gefengpei@bupt.edu.cn

Abstract. The existing methods for predicting Easily Confused Charges (ECC) primarily rely on factual descriptions from legal cases. However, these approaches overlook some key information hidden in these descriptions, resulting in an inability to accurately differentiate between ECC. Legal domain knowledge graphs can showcase personal information and criminal processes in cases, but they primarily focus on entities in cases of insolation while ignoring the logical relationships between these entities. Different relationships often lead to distinct charges. To address these problems, this paper proposes a charge prediction model that integrates a Criminal Behavior Knowledge Graph (CBKG), called Charge Prediction Knowledge Graph (CP-KG). Firstly, we defined a diverse range of legal entities and relationships based on the characteristics of ECC. We conducted fine-grained annotation on key elements and logical relationships in the factual descriptions. Subsequently, we matched the descriptions with the CBKG to extract the key elements, which were then encoded by Text Convolutional Neural Network (TextCNN). Additionally, we extracted case subgraphs containing sequential behaviors from the CBKG based on the factual descriptions and encoded them using a Graph Attention Network (GAT). Finally, we concatenated these representations of key elements, case subgraphs, and factual descriptions, collectively used for predicting the charges of the defendant. To evaluate the CP-KG, we conducted experiments on two charge prediction datasets consisting of real legal cases. The experimental results demonstrate that the CP-KG achieves scores of 99.10% and 90.23% in the Macro-F1 respectively. Compared to the baseline methods, the CP-KG shows significant improvements with 25.79% and 13.82% respectively.

Keywords: Charge prediction · Easily confused charges · Knowledge graph · Graph attention network

---

S. Gao and R. Sa—Equal Contribution.

B. Luo et al. (Eds.): ICONIP 2023, LNCS 14452, pp. 408–420, 2024.
https://doi.org/10.1007/978-981-99-8076-5_30

1 Introduction

With the application of information technologies such as Artificial Intelligence (AI) and big data in various scenarios, there is a growing demand for judicial services. To meet the evolving needs of legal judgment by the general public and keep up with the development of the times, researchers have attempted to integrate AI into the field of legal to achieve fairness, intelligence, and efficiency in legal judgments. In China, to promote AI applications such as intelligent information retrieval and Natural Language Processing (NLP) in the legal domain, the Supreme People's Court and the Chinese Information Processing Society of China jointly organized the Challenge of AI in Law (CAIL) competition. This competition has been held continuously for five years since 2018 and has become an important platform for academic exchanges in the field of legal AI. In CAIL 2018, three tasks were set: legal article recommendation, charge prediction, and sentence prediction.

This study focuses on charge prediction within the CAIL competition. Charge prediction is a crucial task in the intelligent judiciary, which involves analyzing factual descriptions in legal cases to predict the charges for defendants. The predicted results can serve as references for judicial personnel, helping to correct the subjective biases of judges and reduce negative impacts caused by intuition and other subjective factors. Predicting Easily Confused Charges (ECC) has always been a research hotspot and challenge in charge prediction. ECC often share similar criminal processes, but they differ in individual criminal behaviors and outcomes. Different charges often lead to different legal judgments and punishments. The examples of ECC are illustrated in Fig. 1. Among them, blue represents the same elements in the two legal cases, and red represents different elements. Although the behavior information and sequence of behaviors of the two defendants were the same, the behavior on the right also resulted in death. The charges and punishments in the two legal cases are completely different. Therefore, differentiating these key elements and the order of elements is crucial for predicting ECC accurately. Existing methods mostly apply mature AI to predict ECC, which often fails to meet many specialized needs of legal professionals. For example, when traditional text classification models like Long Short-Term Memory (LSTM) [1] are applied to charge prediction, they merely encode and process factual descriptions of legal cases, lacking the fine-grained recognition and understanding of tools and behavioral elements within these descriptions. They do not fully utilize the information presented in legal cases.

Charge: **Dangerous Driving**	Charge: **Traffic Accident**
Case Facts: The defendant, *A*, drove a car with license plate number Ji CXXX along the West Expressway, traveling in the opposite direction from north to south at the segment of Cuigezhuang Road. The defendant's vehicle collided with an oncoming car, resulting in a traffic accident and damage to both vehicles. After the accident, *A* fled the scene. According to the examination, *A*′s blood alcohol concentration was 140.30 mg/100ml, indicating drunk driving.	**Case Facts:** The defendant, *B*, drove a car with license plate number Ji AXXX along the 393 road, traveling from east to west. At the intersection near Sunjiazhuang Village, the defendant's vehicle collided with a three-wheeler driven by *C*, resulting in a traffic accident where *C* sustained fatal injuries despite rescue efforts and both vehicles were damaged. After the accident, *B* fled the scene. It was determined that *B* bears full responsibility for this accident.

Fig. 1. Easily confused charges comparison

Knowledge Graphs (KGs) are graph-structured data that describe entities and their relationships. Due to their intuitive and rich knowledge representation, KGs have been widely applied in NLP tasks. KGs can generally be categorized into two types: general-purpose KGs and domain-specific KGs. The former contains extensive knowledge, mainly consisting of common-sense knowledge and coarse-grained knowledge. The latter focuses on a specific domain, emphasizing deeper and more specialized knowledge. It incorporates expert experience and industry-specific information, covering fine-grained knowledge. Compared to text, KGs provide more explicit and logical representations, playing a supportive and enabling role in intelligent judiciary. In the legal domain, general-purpose KGs provide limited knowledge. Therefore, researchers have begun to construct domain-specific KGs for addressing legal domain tasks. However, existing legal domain KGs often focus on the entities within legal cases and often overlook the logical relationships between legal entities, that different behavioral relationships between entities can lead to different charges and varying punishments. Therefore, for the task of predicting ECC, it is necessary to construct a novel domain-specific KG that highlights key elements and sequential information of legal cases.

To address these challenges, this paper proposes a Charge Prediction method called CP-KG, which integrates a Crime Behavior Knowledge Graph (CBKG). The contributions of this work are summarized as follows:

(1) We constructed a novel domain-specific knowledge graph in the field of legal, known as the CBKG. It encompasses a wealth of legal entities and relationships, enabling a clear depiction of the criminal process associated with ECC.
(2) We extracted Key Elements (KE) and Case Subgraphs (CS) from legal cases, which supplements the model with fine-grained legal knowledge and enhances its ability to comprehend ECC.
(3) Experimental results on two charge prediction datasets consisting of real legal cases demonstrate that the CP-KG achieves Macro-F1 scores of 99.10% and 90.23% respectively. Moreover, it significantly outperforms the baseline methods, achieving approximately 25.79% and 13.82% improvement in the Macro-F1 metric.

2 Related Work

2.1 Charge Prediction

With the development of deep learning, it has achieved successful applications in NLP in recent years. Inspired by this, researchers have attempted to apply deep neural networks to the task of charge prediction, aiming to improve accuracy by extracting deep semantic information from legal cases. Jiang et al. [2] treated charge labels as supervision and used deep reinforcement learning to extract key factual fragments from case facts, enhancing case representations and improving model performance. Zhong et al. [3] proposed a topological multi-task learning framework that models the explicit dependency relationships among three sub-tasks: legal article recommendation, charge prediction, and sentence prediction. Yang et al. [4] introduced the Multi-Perspective Bi-directional Feedback Network (MPBFN) with word collocation attention mechanism to fully utilize the topological dependencies among multiple task results, effectively improving judgment prediction. Zhao et al. [5] employed a reinforcement learning method to extract sentences containing criminal elements in cases, simulating the process of judgment in real-world scenarios.

2.2 Legal Domain Knowledge Graph

The construction of a legal domain KGs based on legal cases is beneficial for various important legal tasks. Chen et al. [6] treated criminal charges and keywords from charge description articles as entity nodes and defined four types of relationships. They constructed a charge knowledge graph by using a relationship classifier to determine the relationships between entities. Chen et al. [7] addressed the issue of dispersed knowledge and inconvenient queries in the judicial domain by constructing a knowledge graph based on a legal case. They preprocessed the legal case using the Language Technology Platform (LTP). Then, they organized and compiled it using the Neo4j graph database to build a case knowledge graph. Chen [8] proposed a knowledge graph construction method focused on criminal behavior. They utilized the LTP and a pre-built legal dictionary to extract elements from legal cases and then extracted triples of criminal behavior entity relationships with the assistance of Chinese grammar rules. Guo [9] introduced a causation graph based on legal cases, using dependency syntactic analysis and regular expression matching to obtain a relationship between events. Chen et al. [10] developed an information extraction model specifically for criminal cases to construct a drug-related case graph. Hong et al. [11] focused on judicial charges of "motor vehicle traffic accident liability disputes" and defined 20 entity types and 9 relationship types. They used deep learning to extract entities and relationships and construct a knowledge graph for traffic cases. To address the interpretability issue in sentencing prediction, Wang et al. [12] focused on drug trafficking cases as their research subject. Under the guidance of domain experts, they designed concepts and relationships within the case factual descriptions and extracted triples from the case facts based on the knowledge graph ontology.

Although these methods have made some progress in the task of charge prediction, there are still challenges that need to be addressed. Firstly, deep learning cannot accurately identify behavior within factual descriptions at a fine-grained level. This limitation results in poor model understanding for ECC. Secondly, existing KGs tailored to the legal domain often focus on the entity within factual descriptions while neglecting the behavioral relationship between legal entities. This limitation hinders the accurate identification of the criminal process associated with ECC, leading to lower prediction performance.

3 Methods

In this section, we detail the construction process of the CBKG and the various components of the CP-KG. Firstly, in Sect. 3.1, we introduce the process of constructing the CBKG. Next, in Sect. 3.2, we discuss the extraction process of KE and CS, while Sect. 3.3 covers the encoding process of KE and CS. Moving forward, Sect. 3.4 presents the encoding process of Fact Descriptions (FD) in legal cases. Finally, in Sect. 3.5, we describe the process of feature fusion and predict charge.

3.1 Overview

The CP-KG model extracts features and fuses them based on the input legal cases. Then, the resulting information is fed into a Fully Connected Layer to predict the defendant's charges. The specific process is illustrated in Fig. 2. Firstly, the FD of the legal cases and the CBKG are used as input data and passed to the Extract Module, which obtains KE from the FD and CS of the CBKG. Subsequently, TextCNN and GAT are employed to encode KE and CS. Finally, the representations of the FD, KE, and CS are fused to predict the charges. Detailed descriptions of these modules will be provided in the following sections.

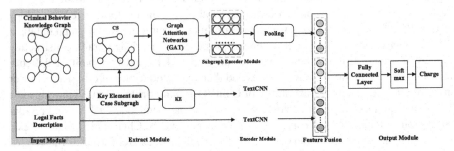

Fig. 2. The structure of the CP-KG model

3.2 Criminal Behavior Knowledge Graph (CBKG)

Collect Legal Cases. The data used for constructing the CBKG is sourced from the criminal legal cases available on China Judgments Online[1]. We have selected eight categories of ECC, namely: intentional injury, intentional homicide, dangerous driving, traffic accident, theft, fraud, robbery, and snatching. For each category, we collected 600 legal cases, resulting in a total of 4,800 legal cases. (Note: The collected legal cases for each charge are from the courts in the same region and only include criminal first-instance cases.)

Define Legal Entities and Relationships. We have redefined 19 categories of legal entities and 11 categories of legal relationships to address the characteristics of cases involving ECC. By utilizing diverse entity and relationship information, we can accurately construct the criminal processes of legal cases and differentiate the subtle differences between ECC. The detailed legal entities and legal relationships are presented in Table 1.

We following the defined entities and relationships as presented in Table 1, with the annotation and verification of legal experts, we ultimately constructed a high-quality dataset for **Judicial Long** Text **Triple** Extraction (JLT). Detailed statistics regarding the dataset will be presented in Sect. 4.

---

[1] https://wenshu.court.gov.cn/.

Table 1. Definition of legal entities and relationships

	Information						
Entity	Case	Defendant	Victim	Cause	Behavior	Tool	Injury
	Primary Culprit	Accessory	Recidivism	Surrender	Forgiveness	Amount	Total Amount
	Law	Charge	Prison Term	Penalty	Truthful Confession		
Relationship	Include	Involve	Crime Cause	Behavior Description	Use	Total	Injury Classification
	Crime Type	Sentencing	Violate	Judgment Information			

Defining Criminal Behavior Knowledge Graph. For predicting ECC tasks, the sequential information of criminal behaviors within criminal events is of paramount importance. Consequently, following the defined legal entities and relationships, we extracted the most relevant criminal behaviors of the defendant and their respective sequences. Criminal behaviors constitute KE, while sequence information constitutes CS. The extracted relevant entities include Crime Cause, Behavior, Tool, Injury Classification, and Charge. The extracted relevant relationships include Use, Cause, Favour, Subsequence, Lead to, and Constitute. From the JLT, we selected entities and relationships related to criminal behaviorsX and constructed CBKG using NetworkX[2]. At this stage, CBKG represents a large graph containing information about criminal behaviors. Subsequently, specific CS are extracted based on the behaviors corresponding to each legal case. The CBKG is illustrated in Fig. 3.

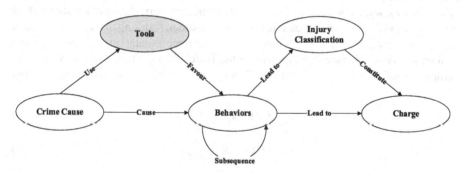

Fig. 3. An example of CBKG

3.3 Extract Module

In criminal cases in China, defendants often have a series of criminal behaviors that occur in a sequence from cause to outcome. This sequential information on criminal

---

[2] https://github.com/networkx/networkx.

behaviors can reflect the order in which the defendant engaged in criminal activities and provide additional perspectives for legal professionals in their judgment. Within a case, multiple behaviors of the defendant can be extracted, and then multiple paths related to these behaviors can be captured from the CBKG. By combining the captured paths, we can obtain a CS that reflects the differences in criminal behaviors between different charges, providing distinguishing features of the criminal process.

The Extract Module takes FD and CBKG as input data and is responsible for extracting KE and CS. Initially, all node information from CBKG is retrieved, and a string matching is conducted with the FD. Successfully matched nodes are identified as KE within the FD. Additionally, multiple paths related to behaviors are extracted from CBKG based on the sequence of node appearances, and these paths are combined to construct CS for criminal behaviors.

3.4 Encoder Module

The FD and KE both fall under textual information. In order to obtain rich features, we choose to encode them using a TextCNN. Specifically, the representation of the KE can provide the model with additional fine-grained semantic information.

For a legal case, denoted as $d_k = [w_1, w_2, ..., w_t, ...w_T]$, where T represents the number of words in the k-th case, and w_t represents the t-th word in the case facts d_k. For the KE extracted from the FD, denoted as $E_k = [e_1, e_2, ..., e_i, ..., e_I]$, where I represent the number of KE in the k-th case, and e_i represents the i-th KE in the KE sequence E_k. In this study, GloVe [13] embeddings are used to obtain word representations for each word in d_k and each element in E_k, represented as $w_t \in R^m$ and $e_t \in R^m$, respectively, where m is the dimensionality of the vectors. The representations of all the words in d_k are concatenated in the order they appear in the text, resulting in the representation $X_{dk} = w_{1:T} = [w_1, w_2, ..., w_t, ..., w_T]$ for the case facts. Similarly, the representations of all the elements in E_k are concatenated in the sequence order, resulting in the representation $X_{Ek} = e_{1:T} = [e_1, e_2, ..., e_i, ..., e_I]$ for the KE. These representations, X_{dk} and X_{Ek}, are then inputted into the convolutional layer of the TextCNN. Different-sized convolutional kernels are used to extract text features. This process can be represented as Eqs. (1) and (2).

$$c = f\left(W_1 * X_{d_k} + b_1\right) \tag{1}$$

$$l = f\left(W_2 * X_{E_k} + b_2\right) \tag{2}$$

where $W_1, W_2 \in R^{h \times m}$ represent the sizes of the convolutional kernels, b_1, b_2 are biases, and $f(\cdot)$ represents a non-linear function. By using convolutional kernels of different sizes, feature sets are obtained for the case facts d_k and the KE sequence E_k, represented as $C = [c_1, c_2, ..., c_{T-h+1}]$ and $L = [l_1, l_2, ..., l_{I-h+1}]$, respectively. T and I represent the number of words in the FD and KE respectively. Finally, the maximum pooling is applied to obtain the text representation C_{dk} for the case facts d_k and the representation L_{Ek} for the KE E_k, capturing representative features. This process can be represented as Eqs. (3) and (4).

$$C_{d_k} = [\max(C)] \tag{3}$$

$$LE_k = [\max(L)] \tag{4}$$

3.5 Subgraph Encoder Module

Graph Convolutional Neural Networks (GCNs) typically aggregate neighboring node information using equal or pre-defined edge weights. In contrast, GAT can assign different weights to different nodes in the graph, allowing for varying importance levels of each neighboring node. Furthermore, GATs do not require the utilization of the entire graph; they only rely on first-order neighboring node information. This addresses some of the limitations of GCNs. Therefore, we choose GAT to encode the extracted CS.

Assuming that the CS contains N nodes, we denote the node vectors of graph as $h = \{h_1, h_2, ..., h_N\}$, $h_i \in R^F$ where F represents the feature dimension of input nodes. Next, the CS is fed into the GAT. When aggregating information between nodes, GAT incorporates an attention mechanism. The formula for calculating the attention coefficients e_{ij} is shown in Eq. (5).

$$e_{ij} = a(Wh_i, Wh_j) \tag{5}$$

where W represents the weight matrix that can be shared, and a denotes the shared attention mechanism. E_{ij} represents the relevance of node I to node j, where $j \in N_i$ and N_i represents all the first-order neighboring nodes of node i. Next, the attention obtained is normalized using the Softmax function. This normalization allows for easy comparison of attention weights between nodes. This process can be represented by Eq. (6).

$$\alpha_{ij} = softmax(e_{ij}) = \frac{\exp\left(LeakyReLU\left(a^T\left[Wh_i \| Wh_j\right]\right)\right)}{\sum_{k \in N_i} \exp\left(LeakyReLU\left(a^T\left[Wh_i \| Wh_k\right]\right)\right)} \tag{6}$$

where the attention mechanism is implemented as a single-layer feed-forward neural network with parameters denoted as a. The activation function used is the LeakyReLU. W represents trainable parameters and ‖ denotes the concatenation operation.

In order to stably represent nodes, we extend the attention mechanism to a multi-head attention mechanism, aiming to improve model stability. M is used to represent the number of attention heads. The final node representation is obtained by averaging the representations obtained through M attention heads. The output can be expressed as shown in Eq. (7).

$$h_{i'} = \sigma\left(\frac{1}{M}\sum_{m=1}^{M}\sum_{j \in N_i} \alpha_{ij}^M W^M h_j\right) \tag{7}$$

where $\sigma(\cdot)$ represents the activation function. α_{ij}^M represents the value computed by the m-th attention head. W^M is the linear matrix used for the linear transformation of the input vector. h_i' represents the feature obtained after feature extraction through the multi-head GAT, which captures the aggregated semantic information of neighboring nodes.

Next, applies max pooling to all the nodes h_i' in the subgraph of the *k-th* case, resulting in a CS representation G_{d_k} that captures the sequential information of criminal behavior. This can be expressed as shown in Eq. (8).

$$G_{d_k} = \text{MaxPooling}(h') \tag{8}$$

3.6 Feature Fusion

The KE of the case, as well as the relevant CS representation, can provide additional information for representing the case facts. Therefore, in this paper, the CS representation G_{dk}, KE representation L_{Ek}, and FD representation Cd_k are concatenate and fed into a fully connected layer, resulting in a representation denoted as p, as shown in Eq. (9). Finally, this representation is inputted into a softmax to predict the involved charges. This process is represented by Eq. (10).

$$p = concat\left[L_{E_k}, C_{d_k}, G_{d_k}\right] \tag{9}$$

$$z = softmax(p) \tag{10}$$

4 Experimental Setup

4.1 Dataset

This study conducted experiments on two charge prediction datasets consisting of real legal cases, JLT and CAIL-8. The JLT includes 8 ECC and comprises 3,842 cases. The CAIL-2018 is an official charge prediction task dataset released by CAIL and contains 202 charges. The CAIL-8 dataset is a subset of CAIL-2018, consisting of legal cases with the same charges as the JLT dataset. (Note: Extracting legal cases with the same charges as the JLT dataset aims to fully validate the effectiveness of the CBKG. In the experiments, CS was extracted from the CBKG, while KE and FD corresponded to the content of the respective dataset.) We conducted a detailed analysis of the case length, and the statistical data is presented in Table 2.

4.2 Implementation Details

The baseline and CP-KG were trained and tested on NVIDIA Tesla V100. According to the analysis in Table 2, the average length of a legal case fact description when using characters as the smallest semantic unit is 497, while it is 371 when using words as the smallest semantic unit. Therefore, in the experiments, the fixed length for input text sequences was set to 400, and the fixed length for KE was set to 50. The experiments utilized word embeddings trained by the GloVe with a dimension of 200 for parameter settings. The convolutional kernel sizes were set to 2, 3, 4, and 5. The training was performed for 20 epochs, and the Adam optimizer was used. The dropout rate was set to 0.5, and the learning rate was set to 1e−3.

Table 2. Dataset statistics information

	JLT-Train	JLT-Test	JLT-Valid	CAIL-8-Train	CAIL-8-Test	CAIL-8-Valid
Total cases	2304	769	769	8243	5227	2441
Average words	391.08	398.76	396.08	281.00	263.48	269.99
Average characters	530.87	564.76	555.82	352.99	331.30	340.56

4.3 Metric

The performance of the CP-KG in this experiment was evaluated using Accuracy (Acc), Macro-Precision (Mac-P), Macro-Recall (Mac-R), and Macro-F1 (Mac-F1) as metrics.

4.4 Baseline Methods

To evaluate the experimental performance of the CP-KG, the following methods were chosen as baseline. All methods used the default settings from the original papers.

- **TFIDF-SVM** [14] is a baseline model provided in the CAIL2018 competition. It uses TF-IDF to extract text features and employs a Support Vector Machine (SVM) for case fact classification.
- **TextCNN** [15] utilizes multiple convolutional layers with different kernel sizes followed by max pooling to encode case facts and predict charges.
- **Bi-GRU** employs a Bi-directional Gated Recurrent Unit to capture text features.
- **TOPJUDGE** [3] is a topological multitask framework that captures the topological dependencies among three judgment prediction subtasks: charge prediction, law article recommendation, and sentence prediction. It uses a Directed Acyclic Graph (DAG) structure.
- **NeurJudge** [16] primarily uses the Bi-GRU approach to encode texts and construct charge graphs and legal provision graphs using charge definition and legal provision. It utilizes graph decomposition to distinguish confusing legal provisions and charge categories and combines the obtained label semantic information with case facts for prediction.

5 Experimental Results

5.1 Analyze

From the experimental results in Table 3, it can be observed that the results of deep learning are generally higher than those of the TFIDF-SVM. This is because the length of case facts is not uniform, and there is a considerable amount of content in the FD. The use of the Bi-GRU leads to the issue of content forgetting, resulting in poor performance in charge prediction. TextCNN effectively captures local features of the text, thus demonstrating better performance in charge prediction. TOPJUDGE, also based on a CNN, differs in associating the three judgment prediction subtasks, leading to superior performance. NeurJudge is based on the Bi-GRU and also establishes some degree of correlation in the three subtasks. However, the graph decomposition operation in this method is more suitable for multiple classes of charges, making it difficult to effectively distinguish distinctive features and resulting in lower prediction performance than other models. The proposed CP-KG combines CS, KE, and FD for feature fusion and prediction. Experimental results demonstrate that CP-KG achieves State-Of-The-Art (SOTA) performance among the baseline. This is primarily due to the fine-grained semantic information provided by the KE in the cases, as well as the criminal process of the defendant and the relationships between elements provided by the CS, which enable a deep understanding of the FD.

Table 3. Comparison of experimental results

Dataset	JLT				CAIL-8			
Methods	Acc	Mac-P	Mac-R	Mac-F1	Acc	Mac-P	Mac-R	Mac-F1
TFIDF-SVM	0.7757	0.7722	0.7749	0.7331	0.8041	0.8191	0.8265	0.7641
TextCNN	0.9831	0.9813	0.9784	0.9796	0.9346	0.9044	0.8782	0.8773
Bi-GRU	0.8388	0.8145	0.8142	0.8090	0.9038	0.8611	0.8791	0.8569
TOPJUDGE	0.9863	0.9897	0.9784	0.9839	0.8719	0.8581	0.8608	0.8334
NeurJudge	0.9714	0.9697	0.9672	0.9683	0.9066	0.8644	0.8967	0.8699
CP-KG	**0.9922**	**0.9917**	**0.9903**	**0.9910**	**0.9474**	**0.9165**	**0.8975**	**0.9023**

5.2 Ablation Study

We conducted an extensive ablation study on CP-KG to validate the effectiveness of the KE and CS. In these experiments, the names of methods indicate the encoder and the encoded object.

From Table 4, it can be observed that incorporating the KE and CS along with the FD leads to improved prediction performance compared to using only the FD, KE, or CS. Mac-F1 increases by an average of 2%. Furthermore, the results show that incorporating the CS yields higher performance than incorporating the KE. This indicates that the sequential information of the elements in the CBKG provides additional semantics, which helps differentiate between ECC. The ablation study results on the CAIL-8 demonstrate that the CBKG is applicable not only to the JLT but also to different cases with the same charges. This validates the generalizability of the CBKG and the effectiveness of the CP-KG.

Table 4. Ablation study results (For example, TextCNN w/KE means that TextCNN was used to encode the Key Elements).

Dataset	JLT				CAIL-8			
Methods	Acc	Mac-P	Mac-R	Mac-F1	Acc	Mac-P	Mac-R	Mac-F1
TextCNN w/FD	0.9831	0.9813	0.9784	0.9796	0.9346	0.9044	0.8782	0.8773
TextCNN w/KE	0.9727	0.9709	0.9727	0.9716	0.9045	0.8545	0.8839	0.8603
GAT w/CS	0.9493	0.9503	0.9432	0.9458	0.9013	0.8505	0.8676	0.8547
TextCNN w/FD+KE	0.9896	0.9879	0.9876	0.9877	0.9334	0.8976	0.8865	0.8821
TextCNN w/FD+CS	0.9909	0.9899	0.9874	0.9885	0.9336	0.8966	0.8903	0.8867
CP-KG	**0.9922**	**0.9917**	**0.9903**	**0.9910**	**0.9474**	**0.9165**	**0.8975**	**0.9023**

6 Conclusion

In this paper, we propose the CP-KG model that integrates a Criminal Behavior Knowledge Graph (CBKG) to address the problem of overlooking crucial elements and sequential information in legal cases in the context of charge prediction methods. The CP-KG first extracts Key Elements (KE) and Case Subgraphs (CS) from the CBKG and the Fact Descriptions (FD). Subsequently, these KE and CS are encoded by a combination method of TextCNN and GAT. Finally, KE, CS, and FD representations are fused to predict the defendant's charges. Experimental results demonstrate that CP-KG outperforms the baseline and achieves state-of-the-art performance, with 25.79% and 13.82% improvements in the Macro-F1 metric. Moreover, the ablation study validates the effectiveness of KE and CS, as they collectively enhance the predictive performance of the CP-KG. Additionally, the CBKF constructed in this paper provides fine-grained semantic information for charge prediction and exhibits generalizability, making it transferable to related legal tasks.

Acknowledgments. This paper was supported by the National Natural Science Foundation of China (12204062, 61806103, 61562068), National Natural Science Foundation of Inner Mongolia, China (2022LHMS06001), Basic Scientific Research Business Project of Inner Mongolia Normal University (2022JBQN106, 2022JBQN111).

References

1. Hochreiter, S., Schmidhuber, J.: Long short-term memory. Neural Comput. **9**(8), 1735–1780 (1997)
2. Jiang, X., Ye, H., Luo, Z., Chao, W., Ma, W.: Interpretable rationale augmented charge prediction system. In: Proceedings of the 27th International Conference on Computational Linguistics: System Demonstrations, pp. 146–151 (2018)
3. Zhong, H., Guo, Z., Tu, C., Xiao, C., Liu, Z., Sun, M.: Legal judgment prediction via topological learning. In: Proceedings of the 2018 Conference on Empirical Methods in Natural Language Processing, pp. 3540–3549 (2018)

4. Yang, W., Jia, W., Zhou, X., Luo, Y.: Legal judgment prediction via multi-perspective bi-feedback network. In: Twenty-Eighth International Joint Conference on Artificial Intelligence (2019)

5. Zhao, J., Guan, Z., Xu, C., Zhao, W., Chen, E.: Charge prediction by constitutive elements matching of crimes. In: Proceedings of the Thirty-First International Joint Conference on Artificial Intelligence, vol. 22(23–29), pp. 4517–4523 (2022)

6. Chen, S., Wang, P., Fang, W., Deng, X., Zhang, F.: Learning to predict charges for judgment with legal graph. In: Artificial Neural Networks and Machine Learning, pp. 240–252 (2019)

7. Chen, J.X., Huang, Y.J., Cao, G.J., Yang, F., Li, C., Ma, Z.B.: Research and implementation of judicial case visualization based on knowledge graph. J. Hubei Univ. Technol. **34**(05), 72–77 (2019)

8. Chen, W.Z.: Research on Legal Text Representation Method Fused with Knowledge Graph. GuiZhou University (2020)

9. Guo, J.: Research and Implementation of Auxiliary Judgment Technology Based on Affair Graph. Beijing University of Posts and Telecommunications (2021)

10. Chen, Y.G.: Research on Entity Relationship Extraction Algorithm for Legal Documents. Dalian University of Technology (2021)

11. Hong, W.X., Hu, Z.Q., Weng, Y., Zhang, H., Wang, Z., Guo, Z.X.: Automatic construction of case knowledge graph for judicial cases. J. Chin. Inf. Process. **34**(01), 34–44 (2020)

12. Wang, Z.Z., et al.: Sentencing prediction based on multi-view knowledge graph embedding. Pattern Recogn. Artif. Intell. **34**(07), 655–665 (2021)

13. Pennington, J., Socher, R., Manning, C.D.: Glove: global vectors for word representation. In: Proceedings of the 2014 Conference on Empirical Methods in Natural Language Processing (EMNLP), pp. 1532–1543 (2014)

14. Xiao, C., Zhong, H., Guo, Z., et al.: CAIL2018: a large-scale legal dataset for judgment prediction. arXiv preprint (2018)

15. Kim, Y.: Convolutional neural networks for sentence classification. arXiv preprint (2014)

16. Yue, L., Liu, Q., Jin, B., et al.: NeurJudge: a circumstance-aware neural framework for legal judgment prediction. In: Proceedings of the 44th International ACM SIGIR Conference on Research and Development in Information Retrieval, pp. 973–982 (2021)

17. Ma, L., et al.: Legal judgment prediction with multi-stage case representation learning in the real court setting. In: Proceedings of the 44th International ACM SIGIR Conference on Research and Development in Information Retrieval, pp. 993–1002 (2021)

18. Lyu, Y., et al.: Improving legal judgment prediction through reinforced criminal element extraction. Inf. Process. Manage. **59**(1), 102780 (2022)

19. Dong, Q., Niu, S.: Legal judgment prediction via relational learning. In: Proceedings of the 44th International ACM SIGIR Conference on Research and Development in Information Retrieval, pp. 983–992 (2021)

20. Feng, Y., Li, C., Ng, V.: Legal judgment prediction via event extraction with constraints. In: Proceedings of the 60th Annual Meeting of the Association for Computational Linguistics, pp. 648–664 (2022)

CMFN: Cross-Modal Fusion Network for Irregular Scene Text Recognition

Jinzhi Zheng[1,3], Ruyi Ji[2(✉)], Libo Zhang[1,4], Yanjun Wu[1,4], and Chen Zhao[1,4]

[1] Intelligent Software Research Center, Institute of Software Chinese Academy of Sciences, Beijing 100190, China
{jinzhi2018,Libo,yanjun,zhaochen}@iscas.ac.cn
[2] The Laboratory of Cognition and Decision Intelligence for Complex Systems, Institute of Automation, Chinese Academy of Sciences, Beijing 100190, China
jrylovezd@gmail.com
[3] University of Chinese Academy of Sciences, Beijing, China
[4] State Key Laboratory of Computer Science, Institute of Software Chinese Academy of Sciences, Beijing 100190, China

Abstract. Scene text recognition, as a cross-modal task involving vision and text, is an important research topic in computer vision. Most existing methods use language models to extract semantic information for optimizing visual recognition. However, the guidance of visual cues is ignored in the process of semantic mining, which limits the performance of the algorithm in recognizing irregular scene text. To tackle this issue, we propose a novel cross-modal fusion network (CMFN) for irregular scene text recognition, which incorporates visual cues into the semantic mining process. Specifically, CMFN consists of a position self-enhanced encoder, a visual recognition branch and an iterative semantic recognition branch. The position self-enhanced encoder provides character sequence position encoding for both the visual recognition branch and the iterative semantic recognition branch. The visual recognition branch carries out visual recognition based on the visual features extracted by CNN and the position encoding information provided by the position self-enhanced encoder. The iterative semantic recognition branch, which consists of a language recognition module and a cross-modal fusion gate, simulates the way that human recognizes scene text and integrates cross-modal visual cues for text recognition. The experiments demonstrate that the proposed CMFN algorithm achieves comparable performance to state-of-the-art algorithms, indicating its effectiveness.

Keywords: Scene Text Recognition · Scene Text Understanding · Neural Networks · OCR

1 Introduction

Scene text recognition, whose main task is to recognize text in image blocks [21], remains a research hotspot in the field of artificial intelligence because of

B. Luo et al. (Eds.): ICONIP 2023, LNCS 14452, pp. 421–433, 2024.
https://doi.org/10.1007/978-981-99-8076-5_31

its wide application scenarios [25], such as intelligent driving, visual question answering, image caption. This task still faces great challenges, mainly due to the complexity and diversity of scene text [22].

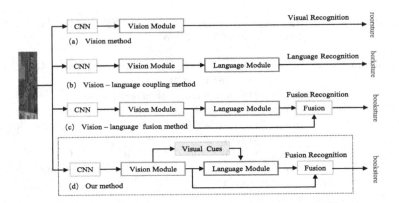

Fig. 1. Comparison of different scene text recognition methods related to our algorithm. (a) Visual recognition methods. (b) Recognition method of visual module series language module. (c) The method of visual module recognition is modified by the language module. (D) Our scene text recognition method(CMFN). Our CMFN fuses visual cues in the language module when mining semantic information.

Most existing methods mainly consider scene text recognition as a sequence generation and a prediction task [1,11]. In the early methods [1,11,16], CNN extracts visual features from the scene image. The vision module decodes the visual features and predicts the scene text (Fig. 1(a)). For example, TRBA [1], MORAN [11], SATRN [9] and VisionLAN [20] use LSTM or transformer module to decode the visual features extracted by CNN. This kind of vision method decodes the visual features but encodes less textual semantic information.

The text carries rich semantic information, which is of great significance in improving the accuracy of text recognition. To exploit the semantic information of text in the recognition process, several visual-language coupling methods [2,14] have been proposed. This kind of method connects the language module after the vision module(Fig. 1(b)). For example, a language decoder set up in PIMNet [14] extracts semantic information from previous predictive text for semantic recognition. The advantage is that the text semantic information can be mined in the recognition process, but the disadvantage is that the recognition result only depends on the performance of the language module, and visual recognition is only the intermediate result. In order to overcome this problem, some visual-language fusion methods [3,21–23] have been gradually proposed(Fig. 1 (c)). After the language module, the fusion gate fuses the visual recognition of the vision module with the semantic recognition of the language module and outputs the fused recognition. The language module in these algorithms only excavates semantic information from the recognized text and ignores visual clues, which means that even if interesting visual clues are found in images or videos, the language module cannot use them to enhance its understanding of semantics.

When humans read scene text, if the visual features are not enough to recognize the text, they will extract semantic information based on the previous recognition. The process of extracting semantic information, not only relies on the context of previous recognition but also incorporates visual cues. It is important to note that the visual cues here are slightly different from explicit visual information such as color, shape, brightness, etc. They refer to more abstract signals or hints perceived from this explicit visual information. Inspired by this, this paper proposes a novel cross-modal fusion network(CMFN) to recognize irregular scene text. The contributions of this paper can be summarized as follows:

- We propose a novel cross-modal fusion network that divides the recognition process into two stages: visual recognition and iterative semantic recognition, with cross-modal fusion in the iterative semantic recognition process.
- We design a position self-enhanced encoder to provide more efficient position coding information for the visual recognition branch and iterative semantic recognition branch.
- In the iterative semantic recognition branch, we design a language module that fuses visual cues. It can alleviate the problem of over-reliance on visual recognition when language the module mining semantic information.
- To verify the effectiveness of the proposed algorithm, abundant experiments are carried out on publicly available datasets.

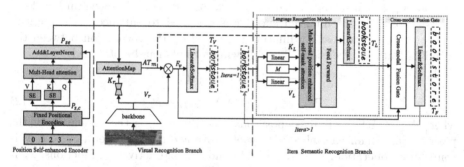

Fig. 2. The overall architecture of CMFN, comprises a position self-enhanced encoder, a visual recognition branch, and an iterative semantic recognition branch. The dashed arrow indicates the direction of attention maps AT_m as visual cues transmission. The blue arrows represent the iterative process. (Color figure online)

2 Methodology

The overall structure of CMFN is shown in Fig. 2. Given scene text image I and pre-defined maximum text length T, the recognition process of scene text consists of three steps. Firstly, the position self-enhanced encoder encodes the character sequence information of the text and outputs position self-enhanced embeddings P_{se}. The character sequence information is an increasing sequence starting from 0 to T-1. Then, the visual branch extracts text visual features from the scene image I and performs text visual recognition based on the visual features and the position self-enhanced embeddings P_{se} and outputs the text prediction probability as text visual recognition T_V. In addition, attention map AT_m is also output as visual cues. Finally, the iterative semantic recognition branch mines semantic information in an iterative way. The iterative semantic recognition branch consists of a language recognition module and a cross-modal fusion gate. The language recognition module integrates visual cues to mine semantic information and outputs the text language prediction probability as text language recognition T_L. The cross-modal fusion gate fuses visual and language features to output cross-modal text fusion prediction probability as text fusion recognition T_F. At the end of the last iteration, the text fusion recognition T_F is output as the final recognition result.

2.1 Position Self-enhanced Encoder

While the structure of learnable positional encoding [4] is relatively simple, a large dataset is required for training. The fixed positional encoding [17] uses constant to express position information, which can meet the requirements in some tasks with relatively low sensitivity to the position. However, in scene text recognition tasks, there are limited training samples and existing models are sensitive to character position information. We are inspired by [3,17,22] to propose a position self-enhanced encoder whose structure is shown in Fig. 2.

The position self-enhanced encoder is designed to enhance the expression ability of character position information based on the correlation between the encoding feature dimensions. The following formula can be obtained:

$$E_\delta = \delta \left(CNN \left(pool \left(P_{s,c} \right) \right) \right) \tag{1}$$

$$P_e = P_{s,c} * \sigma \left(CNN \left(E_\delta \right) \right) + P_{s,c} \tag{2}$$

where $P_{s,c} \in \mathbb{R}^{T \times C_p}$ is fixed positional encoding, C_p is the dimension of the position self-enhanced embedding, *pool* is average pooling operation, CNN is convolution operation, δ and σ are relu and sigmoid activation functions, P_e is the output of the self-enhance (SE) block. The formulaic expression of position self-enhanced encoder coding process:

$$M_p = softmax \left(\frac{QK^{\mathsf{T}}}{\sqrt{C_p}} \right) V + P_{s,c} \tag{3}$$

$$P_{se} = LayNorm \left(M_p \right) \in \mathbb{R}^{T \times C_p} \tag{4}$$

where query Q is from the fixed positional encoding $P_{s,c}$, key K and value V are from two self-enhanced block outputs P_e, respectively. K^{T} is the transpose of key K. $LayNorm$ represents layer normalization and P_{se} is the output of position self-enhanced encoder.

2.2   Visual Recognition Branch

The branch encodes visual features of the image according to the position self-enhanced embedding P_{se}, recognizes the scene text, and outputs the attention map AT_m as visual cues. In this paper, ResNet50 is used as the backbone network to extract visual features. Given scene text image $I \in \mathbb{R}^{H \times W \times 3}$, the scene text recognition process of the visual recognition branch can be expressed as:

$$V_r = Res\left(I\right) \in \mathbb{R}^{\frac{H}{4} \times \frac{W}{4} \times C} \tag{5}$$

$$AT_m = softmax\left(\frac{P_{se}K_r^\mathsf{T}}{\sqrt{C}}\right) \in \mathbb{R}^{\frac{H}{4} \times \frac{W}{4} \times 1} \tag{6}$$

$$F_v = AT_m * V_r \tag{7}$$

where C is the dimension of the feature channel, Res stands for $Resnet50$, $K_r = U(V_r)$, and U is $U - Net$. Based on the feature F_v, the recognition of the visual recognition branch can be formalized:

$$T_V = softmax(fully\left(F_v\right)) \in \mathbb{R}^{T \times cls} \tag{8}$$

where cls is the number of character classes, $fully$ is the fully connected layer and $softmax$ is activation function.

2.3   Iterative Semantic Recognition Branch

Language Recognition Module. The language recognition module integrates visual cues to mine language information from the currently recognized text. The structure of the multi-head position enhanced self-mask attention module is shown in Fig. 3. Inspired by [3, 22], language models treat character reasoning as a fill-in-the-blank task. In order to prevent the leakage of its own information, the mask matrix $M \in R^{T \times T}$ is set. In this matrix, the elements on the main diagonal are negative infinity and the other elements are 0. Text semantic coding features can be obtained by position self-enhanced embedding P_{se}, visual recognition T_v and attention map AT_m:

$$V_L = K_L = fully(T_V) \in \mathbb{R}^{T \times C} \tag{9}$$

$$F_L = softmax\left(\frac{P_{se}K_L^\mathsf{T}}{\sqrt{C}} + M\right)V_L \tag{10}$$

where $fully$ is the fully connected layer and $softmax$ is activation function. The language recognition can be expressed as:

$$L_{vc} = AT_m * P'_{s,c} \in \mathbb{R}^{T \times C} \tag{11}$$

$$F_{vL} = F_L + L_{vc} \tag{12}$$

$$T_L = softmax(fully\left(F_{vL}\right)) \in \mathbb{R}^{T \times cls} \tag{13}$$

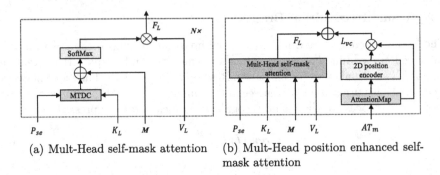

(a) Mult-Head self-mask attention   (b) Mult-Head position enhanced self-mask attention

Fig. 3. The structure of the Mult-Head self-mask attention [3] and Mult-Head position enhanced self-mask attention. MTDC is short for Matrix multiplication, Transpose, Division, Channel square root. AT_m comes from the visual recognition branch, and L_{vc} is the representation of visual cues in the semantic space of the text.

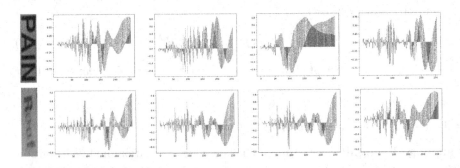

Fig. 4. Visualization of L_{vc} for two scene text instances ("PAIN" and "Root"). The first figure in each row represents a scene text example, followed by the visual cues L_{vc} representation corresponding to each character in the text. In the visualization diagram, the horizontal axis represents feature dimensions and the vertical axis represents the corresponding feature values.

where $P'_{s,c} \in \mathbb{R}^{\frac{H}{4} \times \frac{W}{4} \times C}$ is the sine and cosine position code in two-dimensional space [10,17], and the sizes of its first two dimensions are consistent with the first two dimensions of AT_m. T_L is visual text recognition results.

L_{vc} is considered as the representation of visual cues in the semantic space of the text, so can also refer to visual cues if not otherwise specified. The visualization of visual cues L_{vc} for each character of two text instances "PAIN" and "Root" as shown in Fig. 4.

As can be seen from Fig. 4, on the one hand, the visual cues L_{vc} of different characters in the text have obvious differences. On the other hand, the L_{vc} of the same character in a text tends to have similar expressions. However, due to the different positions, the expression also has a certain difference. This finding suggests that the use of visual cues can enhance the expression of text features.

Cross-Modal Fusion Gate. In order to integrate the recognition of the visual recognition branch and language recognition module, this paper uses a simple fusion gate [3,22]. The cross-modal fusion process can be expressed as:

$$F_g = \sigma([F_v, F_{vL}]Wg) \tag{14}$$

$$T_F = F_v * (1 - F_g) + F_{vL} * F_g \tag{15}$$

where W_g and F_g are trainable superparameters and fusion weights, respectively.

2.4 Training Objective Function

In the training process, the recognition results of multiple modules need to be optimized, so we set a multi-objective loss function:

$$L = \gamma_v L_{Tv} + \frac{1}{N} \sum_{i=1}^{N} (\gamma_l \times L_{TL}^i + \gamma_f \times L_{TF}^i) \tag{16}$$

where L_{Tv} is the standard cross-entropy loss between the predicted probability T_V of the visual recognition branch and ground truth labels for the text. L_{TL}^i and L_{TF}^i are standard cross-entropy loss corresponding to the predicted probability T_L of the language recognition module and T_F of cross-modal fusion gate at the ith iteration. N is the number of iterations. γ_v, γ_l and γ_f are balanced factors. All of these balance factors are set to 1.0 in the experimental part of this paper.

3 Experiments

3.1 Datasets

Synthetic Datasets. MJSynth [6] is a dataset generated by rendering text in scene images. It contained 9M texts, each of which is generated from a dictionary containing 90,000 English words, covering 1,400 Google fonts. SynthText [5] dataset is originally proposed for the scene text detection task. In scene text recognition tasks, the dataset contains 8M images cut from detection task sets.

Regular Datasets. The three regular text datasets are ICDAR2013 (IC13) [8], Street View Text (SVT) [19] and IIIT5k-words (IIIT) [12]. IC13 consists of a training subset with 848 images and a validation subset with 1015 images. SVT comprises images sourced from Google Street View, which includes a training set of 257 samples and a validation set of 647 samples. IIIT includes a training subset of 2000 samples and a validation subset of 3000 samples.

Irregular Datasets. Three irregular scene text datasets are ICDAR2015 (IC15) [7], SVT Perspective (SVTP) [13] and CUTE80 (CUTE) [15]. IC15 contains 4468 train samples and 2077 validation samples. SVTP is from Google Street View, which includes 238 images. 645 image blocks containing text are cropped from this dataset for scene text recognition tasks. The images in the CUTE dataset are collected from digital cameras or the Internet, with a total of 288 images, usually with high resolution and irregularly curved text. All the images are used to verify the algorithm.

Table 1. Text recognition accuracy comparison with other methods on six datasets. The current best performance on each dataset is shown in bold.

Methods	Years	Regular			Irregular			Params
		IC13	SVT	IIIT	IC15	SVTP	CUTE	
MORAN [11]	2019	92.4	88.3	91.2	68.8	76.1	77.4	–
TRBA [1]	2019	92.3	87.5	87.9	77.6	79.2	74.0	49.6M
Textscanner [18]	2020	92.9	90.1	93.9	79.4	84.3	83.3	56.8M
RobustScanner [22]	2020	94.8	88.1	95.3	77.1	79.5	90.3	–
SRN [21]	2020	95.5	91.5	94.8	82.7	85.1	87.8	49.3 M
PIMNet [14]	2021	95.2	91.2	95.2	83.5	84.3	84.8	–
VisionLAN [20]	2021	95.7	91.7	95.8	83.7	86.0	88.5	33M
ABINet [3]	2021	97.4	93.5	96.2	86.0	89.3	89.2	36.7M
SGBANet [25]	2022	95.1	89.1	95.4	78.4	83.1	88.2	–
S-GTR [23]	2022	96.8	94.1	95.8	84.6	87.9	92.3	42.1M
ABINet-ConCLR [24]	2022	97.7	**94.3**	96.5	85.4	89.3	91.3	–
CMFN	Ours	**97.9**	94.0	**96.7**	**87.1**	**90.1**	**92.0**	37.5M

3.2 Implementation Details

In order to make the comparison with other SOTA algorithms [3,24] as fair as possible, this paper adopts the experimental setup that is as close to these algorithms as possible. Before the scene image is input into the model, the size is adjusted to 32 × 128, and data enhancement pre-processings are adopted, such as random angle rotation, geometric transformation, color jitter, etc. In the experiment, the maximum length T of the text is set to be 26. The text characters recognized are 37 classes, including 26 case-insensitive letters, 10 digits, and a token. The multi-head attention unit in the visual recognition branch is set to 1 layer. The language recognition module transformer is set as 4 layers and 8 heads. The experiment is implemented based on Pytorch. Two NVIDIA TITAN RTX graphics cards are used, each with 24GB of space. When MJSynth and SynthText are used for training, the training model provided by ABINet [3] is used for initialization, the batch size is set to 200, the initial learning rate is $1e^{-4}$, and the optimizer is ADAM. The model is trained for a total of 10 epochs. The learning rate decays by one-tenth with each increase of epoch after 5 epochs.

3.3 Comparisons with State-of-the-Arts

In fact, due to the differences in training strategies and backbone, it is difficult to make an absolutely fair comparison between different methods. To be as fair as possible, only the results using the same or similar training strategies are analyzed and compared in this experiment. Specifically, each algorithm model is supervised learning, and the same datasets are used for training.

The statistical results of our CMFN and other published state-of-the-art methods are shown in Table 2. Compared with ABINet-ConCLR [24], our CMFN improved 1.7%, 0.8%, and 0.7% on the three irregular datasets IC15, SVTP, and CUTE, respectively. Compared to irregular datasets, the improvement on regular datasets is relatively small. This is because the accuracy of regular scene text recognition is already quite high (for example, the recognition accuracy of ABINet-ConCLR [24] on the IC13 dataset has reached 97.7%), leaving a relatively limited space for improvement.

In conclusion, our CMFN outperforms state-of-the-art methods in recognizing irregular scene text and achieves comparable results for regular scene text recognition.

3.4 Ablation Study

Analysis of Iteration Number. The visualization of text recognition results under different iterations number is shown in Fig. 5. The overall recognition accuracy reaches 88.3% on three irregular text datasets and 96.5% on three regular datasets when the iteration number is set to 3. Further increasing the iteration number does not lead to significant improvement in accuracy. Therefore, the default iteration number for subsequent experiments is set to 3.

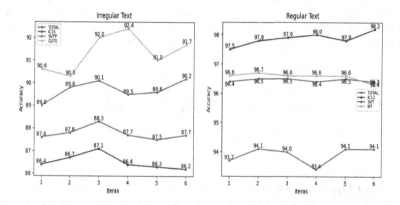

Fig. 5. Text recognition accuracy of different iteration numbers. TOTAL indicates the statistic result of the three corresponding scene text datasets as a whole.

Analysis of the Position Self-enhanced Encoder and Visual Cues. To verify the effectiveness of the position self-enhanced encoder, we compared its performance with learnable positional encoding [4] and fixed positional encoding [17], as shown in Table 1. The position self-enhanced encoder improved by 0.6%, 1.3%, and 1.0% compared to fixed positional encoding, and by 1.0%, 1.1%, and 1.7% compared to learnable positional encoding, on IC15, SVTP, and CUTE datasets. This indicates that our position self-enhanced encoder can bring higher recognition performance. Compared to the language module without visual cues, the inclusion of visual cues in the language module improves performance by 0.6%, and 0.8% on the IC15 and CUTE datasets. This demonstrates the effectiveness of the fusion of visual cues proposed in this paper (Table 3).

Table 2. Ablation study of position self-enhanced encoder. PSE, LPE, VPE are abbreviations for position self-enhanced encoder, learnable positional encoding and fixed positional encoding, respectively. TV, TL, TLn and TF represent the visual recognition branch, language recognition module with visual cues, language recognition module with visual cues exclude visual cues and the cross-modal fusion gate, respectively.

Module	IC15	SVTP	CUTE
VPE+TV+TLn+TF	85.9	88.7	89.2
VPE+TV+TL+TF	86.5	88.8	91.0
LPE+TV+TL+TF	86.1	89.0	90.3
PSE+TV+TL+TF	**87.1**	**90.1**	**92.0**

Table 3. Ablation study of different modules. PSE represents the position self-enhanced encoder. TV(text of visual recognition), TL(text of language recognition) and TF(text of fusion gate recognition) represent the visual recognition branch, language recognition module, and the cross-modal fusion gate, respectively.

Module	IC15	SVTP	CUTE
PSE+TV	84.6	85.4	88.9
PSE+TV+TL	84.3	89.8	89.9
PSE+TV+TL+TF	**87.1**	**90.1**	**92.0**

Analysis of Different Module. The ablation data of each module are shown in Table 1. The visual branch achieved 84.6%, 85.4%, and 88.9% recognition accuracy on the IC5, SVTP, and CUTE. Compared to the visual recognition branch, the text recognition accuracy of the language recognition module decreased by 0.3% on IC15 but increased by 4.4% on SVTP and 1.0% on CUTE. The cross-modal fusion gate achieved optimal recognition performance on IC15, SVTP, and CUTE, with improvements of 2.5%, 4.7%, 3.1% compared to the visual recognition branch, and 2.8%, 0.3%, 2.1% compared to the language recognition branch. This further shows that the modules in our model are valid.

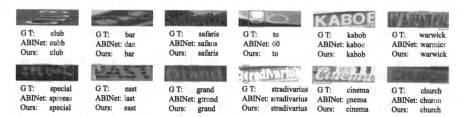

G T: club	G T: bar	G T: safaris	G T: to	G T: kabob	G T: warwick
ABINet: cubb	ABINet: dan	ABINet: safans	ABINet: 60	ABINet: kaboc	ABINet: warmicr
Ours: club	Ours: bar	Ours: safaris	Ours: to	Ours: kabob	Ours: warwick
G T: special	G T: east	G T: grand	G T: stradivarius	G T: cinema	G T: church
ABINet: spreeas	ABINet: last	ABINet: gtrnnd	ABINet: srradivarius	ABINet: gnema	ABINet: churon
Ours: special	Ours: east	Ours: grand	Ours: stradivarius	Ours: cinema	Ours: church

Fig. 6. Examples of successful recognition of irregular text. GT are the ground truth. ABINet and Ours are the recognition results corresponding algorithms, respectively.

3.5   Qualitative Analysis

For qualitative analysis, some recognition cases are visualized as shown in Fig. 6. Compared with ABINet, CMFN has stronger recognition performance. For example, the visual features of "ri" in "safaris" are similar to those of "n", but "safans" is not a meaningful text, and our algorithm can correct "ri" to "n" according to the semantic relation, to correctly recognize the text. In summary, our CMFN has stronger text visual expression ability. For scene texts with insufficient visual features, the text also can be correctly recognized through semantic mining. Even for some scene texts that are difficult for human eyes to recognize, such as "church" and "grand", CMFN can also recognize them correctly.

4   Conclusion

Inspired by human recognition of scene text, this paper proposes a cross-modal fusion network(CMFN) for irregular scene text recognition. The network mainly consists of three parts: a position self-enhanced encoder, a visual recognition branch, and an iterative semantic recognition branch. The position self-enhanced encoder encodes the position information of characters in the text. The visual recognition branch decodes the visual recognition text based on the visual features. The iterative semantic recognition branch simulates the way of integrating visual and semantic modes when a human recognizes scene text and improves the recognition performance of irregular texts by fusing visual features with semantic information. The experimental results show our CMFN not only achieves optimal performance in the recognition of irregular scene text but also has certain advantages in the recognition of regular scene text. In future research, we will explore how to design the recognition model based on knowledge reasoning.

References

1. Baek, J., et al.: What is wrong with scene text recognition model comparisons? dataset and model analysis. In: ICCV, pp. 4714–4722 (2019)
2. Bhunia, A.K., Sain, A., Kumar, A., Ghose, S., Nath Chowdhury, P., Song, Y.Z.: Joint visual semantic reasoning: multi-stage decoder for text recognition. In: ICCV, pp. 14920–14929 (2021)

3. Fang, S., Xie, H., Wang, Y., Mao, Z., Zhang, Y.: Read like humans: autonomous, bidirectional and iterative language modeling for scene text recognition. In: CVPR, pp. 7094–7103 (2021)
4. Gehring, J., Auli, M., Grangier, D., Yarats, D., Dauphin, Y.N.: Convolutional sequence to sequence learning. In: ICML, pp. 1243–1252. JMLR. org (2017)
5. Gupta, A., Vedaldi, A., Zisserman, A.: Synthetic data for text localisation in natural images. In: CVPR, pp. 2315–2324 (2016)
6. Jaderberg, M., Simonyan, K., Vedaldi, A., Zisserman, A.: Synthetic data and artificial neural networks for natural scene text recognition. CoRR abs/arXiv: 1406.2227 (2014)
7. Karatzas, D., et al.: Icdar 2015 competition on robust reading. In: ICDAR, pp. 1156–1160 (2015)
8. Karatzas, D., et al.: Icdar 2013 robust reading competition. In: ICDAR, pp. 1484–1493 (2013)
9. Lee, J., Park, S., Baek, J., Oh, S.J., Kim, S., Lee, H.: On recognizing texts of arbitrary shapes with 2d self-attention. In: CVPRW, pp. 2326–2335 (2020)
10. Liao, M., Lyu, P., He, M., Yao, C., Wu, W., Bai, X.: Mask textspotter: an end-to-end trainable neural network for spotting text with arbitrary shapes. IEEE Trans. Pattern Anal. Mach. Intell. **43**(2), 532–548 (2021)
11. Luo, C., Jin, L., Sun, Z.: Moran: a multi-object rectified attention network for scene text recognition. Pattern Recogn. **90**, 109–118 (2019)
12. Mishra, A., Karteek, A., Jawahar, C.V.: Scene text recognition using higher order language priors. In: BMVC, pp. 1–11 (2012)
13. Phan, T.Q., Shivakumara, P., Tian, S., Tan, C.L.: Recognizing text with perspective distortion in natural scenes. In: ICCV, pp. 569–576 (2013)
14. Qiao, Z., et al.: Pimnet: a parallel, iterative and mimicking network for scene text recognition. In: ACM MM, pp. 2046–2055 (2021)
15. Risnumawan, A., Shivakumara, P., Chan, C.S., Tan, C.L.: A robust arbitrary text detection system for natural scene images. Expert Syst. Appl. **41**(18), 8027–8048 (2014)
16. Shi, B., Yang, M., Wang, X., Lyu, P., Yao, C., Bai, X.: Aster: An attentional scene text recognizer with flexible rectification. IEEE Trans. Pattern Anal. Mach. Intell. **41**(9), 2035–2048 (2019)
17. Vaswani, A., et al.: Attention is all you need. In: NIPS, pp. 6000–6010 (2017)
18. Wan, Z., He, M., Chen, H., Bai, X., Yao, C.: Textscanner: reading characters in order for robust scene text recognition. In: AAAI, pp. 12120–12127 (2020)
19. Wang, K., Babenko, B., Belongie, S.: End-to-end scene text recognition. In: ICCV, pp. 1457–1464 (2011)
20. Wang, Y., Xie, H., Fang, S., Wang, J., Zhu, S., Zhang, Y.: From two to one: a new scene text recognizer with visual language modeling network. In: ICCV, pp. 14194–14203 (2021)
21. Yu, D., et al.: Towards accurate scene text recognition with semantic reasoning networks. In: CVPR, pp. 12110–12119 (2020)
22. Yue, X., Kuang, Z., Lin, C., Sun, H., Zhang, W.: RobustScanner: dynamically enhancing positional clues for robust text recognition. In: Vedaldi, A., Bischof, H., Brox, T., Frahm, J.-M. (eds.) ECCV 2020. LNCS, vol. 12364, pp. 135–151. Springer, Cham (2020). https://doi.org/10.1007/978-3-030-58529-7_9
23. He, Y., et al.: Visual semantics allow for textual reasoning better in scene text recognition. In: AAAI, pp. 888–896 (2022)

24. Zhang, X., Zhu, B., Yao, X., Sun, Q., Li, R., Yu, B.: Context-based contrastive learning for scene text recognition. In: AAAI, pp. 3353–3361 (2022)
25. Zhong, D., et al.: Sgbanet: semantic gan and balanced attention network for arbitrarily oriented scene text recognition. In: ECCV, pp. 464–480 (2022). https://doi.org/10.1007/978-3-031-19815-1_27

Introducing Semantic-Based Receptive Field into Semantic Segmentation via Graph Neural Networks

Daixi Jia[1,2], Hang Gao[1,2(✉)], Xingzhe Su[1,2], Fengge Wu[1,2],
and Junsuo Zhao[2]

[1] Institute of Software Chinese academy of Sciences, No. 4 South Fourth Street,
Zhongguancun, Haidian District, Beijing, China
{jiadaixi2021,gaohang,xingzhe2018,fengge}@iscas.ac.cn
[2] University of Chinese Academy of Sciences, No. 1 Yanqi Lake East Road, Huairou District,
Beijing, China
junsuo@iscas.ac.cn

Abstract. Current semantic segmentation models typically use deep learning models as encoders. However, these models have a fixed receptive field, which can cause mixed information within the receptive field and lead to confounding effects during neural network training. To address these limitations, we propose the "semantic-based receptive field" based on our analysis in current models. This approach seeks to improve the segmentation performance by aggregate image patches with similar representation rather than their physical location, aiming to enhance the interpretability and accuracy of semantic segmentation models. For implementation, we utilize Graph representation learning (GRL) approaches into current semantic segmentation models. Specifically, we divide the input image into patches and construct them into graph-structured data that expresses semantic similarity. Our *Graph Convolution Receptor* block uses graph-structured data purpose-built from image data and adopt a node-classification-like perspective to address the problem of semantic segmentation. Our GCR module models the relationship between semantic relative patches, allowing us to mitigate the adverse effects of confounding information and improve the quality of feature representation. By adopting this approach, we aim to enhance the accuracy and robustness of the semantic segmentation task. Finally, we evaluated our proposed module on multiple semantic segmentation models and compared its performance to baseline models on multiple semantic segmentation datasets. Our empirical evaluations demonstrate the effectiveness and robustness of our proposed module, as it consistently outperformed baseline models on these datasets.

Keywords: Semantic Segmentation · Computer Vision · Graph representation learning

1 Introduction

As a fundamental task in computer vision, semantic segmentation strives to give a rich per-pixel classification of image data. These methods for interpreting visual input have a variety of applications including scene comprehension, object recognition and augmented reality. Existing semantic segmentation methods are mostly end-to-end, deep-learning-based methods [23] that use Convolutional Neural Networks (CNNs) [19] or transformer-based models [32] as encoder, while generating prediction mask with a decode neural network. These current models have already achieved excellent performance, but still facing some problems.

The "neurons" within deep neural network models possess receptive fields, which are defined as the spatial extent of image that a set of weights can access. At lower levels of processing, both convolutional neural network (CNN) and current vision transformer encoders have local receptive fields with fixed sizes [28]. This characteristic can lead to the loss of crucial global information that underpins semantic segmentation, as well as the confounding interference of information within receptive field. Consequently, these limitations can impede the training and prediction of the model and make it difficult for models to learn the true pattern in the image.

The fixed-shape receptive field of CNNs ensures the network's inductive biases, such as translation invariance and locality. However, the fixed nature of the receptive field shape can result in the inclusion of information that is irrelevant to the target task, which may interfere with the neuron's ability to accurately perceive the target information. This information has the potential to negatively impact the performance and generalization ability of the neural network. To address this issue, research efforts have focused on developing techniques to effectively handle confounding information, including approaches such as increasing network depth [12] and incorporating attention-based mechanisms [9].

(a) The receptive field of CNNs. (b) The receptive field of pyramid ViTs. (c) The semantic-based receptive field.

Fig. 1. Illustration of the receptive fields. As illustrated in (a), the receptive field of CNNs have a fixed size, information from multiple categories are included in it. The receptive field of ViT (b) is even better, as it can change the weights of internal vectors to focus attention on key parts. The semantic-based receptive field (c), on the contrary, covers only those parts that can be classified as "goat" and ignores the irrelevant part for segmentation.

Vision transformers regards an image as a sequence, and achieve cognition by performing attention-weighting on every position in the image, leading to the global receptive field [8]. However, this global receptive field may have limitations in segmentation tasks. In the context of semantic segmentation, the global receptive field of vision transformers has some inherent problems such as the loss of crucial local information and inevitably introduce confounding information that interferes with model convergence and leads to performance degradation. To address this issue, contemporary vision transformers address the locality issue by employing methods such as sliding windows [21] and limiting the local receptive field [31]. However, this still does not fully resolve the problem of confounding factors within the receptive field for the model still process information from irrelevant pixels. Because the features at each position are obtained by weighting information from all pixels in the entire image or a certain set of image patches.

In light of the limited success of previous works involving CNNs and vision transformers in addressing the issue of confounding information, we propose the "semantic-based receptive field". As illustrated in Fig. 1(c), the semantic-based receptive field has a more flexible form depending on semantic information, whereas the receptive field of CNNs and transformers has a fixed size. Learning direct effect from high-relative part of image can allow specific patches to aggregate only with those patches that are related to their representation. However, while CNN and transformer-based models both use a predefined paradigm to learn how to represent images, neither of them can implement the suggested idea. Notice that Graph Representation Learning (GRL) are a more flexible approach to processing, we introduce GRL to address this issue. GRL methods are usually used to process graph data with topological forms. To represent images as graph structure for processing, before each time that data is input into the model, we first divide the image into multiple patches as graphs nodes. Then, based on the similarity between these patches, they are constructed as graphs. Finally, we employ GNNs to extract representations from the constructed graphs. The advantage of such design is that similarity, rather than spatial information, is used to bind the patches together. Patches with more similar information tend to be connected as training goes on, which is akin to the receptive field being deformed in accordance with the semantic information. However, this design also has a disadvantage. That is, it also lacks important inductive biases like locality and translation invariance.

So, for implementation, we propose our *Graph Convolution Receptor* module as a plugin to current semantic segmentation models. By replacing a portion of the CNN or transformer channels, we are able to enhance the model's performance without introducing additional parameters. Empirically, we evaluate our proposed structure on three different datasets. Extensive experiments show that model GCR outperforms state-of-the-art graph representation learning methods on semantic segmentation tasks. Furthermore, models adding our GCR module show better robustness on naturally and artificially corrupted data. Our contributions are as follows:

- We analyze the receptive field of current semantic segmentation models and propose the semantic-based receptive field which is crucial for semantic segmentation tasks.
- We propose a novel block, GCR, which uses Graph representation learning to enhance the performance of semantic segmentation models.
- We conduct multiple experiments to verify the effectiveness of our method. Our model also shows advantages in robustness and convergence speed.

2   Related Works

2.1   Semantic Segmentation

Semantic information holds wide applications within machine learning tasks [1, 16], and semantic segmentation utilizes semantic information among images to forecasts a pixel-level categorization. Such task can be viewed as a more challenging, fine-grained form of image classification. The relationship is pointed out and studied systematically in a seminal work [23] where authors designed fully convolutional networks (FCNs) for semantic segmentation. Since then, the FCN model has become a standard design paradigm for dense prediction and has inspired many follow-up works. Works after mainly focused on improving FCNs from aspects such as using deeper neural network [41], enlarging the receptive field [38], introducing boundary information [7]. These architectures improved the performance of semantic segmentation significantly. Furthermore, transformer models introduced from natural language processing (NLP) also achieved competitive performance on semantic segmentation tasks. The application of semantic segmentation is not limited to 2D images, but point clouds [18], 3D scenes [27] and videos.

2.2   Graph Neural Networks (GNNs)

GNNs conduct graph representation learning by propagating information among neighbor nodes. The learned representation have various applications in downstream tasks. Similar to other artificial neural networks, multiple variants of GNNs were developed, such as Graph Convolution Networks (GCNs) [17], which utilizes convolution for graph learning, Graph Attention Networks (GATs) [33], which introduces the attention mechanism, and Graph Isomorphism Networks (GINs) [36], which proposes a graph learning architecture that is as powerful as the Weisfeiler-Lehman test. Based on these networks, relevant researchers also propose many improved architectures according to application domains, including social networks [37], natural language processing [25], molecular biology [10], chemistry [2], and physics [26].

Some methods also adopt GNNs for computer vision tasks. In these tasks, GNNs were mainly used to learn data that naturally possess a graphic structure, e.g., 3D point cloud segmentation [18] and human action recognition [15]. However, recent works applied GNNs for image classification [11] and proved their potential in processing digital images.

3   Method

Base on our analysis of the prediction generation process for semantic segmentation, we introduce the GCR block. As shown in Fig. 2, our model consists of two major components: 1) standard pyramid semantic segmentation model that provides useful inductive biases, 2) a graph convolution receptor implemented using graph representation learning, is used to replace part of the channels of the pyramid encoder to achieve semantic-based receptive field.

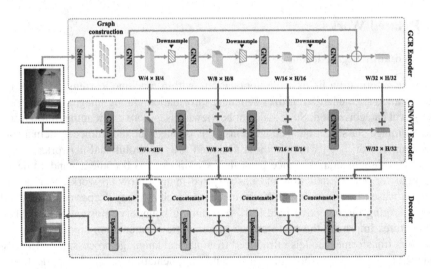

Fig. 2. The overview of our model structure.

3.1 Graph Convolution Receptor

We use a GNN to implement our semantic-based receptive field by replacing a set of CNN or ViT channels to GNN channels. This module can enhance the performance of existing models, allowing neurons to more prominently attend to image blocks that have semantic relationships with the position being considered. Different from CNNs and ViTs, our graph convolution receptor possesses a changeable form that based on learned information. As shown in right in Fig. 3, the semantic-based receptor selects semantically related neurons from the feature map for aggregation, rather than according to the physical location like CNNs. The form of the semantic-based receptive field of each neuron is defined based on representation of the input visual data. To generate such receptive field, a GNN architecture was introduced into our encoder.

The input image, denoted as I, with dimensions of $H \times W \times C$, is transformed into a graph-based representation, $G = \{V, E\}$, where V represents the node set and E represents the edge set. The image is divided into N patches, each of which is represented as a vector $x_i \in \mathbb{R}^D$, where D is the dimension of the patch vector.

In order to mitigate the potential loss of information due to direct slicing of an image into patches, a stem architecture utilizing convolutional layers is implemented. The stem architecture downsamples the input image I to a feature map with a size of $H/4 \times W/4$, providing each patch with overlapped data. These patch vectors are then utilized as node features within the set of nodes, V. Edges within the set of edges, E, are established based on similarity, as calculated from the semantic representation. Specifically, an edge $e_{uv} \in E$ is created between nodes v and u if node u is among the k-nearest neighbors of node v in terms of similarity.

The encoder employs a neighborhood aggregation strategy, in which the representation of node $h_v^{(k)}$ is iteratively updated at the k-th layer. Through k rounds of aggregation, the information of the k-hop neighborhood of node v is propagated to itself. The architecture of the k-th layer of the GCR module can be mathematically formulated as follows:

$$a_v^{(k)} = f^{(k)}(\{h_u^{(k-1)} : u \in \mathcal{N}(v)\}), \tag{1}$$

$$h_v^{(k)} = \phi^{(k)}(h_v^{(k-1)}, a_v^{(k)}). \tag{2}$$

where $h_v^{(k)}$ is the vector representation of node v in the k-th layer of the network. The aggregation function $f(\cdot)$ and combine function $\phi(\cdot)$ are critical operators for GNNs. There exists multiple choices for these operators, as many kinds of aggregation functions with different characteristics have been proposed. Here, we model the pairwise distances between these nodes as edge weights. Specifically, the weight of the edge between any two nodes (i.e., two patch representations) is positively correlated with the proximity of their respective distances. Formally, the aggregation function $f(\cdot)$ can be formulated as:

$$f^{(k)} = \frac{\mathcal{D}_{uv}(\{h_v^{(k-1)} - h_u^{(k-1)} | u \in \mathcal{N}(v)\})}{N} \tag{3}$$

where $\mathcal{D}_{uv}$ represents the pairwise distance between feature u and v and N represents the number of neighbours of node v. As for the combine function $\phi(\cdot)$, the formal representation is as follows:

$$h_v^{(k)} = \phi^{(k)}(h_v^{(k-1)}, f^{(k)}(h_v^{(k-1)})) \tag{4}$$

With the transformed data and the GCR module, we are able to implement semantic-based receptive field for our method. As the graph generation do not rely on physical position to select neighbors but based on similarity, the data propagation will be committed among semantically related nodes, which yields a semantic-based respective field as result. Figure 2 gives an illustration for such semantic-based receptive field. Such mechanism also provides an explicit structure for the aggregation of high-level semantics. Experimental results of Sect. 4.3 demonstrate that models with GCR module converges significantly faster than other models. Moreover, the GCR module encodes relative value of each nodes generated from image patches, thus it shows better zero-shot robustness especially on corruptions like blur, fog and color casts.

3.2 Over-Smoothing Alleviation

Our GCR module utilizes data aggregation among nodes to learn the representation of graph nodes. However, this mechanism can lead to the over-smoothing phenomenon, which results in the decrease of distinctiveness in node features and ultimately results in all nodes possessing the same vector representation. This finally leads to degradation of performance in the segmentation task. This can negatively impact performance in the segmentation task. To mitigate this issue, we propose the implementation of the following modifications to our GCR module:

Diversity Elevation. Alike our baseline methods [11,21,22], we apply a linear layer before and after the graph convolution layer to increase the diversity and limit the dimensionality of features. A nonlinear layer is added after the graph convolution layer. Given a feature map $H^{(k)} = \{h_1^{(k)}, h_2^{(k)}, ..., h_v^{(k)}\} \in \mathbb{R}^{N^{(k)}} \times D^{(k)}$ of layer k, and define the graph process mentioned above as $H^{(k)} = \Phi(H^{(k-1)})$, the modified graph procession module of layer k can be formalized as:

$$H_g^{(k)} = \sigma(\Phi(H^{(k)})W_{in}^{(k)})W_{out}^{(k)} + H^{(k)} \tag{5}$$

where σ represents nonlinear activate function for diversity enhancement, W_{in} and W_{out} denote the learnable parameters of the linear layers.

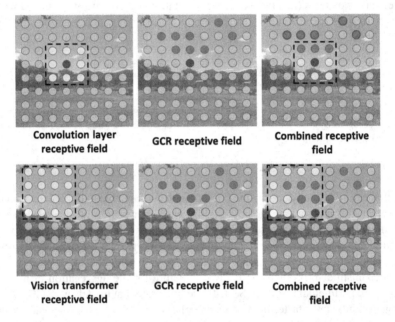

Fig. 3. The receptive field of different models. Our method improves the receptive field of CNN or ViT, enabling the models to perceive global semantic information without losing important inductive biases.

Transformation Elevation. To further alleviate the over-smoothing phenomenon, we use a simple MLP (multilayer perceptrons) on each node of every graph node for more transformation on vector space, like the method used in our baseline models [11,21,22]. This can be described as:

$$H_f^{(k)} = MLP(H_g^k) \tag{6}$$

where $H_f^{(k)} \in \mathbb{R}^{N^{(k)} \times D^{(k)}}$ is the final output of our GCR module in the k-th layer. The hidden dimension of MLP is usually greater than $D^{(k)}$ by an augmentation factor called expansion ratio. Following the implementation of all fully-connected layers, a batch normalization operation is applied as a standard practice, although it has been omitted in this concise description. This structure forms the core component of the GCR module.

Graph Shortcut. Inspired by the idea of residual connection, we apply a strategy called Graph Shortcut which effectively mitigates the issue of over-smoothing while introducing a minimal number of additional parameters. Formally, our Graph Shortcut is defined as:

$$H_f^{l'} = \alpha \mathcal{D} H_f^0 + (1 - \alpha) H_f^l \tag{7}$$

where H_f^l represents the last layer's output of the GCR module and α is a hyperparameter that will be discussed later. The function $\mathcal{D}$ is a simple MLP that downsamples the scale of H_f^0 to be the same as H_f^l.

3.3 Baseline Encoder

The GCR module takes image patches as unordered graph data, which causes the loss of inductive biases like translation invariance and localization. Based on the analysis above, we build our GCR module in parallel to any typical pyramid backbones. The baseline encoder works in parallel with the GCR module, providing basic inductive biases for image understanding. As shown in Fig. 2, our GCR is a plugin to current models that process in parallel to baseline encoder.

To assess the efficacy of our proposed GCR method, we evaluate its performance using commonly used state-of-the-art transformer and CNN models as baseline. By comparing our method to these well-established baselines, we aim to demonstrate the superiority of our approach and establish its effectiveness in improving the overall quality and accuracy of semantic segmentation results.

3.4 Decoder

Pyramid architecture [12] considers the multi-scale property of images by extracting features with gradually smaller spatial size and greater dimension as network goes deeper and has become a popular paradigm for semantic segmentation. Larger numbers of empirical results have shown that pyramid architecture is effective for visual tasks especially semantic segmentation task which needs both global semantics and fine-grained forecasting. The representation of a typical semantic segmentation decoder can be described formally as:

$$Output^o = Concat(H_f^o, C^o) \tag{8}$$

where $o \in \{1, 2, ..., n\}$ represents the n output stages of pyramid encoder. C^o is the output of CNN at stage o, and H_f^o is the output of GNN at stage o. The $Concat(\cdot)$ is the concatenate operator. The output scales of encoder are $H/4 \times W/4$, $H/8 \times W/8$, $H/16 \times W/16$ and $H/32 \times W/32$ of an image with the scale of $H \times W$. An UperNet [35] decoder that takes the pyramid feature maps as input and then generating per-pixel prediction is applied.

4 Experiments

In this section, we present a series of experiments to evaluate the performance of our proposed GCR module and examine their structures.

4.1 Datasets

We adopt ADE20k and Cityscapes datasets for semantic segmentation. ADE20k [42] is a widely-used benchmark which has 20,000 well-labeled training images and 2,000 validation images which belong to 150 categories. For the licenses of ADE20k dataset, please refer to https://groups.csail.mit.edu/vision/datasets/ADE20k/. Most images in ADE20k belong to indoor scene segmentation, while Cityscapes [4] mainly focuses on outdoor scenes on roads. Cityscapes contains 19 categories of labels and has 2,975 training images along with 500 validation images. For the licenses of the Cityscapes dataset, please refer to https://www.cityscapes-dataset.com/.

4.2 Experimental Settings

For all our models, we pretrain the encoder blocks on ImageNet1k or ImageNet22k [6] dataset. For the pretraining strategies, we follow the widely-used training strategies proposed in DeiT [30] for fair comparison. The data augmentation techniques we used includes RandAugment [5], Mixup [40], CutMix [39], repeated augment [13]and stochastic path [14]. To ensure fairness in comparison, a widely-utilized Uper-Net decoder structure is adopted in the decoder without any additional modifications. The models are trained using a batch size of 16 for the ADE20k dataset and 8 for the Cityscapes dataset on 8 RTX5000 GPUs. The implementation of our method is based on MMsegmentation [3] codebase and during training, data augmentation techniques such as random resize with a ratio of 0.5–2.0, horizontal flipping, and random cropping to 512×512 and 1024×1024 for ADE20k and Cityscapes respectively are applied. The models are trained using the AdamW optimizer for 160,000 iterations on both ADE20k and Cityscapes datasets, with an initial learning rate of 0.00006 and a "poly" learning rate schedule with a default factor of 1.0. Inference is performed using a sliding window approach by cropping 1024×1024 windows for Cityscapes.

4.3 Comparison with Baseline Methods

In our evaluation, we compare our proposed method against several baseline methods from three key aspects: performance, robustness, and convergence speed. Specifically, we conduct a comprehensive analysis of the performance of our approach and the baselines in terms of multiple metrics. We also investigate the robustness of the methods by examining their performance under various perturbations, such as noise and occlusion, and assess their ability to handle input variations. Furthermore, we analyze the convergence speed of the methods and compare the time taken to reach a satisfactory level of performance.

Table 1. The main results on ADE20k and Cityscapes. Results with '*' mark represents the model is pretrained on ImageNet22k dataset.

Method	Encoder	Decoder	ADE20k (Val)		CityScapes (Val)		Parameters (M)
			mIoU	aACC	mIoU	aACC	
UperNet-r50	ResNet-50	UperNet	41.65	80.14	-	-	66.52
UperNet-r101	ResNet -101	UperNet	43.21	80.81	-	-	85.51
ConvNeXt-Ti	ConvNext-Ti	UperNet	46.41	81.77	80.78	96.57	60.24
ConvNeXt-S	ConvNext-S	UperNet	48.33	82.44	81.93	96.77	81.88
ConvNeXt-B	ConvNext-B	UperNet	*51.44	*84.27	82.01	96.79	122.1
UperNet-r50(with GCR)	ResNet-50+GCR	UperNet	42.71	80.3	-	-	66.49
UperNet-r101(with GCR)	ResNet -101+GCR	UperNet	44.13	80.96	-	-	85.47
ConvNeXt-T(with GCR)	ConvNext-Ti+GCR	UperNet	**46.63**	**81.85**	**80.84**	**96.5**	60.19
ConvNeXt-S(with GCR)	ConvNext-S+GCR	UperNet	**48.59**	**82.49**	**81.99**	**96.79**	81.81
ConvNeXt-B(with GCR)	ConvNext-B+GCR	UperNet	***51.64**	***84.62**	**82.24**	**96.89**	121.9
SwinTransformer-T	SwinTransformer-Ti+GCR	UperNet	44.53	81.17	78.66	96.21	59.94
SwinTransformer-S	SwinTransformer-S	UperNet	47.99	82.58	79.62	96.4	81.26
SwinTransformer-B	SwinTransformer-B	UperNet	*51.27	*84.27	80.42	96.53	121.28
SegFormer-b5	MiT	MLP	48.7	82.51	80.35	96.53	82.01
SwinTransformer-T(with GCR)	SwinTransformer-Ti	UperNet	**45.12**	**81.41**	**78.16**	**96.01**	59.89
SwinTransformer-S (with GCR)	SwinTransformer-S+GCR	UperNet	**48.31**	**82.65**	**79.75**	**96.46**	81.19
SwinTransformer-B (with GCR)	SwinTransformer-B+GCR	UperNet	***51.52**	***84.46**	**80.53**	**96.62**	121.08
SegFormer-b5 (with GCR)	MiT+GCR	MLP	49.08	82.73	80.41	96.57	85.17

Results on Performance. The experimental results on the ADE20k and Cityscapes datasets are demonstrated in Table 1. For metrics, we evaluate semantic segmentation performance using mean Intersection over Union (mIoU) and average Accuracy (aACC). We observed that all the methods achieved performance improvement after incorporating our module. We attribute such performance to the characteristics of our GCR that can perceive the global semantic information more effectively.

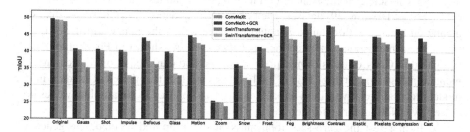

Fig. 4. Experimental results on ADE-C. Among these blurs and noises, 'Gauss' represents Gaussian noise; 'Shot' represents shot noise; 'Impulse' represents impulse noise; 'Glass' represents glass blur. Models added GCR mainly shows more prominent robustness in color cast, fog and defocus blur.

Fig. 5. Examples tested for different models in Foggy-Driving dataset without additional training. Models with GCR can predict segmentation edges more accurately in the foggy weather.

Results on Robustness. Compared with curated data in well-labeled datasets, images collected in the natural world face various noises. Examples include clouds, Gaussian noise, blur, and smudges. Our model introduces the idea of semantic-based receptive field and implements it using graph representation learning, making the model robust to these perturbations. In each layer, GNN selects neighbors dynamically according to the similarity for aggregating. It also processes the difference between feature vectors rather than their absolute values, which leads to stronger ability to deal with interference. We, therefore, design experiments to test the performance of our models on noisy data.

We experimentally demonstrate that our method is structurally robust. We train the model on Cityscapes using standard training pipeline and test the performance directly on the Foggy-Driving [29] dataset. The Foggy-Driving dataset is a Cityscapes-like segmentation dataset with detailed annotations collected in the real world under fog. It shares the same annotation categories as Cityscapes. As shown in Fig. 5 and Table 2, compared to CNN or transformer models, our method can predict more accurate segmentation edges under fog. This is likely due to the fact that the GCR module in our model processes relative values of patch vectors, as opposed to the absolute values processed by CNNs and transformers.

To further test the zero-shot robustness of our proposed method, We generated ADE-C based on [24], which expands the validation set by 16 kinds of algorithmically generated corruptions. ADE-C is used to test the robustness of Swin-transformer, ConvNeXt with or without GCR module. The results, as depicted in Fig. 4, demonstrate that our method exhibits a high level of robustness, particularly in regards to color cast and blur distortions. It should be noted that all models tested in this section have a similar parameter size to that of the ConvNeXt Base-size model.

Table 2. Test result on Foggy-Driving dataset, all models are trained on Cityscapes dataset and directly tested without other training procedure.

Model	mIoU
ConvNeXt	48.54
ConvNeXt+GCR	52.83
Swin-Transformer	47.22
Swin-Transformer+GCR	49.98

Visualization. The visualization results presented in Fig. 6 demonstrate the efficacy of our proposed method. Specifically, we used the shallow layers of the neural network to generate CAM maps for visualization. Clearly, our model is capable of accurately attending to category-specific edge information in images at shallow levels, which provides an experimental explanation for the good performance of our model. This observation provides compelling evidence of the enhanced performance achieved by our method.

Fig. 6. Visualization examples of Class Activation Mapping (CAM) for Swin-Transformer with or without our GCR module at low stage. The tested activation class is 'dog', 'llama', 'frog', 'cat' respectively.

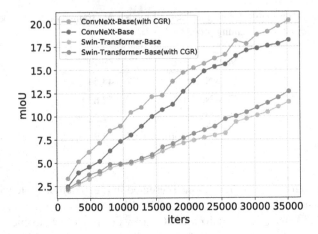

Fig. 7. The convergence speed of different models training on ADE20k directly without any pre-training, all the results were tested on the validation set. This figure shows the early 34000 iterations with the batchsize of 16.

Results on Convergence Speed. In contrast to CNNs and transformers, our GCR module employs a novel strategy for semantic segmentation tasks by selecting neighbors for aggregation based on similarity. This graph convolution module allows the model to aggregate global information at shallow layers, resulting in faster convergence as observed through experimental results. The utilization of GCR module is believed to contribute to the improved performance of the segmentation models.

In order to accentuate the distinction, we refrain from utilizing pre-trained parameters and instead opt for direct training on the ADE20k dataset. Empirical evidence demonstrates that our proposed GCR method exhibits a faster convergence rate during the initial training iterations when compared to other models. The experimental results are shown in Fig. 7.

4.4 Ablation Study

The Necessity of GCR Module. To demonstrate the effectiveness of the GCR module, we tested the performance of model by replacing the GCR module with other CNNs and transformers with corresponding amount of parameters. In Table 3 we show evaluate results of models with parameter size around ConvNeXt-Base.

Influence of Different GNN Types. In this experiment, we aimed to demonstrate the robustness of our model by evaluating its performance when utilizing different types of graph aggregation operator. Specifically, we modified the GNN structure in our encoder while maintaining the same pre-training techniques and training procedures on ImageNet1k and ADE20k datasets while using ConvNeXt-B as our baseline. Our results, as presented in Table 4, indicate that the graph convolution we applied yielded a better performance and other graph aggregators has also achieved competitive capacity.

Table 3. mIoU on ADE20k of different network structure concatenated on ConvNeXt. '+C', '+S', '+N' represents concatenating GCR, Swin-Transformer, ConvNeXt to network, respectively.

Model	mIoU	Model	mIoU
ConvNeXt	49.25	Swin-Transformer	48.75
ConvNeXt+C	49.63	Swin-Transformer+C	49.17
ConvNeXt+S	49.01	Swin-Transformer+S	48.42
ConvNext+N	48.84	Swin-Transformer+N	48.67

This experiment particularly demonstrated the structural rationality and potential of our method, as it achieved good performance when using various graph operators.

Table 4. ADE20k segmentation results using different type of GNN operators. The FLOPS is calculated with a image scale of $3 \times 512 \times 512$.

GNN type	Parameters(M)	GFLOPS	mIoU
Ours	121.9	306.7	49.63
Max-Relative [20]	121.58	301.86	49.53
GIN [36]	121.24	301.57	48.81
Edge Conv [34]	121.9	306.7	49.56

Influence of the Shortcut Coefficient. The utilization of the shortcut coefficient in our GCR module is crucial due to the potential risk of over-smoothing within a GNN with a deep structure. To evaluate the effect of the hyperparameter α in Eq. 7 on model performance, we conducted experiments using GCR-ConvNeXt-B and ADE20k dataset while pretraining on ImageNet1k. The results demonstrate that the model performs optimally when α is approximately 0.5, as illustrated in Fig. 8.

Influence of the Downsample Layer Type. In order to generate features with a pyramid structure in GCR module, a downsampling layer is required as GNNs do not possess the same convenient downsampling method as CNNs do through pooling. By default, a single convolution layer is utilized as the downsampling layer. In order to demonstrate that the type of downsample layer is inconsequential for the GNN model, an experiment was conducted utilizing various implementations of the downsample layer on the GCR-ConvNeXt-B model. The results of these different downsample methods were then compared on the ADE20k dataset (as shown in Table 5). The results of this experiment indicate that the use of different types of downsample layers has minimal impact on the performance of GNN.

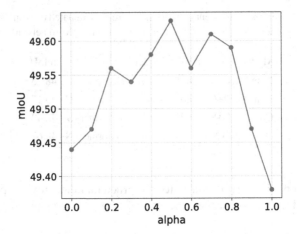

Fig. 8. The influence of α on mIoU.

Table 5. ADE20k Segmentation results using different types of downsample layer of GNN.

Downsample	Conv layer	FC layer
mIoU	49.63	49.51

5   Conclusion

This paper presents a novel module, GCR to address the semantic segmentation problem with a semantic-based receptive field. We put up the idea of such receptive field with a semantic foundation that adhere to the fundamental tenets of semantic segmentation, and implement it with graph representation learning. Furthermore, we adopt a paralleled CNN or ViT encoder to ensure the beneficial inductive bias is also accessible for our model. Experimental results verify the effectiveness of our proposed model, along with other advantages. Future work will focus on integrating the research intuition of graph representation learning more deeply into various tasks of image segmentation. Specifically, for open vocabulary segmentation and scene understanding.

Acknowledgements. This work was supported by the CAS Project for Young Scientists in Basic Research, Grant No. YSBR-040.

References

1. Chen, L.C., Papandreou, G., Kokkinos, I., Murphy, K.P., Yuille, A.L.: DeepLab: semantic image segmentation with deep convolutional nets, atrous convolution, and fully connected CRFs. IEEE Trans. Pattern Anal. Mach. Intell. **40**, 834–848 (2016)
2. Coley, C.W., et al.: A graph-convolutional neural network model for the prediction of chemical reactivity. Chem. Sci. **10**(2), 370–377 (2019)
3. MMS Contributors: MMSegmentation: OpenMMLab semantic segmentation toolbox and benchmark (2020). https://github.com/open-mmlab/mmsegmentation

4. Cordts, M., et al.: The cityscapes dataset for semantic urban scene understanding. In: Proceedings of the IEEE Conference on Computer Vision and Pattern Recognition (CVPR) (2016)

5. Cubuk, E.D., Zoph, B., Shlens, J., Le, Q.V.: RandAugment: practical automated data augmentation with a reduced search space. In: 2020 IEEE/CVF Conference on Computer Vision and Pattern Recognition, CVPR Workshops 2020, Seattle, WA, USA, 14–19 June 2020, pp. 3008–3017. Computer Vision Foundation/IEEE (2020). https://doi.org/10.1109/CVPRW50498.2020.00359. https://openaccess.thecvf.com/content_CVPRW_2020/html/w40/Cubuk_Randaugment_Practical_Automated_Data_Augmentation_With_a_Reduced_Search_Space_CVPRW_2020_paper.html

6. Deng, J., Dong, W., Socher, R., Li, L.J., Li, K., Fei-Fei, L.: ImageNet: a large-scale hierarchical image database. In: CVPR 2009 (2009)

7. Ding, H., Jiang, X., Liu, A.Q., Magnenat-Thalmann, N., Wang, G.: Boundary-aware feature propagation for scene segmentation. In: 2019 IEEE/CVF International Conference on Computer Vision, ICCV 2019, Seoul, South Korea, 27 October–2 November 2019, pp. 6818–6828. IEEE (2019). https://doi.org/10.1109/ICCV.2019.00692

8. Dosovitskiy, A., et al.: An image is worth 16×16 words: transformers for image recognition at scale. In: 9th International Conference on Learning Representations, ICLR 2021, Virtual Event, Austria, 3–7 May 2021. OpenReview.net (2021). https://openreview.net/forum?id=YicbFdNTTy

9. Du, Y., Yuan, C., Li, B., Zhao, L., Li, Y., Hu, W.: Interaction-aware spatio-temporal pyramid attention networks for action classification. CoRR abs/1808.01106 (2018). http://arxiv.org/abs/1808.01106

10. Gilmer, J., Schoenholz, S.S., Riley, P.F., Vinyals, O., Dahl, G.E.: Neural message passing for quantum chemistry. In: International Conference on Machine Learning, pp. 1263–1272. PMLR (2017)

11. Han, K., Wang, Y., Guo, J., Tang, Y., Wu, E.: Vision GNN: an image is worth graph of nodes. CoRR abs/2206.00272 (2022). https://doi.org/10.48550/arXiv.2206.00272

12. He, K., Zhang, X., Ren, S., Sun, J.: Deep residual learning for image recognition. In: 2016 IEEE Conference on Computer Vision and Pattern Recognition, CVPR 2016, Las Vegas, NV, USA, 27–30 June 2016, pp. 770–778. IEEE Computer Society (2016). https://doi.org/10.1109/CVPR.2016.90

13. Hoffer, E., Ben-Nun, T., Hubara, I., Giladi, N., Hoefler, T., Soudry, D.: Augment your batch: improving generalization through instance repetition. In: 2020 IEEE/CVF Conference on Computer Vision and Pattern Recognition, CVPR 2020, Seattle, WA, USA, 13–19 June 2020, pp. 8126–8135. Computer Vision Foundation/IEEE (2020). https://doi.org/10.1109/CVPR42600.2020.00815. https://openaccess.thecvf.com/content_CVPR_2020/html/Hoffer_Augment_Your_Batch_Improving_Generalization_Through_Instance_Repetition_CVPR_2020_paper.html

14. Huang, G., Sun, Yu., Liu, Z., Sedra, D., Weinberger, K.Q.: Deep networks with stochastic depth. In: Leibe, B., Matas, J., Sebe, N., Welling, M. (eds.) ECCV 2016, Part IV. LNCS, vol. 9908, pp. 646–661. Springer, Cham (2016). https://doi.org/10.1007/978-3-319-46493-0_39

15. Jain, A., Zamir, A.R., Savarese, S., Saxena, A.: Structural-RNN: deep learning on spatio-temporal graphs. In: 2016 IEEE Conference on Computer Vision and Pattern Recognition, CVPR 2016, Las Vegas, NV, USA, 27–30 June 2016, pp. 5308–5317. IEEE Computer Society (2016). https://doi.org/10.1109/CVPR.2016.573

16. Jin, Y., Li, J., Lian, Z., Jiao, C., Hu, X.: Supporting medical relation extraction via causality-pruned semantic dependency forest. In: Proceedings of the 29th International Conference on Computational Linguistics, COLING 2022, Gyeongju, Republic of Korea, 12–17 October 2022, pp. 2450–2460. International Committee on Computational Linguistics (2022)

17. Kipf, T.N., Welling, M.: Semi-supervised classification with graph convolutional networks. In: 5th International Conference on Learning Representations, ICLR 2017, Conference Track Proceedings, Toulon, France, 24–26 April 2017. OpenReview.net (2017). https://openreview.net/forum?id=SJU4ayYgl

18. Landrieu, L., Simonovsky, M.: Large-scale point cloud semantic segmentation with super-point graphs. In: 2018 IEEE Conference on Computer Vision and Pattern Recognition, CVPR 2018, Salt Lake City, UT, USA, 18–22 June 2018, pp. 4558–4567. Computer Vision Foundation/IEEE Computer Society (2018). https://doi.org/10.1109/CVPR.2018.00479. http://openaccess.thecvf.com/content_cvpr_2018/html/Landrieu_Large-Scale_Point_Cloud_CVPR_2018_paper.html

19. LeCun, Y., Bottou, L., Bengio, Y., Haffner, P.: Gradient-based learning applied to document recognition. Proc. IEEE **86**(11), 2278–2324 (1998)

20. Li, G., Müller, M., Thabet, A.K., Ghanem, B.: DeepGCNs: can GCNs go as deep as CNNs? In: 2019 IEEE/CVF International Conference on Computer Vision, ICCV 2019, Seoul, South Korea, 27 October–2 November 2019, pp. 9266–9275. IEEE (2019). https://doi.org/10.1109/ICCV.2019.00936

21. Liu, Z., et al.: Swin transformer: hierarchical vision transformer using shifted windows. CoRR abs/2103.14030 (2021). https://arxiv.org/abs/2103.14030

22. Liu, Z., Mao, H., Wu, C., Feichtenhofer, C., Darrell, T., Xie, S.: A ConvNet for the 2020s. In: IEEE/CVF Conference on Computer Vision and Pattern Recognition, CVPR 2022, New Orleans, LA, USA, 18–24 June 2022, pp. 11966–11976. IEEE (2022). https://doi.org/10.1109/CVPR52688.2022.01167

23. Long, J., Shelhamer, E., Darrell, T.: Fully convolutional networks for semantic segmentation. CoRR abs/1411.4038 (2014). http://arxiv.org/abs/1411.4038

24. Michaelis, C., et al.: Benchmarking robustness in object detection: autonomous driving when winter is coming. arXiv preprint arXiv:1907.07484 (2019)

25. Prado-Romero, M.A., Prenkaj, B., Stilo, G., Giannotti, F.: A survey on graph counterfactual explanations: definitions, methods, evaluation, and research challenges. ACM Comput. Surv. (2023). https://doi.org/10.1145/3618105

26. Qasim, S.R., Kieseler, J., Iiyama, Y., Pierini, M.: Learning representations of irregular particle-detector geometry with distance-weighted graph networks. Eur. Phys. J. C **79**(7), 1–11 (2019)

27. Qi, X., Liao, R., Jia, J., Fidler, S., Urtasun, R.: 3D graph neural networks for RGBD semantic segmentation. In: IEEE International Conference on Computer Vision, ICCV 2017, Venice, Italy, 22–29 October 2017, pp. 5209–5218. IEEE Computer Society (2017). https://doi.org/10.1109/ICCV.2017.556

28. Raghu, M., Unterthiner, T., Kornblith, S., Zhang, C., Dosovitskiy, A.: Do vision transformers see like convolutional neural networks? CoRR abs/2108.08810 (2021). https://arxiv.org/abs/2108.08810

29. Sakaridis, C., Dai, D., Van Gool, L.: Semantic foggy scene understanding with synthetic data. Int. J. Comput. Vis. **126**(9), 973–992 (2018). https://doi.org/10.1007/s11263-018-1072-8

30. Touvron, H., Cord, M., Douze, M., Massa, F., Sablayrolles, A., Jégou, H.: Training data-efficient image transformers & distillation through attention. CoRR abs/2012.12877 (2020). https://arxiv.org/abs/2012.12877

31. Touvron, H., Cord, M., Douze, M., Massa, F., Sablayrolles, A., Jegou, H.: Training data-efficient image transformers & distillation through attention. In: International Conference on Machine Learning, July 2021, vol. 139, pp. 10347–10357 (2021)

32. Vaswani, A., et al.: Attention is all you need. In: Advances in Neural Information Processing Systems 30: Annual Conference on Neural Information Processing Systems 2017, Long Beach, CA, USA, December 2017, pp. 4–9, pp. 5998–6008 (2017). https://proceedings.neurips.cc/paper/2017/hash/3f5ee243547dee91fbd053c1c4a845aa-Abstract.html

33. Veličković, P., Cucurull, G., Casanova, A., Romero, A., Lio, P., Bengio, Y.: Graph attention networks. arXiv preprint arXiv:1710.10903 (2017)

34. Wang, Y., Sun, Y., Liu, Z., Sarma, S.E., Bronstein, M.M., Solomon, J.M.: Dynamic graph CNN for learning on point clouds. ACM Trans. Graph. **38**(5), 146:1–146:12 (2019). https://doi.org/10.1145/3326362

35. Xiao, T., Liu, Y., Zhou, B., Jiang, Y., Sun, J.: Unified perceptual parsing for scene understanding. CoRR abs/1807.10221 (2018). http://arxiv.org/abs/1807.10221

36. Xu, K., Hu, W., Leskovec, J., Jegelka, S.: How powerful are graph neural networks? In: International Conference on Learning Representations (2018)

37. Ying, R., He, R., Chen, K., Eksombatchai, P., Hamilton, W.L., Leskovec, J.: Graph convolutional neural networks for web-scale recommender systems. In: Proceedings of the 24th ACM SIGKDD International Conference on Knowledge Discovery & Data Mining, pp. 974–983 (2018)

38. Yu, F., Koltun, V.: Multi-scale context aggregation by dilated convolutions. In: Bengio, Y., LeCun, Y. (eds.) 4th International Conference on Learning Representations, ICLR 2016, Conference Track Proceedings, San Juan, Puerto Rico, 2–4 May 2016 (2016). http://arxiv.org/abs/1511.07122

39. Yun, S., Han, D., Chun, S., Oh, S.J., Yoo, Y., Choe, J.: CutMix: regularization strategy to train strong classifiers with localizable features. In: 2019 IEEE/CVF International Conference on Computer Vision, ICCV 2019, Seoul, South Korea, 27 October–2 November 2019, pp. 6022–6031. IEEE (2019). https://doi.org/10.1109/ICCV.2019.00612

40. Zhang, H., Cissé, M., Dauphin, Y.N., Lopez-Paz, D.: mixup: beyond empirical risk minimization. In: 6th International Conference on Learning Representations, ICLR 2018, Conference Track Proceedings, Vancouver, BC, Canada, 30 April–3 May 2018. OpenReview.net (2018). https://openreview.net/forum?id=r1Ddp1-Rb

41. Zhao, H., Shi, J., Qi, X., Wang, X., Jia, J.: Pyramid scene parsing network. In: 2017 IEEE Conference on Computer Vision and Pattern Recognition, CVPR 2017, Honolulu, HI, USA, 21–26 July 2017, pp. 6230–6239. IEEE Computer Society (2017). https://doi.org/10.1109/CVPR.2017.660

42. Zhou, B., Zhao, H., Puig, X., Fidler, S., Barriuso, A., Torralba, A.: Scene parsing through ADE20k dataset. In: 2017 IEEE Conference on Computer Vision and Pattern Recognition, CVPR 2017, Honolulu, HI, USA, 21–26 July 2017, pp. 5122–5130. IEEE Computer Society (2017). https://doi.org/10.1109/CVPR.2017.544

Transductive Cross-Lingual Scene-Text Visual Question Answering

Lin Li[1]([✉]), Haohan Zhang[1], Zeqin Fang[1], Zhongwei Xie[2], and Jianquan Liu[3]

[1] Wuhan University of Technology, Wuhan, China
{cathylilin,zhaha,zeqinfang}@whut.edu.cn
[2] Huawei Technologies Co., Ltd., Shenzhen, China
xiezhongwei3@huawei.com
[3] NEC Corporation, Tsukuba, Japan
jqliu@nec.com

Abstract. Multilingual modeling has gained increasing attention in recent years, as the cross-lingual Text-based Visual Question Answering (TextVQA) are requried to understand questions and answers across different languages. Current researches mainly work on multimodal information assuming that multilingual pretrained models are effective to encode questions. However, the semantic comprehension of a text-based question varies between languages, creating challenges in directly deducing its answer from an image. To this end, we propose a novel multilingual text-based VQA framework suited for cross-language scenarios(CLVQA), transductively considering multiple answer generating interactions with questions. First, a question reading module densely connects encoding layers in a feedforward manner, which can adaptively work together with answering. Second, a multimodal OCR-based module decouples OCR features in an image into visual, linguistic, and holistic parts to facilitate the localization of a target-language answer. By incorporating enhancements from the above two input encoding modules, the proposed framework outputs its answer candidates mainly from the input image with a object detection module. Finally, a transductive answering module jointly understands input multimodal information and identified answer candidates at the multilingual level, autoregressively generating cross-lingual answers. Extensive experiments show that our framework outperforms state-of-the-art methods for both of cross-lingual (English<->Chinese) and mono-lingual (English<->English and Chinese<->Chinese) tasks in terms of accuracy based metrics. Moreover, significant improvements are achieved in zero-shot cross-lingual settings(French<->Chinese).

Keywords: Cross-Lingual Scenario · Multimodality · TextVQA

This work is partially supported by NSFC, China (No. 62276196).

B. Luo et al. (Eds.): ICONIP 2023, LNCS 14452, pp. 452–467, 2024.
https://doi.org/10.1007/978-981-99-8076-5_33

1   Introduction

It is widely recognized that Images with rich text information, such as product descriptions and advertising images. Texts in such images generally convey valuable information and thus are of critical importance in visual understanding tasks such as Visual Question Answering (VQA) [1].

It requires analyzing both natural language questions and the visual content of images, and answering text questions based on images, as a comprehensive problem involving natural language processing and computer vision. Existing VQA methods [2–6] tend to capture the relationships between visual concepts directly through sophisticated visual attention mechanisms. These methods pay few attention on reading the text in images and suffer performance degradation when answering text-based visual questions (TextVQA) [7]. Compared with ordinary VQA, TextVQA [7–10] is more practical because of its ability to help visually impaired users better identify information about the surrounding physical world, such as time, date, temperature, brand name, etc.

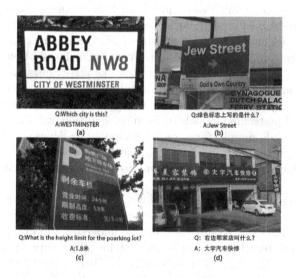

Fig. 1. Our work includes both mono-lingual and cross-lingual tasks. The former represents question-answer pairs in a same language (Chinese and English in our case, sample a and d). The latter represents that questions and answers are not given in the same language (sample b and c). Examples are from the EST-VQA dataset.

With the internationalization of human activities, multi-language TextVQA can well meet people's needs in various scenarios, for example, Be My Eyes APP. As shown in Fig. 1, there are mono-lingual and cross-lingual cases. (1) When questions are given in a specific language, the TextVQA model can provide answers in the same language as the given question. Examples *a* and *d* show that we can ask questions based on images in English or Chinese, and the Text VQA model should give predicted answers in the same language, which represents mono-lingual TextVQA. In this cases, question-answer pairs can cover multiple languages. (2) On the other hand, the asked question and the answer given are not in the same language as example b, the question is in Chinese, "绿色标志上写的是什么？" (What does the green sign say?). Since the answers are given in English, it belongs to cross-lingual TextVQA.

Based on the languages used in questions and answers, recent multi-lingual TextVQA methods are classified into three types. (1) **Monolingual task with one fixed lanaguage** includes: LoRRA [7], RuArt [11], MM-GNN [12], CRN [13], SMA [9], M4C [10], LaAPNet [14], LaTr [15], KTVQA [16], and TAP [17], etc. In terms of text encoding, pre-trained models such as Bert [18], Faster R-CNN [19], and T5 [20] are commonly used. (2) **Monolingual task covering several languages**: [21] is the first attempt to Multilingual TextVQA, which translates the questions in ST-VQA dataset to 3 languages(Catalan, Spanish, and Chinese). The latest attempt multilingual task is MUST-VQA [22]. MUST-VQA translates all the questions in ST-VQA and Text-VQA datasets to 5 languages with 3 scripts; namely Spanish, Catalan, Chinese, Italian and Greek. Compared with monolingual tasks, multilingual tasks usually use state-of-the-art multilingual pre-trained models, such as mBert [23], XLM-R [24], mT5 [25], etc. Current researches mainly work on multimodal information assuming that those multilingual pretrained models are effective to encode questions. (3) **Cross-lingual task**: current multilingual TextVQA like [22] translates questions into six languages, and the target language is only English. Therefore, it is not the true many-to-many cross-lingual evaluation. In summary, current researches utilize multilingual pretrained models to learn the question embeddings of different languages and while mainly focus on multimodalities. However, the process of semantically understanding a text-based question varies depending on the language used, which create challenges when attempting to directly infer its answer from an image.

In cross-language TextVQA, Questions are stated in the source language, with answers in the target language expected to come from the accompanying image. In some case, the question can be understood in one go while others require multiple interactions with the question for fully comprehend it. As shown in Fig. 1(c), TextVQA directly understands the "height limit" and find the corresponding Chinese "限制高度" (height limit). On the other hand, in Fig. 1(b), TextVQA initial focus is "写的什么" (what is written), and after TextVQA understands the question again, it will find that "绿色标志上" (on the green sign) is equally important. Thus, aligning their semantic information and interacting with the question multiple times becomes crucial to obtain accurate answers.

To address this end, we propose a cross-lingual Scene-TextVQA framework called CLVQA with Question Reading, Multimodal OCR, Object Detection and Transductive Answering. The Question Reading module pays attention to the rich language information at different linguistic levels given by a text question, which can adaptively work together with answering. The Multimodal OCR module firstly decouples the OCR modal features in the corresponding image and divides them into three parts: visual, textual, and holistic features. Moreover, redundant and irrelevant features are softly filtered through the cross-modal attention module. The Object Detection module is used to extract visual object features. Then, the three modality features are trained by Transformer to learn a joint semantic embedding. The Transductive Answering module utilizes a mixture of answer vocabulary and OCR tokens to obtain answer candidates, which is transductive conjunction with input multimodal features during iteratively autoregressive decoding to generate a refined answer. This approach helps to enhance the mutual understanding of questions and answers at the cross-lingual level, resulting in effective decoded answers.

Our contributions are summarized as follows: 1) Our proposed CLVQA framework can handle cross-language scenarios, consisting of Question Reading, Multimodal OCR, Object Detection and Transductive Answering. 2) The Transductive Answeringmodule is capable of simultaneously learning multimodal and multi-lingual information while jointly understanding the question and its answer at the cross-lingual level. 3) To the best of our knowledge, it is the first empirical analysis conducted on datasets with different target languages, i.e., English<->Chinese and English<->French, and our proposed framework outperforms baselines in terms of Accuracy and ANLS, at both of supervised and zero-shot settings.

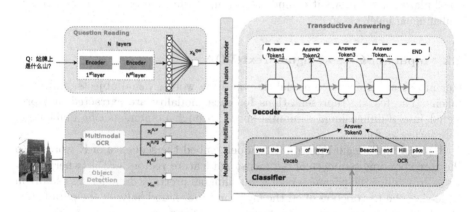

Fig. 2. Our Framework For Cross-Lingual Scene-Text VQA(CLVQA).

2   Our Framework

As shown in Fig. 2, our proposed framework is designed for cross-lingual text based visual question answering. In the remainder of this paper, all W are learned linear transformations, with different symbols to denote independent parameters, e.g., W_{fr}. LN is Layer Normalization [26]. o represents an element-wise product.

2.1   The Overview of Our Framework

In Fig. 2, given a textual question in a source language and an image where OCR is a target language, feature representations are extracted from three modalities, i.e., the question text reading (in the upper and left part of Fig. 2), the visual objects in the image(in the bottom and left part of Fig. 2), and the text tokens identified from the image OCR parts (in the middle and left part of Fig. 2). These three modalities are represented respectively as a list of question words embeddings, a list of visual object features from an off-the-shelf object detector, and a list of OCR token features based on an external OCR system.

Question Reading and Multimodal OCR are part of our concerns since they contain rich linguistic information. For question words, we dynamically weight the output of 12-layer transformer encoders and the weight of each layer is a learnable parameter. In this way, it is effective to obtain the question embedding containing different levels of semantic information for each question token. The identified OCR features with rich modality representations will be decoupled into three parts, linguistic, visual and holistic part. Our main concern is the transductive answering module, which predicts the answer through iterative decoding (in the right part of Fig. 2). During decoding, it feeds in the previous output to predict the next answer component in an autoregressive manner. At each step, it either locates an OCR token from the image or selects a word from its answer vocabulary, which allows it to understand input information and answers at the multilingual level.

2.2   Inputs of Cross-Lingual Text-VQA

Our inputs are from three modalities–question words, visual objects, and OCR tokens. The feature representations for each modality are extracted and projected into d-dimensional semantic space through domain-specific embedding approaches as follows:

Using a multilingual pretrained model, embed the sequence of K words into the corresponding sequence of d-dimensional feature vectors $\{x_k^{qw}\}$ (where $k = 1, \cdots, K$). Since the parameters of a multilingual pretrained model are usually in large-size, the parameters of all the layers are frozen and only let the parameters of the weight fusion layer be fine-tuned during training.

To extract a set of M visual objects from an image, a pretrained detector is used and called Faster R-CNN. Through it we can get visual object features $\{x_m^{fr}\}$ and location feature $\{x_m^b\}$. The final object embedding $\{x_m^{oi}\}$ is obtained by projecting the visual and location features into a d-dimensional space with two learned linear layers and then summing them up.

$$x_m^{oi} = \mathrm{LN}\left(W_1 x_m^{fr}\right) + \mathrm{LN}\left(W_2 x_m^b\right) \tag{1}$$

we categorize the features of N OCR tokens into two types: Visual and Linguistic. We also include an external OCR cross-modal pretrained embedding as an additional feature to aid in recognition. The final linguistic embedding

$x_i^{o,l}$ is obtained by projecting the pre-trained word embedding and character-level pyramidal histogram feature [27] with a linear layers. Off-the-shelf visual encoder can be used to extract object features and spatial features such as VIT and VGG-16, etc. Merge them into the final features $x_i^{o,v}$ using a linear layer. In our approach, we introduce the Recog-CNN [28] feature to enhance cross-modal features. $\mathbf{x}_i^{o,rg}$ as a kind of holistic feature to enhance the text and visual representation learning.

Finally, the Recog-CNN feature, OCR visual and linguistic part features will enter the attention based multimodal multilingual fusion encoder, and please refer to [29] for details.

2.3   Question Reading

Question Reading in Source Language. Multilingual pre-trained models can encode rich linguistic level information, the surface information features are in the bottom network, the syntactic information features are in the middle layer network, and the semantic information features are in the high-level network. Given a question, different languages have their linguistics usage. Some may like the lower-level information and others may consider the higher-level semantics. Thus our framework adopts the multi-layer connection to capture different levels of semantic information [30,31] by using different hierarchical features (for example, 12 layers in our case).

As shown in Fig. 3 and Eq. 2, in this way, the hierarchical information of the pre-trained model can be better utilized, and the information beneficial to the current text can be mined. unit $= 20$ represents a question that contains up to 20 tokens. ∂_i represents the output of a certain encoder layer. W_k^h is the weight score of ∂_i, k represents the kth token in the question, and the h is the current number of layers of the encoder. And the objective function of our framework is based on cross-entropy defined in Eq. 3.

$$x_k^{ques} = Dense_{unit=20} \left(\sum_{i=1}^{n} \partial_i \cdot W_k^h \right) \tag{2}$$

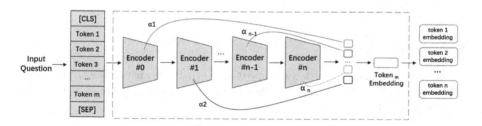

Fig. 3. Question Reading Network

The above output the d-dimensional embedding of question words, visual objects, and OCR tokens from each modality. A multi-layer Transformer encoder is applied to fuse the three modalities, and the L stacked decoder outputs the final embedding of $K + M + N$ entities. With the multi-head self-attention mechanism in the Transformer, each entity can jointly attend to all other entities freely, regardless of their modality. This enables us to model both inter-modal and intra-modal relationships uniformly with the same multi-layers parameters. The output of our multi-modal decoder is a list of d-dimensional feature vectors for the entities in each modality.

2.4 Transductive Answering

During prediction, the argmax is done on the concatenation $y_t^{all} = [y_t^{voc}; y_t^{ocr}]$ of fixed answer vocabulary scores and dynamic OCR-copying scores, selecting the top scoring element (either a vocabulary word or an OCR token) from all $V + N$ candidates. In our iterative auto-regressive decoding procedure, if the prediction at decoding time-step t is an OCR token, its OCR representation is feed to x_n^{ocr} as the Transformer input x_{t+1}^{dec} to the next prediction step $t + 1$. Otherwise, if the previous prediction is a word from the fixed answer vocabulary, its corresponding weight vector w_i^{voc} is the next step's input x_{t+1}^{dec}. In addition, two extra d-dimensional vectors are added to our inputs. One is a positional embedding vector corresponding to step t, and the other is a type embedding vector corresponding to whether the previous prediction is a fixed vocabulary word or an OCR token.

Similar to machine translation, our answer vocabulary is augmented with two special tokens, $< begin >$ and $< end >$. Here $< begin >$ is used as the input to the first decoding step, and the decoding process is stopped after $< end >$ is predicted. To ensure causality in answer decoding, the attention weights are masked in the self-attention layers of the Transformer architecture such that question words, detected objects and OCR tokens cannot attend to any decoding steps. All decoding steps can only attend to previous decoding steps in addition to question words, detected objects and OCR tokens. In addition, a random probability is adopted when it is greater than 0.5, the input of the next step is kept unchanged; when the probability is less than 0.5, the token representation of the corresponding position of ground truth is used as the input of the next step instead.

2.5 Training Loss

Given that an answer word can appear in both fixed answer vocabulary and OCR tokens, multi-label sigmoid loss (instead of softmax loss) is applied, as defined in Eq. 3.

$$\mathcal{L}_{bce} = -y_{gt} \log(\, sig(y_{pred})) - (1 - y_{gt}) \log(1 - sig(y_{pred})) \tag{3}$$

where y_{pred} is prediction and y_{gt} is ground-truth target.

3 Experiments

3.1 EST-VQA Dataset and Baselines

Dataset. To the best of our knowledge, the EST-VQA proposed in 2020 is the first bilingual scene text VQA dataset [32] covering English<->Chinese where target languages are different. Further discussions on other related datasets are in Sec4.2 where most ones are with one fixed target language. As shown in Table 1, the training set of this dataset contains 17102 pictures, each picture corresponds to a question, of which 9515 pictures are asked in Chinese and 7532 are asked in English. The test set contains 4000 pictures, of which 2530 are asked in Chinese.

Table 1. Statistics of EST-VQA dataset.

Set	English		Chinese		All	
	#Image	#Question	#Image	#Question	#Image	#Question
Train	6,500	6,500	8,000	8,000	14,500	14,500
Val	500	500	750	750	1,250	1,250
Test	587	587	765	765	1,352	1,352

Zero-Shot. To verify the language adaption of our trained framework, we construct an additional set of Chinese<->French question answering pairs to the test set for the zero-shot setting of Cross-language Text-VQA. Following [22] and [33], Baidu-translation-API[1] is used to translate English questions and answers in EST-VQA into French questions and answers.

Evaluation Methodology. Since the original test set of EST-VQA is not public now, the original training set is randomly divided into our training set, validation set and test set in a ratio of 8:1:1, following EST-VQA's official splitting. The EST-VQA dataset contains multilingual question-answer pairs, with human-written questions asking to reason about the text in the image. Each question in the dataset has 10 human annotated answers. Following previous works [7,9,10,22,29], the main metrics in our experiment is Accuracy that counts the proportion of successful matching with the real answers. Final accuracy is measured by averaging the 10 answers. At the same time, edit-distanced based metric is also considered, i.e., ANLS, which is more relaxed than Accuracy.

Baselines. LORRA [7], SMA [9] and M4C [10] are recent works and those baselines deal well with monolingual tasks. Their question encoders are replaced by the multilingual mBERT [23] while reproducing their overall frameworks. The latest multi-lingual works include MUST-VQA [22] and AttText-VQA [29]. The former includes the testing of multiple Text-VQA models, among which the best is LaTr [25] that replaces the T5 language model with the mT5 language model

---

[1] https://api.fanyi.baidu.com.

to better adapt to multilingual tasks. The latter language model uses mBERT [23]. For fair comparisons, AttText-VQA [29] is reproduced as our baseline.

3.2 Implementation Details

Our framework is implemented based on PyTorch. For the visual modality, the settings of M4C are used where Faster R-CNN [19] fc7 features of 100 top-scoring objects in the image pretrained on the Visual Genome dataset. The fc7 layer weights are fine-tuned during training. For the text modality, we use two iFLYTEK OCR systems[2] to recognize scene text. A trainable 12-layer mBERT or XLM-R is used [24] for text representation. Pyramidal histogram of characters (PHOC) representations [34] are applied on OCR text. VIT [35] extracts visual features in OCR and Recog-CNN [28] extracts holistic features in OCR. Adam is used as the optimizer, and its learning rate is set to 1e-4, and the epsilon value is 1e-8. For the training strategy, L2 regularization, warmup learning, gradient clipping, early stopping, learning rate decay, dropout, etc. are used in our experiment. $L = 4$ layers of multimodal transformer with 12 attention heads are our encoders. As for the decoder, a maximum decoding stride of $T = 12$ is adopted since it is sufficient to cover almost all answers unless otherwise stated.

3.3 Overall Performance

The overall experimental results are reported in Table 2, * represents that modality feature processing is followed with self-attention on each of the three modalities of question, detected object, and OCR. In addition, for fair comparisons, those baselines and ours use XLM-R for multi-lingual text encoding.

Table 2. Supervised Setting covering Chinese<->English.

#	Method	Acc on val	Acc on test	ANLS on val	ANLS on test
1	LORRA(*)	0.3842	0.3906	0.4792	0.4876
2	SMA(*)	0.4092	0.4168	0.4961	0.5034
3	M4C(*)	0.4048	0.4122	0.5014	0.4985
4	AttText-VQA	0.4326	0.4332	0.5368	0.5294
5	**Ours**	**0.4614**	**0.4628**	**0.5708**	**0.5652**

Experimental results show that our proposed framework with question reading, multimodal OCR, oject detection and transductive answering significantly outperforms prior works. Specifically, our framework achieved the highest performance on the test set with an accuracy of 0.4628 and an ANLS value of 0.5652.

---

[2] https://www.xfyun.cn/services/common-ocr.

The first three experiments in Table 2 show that it is not effective if just replacing the mono-lingual pretrained embedding of baselines with the multilingual one. On the test set, our answering accuracy is improved by nearly by 12.27% compared to M4C [10], released as the star codes of TextVQA Challenge 2020. M4C fuses different modalities homogeneously by embedding them into a common semantic space where self-attention is applied to model inter- and intra- modality context. In our reproduction, all entities of the three modalities are fused based on the transformer, and the question encoder is replaced with XLM-R.

Our framework is improved by 6.9% compared to the recent multi-lingual Text-VQA work of AttText-VQA [29]. AttText-VQA works with Transformer by using a set of cross-modal attentions. Under ANLS metric, our framework is improved by 15.91%, 12.27%, 13.38%, and 6.76%, respectively, further highlighting the superiority of our framework.

To verify the language adaption of our trained framework, we construct an additional set of Chinese<->French question answering pairs to the test set for the zero-shot setting. The specific results are shown in Table 3.

Table 3. Zero-shot Setting for Chinese<->French.

#	Method	Acc on test (Fr)	ANLS on test (Fr)
1	LORRA(*)	0.3523	0.4398
2	SMA(*)	0.3877	0.4762
3	M4C(*)	0.3849	0.4826
4	AttText-VQA	0.4216	0.5134
5	**Ours**	**0.4504**	**0.5428**

In this scenario, the accuracy performance of ours has been improved by 23.41%, 13.98%, 12.47%, and 5.72% and the ANLS performance has been improved by 23.41%, 13.98%, 12.47%, and 5.72% respectively compared with the four baselines. At the same time, compared with Table 2, our proposed model has only a small performance loss, while the results of those baselines have been dropped a lot. This shows that our proposed framework is more robust to Text-VQA problems in cross-lingual situations.

3.4 Multi-lingual and Multi-modal Embeddings

This part explains and analyzes the influence of different multilingual models and different visual feature extractors on our proposed framework. The experimental results are reported in Table 4 and Table 5.

(1) Experimentingon two mainstream multilingual pre-training models mBERT and XLM-R, we found that the XLM-R model is more applicable to this task and can better extract information from texts in different languages. As

listed in Table 4, XLM-R reaches the Accuracy 0.5444, i.e., 4.4% improvement over mBERT on English->Chinese. Ours w/o means that the different levels of semantics of a question is not jointly worked with answer generating. From Table 4, its Accuracy is decreased from 0.5440 to 0.4482. This observation underscores the ability of our framework to perform well in cross-lingual textual question answering.

(2) Advanced visual encoding models are effective to extract features. As shown in Table 5, when our framework uses Faster R-CNN as the target detector and extracts visual features and uses the VIT model to extract OCR marked visual features, it can be found that the Accuracy and ANLS indicators have been further improved on both of dev and test sets. Therefore, paying attention to new technical approaches and putting them into practice is an effective strategy.

Table 4. Muli-lingual Pretrained Models

Setting	Cross-lingual		Multilingual		All data
Type	En ->CH	Ch->EN	CH ->CH	En ->En	
Ours w/o (mBERT)	0.4208	0.4426	0.4149	0.4192	0.4282
Ours w/o (XLM-R)	0.4482	0.4680	0.4267	0.4236	0.4306
Ours (mBERT)	0.5212	**0.5328**	0.4354	0.4492	0.4572
Ours (XLM-R)	**0.5440**	0.5262	**0.4396**	**0.4547**	**0.4628**

Table 5. Visual Feature Extractors

Method	visual features	OCR visual features	acc/%		ANLS	
			dev	test	dev	test
Ours	Faster R-CNN	VGG16	46.14	46.28	0.5708	0.5652
Ours	Faster R-CNN	VIT	**46.28**	**46.75**	**0.5783**	**0.5726**
Ours	VIT	VGG16	45.64	46.04	0.5592	0.5635
Ours	VIT	VIT	45.92	45.36	0.5646	0.5598

4   Related Work

4.1   Multilingual Language Models

This paper works on cross-lingual information processing in the field of Text VQA. Multi-lingual language models are basic to handle different languages. Common multi-lingual pre-training models include mBert [23]], LASER [36], MultiFiT [37], XLM-R [24], etc. In current cross-language tasks, mBert and XLM-R are the mainstream methods with better effects. XLM-R is a scaled cross-language sentence encoder derived from 2.5T corpus data in hundreds of languages. mBert (Multilingual BERT), pre-trained on the large Wikipedia, is a multilingual extension of BERT that provides word and sentence representations for 104 languages. Both of mBert and XLM-R have been shown to cluster polysemy into the different regions of embedding space.

4.2   Text-VQA Datasets

Although the importance of involving scene texts in visual question answering tasks is originally emphasized by [38], due to the lack of available large-scale datasets, early development of question answering tasks related to embedded text understanding is limited in narrow domains such as bar-charts or diagrams [39, 40]. TextVQA [7] is the first large-scale open-domain dataset of the Text-VQA task, followed by TVQA [8], OCR-VQA [21, 41] is the first attempt to do Multilingual Text VQA, but the question and answer of the above data set is only in English, which cannot be applied to this task. Moreover, MUST-VQA [22] uses the Google translation api to translate the questions of ST-VQA and TestVQA into 6 languages as a multilingual scene. Similar to MUST-VQA, xGQA [33] (not focusing on Text VQA) claims to be a cross-language task, but the target language is fixed as English. Therefore, for our cross-lingual task, none of the above datasets really covered cross-language until the EST-VQA [32] appeared. It is the first public dataset for bilingual scene Text-VQA with English and Chinese text-VQA pairs.

4.3   Text-VQA Methods

Recent studies [7, 9, 10, 17, 42–44] have proposed several models and network architectures for the Text VQA task. As previous approaches struggled to answer questions requiring reading text from images, as the first time, As the first work, LoRRA [7] based on Pythia [45] studies an OCR branch to encode text information in images. MMGAN [43] constructs three sub-graphs for visual, semantic and numeric modalities. M4C [10] fuses multimodal information from questions, visual objects and OCR tokens with Transformer, and iteratively decodes answers with a dynamic pointer network. Since remarkable performance has been achieved, M4C is released as the star codes of TextVQA Challenge 2020. Subsequently, a series of M4C-derived approaches have been reported. For example, SMA [9] embeds a Graph Convolution Network (GCN) before the multimodal feature fusion module to encode object-object, object-text and text-text relations. LaAP-Net [14] thinks the view that OCR tokens play a more important role than visual objects, so it only takes questions and OCR tokens into consideration when encoding and decoding multimodal information. On the other hand, SA-M4C [42] takes spatial relations more than semantic relations. It has handled 12 types of spatial relations with the multiple heads of Transformer, and meanwhile, intentionally suppressed the semantic relations of text and visual objects. Recent SSbaseline [46] has achieved state-of-the-arts by splitting OCR token features into separate visual and linguistic attention branches with a simple attention mechanism. TAP [17] proposes to pretrain the model on several auxiliary tasks such as masked language modeling and relative position prediction.

5   Conclusions and Future Work

In this paper, we propose a transductive cross-lingual scene-text VQA framework with Question Reading, Mutilmodal OCR, Object Detection and Transductive Answering, which aims to delve into the current state-of-the-art studies to cross-lingual vision and language learning. A series of analytical and empirical comparisons is conducted to show its effectiveness. In future, OCR system, advanced language inference module, and generator head will further improve the performance of cross-lingual Text-VQA.

In summary, the most recent monolingual TextVQA approaches are derived from M4C [10], and these approaches highlight the role of OCR blocks and spatial relations. Current studies pay less attentions at the semantic understanding of questions in cross-lingual scenarios by assuming multi-language pretrained models are good enough. Thanks to the release of EST-VQA [32], our idea can be evaluated to show its effectiveness and the importance of aggressively considering question reading and answer generating.

References

1. Kafle, K., Price, B., Cohen, S., Kanan, C.: Dvqa: understanding data visualizations via question answering. In: Proceedings of the IEEE Conference on Computer Vision and Pattern Recognition, pp. 5648–5656 (2018)
2. Anderson, P., et al.: Bottom-up and top-down attention for image captioning and visual question answering. In: Proceedings of the IEEE Conference on Computer Vision and Pattern Recognition, pp. 6077–6086 (2018)
3. Dong, X., Zhu, L., Zhang, D., Yang, Y., Wu, F.: Fast parameter adaptation for few-shot image captioning and visual question answering. In: Proceedings of the 26th ACM International Conference on Multimedia, pp. 54–62 (2018)
4. Hu, R., Rohrbach, A., Darrell, T., Saenko, K.: Language-conditioned graph networks for relational reasoning. In: Proceedings of the IEEE/CVF International Conference on Computer Vision, pp. 10294–10303 (2019)
5. Liu, F., Liu, J., Hong, R., Lu, H.: Erasing-based attention learning for visual question answering. In: Proceedings of the 27th ACM International Conference on Multimedia, pp. 1175–1183 (2019)
6. Peng, L., Yang, Y., Wang, Z., Wu, X., Huang, Z.: Cra-net: composed relation attention network for visual question answering. In: Proceedings of the 27th ACM International Conference on Multimedia, pp. 1202–1210 (2019)
7. Singh, A., et al.: Towards vqa models that can read. In: Proceedings of the IEEE/CVF Conference on Computer Vision and Pattern Recognition, pp. 8317–8326 (2019)
8. Biten, A.F., et al.: Scene text visual question answering. In: Proceedings of the IEEE/CVF International Conference on Computer Vision, pp. 4291–4301 (2019)

9. Gao, C., Zhu, Q., Wang, P., Li, H., Liu, Y., van den Hengel, A., Wu, Q.: Structured multimodal attentions for textvqa. CoRR abs/ arXiv: 2006.00753 (2020)
10. Hu, R., Singh, A., Darrell, T., Rohrbach, M.: Iterative answer prediction with pointer-augmented multimodal transformers for textvqa. In: Proceedings of the IEEE/CVF Conference on Computer Vision and Pattern Recognition, pp. 9992–10002 (2020)
11. Jin, Z., et al.: Ruart: a novel text-centered solution for text-based visual question answering. IEEE Trans. Multim. **25**, 1–12 (2023). https://doi.org/10.1109/TMM.2021.3120194
12. Gao, D., Li, K., Wang, R., Shan, S., Chen, X.: Multi-modal graph neural network for joint reasoning on vision and scene text. In: 2020 IEEE/CVF Conference on Computer Vision and Pattern Recognition, CVPR 2020, Seattle, WA, USA, 13–19 June 2020, pp. 12743–12753. Computer Vision Foundation/IEEE (2020). https://doi.org/10.1109/CVPR42600.2020.01276
13. Liu, F., Xu, G., Wu, Q., Du, Q., Jia, W., Tan, M.: Cascade reasoning network for text-based visual question answering. In: Chen, C.W., et al. (eds.) MM 2020: The 28th ACM International Conference on Multimedia, Virtual Event / Seattle, WA, USA, 12–16 October 2020, pp. 4060–4069. ACM (2020). https://doi.org/10.1145/3394171.3413924,https://doi.org/10.1145/3394171.3413924
14. Han, W., Huang, H., Han, T.: Finding the evidence: Localization-aware answer prediction for text visual question answering. arXiv preprint arXiv:2010.02582 (2020)
15. Biten, A.F., Litman, R., Xie, Y., Appalaraju, S., Manmatha, R.: Latr: layout-aware transformer for scene-text VQA. In: IEEE/CVF Conference on Computer Vision and Pattern Recognition, CVPR 2022, New Orleans, LA, USA, 18–24 June 2022, pp. 16527–16537. IEEE (2022). https://doi.org/10.1109/CVPR52688.2022.01605, https://doi.org/10.1109/CVPR52688.2022.01605
16. Dey, A.U., Valveny, E., Harit, G.: External knowledge augmented text visual question answering. CoRR abs/ arXiv: 2108.09717 (2021)
17. Yang, Z., et al.: Tap: text-aware pre-training for text-vqa and text-caption. In: Proceedings of the IEEE/CVF Conference on Computer Vision and Pattern Recognition, pp. 8751–8761 (2021)
18. Devlin, J., Chang, M., Lee, K., Toutanova, K.: BERT: pre-training of deep bidirectional transformers for language understanding. In: Burstein, J., Doran, C., Solorio, T. (eds.) Proceedings of the 2019 Conference of the North American Chapter of the Association for Computational Linguistics: Human Language Technologies, NAACL-HLT 2019, Minneapolis, MN, USA, 2–7 June 2019, Volume 1 (Long and Short Papers), pp. 4171–4186. Association for Computational Linguistics (2019). https://doi.org/10.18653/v1/n19-1423
19. Ren, S., He, K., Girshick, R.B., Sun, J.: Faster R-CNN: towards real-time object detection with region proposal networks. IEEE Trans. Pattern Anal. Mach. Intell. **39**(6), 1137–1149 (2017). https://doi.org/10.1109/TPAMI.2016.2577031
20. Raffel, C., et al.: Exploring the limits of transfer learning with a unified text-to-text transformer. J. Mach. Learn. Res. **21**, 140:1–140:67 (2020). http://jmlr.org/papers/v21/20-074.html
21. i Pujolràs, J.B., i Bigorda, L.G., Karatzas, D.: A multilingual approach to scene text visual question answering. In: Uchida, S., Smith, E.H.B., Eglin, V. (eds.) Document Analysis Systems - 15th IAPR International Workshop, DAS 2022, La Rochelle, France, 22–25 May 2022, Proceedings. LNCS, vol. 13237, pp. 65–79. Springer (2022). https://doi.org/10.1007/978-3-031-06555-2_5

22. Vivoli, E., Biten, A.F., Mafla, A., Karatzas, D., Gómez, L.: MUST-VQA: multilingual scene-text VQA. In: Karlinsky, L., Michaeli, T., Nishino, K. (eds.) Computer Vision - ECCV 2022 Workshops - Tel Aviv, Israel, 23–27 October 2022, Proceedings, Part IV. LNCS, vol. 13804, pp. 345–358. Springer (2022). https://doi.org/10.1007/978-3-031-25069-9_23

23. Pires, T., Schlinger, E., Garrette, D.: How multilingual is multilingual bert? arXiv preprint arXiv:1906.01502 (2019)

24. Conneau, A., et al.: Unsupervised cross-lingual representation learning at scale. arXiv preprint arXiv:1911.02116 (2019)

25. Xue, L., et al.: mt5: a massively multilingual pre-trained text-to-text transformer. In: Toutanova, K. (eds.) Proceedings of the 2021 Conference of the North American Chapter of the Association for Computational Linguistics: Human Language Technologies, NAACL-HLT 2021, Online, June 6–11, 2021, pp. 483–498. Association for Computational Linguistics (2021). https://doi.org/10.18653/v1/2021.naacl-main.41

26. Ba, J.L., Kiros, J.R., Hinton, G.E.: Layer normalization. arXiv preprint arXiv:1607.06450 (2016)

27. Ghosh, S.K., Valveny, E.: R-phoc: segmentation-free word spotting using cnn. In: 2017 14th IAPR International Conference on Document Analysis and Recognition (ICDAR), vol. 1, pp. 801–806. IEEE (2017)

28. Yang, L., Wang, P., Li, H., Li, Z., Zhang, Y.: A holistic representation guided attention network for scene text recognition. Neurocomputing **414**, 67–75 (2020)

29. Fang, Z., Li, L., Xie, Z., Yuan, J.: Cross-modal attention networks with modality disentanglement for scene-text VQA. In: IEEE International Conference on Multimedia and Expo, ICME 2022, Taipei, Taiwan, 18–22 July 2022, pp. 1–6. IEEE (2022). https://doi.org/10.1109/ICME52920.2022.9859666

30. Huang, G., Liu, Z., Van Der Maaten, L., Weinberger, K.Q.: Densely connected convolutional networks. In: Proceedings of the IEEE Conference on Computer Vision and Pattern Recognition, pp. 4700–4708 (2017)

31. Nai, P., Li, L., Tao, X.: A densely connected encoder stack approach for multi-type legal machine reading comprehension. In: Huang, Z., Beek, W., Wang, H., Zhou, R., Zhang, Y. (eds.) WISE 2020. LNCS, vol. 12343, pp. 167–181. Springer, Cham (2020). https://doi.org/10.1007/978-3-030-62008-0_12

32. Wang, X., et al.: On the general value of evidence, and bilingual scene-text visual question answering. In: Proceedings of the IEEE/CVF Conference on Computer Vision and Pattern Recognition, pp. 10126–10135 (2020)

33. Pfeiffer, J., etal.: xgqa: cross-lingual visual question answering. In: Muresan, S., Nakov, P., Villavicencio, A. (eds.) Findings of the Association for Computational Linguistics: ACL 2022, Dublin, Ireland, 22–27 May 2022, pp. 2497–2511. Association for Computational Linguistics (2022). https://doi.org/10.18653/v1/2022.findings-acl.196

34. Almazán, J., Gordo, A., Fornés, A., Valveny, E.: Word spotting and recognition with embedded attributes. IEEE Trans. Pattern Anal. Mach. Intell. **36**(12), 2552–2566 (2014)

35. Dosovitskiy, A., et al.: An image is worth 16x16 words: Transformers for image recognition at scale. In: 9th International Conference on Learning Representations, ICLR 2021, Virtual Event, Austria, 3–7 May 2021. OpenReview.net (2021). https://openreview.net/forum?id=YicbFdNTTy

36. Feng, F., Yang, Y., Cer, D., Arivazhagan, N., Wang, W.: Language-agnostic bert sentence embedding. arXiv preprint arXiv:2007.01852 (2020)

37. Eisenschlos, J.M., Ruder, S., Czapla, P., Kardas, M., Gugger, S., Howard, J.: Multifit: efficient multi-lingual language model fine-tuning. arXiv preprint arXiv:1909.04761 (2019)
38. Bigham, J.P., et al.: Vizwiz: nearly real-time answers to visual questions. In: Proceedings of the 23nd Annual ACM Symposium on User Interface Software and Technology, pp. 333–342 (2010)
39. Kembhavi, A., Salvato, M., Kolve, E., Seo, M., Hajishirzi, H., Farhadi, A.: A diagram is worth a dozen images. In: Leibe, B., Matas, J., Sebe, N., Welling, M. (eds.) ECCV 2016. LNCS, vol. 9908, pp. 235–251. Springer, Cham (2016). https://doi.org/10.1007/978-3-319-46493-0_15
40. Kembhavi, A., Seo, M., Schwenk, D., Choi, J., Farhadi, A., Hajishirzi, H.: Are you smarter than a sixth grader? textbook question answering for multimodal machine comprehension. In: Proceedings of the IEEE Conference on Computer Vision and Pattern Recognition, pp. 4999–5007 (2017)
41. Mishra, A., Shekhar, S., Singh, A.K., Chakraborty, A.: Ocr-vqa: visual question answering by reading text in images. In: 2019 International Conference on Document Analysis and Recognition (ICDAR), pp. 947–952. IEEE (2019)
42. Kant, Y., et al.: Spatially aware multimodal transformers for TextVQA. In: Vedaldi, A., Bischof, H., Brox, T., Frahm, J.-M. (eds.) ECCV 2020. LNCS, vol. 12354, pp. 715–732. Springer, Cham (2020). https://doi.org/10.1007/978-3-030-58545-7_41
43. Gao, D., Li, K., Wang, R., Shan, S., Chen, X.: Multi-modal graph neural network for joint reasoning on vision and scene text. In: Proceedings of the IEEE/CVF Conference on Computer Vision and Pattern Recognition, pp. 12746–12756 (2020)
44. Liu, F., Xu, G., Wu, Q., Du, Q., Jia, W., Tan, M.: Cascade reasoning network for text-based visual question answering. In: Proceedings of the 28th ACM International Conference on Multimedia, pp. 4060–4069 (2020)
45. Singh, A., et al.: Pythia-a platform for vision & language research. In: SysML Workshop, NeurIPS, vol. 2018 (2018)
46. Zhu, Q., Gao, C., Wang, P., Wu, Q.: Simple is not easy: a simple strong baseline for textvqa and textcaps. In: Thirty-Fifth AAAI Conference on Artificial Intelligence, AAAI 2021, Thirty-Third Conference on Innovative Applications of Artificial Intelligence, IAAI 2021, The Eleventh Symposium on Educational Advances in Artificial Intelligence, EAAI 2021, Virtual Event, 2–9 February 2021, pp. 3608–3615. AAAI Press (2021). https://ojs.aaai.org/index.php/AAAI/article/view/16476

Learning Representations for Sparse Crowd Answers

Jiyi Li[✉][iD]

University of Yamanashi, Kofu, Japan
jyli@yamanashi.ac.jp

Abstract. When collecting answers from crowds, if there are many instances, each worker can only provide the answers to a small subset of the instances, and the instance-worker answer matrix is thus sparse. The solutions for improving the quality of crowd answers such as answer aggregation are usually proposed in an unsupervised fashion. In this paper, for enhancing the quality of crowd answers used for inferring true answers, we propose a solution with a self-supervised fashion to effectively learn the potential information in the sparse crowd answers. We propose a method named CROWDLR which first learns rich instance and worker representations from the crowd answers based on two types of self-supervised signals. We create a multi-task model with a Siamese structure to learn two classification tasks for two self-supervised signals in one framework. We then utilize the learned representations to complete the answers to fill the missing answers, and can utilize the answer aggregation methods to the complete answers. The experimental results based on real datasets show that our approach can effectively learn the representations from crowd answers and improve the performance of answer aggregation especially when the crowd answers are sparse.

Keywords: Crowdsourcing · Answer Aggregation · Representation Learning

1 Introduction

Quality control is a crucial issue for the answers collected by crowdsourcing because the ability and diligence of the crowd workers are diverse. A typical solution is building redundancy in the collected answers, i.e., people ask multiple workers to assign the answers to an instance and then aggregate the answers of the instances to improve the quality of the crowd answers and discover the truths.

Because it is practical to assume that the ground truths are always not available before collecting the crowd answers, many answer aggregation approaches are proposed in an unsupervised fashion [2,4,16,25,29,30]. However, when there are many instances, each worker can only provide the answers to a small subset of the instances; many workers are required; the instance-worker answer matrix is usually sparse in such cases, which results in that the existing answer aggregation models may be not well-learned. In this paper, for enhancing the quality of crowd answers used for inferring true answers, we propose a solution in a self-supervised fashion to effectively learn the

B. Luo et al. (Eds.): ICONIP 2023, LNCS 14452, pp. 468–480, 2024.
https://doi.org/10.1007/978-981-99-8076-5_34

potential information in the sparse crowd answers. We then utilize the learned representations to fill the missing answers. The answers with enhanced quality are then used by label aggregation methods and improve the performance of label aggregation.

To learn rich instance and worker representations from the sparse crowd answers, we leverage two types of self-supervised signals in the crowd answers to capture the potential relations among instances and workers. We propose a multi-task self-supervised method based on a neural network to learn the representations, named CROWDLR (Learning Representations for CROWD answers). Self-supervised methods have recently been concentrated on for learning the feature representations from the data with unknown ground truths [3,8,9,26]. One popular type of self-supervised methods is the task-related methods which tend to create supervised pretext tasks from the raw data, e.g., the context prediction tasks in computer vision [6] and natural language processing [5], the rotation prediction tasks in computer vision [10].

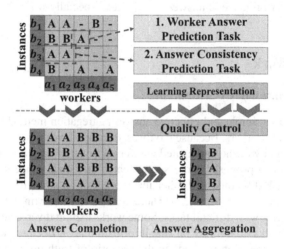

Fig. 1. Our solution for quality control in crowdsourcing by self-supervised representation learning on sparse crowd answers. There are two self-supervised tasks and two quality control tasks.

In our work, we design two self-supervised tasks. One intuitive task is predicting the answer of a worker to an instance. To better understand the relations among the workers, instances, and answers in the collection, another task is predicting whether two workers assign the same answers to an instance, which can detect the additional instance-wise relation information among the workers. Both two self-supervised tasks can learn different and partial information. We thus propose a multi-task model with a Siamese structure to learn these two tasks in one framework.

After pre-training the proposed self-supervised model and obtaining the representations, we can utilize them in the tasks related to quality control in crowdsourcing, i.e., answer aggregation. We first make an answer completion, which predicts the answers of all workers and instances, can fill the missing answers of the instances that a worker does not annotate. It can naturally be implemented by the module of worker answer prediction in the trained CROWDLR. After that, we can utilize various answer aggregation approaches such as [2,4,29] on the complete answers for answer aggregation. In this paper, we utilize the Majority Voting (MV) method as the backbone answer aggregation method, which is one of the most typical answer aggregation method always used by people who are not major in the crowdsourcing research and are working on diverse practical scenarios. Figure 1 summarizes the framework of our solution for quality control. The contributions of this paper can be addressed as follows.

- We propose a method named CROWDLR to learn worker and instance representations with two self-supervised prediction tasks on the crowd answers.
- We propose a multi-task architecture with Siamese structure, comparison-based learning for efficiently learning the rich information from the raw crowd answers.
- The trained model and learned representations can be utilized for quality control task of answer aggregation. The experimental results based on real datasets verify that our approach can effectively learn the representations from crowd answers and improve the performance of answer aggregation especially when the crowd answers are sparse.

2 Related Work

2.1 Quality Control in Crowdsourcing

Majority voting is a simple and effective answer aggregation method but only assigns equal weights to all workers. Because it only assigns equal weights to all crowd workers and labels, the quality of the aggregated answers is not stable because of diverse quality. Researchers also proposed more sophisticated statistical models which consider worker ability, task difficulty, and other uncertainties and strengthen the opinions of the majority [2,4,17,21,22,25,29,30]. There are also some existing works focusing on few-expert scenarios with difficult tasks. Some works exploit diverse auxiliary information [11]; some works do not rely on side information [16]. Zheng et al. [28] surveyed existing answer aggregation methods in the scenario of truth discovery. Furthermore, besides categorical labels, there are some works for other types of crowd labels such as pairwise similarity or preference comparison labels [1,15,18,27,31], triplet preference comparison labels [19,24] and text sequence [14,20]. Besides aggregating the crowd labels, there are some works that trains the classification models by using the noisy crowd labels directly, e.g., [12]. There are also some works on reducing the budget while the utility of the aggregated labels is preserved, e.g., [13]. In this paper, answer aggregation is a downstream task; our method incorporates the existing answer aggregation approaches by executing them on the complete worker answers predicted by the pre-trained CROWDLR model.

2.2 Self-supervised Representation Learning

Self-supervised methods have obtained many attractions in recent years to learn the rich feature representations from the data with unknown ground truths [3,8,9,26]. There are many types of self-supervised methods. One popular type is the task-related methods, which tend to create supervised auxiliary tasks based on the self-supervised signals from the raw data, e.g., the context prediction tasks in computer vision [6] and natural language processing [5], the rotation prediction tasks in computer vision [10]. In this paper, we focus on proposing a task-related self-supervised method for our scenario of crowd worker answers. We propose specific self-supervised tasks and the corresponding model.

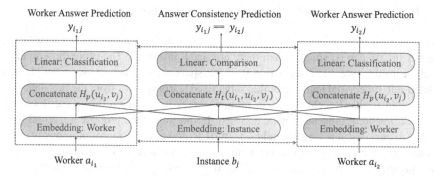

Fig. 2. The architecture of our self-supervised representation learning model CROWDLR. It has two modules for the worker answer prediction task and one module for the answer consistency prediction task in a multi-task architecture. It has a Siamese structure, i.e., the two modules of worker answer prediction share the parameters.

3   Problem Setting

We focus on the multi-choice-one-answer questions in crowdsourcing which is a typical form for collecting the answers from crowd workers. Workers are asked to select one answer from multiple candidates for an instance, e.g., a label of an image or an answer to a scientific question. We define a question assigned to a worker as an *instance*. For each instance, the potential answer set is $C = \{c_k\}_{k=1}^{K}$. We assume that K is the same for all questions that is a typical setting. We define the set of workers as $A = \{a_i\}_{i=1}^{N}$, the set of instances as $B = \{b_j\}_{j=1}^{M}$. We define y_{ij} as the answer given by worker a_i to instance b_j. We denote the set of all answers as $Y = \{y_{ij}\}_{i,j}$, the set of answers given to b_j as $Y_{*j} = \{y_{ij} | a_i \in A\}$ and the set of answers given by a_i as $Y_{i*} = \{y_{ij} | b_j \in B\}$. Because the number of instances can be large, each worker only needs to annotate a subset of them. The worker representations learned by the self-supervised model are defined as $U = \{u_i\}_{i=1}^{N}, u_i \in \mathbb{R}^d$; the instance representation is defined as $V = \{v_j\}_{j=1}^{M}, v_j \in \mathbb{R}^d$. For simplicity of the model hyperparameter, we use the same number of dimensions for worker and instance representations. The estimated true answers is defined as $\bar{Z} = \{\bar{z}_j\}_{j=1}^{M}$. The true answers are $\hat{Z} = \{\hat{z}_j\}_{j=1}^{M}$ which are unknown to the quality control methods for crowdsourcing. The cold-start problem is out of scope, we assume that each instance has been annotated by some workers, which is always guaranteed in crowdsourcing when collecting the crowd answers. In summary, the problem settings are as follows.

- **Inputs**: Worker set A, instance set B, and answer set Y.

- **Learning Representations**: Learn the worker representation U and instance representations V, and a model that predicts worker answers and whether two workers assign the same answers to an instance.
- **Answer Aggregation**: The outputs are the estimated true answers $\bar{\mathcal{Z}} = \{\bar{z}_j\}_j$.

4   Learning Representations for Crowd Answers

Because each worker can only annotate a small instance subset, the instance-worker answer matrix can be sparse when there are many instances. Therefore, how to effectively learn the potential information from the sparse crowd answers is essential. Task-related self-supervised learning builds supervised learning tasks from the raw data without ground truths. Recent works in multiple areas have shown that it can learn rich potential feature representations, e.g., [5,9]. In this paper, we propose a method based on self-supervised learning, namely CROWDLR, to learn the representations of workers and instances, and the relations among workers, instances, and answers. We propose two specific self-supervised tasks and the corresponding model based on these tasks.

4.1   Self-supervised Tasks

Because the raw crowd answers have no ground truths, we need to create supervised tasks and convert the data into a format with ground truth labels. We consider the following two self-supervised tasks. An intuitive task is predicting the answer of a worker to an instance. It can model the relation between instances and workers. It is similar to the cases based on tensor factorization (TF) [7,30]. To better understand the relations among the workers, instances, and answers in the collection, a question is what kinds of additional information we can observe from raw crowd answers. In Fig. 1, from the answer matrix by workers and instances, we can also find that whether two workers agree with each other on an instance is important information to describe the instance-wise relation information among the workers, especially when both of these two workers only annotate a small subset of the instances. It is possible to help the model understand the potential ability relations of two workers, i.e., if two workers agree with each other on many instances, they may have similar abilities; if two workers disagree with each other on many instances, they may have different abilities. Therefore, another self-supervised task leverages this instance-wise worker relation and predicts whether two workers provide the same answers to a given instance. We clarify these two self-supervised tasks: (1) Worker Answer Prediction (WAP) Task; (2) Answer Consistency Prediction (ACP) Task.

4.2   Neural Architecture of CROWDLR

Both of these two tasks can learn different types of information. They partially represent the relations among the workers, instances, and answers. We thus learn them in one neural-based model with a multi-task architecture. Figure 2 shows the architecture of the proposed model. It consists of two modules for the worker answer prediction task and one module for the answer consistency prediction task. It has a Siamese structure, i.e., the two modules of worker answer prediction for two workers share the same parameters.

To generate the samples for training the model, the raw crowd answers are converted into quintuples $(b_j, a_{i_1}, a_{i_2}, y_{i_1j}, y_{i_2j}, y_{i_1j} == y_{i_2j})$. The inputs are the one-hot worker IDs of two workers a_{i_1} and a_{i_2} and one-hot instance ID of one instance b_j. The embedding layer converts them into worker representations $\mathbf{u}_{i_1}$ and $\mathbf{u}_{i_2}$, and instance representation $\mathbf{v}_j$. The parameters in the worker and instance embedding layers are two matrices, i.e., $\mathcal{E}_{\mathcal{A}} \in \mathbb{R}^{|\mathcal{A}|*d}$ for the workers and $\mathcal{E}_{\mathcal{B}} \in \mathbb{R}^{|\mathcal{B}|*d}$ for the instances. The worker representation can be obtained by $\mathbf{u}_i = \mathcal{E}_{\mathcal{A}} \cdot a_i$; the instance representation can be obtained by $\mathbf{v}_j = \mathcal{E}_{\mathcal{B}} \cdot b_j$.

After that, we fuse the worker and instance representations with some computations $\mathcal{H}_p(\mathbf{u}_{i_1}, \mathbf{v}_j)$, $\mathcal{H}_p(\mathbf{u}_{i_2}, \mathbf{v}_j)$ and $\mathcal{H}_t(\mathbf{u}_{i_1}, \mathbf{u}_{i_2}, \mathbf{v}_j)$, where $\mathcal{H}_p$ is for worker answer prediction and $\mathcal{H}_t$ is for answer consistency prediction. There are several alternatives for the computation of $\mathcal{H}_p$ for fusing worker and instance representations. One is based on a dot-product computation on the worker and instance representations, i.e. $\mathcal{H}_p^1(\mathbf{u}_i, \mathbf{v}_j) = \mathbf{u}_i \cdot \mathbf{v}_j$. It can be regarded as a neural-based implementation of tensor factorization (TF) [7,30] while the non-neural-based TF one has no subsequent linear layers. Another option is based on a concatenation operation, i.e. $\mathcal{H}_p^2(\mathbf{u}_i, \mathbf{v}_j) = [\mathbf{u}_i, \mathbf{v}_j]$. It is similar to the neural collaborative filtering (NCF) [7]. We utilize the concatenation operation for $\mathcal{H}_p$ following the neural-based existing work NCF.

For the computation of $\mathcal{H}_t$, we also utilize the concatenation operation. An intuitive method is $\mathcal{H}_t^1 = [\mathbf{u}_{i_1}, \mathbf{u}_{i_2}, \mathbf{v}_j]$. Because the answer consistency prediction implies the comparisons between two workers and utilizing $\mathcal{H}_t^1$ with the linear layer cannot guarantee these comparisons are implicitly computed, explicitly appending the computation of comparison into $\mathcal{H}_t^1$ is potentially better. We thus have a variant of $\mathcal{H}_t$ with explicit computation of the comparisons,

$$\mathcal{H}_t^2 = [\mathbf{u}_{i_1}, \mathbf{u}_{i_2}, |\mathbf{u}_{i_1} - \mathbf{u}_{i_2}|, \mathbf{v}_j]. \tag{1}$$

Finally, a linear layer $\mathcal{F}$ and an activation function are used to compute the output predictions, i.e., the predicted answer

$$\bar{y}_{ij} = \sigma_p(\mathcal{F}(\mathcal{H}_p(\mathbf{u}_{i_1}, \mathbf{v}_j))), \tag{2}$$

and the predicted consistency

$$\bar{y}_{i_1i_2j} = \sigma_t(\mathcal{F}(\mathcal{H}_t(\mathbf{u}_{i_1}, \mathbf{u}_{i_2}, \mathbf{v}_j))). \tag{3}$$

σ is an activation function. σ_p is the Softmax function for the multi-choice-one-answer worker answer prediction and σ_t is the Sigmoid function for the binary answer consistency prediction. It is also possible to instead $\sigma(\mathcal{F}(\cdot))$ by a Multi-Layer Perception (MLP) like that in [7]. Without loss of generality, we didn't select such MLP which needs to tune the number of layers and their dimensions of latent layers; we directly utilize one linear layer in the current implementation. The size of the entire CROWDLR model is small and can be trained at a low time cost.

4.3   Loss Function

The loss function for this multi-task structure is as follows,

$$\mathcal{L} = \lambda_t \mathcal{L}_t + \lambda_{p_1} \mathcal{L}_{p_1} + \lambda_{p_2} \mathcal{L}_{p_2}. \tag{4}$$

$\mathcal{L}_{p1}$ and $\mathcal{L}_{p2}$ are the losses of worker answer prediction tasks and $\mathcal{L}_t$ is the loss of answer consistency prediction task. These two types of tasks can be regarded as the constraints to avoid the overfitting of each other and have the generalization effect. λ_t, λ_{p1} and λ_{p2} are hyperparameters that can be tuned, in our experiments, we set them as general values, i.e., $\lambda_t = \lambda_{p_1} = \lambda_{p_2} = 1$, which is always used when the multiple tasks have same importance and it is not a main-auxiliary multi-task model. This multi-task neural network model can be optimized by a standard optimizer such as Adam.

When people collect answers, people can set the same number of answers to each instance, but the numbers of answers that a worker provides are diverse, the numbers of training samples for the workers are imbalanced. Therefore, we propose the loss functions that consider the imbalanced sample numbers of workers, rather than the vanilla binary cross-entropy, to train the model more effectively. We refer the focal loss [23] to propose our loss functions. For the detailed loss functions, our loss $\mathcal{L}_{p1}$ ($\mathcal{L}_{p2}$) for the worker answer prediction task is,

$$\mathcal{L}_{p_{ij}} = -\sum_k \sum_{r\in\{0,1\}} y_{ij}^{c_k r}(1 - \bar{y}_{ij}^{c_k r})^\gamma \log \bar{y}_{ij}^{c_k r},$$

$$\mathcal{L}_{p_1} = \sum_{(i_1,j)} \mathcal{L}_{p_{ij}}, \quad \mathcal{L}_{p_2} = \sum_{(i_2,j)} \mathcal{L}_{p_{ij}}. \tag{5}$$

It has a dynamic weight based on $(1 - \bar{y}^{c_k r})$ so that it promotes the importance of the hard negative samples and decreases the strong influences of the easy negative samples. On the other hand, the loss $\mathcal{L}_t$ for the worker answer prediction task is,

$$\mathcal{L}_{t_{i_1 i_2 j}} = -\sum_{r\in\{0,1\}} y_{i_1 i_2 j}^r(1 - \bar{y}_{i_1 i_2 j}^r)^\gamma \log \bar{y}_{i_1 i_2 j}^r. \tag{6}$$

4.4   Answer Aggregation

After we obtain the pre-trained model CROWDLR as well as the worker and instance representations, we can utilize them for the quality control task of answer aggregation. First, we utilize the worker answer prediction module to estimate worker answers $\bar{\mathcal{Y}} = \{\bar{y}_{ij}\}_{i,j}$ for all workers and instances in the instances-worker answer matrix. After that, we can utilize the existing answer aggregation approaches (e.g., [2,4,29]) on the complete answers $\bar{\mathcal{Y}}$ to estimate the aggregated labels. In this paper, we utilize the Majority Voting (MV) method as the backbone answer aggregation method, which is one of the most typical answer aggregation method, because it is easy to implement and always used by people who are not major in the crowdsourcing research and are working on diverse practical scenarios.

In addition, for the quality control task of worker ability estimation, we can obtain the results from some answer aggregation methods (e.g., DARE). Furthermore, in the crowdsourcing context, most of the datasets have diverse instances and no-overlap workers. It is generally incapable to transfer the presentations learned from one crowd dataset by the pre-training tasks to the quality control tasks on another dataset. Therefore, in this paper, the quality control tasks need to utilize the trained model and learned representations obtained from the same dataset with the same instances and workers.

Table 1. Statistics of the real datasets with diverse statistical factors. $\mathcal{K}$: # of candidate answers of a question; $|\mathcal{B}|$: # of instances; $|\mathcal{A}|$: # of workers; $|\mathcal{Y}|$: # of answers; AR: Answer Redundancy ($|\mathcal{Y}|/|\mathcal{B}|)$); MWA: Mean Worker Accuracy.

Dataset	$\mathcal{K}$	$|\mathcal{B}|$	$|\mathcal{A}|$	$|\mathcal{Y}|$	AR	MWA
ENG	5	30	63	1890	63.00	0.256
CHI	5	24	50	1200	50.00	0.374
ITM	4	25	36	900	36.00	0.537
MED	4	36	45	1650	45.00	0.475
POK	6	20	55	1100	55.00	0.277
SCI	5	20	111	2220	111.00	0.295

5 Experiments

5.1 Experimental Settings

We utilize some real crowdsourcing datasets proposed in existing work [16] to verify our approach. Table 1 lists the statistical factors of these datasets. These datasets contain the ground truths that are only used for evaluation and not used by the approaches. Mean Worker Accuracy (WMA) is the mean of accuracy of answer set $\mathcal{Y}_{i*}$ of each worker a_i to the ground truths, which shows the instance difficulties in these datasets.

There are six datasets in total, i.e., ENG, CHI, MED, POK, and SCI. A description of the crowd answers of these datasets are as follows.

- ENG (English): the most analogically similar word pair to a word pair;
- CHI (Chinese): the meaning of Chinese vocabularies;
- ITM (Information Technology): basic knowledge of information technology;
- MED (Medicine): about medicine efficacy and side effects;
- POK (Pokémon): the Japanese name of a Pokémon with English name;
- SCI (Science): intermediate knowledge of chemistry and physics.

The type of questions of the instances in these datasets are heterogeneous questions, i.e., the answers for the questions have different contents. For example, an answer "A. creek:river" for a question "Select the analogous pair for word pair hill:mountain" and an answer "A. posse:crowd" for another question "Select the analogous pair for word pair school:fish" are different. In these datasets, each worker annotates all instances. To verify our method, we generate datasets with missing answers by randomly removing the answers. Given a sampling rate r, for each instance in the dataset, we randomly select $\lfloor |\mathcal{A}| * r \rceil$ answers where $|\mathcal{A}|$ is the number of workers. We set the sampling rate of r from the set of $\{0.1, 0.2\}$. The new datasets are named using r at the suffix, e.g., ENG.10% for ENG dataset with $r = 0.1$. For each dataset and r, we run the generation with five trials. We evaluate the average performance of the five trials in the experiments. In addition, we define Answer Completeness Rate (ACR), which shows the sparsity of the answer matrix of workers and instances, i.e., ACR $= |\mathcal{Y}|/(|\mathcal{A}||\mathcal{B}|)$.

These sampled subsets have relatively low ACR ($\leq 20\%$) and low WMA. They are sparse and difficult subsets.

We utilize Tensorflow to implement CROWDLR. We set some of the hyperparameters as the general values, i.e., the weights of the multi-task losses in Eq. (4) are equal, $\lambda_t = \lambda_{p_1} = \lambda_{p_2} = 1$. γ of the loss in Eq. (5) and (6) are 1 which follows the existing works such as [23]. Besides, we utilize the Adam algorithm for optimization with the default learning parameters such as learning rate 0.001. The batch size is set as 2,000. We set 10,000 training epochs with early stopping. We set the dimension of embeddings $d = 1024$ for CROWDLR. Because the size of a training sample and the size of the neural network of CROWDLR is small, it does not need too much memory cost and training time cost. We convert all raw crowd answers into the training quintuples and use all of them to train the self-supervised model.

For the existing works used as baselines for comparisons in the answer aggregation task, besides MV method, we also compare with typical answer aggregation methods D&S [4], DARE [2] and Minimax entropy (MINIMAX) [29]. We implement Majority Voting (MV) by using Python and Scipy. We utilize the the public codes of D&S, DARE and MINIMAX. For MV method, because different rules can be used to break the ties, i.e., the rule of how to select the answer when the candidate choices have equal estimated probability, the reported results of the MV method of a dataset may have minor differences among different existing works as well as our paper. We always select the first answer in the answers with equal estimated probability if there is a tie in the answers of an instance.

Table 2. Experimental Results. Bold values show the best results; underlined values show the results of CROWDLR are better than or equal to that of MV, while CROWDLR utilize MV as the backbone answer aggregation method.

Dataset	CrowdLR	w/o ACP Task	MV	D&S	DARE	MINIMAX
ENG.10%	**0.2933**	0.1867	0.2600	0.2133	0.2467	0.2600
CHI.10%	**0.4500**	0.2750	0.4083	0.3750	0.4417	0.4333
ITM.10%	**0.5680**	0.3760	0.5440	0.4080	**0.5680**	0.5520
MED.10%	**0.5167**	0.3111	0.5056	0.4389	0.5111	0.4833
POK.10%	**0.2700**	0.1500	0.2200	0.2400	0.2500	**0.2700**
SCI.10%	0.4700	0.3600	0.4000	0.3700	**0.4800**	0.4000
ENG.20%	**0.4133**	0.2933	0.3867	0.3000	0.3600	0.2733
CHI.20%	0.5083	0.4000	0.5083	0.4333	**0.5500**	0.4583
ITM.20%	**0.6560**	0.4480	0.6480	0.5680	0.6160	0.5760
MED.20%	0.6333	0.4500	0.6222	**0.6667**	0.6389	0.5611
POK.20%	**0.4600**	0.3600	0.3700	0.3400	0.4400	0.4100
SCI.20%	0.4500	0.2800	0.4700	0.5000	0.4800	**0.5000**

5.2   Experimental Results

We executed the answer aggregation approach MV on the complete worker answers $\bar{\mathcal{Y}}$ predicted by the pre-trained model of CROWDLR. We compared these aggregation results with those from the raw answers $\mathcal{Y}$ by the existing answer aggregation methods. Table 2 lists the results. First, the underlined values in Table 2 show the results of CROWDLR are better than or equal to that of MV. Comparing the results in the "CROWDLR" and "MV" columns, it shows that the performance in the columns of CROWDLR is better than or equal to that of MV in all of the cases. CROWDLR can effectively learn the representations from crowd answers and improve the performance of the answer aggregation.

Second, comparing the results in the column "CROWDLR" with the columns "MV", "D&S", "DARE" and "MINIMAX", the proposed CROWDLR outperforms the baselines in 8 of 12 cases. Considering the backbone answer aggregation method of CROWDLR for estimating the answers is the simple MV, i.e., applying MV on the complete answers by CROWDLR, it also shows that CROWDLR is effective for improving the performance of answer aggregation.

Third, comparing the results between the rows of subsets with $r = 0.1$ and the rows of subsets with $r = 0.2$. Especially, CROWDLR outperforms the baselines in 5 of 6 cases when $r = 0.1$ and 3 of 6 cases when $r = 0.2$. It shows that the proposed CROWDLR is especially effective when the crowd answers are sparse.

We also investigate how the proposed components of CROWDLR influence the results, i.e., the ablation study. Table 2 also lists the ablation results. If removing the ACP task (w/o ACP Task, equivalent to NCF), the performance of CROWDLR decrease a lot. It shows that the ACP task and the multi-task model with Siamese structure are important for CROWDLR to learn the information from the raw crowd data effectively.

6   Conclusion

In this paper, we propose a task-related neural-based self-supervised method named as CROWDLR to learn rich instance and worker representations from the crowd answers. We design two self-supervised tasks and create a multi-task model with a Siamese structure to learn these two tasks in one framework. The experimental results based on real datasets show that our approach can effectively learn the representations from the crowd answers and improve the performance of answer aggregation especially when the crowd answers are sparse. A potential limitation of the proposed method is that the corrected answers predicted by CROWDLR are on some instances with relatively higher quality worker answers; the quality of worker answers of some instances with relatively lower quality may be not improved. Another potential limitation is that the answer aggregation approaches utilize the correlation among workers, instances, and answers with complex behaviors, correcting some of the worker answers may does not exactly generate the correct aggregated answers. In the future work, we will investigate the relation between increase of worker answer accuracy and the increase of answer aggregation accuracy.

Acknowledgements. This work was partially supported by JKA Foundation and JSPS KAK-ENHI Grant Number 23H03402.

References

1. Baba, Y., Li, J., Kashima, H.: Crowdea: multi-view idea prioritization with crowds. Proc. AAAI Conf. Human Comput. Crowdsourcing (HCOMP) **8**(1), 23–32 (2020). https://doi.org/10.1609/hcomp.v8i1.7460
2. Bachrach, Y., Minka, T., Guiver, J., Graepel, T.: How to grade a test without knowing the answers: a bayesian graphical model for adaptive crowdsourcing and aptitude testing. In: Proceedings of the 29th International Coference on International Conference on Machine Learning (ICML), pp. 819–826 (2012), https://dl.acm.org/doi/abs/10.5555/3042573.3042680
3. Chen, T., Kornblith, S., Norouzi, M., Hinton, G.: A simple framework for contrastive learning of visual representations. In: Proceedings of the 37th International Conference on Machine Learning (ICML) (2020). https://dl.acm.org/doi/abs/10.5555/3524938.3525087
4. Dawid, A.P., Skene, A.M.: Maximum likelihood estimation of observer error-rates using the EM algorithm. J. Royal Stat. Society. Series C (Applied Statistics) **28**(1), 20–28 (1979). https://doi.org/10.2307/2346806
5. Devlin, J., Chang, M.W., Lee, K., Toutanova, K.: BERT: pre-training of deep bidirectional transformers for language understanding. In: Proceedings of the 2019 Conference of the North American Chapter of the Association for Computational Linguistics: Human Language Technologies, pp. 4171–4186 (2019). https://doi.org/10.18653/v1/N19-1423
6. Doersch, C., Gupta, A., Efros, A.A.: Unsupervised visual representation learning by context prediction. In: Proceedings of the 2015 IEEE International Conference on Computer Vision (ICCV), pp. 1422–1430 (2015). https://doi.org/10.1109/ICCV.2015.167
7. He, X., Liao, L., Zhang, H., Nie, L., Hu, X., Chua, T.S.: Neural collaborative filtering. In: Proceedings of the 26th International Conference on World Wide Web (WWW), pp. 173–182 (2017). https://doi.org/10.1145/3038912.3052569
8. Hjelm, R.D., et al.: Learning deep representations by mutual information estimation and maximization. In: Proceedings of the Sixth International Conference on Learning Representations (ICLR) (2018). https://doi.org/10.48550/arXiv.1808.06670
9. Jing, L., Tian, Y.: Self-supervised visual feature learning with deep neural networks: a survey. IEEE Trans. Pattern Anal. Mach. Intell. **43**(11), 4037–4058 (2021). https://doi.org/10.1109/TPAMI.2020.2992393
10. Kolesnikov, A., Zhai, X., Beyer, L.: Revisiting self-supervised visual representation learning. In: Proceedings of the IEEE Conference on Computer Vision and Pattern Recognition (CVPR), pp. 1920–1929 (2019). https://doi.org/10.48550/arXiv.1901.09005
11. Li, H., Zhao, B., Fuxman, A.: The wisdom of minority: discovering and targeting the right group of workers for crowdsourcing. In: Proceedings of the 23rd International Conference on World Wide Web (WWW), pp. 165–176 (2014). https://doi.org/10.1145/2566486.2568033
12. Li, J., Sun, H., Li, J.: Beyond confusion matrix: learning from multiple annotators with awareness of instance features. Mach. Learn. **112**(3), 1053–1075 (2022). https://doi.org/10.1007/s10994-022-06211-x
13. Li, J.: Budget cost reduction for label collection with confusability based exploration. In: Neural Information Processing, pp. 231–241 (2019). https://doi.org/10.1007/978-3-030-36802-9_26
14. Li, J.: Crowdsourced text sequence aggregation based on hybrid reliability and representation. In: Proceedings of the 43rd International ACM SIGIR Conference on Research and Development in Information Retrieval (SIGIR), pp. 1761–1764 (2020). https://doi.org/10.1145/3397271.3401239
15. Li, J.: Context-based collective preference aggregation for prioritizing crowd opinions in social decision-making. In: Proceedings of the ACM Web Conference 2022 (WWW), pp. 2657–2667 (2022). https://doi.org/10.1145/3485447.3512137

16. Li, J., Baba, Y., Kashima, H.: Hyper questions: unsupervised targeting of a few experts in crowdsourcing. In: Proceedings of the 2017 ACM on Conference on Information and Knowledge Management (CIKM), pp. 1069–1078 (2017). https://doi.org/10.1145/3132847.3132971

17. Li, J., Baba, Y., Kashima, H.: Incorporating worker similarity for label aggregation in crowdsourcing. In: Proceedings of the 27th International Conference on Artificial Neural Networks (ICANN), pp. 596–606 (2018). https://doi.org/10.1007/978-3-030-01421-6_57

18. Li, J., Baba, Y., Kashima, H.: Simultaneous clustering and ranking from pairwise comparisons. In: Proceedings of the Twenty-Seventh International Joint Conference on Artificial Intelligence (IJCAI), pp. 1554–1560 (2018). https://doi.org/10.24963/ijcai.2018/215

19. Li, J., Endo, L.R., Kashima, H.: Label aggregation for crowdsourced triplet similarity comparisons. In: Neural Information Processing, pp. 176–185 (2021). https://doi.org/10.1007/978-3-030-92310-5_21

20. Li, J., Fukumoto, F.: A dataset of crowdsourced word sequences: collections and answer aggregation for ground truth creation. In: Proceedings of the First Workshop on Aggregating and Analysing Crowdsourced Annotations for NLP, pp. 24–28 (Nov 2019). https://doi.org/10.18653/v1/D19-5904

21. Li, J., Kashima, H.: Iterative reduction worker filtering for crowdsourced label aggregation. In: Proceedings of the 18th International Conference on Web Information Systems Engineering (WISE), pp. 46–54 (2017). https://doi.org/10.1145/978-3-319-68786-5_4

22. Li, J., Kawase, Y., Baba, Y., Kashima, H.: Performance as a constraint: an improved wisdom of crowds using performance regularization. In: Proceedings of the Twenty-Ninth International Joint Conference on Artificial Intelligence (IJCAI), pp. 1534–1541 (2020). https://doi.org/10.24963/ijcai.2020/213, main track

23. Lin, T., Goyal, P., Girshick, R., He, K., Dollár, P.: Focal loss for dense object detection. In: Proceedings of the IEEE International Conference on Computer Vision (ICCV), pp. 2999–3007 (2017). arXiv:1708.02002

24. Lu, X., Li, J., Takeuchi, K., Kashima, H.: Multiview representation learning from crowdsourced triplet comparisons. In: Proceedings of the ACM Web Conference 2023 (WWW), pp. 3827–3836 (2023). https://doi.org/10.1145/3543507.3583431

25. Venanzi, M., Guiver, J., Kazai, G., Kohli, P., Shokouhi, M.: Community-based Bayesian aggregation models for crowdsourcing. In: Proceedings of the 23rd International Conference on World Wide Web (WWW), pp. 155–164 (2014). https://doi.org/10.1145/2566486.2567989

26. Zhai, X., Oliver, A., Kolesnikov, A., Beyer, L.: S4l: self-supervised semi-supervised learning. In: Proceedings of the IEEE international conference on computer vision (CVPR), pp. 1476–1485 (2019). arXiv:1905.03670

27. Zhang, G., Li, J., Kashima, H.: Improving pairwise rank aggregation via querying for rank difference. In: 2022 IEEE 9th International Conference on Data Science and Advanced Analytics (DSAA), pp. 1–9 (2022). https://doi.org/10.1109/DSAA54385.2022.10032454

28. Zheng, Y., Li, G., Li, Y., Shan, C., Cheng, R.: Truth inference in crowdsourcing: is the problem solved? Proc. VLDB Endow. 10(5), 541–552 (2017). https://doi.org/10.14778/3055540.3055547

29. Zhou, D., Platt, J.C., Basu, S., Mao, Y.: Learning from the wisdom of crowds by minimax entropy. In: Proceedings of the 25th International Conference on Neural Information Processing Systems (NIPS), pp. 2195–2203 (2012). https://dl.acm.org/doi/10.5555/2999325.2999380

30. Zhou, Y., He, J.: Crowdsourcing via tensor augmentation and completion. In: Proceedings of the Twenty-Fifth International Joint Conference on Artificial Intelligence (IJCAI), pp. 2435–2441 (2016). https://dl.acm.org/doi/10.5555/3060832.3060962
31. Zuo, X., Li, J., Zhou, Q., Li, J., Mao, X.: Affecti: a game for diverse, reliable, and efficient affective image annotation. In: Proceedings of the 28th ACM International Conference on Multimedia (MM), pp. 529–537 (2020). https://doi.org/10.1145/3394171.3413744

Identify Vulnerability Types: A Cross-Project Multiclass Vulnerability Classification System Based on Deep Domain Adaptation

Gewangzi Du[1,2], Liwei Chen[1,2(✉)], Tongshuai Wu[1,2], Chenguang Zhu[1,2], and Gang Shi[1,2]

[1] Institute of Information Engineering, Chinese Academy of Sciences, Beijing, China
{dugewangzi,chenliwei,wutongshuai,zhuchenguang,shigang}@iie.ac.cn
[2] School of Cyber Security, University of Chinese Academy of Sciences, Beijing, China

Abstract. Software Vulnerability Detection(SVD) is a important means to ensure system security due to the ubiquity of software. Deep learning-based approaches achieve state-of-the-art performance in SVD but one of the most crucial issues is coping with the scarcity of labeled data in projects to be detected. One reliable solution is to employ transfer learning skills to leverage labeled data from other software projects. However, existing cross-project approaches only focused on detecting whether the function code is vulnerable or not. The requirement to identify vulnerability types is essential because it offers information to patch the vulnerabilities. Our aim in this paper is to propose the first system for cross-project multiclass vulnerability classification. We detect at the granularity of code snippet, which is finer-grained compare to function and effective to catch inter-procedure vulnerability patterns. After generating code snippets, we define several principles to extract snippet attentions and build a deep model to obtain the fused deep features; We then extend different domain adaptation approaches to reduce feature distributions of different projects. Experimental results indicate that our system outperforms other state-of-the-art systems.

Keywords: cyber security · multiclass classification · snippet attention · deep learning · domain adaptation

1 Introduction

Software vulnerabilities dreadfully undermine the security of computer systems due to the inevitability of vulnerabilities. Thousands of vulnerabilities are reported publicly to Common Vulnerabilities and Exposures (CVE) [1] and National Vulnerability Database (NVD) [2] each year. In the field of cyber security, SVD has always been one of the most important problems faced by researchers. Traditional static methods rely experts to define vulnerability patterns [3,4] but failed to achieve decent performance owning to the incomplete and subjective patterns given by experts. Besides, the work of manually defining vulnerability patterns is tedious and laborious. Traditional dynamic analysis such as fuzzing [5,6] inspect vulnerabilities during the program execution but have low code coverage.

With the rapid development of deep neural networks, researchers leverage deep neural networks to relive human experts from the arduous task of manually defining patterns [7–11]. Deep learning-based methods automatically learn vulnerability patterns using rough features as input and achieve huge success. In the mainstream deep neural networks, RNN is naturally designed to work with sequential data such as text. In particular, the bidirectional form of RNN can capture the long-term dependencies of a sequence. Therefore, a large number of studies have applied RNN variants to learn the semantic meaning of vulnerabilities and harvest excellent detection performance.

A practical challenge of deep learning-based SVD is the shortage of labeled data in one software project. Deep models are trained with large amount of labelled data but the process of labeling vulnerable code is very time-consuming. Moreover, models trained from one project cannot be generalized to another since data distribution differ across different projects. One ingenious solution is to adopt transfer learning skills to reduce the distribution discrepancy. In the field of SVD, several researches attempt to draw close data distributions between different domains. After feature-level transfer, labeled data in source domain can then be applied to the target domain [12–15]. However, all these work only focus on binary classification but cannot figure out vulnerability types, which is essential for security analysts to determine the risk level and patch the vulnerabilities. Furthermore, all these existing cross-project studies detect vulnerabilities at function level, which is too coarse-grained and cannot capture inter-procedure vulnerabilities.

To solve these problems, we propose a novel system for cross-project multiclass vulnerability detection at code snippet level. Compared with previous studies, our work is the initial investigation on cross-project multiclass vulnerability detection. Besides, there are three more contributions:

1. Inspired by fine-grained image recognition in computer vision [16,17], we defined several types of statement and several principles to generate snippet attentions by matching these types of statement. In addition, we build a deep model, which is composed of three sub-models to accommodate code snippets/snippet attentions to obtain fused deep features. Our deep feature extracting strategy achieves better experimental results than others.
2. We devise to align multiple source projects before involving the target projects. Also, we extend a Mahalanobis distance metric learning algorithm to draw close distributions between source and target projects. Experimental results indicate that our multi-source alignment method improves the detection performance.

3. We collected and labeled 5 real-world open source projects with two professional security researchers and it took more than 240 h of manual work.
   Equipped with the dataset, we conducted extensive experiments, which verify that our system significantly outperforms other state-of-the-art approaches
   in the following cross-project settings:
   - Scenario 1: Single source to single target project. Source project contains
     all the corresponding vulnerability types with target project, which is the
     most straightforward scenario.
   - Scenario 2: Single source to single target, but the vulnerability types in
     source project are not identical with target project. This scenario mea-
     sures the capability to detect across vulnerability types.
   - Scenario 3: Multi-source to single target. This scenario is very common
     since it is likely that some types of vulnerability are absent in one source
     project (subcase 1), or there's no label absence but the samples of each
     vulnerability type in each source project are small (subcase 2). Thus,
     combining multiple sources together is supposed to achieve better perfor-
     mance.

Paper Organization. The rest of the paper is organized as follows. Section 2
describes the previous work. Section 3 states our proposed method in detail.
Experiment implementation and results analysis are presented in Sect. 4.
Section 5 concludes the paper and prospects for future work.

2   Previous Work

Traditional approaches completely rely on experts to formulate vulnerability
patterns, which always suffers unsatisfied performance. Recent years, more and
more researchers employ deep neural networks to conduct SVD since they obviously outperforms traditional approaches and relief security experts from tedious
manual work. Li et al. proposed VulDeePecker [7], which is the first SVD system based on deep learning. They extracted code slices from API calls though
dataflow propagations. Then Zou et al. proposed a multiclass vulnerability detection system [8] by combining different kinds of features. However, this in-domain
detecting system only works on slices generated from API calls. After that, Li et
al. proposed SySeVR [9], which is a binary classification system and expands the
slicing points, not limited to API calls. Duan et al. [10] embedded features from
code property graph and apply attention mechanism to capture potential vulnerable code lines. Zhu et al. [11] adopted bidirectional self-attention mechanism
and used a pretrained BERT to detect multiclass vulnerabilities. However, all
these approaches split training and test data from a same dataset, which means
deep model built by the training data is able to directly apply to the test data.
    When coping with the scarcity of labeled vulnerabilities in a project,
researchers resort to transfer learning technique and leverage labeled data from
other projects. Lin et al. [12] employed BiLSTM to learn the transferable representation between projects and used a random forest as downstream classifier.

Then they combined the heterogeneous data of entire code text and abstract syntax tree(AST) to learn unified representations of the vulnerability patterns [13]. Nguyen et al. [14] employed adversarial learning framework to learn domain-invariant features that can be transferred from source to target project. Liu et al. [15] adopted RNN to learn high-level features and utilised the metric transfer learning framework to transform these features to transferable ones. However, all these cross-project researches only detect the presence of vulnerabilities, but cannot determine the vulnerability types.

3   Framework

A domain $D = \{X, Y, P(X, Y)\}$ is characterized by a feature space X, a label space Y and a joint probability distribution $P(X, Y)$. $P_s(X, Y) \neq P_t(X, Y)$ where $P_s(X, Y)$ and $P_t(X, Y)$ represent source domain and target domain respectively. Further more, since $P(X, Y) = P(X)P(Y|X)$ and $P_s(Y|X) = P_t(Y|X)$, the discrepancy of joint probability distribution can be reduced by minimizing the discrepancy between $P_s(X)$ and $P_t(X)$ and thus we can train a classifier in source domain and apply it to target domain. Without drawing close these distributions, classifiers trained in one domain suffer over-fitting results in another domain. One project represents a domain since its feature distribution differs with other projects and several researches tried to tackle this problem as mentioned in Sect. 2. The main idea of our work is to learn cross-domain representations and diminish the gap between different domains at the code snippet level. Consequently, classifier trained by source domain representations can be applied to target. The structure of system is shown in Fig. 1, which will be illustrated in details.

3.1   Code Snippet Generation

There are two disadvantages of function-level detection: When a function is detected as vulnerable, it takes experts laborious work to locate vulnerable lines because a real-word software project function always contains abundant lines of code and experts still need other techniques or manual check to figure out the vulnerability locations. Moreover, detecting within function scope fails grasp vulnerability patterns across functions calls. These data propagation-caused vulnerabilities through function calls are pretty common and easily lead to serious security issues. Thus we detect at the granularity of code snippet, which includes significantly less numbers of code lines than functions and also spans across function calls to include potential inter-procedure vulnerabilities.

- **Step1: Key Points Locating.** In the first place, we select key points for the source program. By analyzing the syntax characteristics of the program, we focus on four key points which are API/Library Call, Array Usage, Pointer Usage and Expression. We use the open-source analysis tool Joern [18] to generate the AST and match these four types of syntactic features in the AST, as shown in Fig. 2.

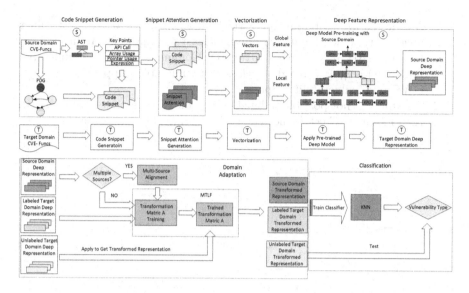

Fig. 1. The framework of our system.

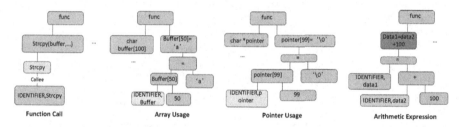

Fig. 2. Four types of key points.

- **Step2: Code Snippets Extracting**. The Program Dependency Graph
  (PDG) of code contains data dependency and control dependency information. We traverse the PDG starting from four types of key points and get
  the traversed code lines as code snippets. Both forward and backward traver-
  sals are considered. For relationships between functions, we extracted code
  lines along function calls and put them ordered. Figure 3 includes the process of generating a code snippet from a function via PDG. We start from the
  Pointer Usage "*pointer*[9] =' \0';" and extract snippet by traversing the PDG.
  Function "*void print(const char *ptr)*" is reached through data dependency.
  Finally, we get the code snippet.
- **Step3: Labeling Code Snippets**. We labeled the ground truth data according to the method in [7] and extended it to figure out the vulnerability types:
  1. We collect projects from the NVD and exam the diff files of all functions.
     If a diff file contains "-", we mark the corresponding lines as vulnerable.
     Professional experts manually mark vulnerable lines if there is no "-".

2. A code snippet containing no marked line is labeled as "0" (i.e., not vulnerable). It is vulnerable if containing at least one marked statement. Since each CVE-ID in NVD has a CWE-ID (Common Weakness Enumeration Identifier) [19], we labeled the classes of vulnerable snippets according to CWE-IDs (vulnerability types).

3. Because it is not possible to label vulnerable code snippets as "0", experts then manually check if the "vulnerable"' snippets are mislabeled.

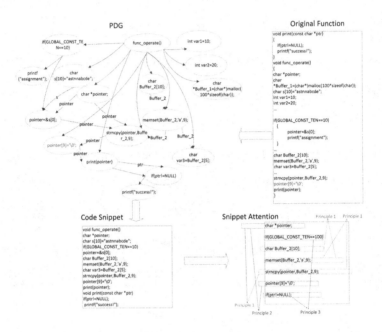

Fig. 3. Process of generating code snippet and snippet attention from a function. In the PDG, red arrows represents control dependency while blue arrows represents data dependency. (Color figure online)

3.2   Snippet Attention Generation

Region attention in image recognition refers to the location in a picture that is able to provide a strong basis for fine-grained classification. For example, mouth shape and hair color are discriminative in determining the types of birds. Inspired by this, we focus on several types of local features such as pointer/array usage statements and API/Library function call statements in our PDG-based code snippet, which we call snippet attention. We believe that our snippet attention can provide more information in our cross-project multiclass detection scenario. For example, pointer usage statements often provide information on many pointer-related vulnerabilities such as null pointer dereference and use after free.

Array usage easily contributes buffer-related vulnerabilities. APIs such as *strcpy()* and *memset()* can cause buffer overflows while *scanf()* can cause improper input validation. To capture information on the dataflow, we also focus on variable definition, formal parameter and bounds-checking operations. Thus, we define variable definition and formal parameter in function definition as "definition" statement type, pointer and array usage as "usage" statement type, API call as "API" statement type, control statement such as "if" or "while" as "control" statement type. The "control" statement may also provide information on vulnerabilities such as infinite loop. Pointer/array usage here has the same meaning with key points in Subsect. 3.1, which means a write operation in the corresponding memory. We further define the following principles to get snippet attentions:

1. Match between "definition" statement and "API" statement: If there is a variable in a "definition" statement which matches an argument in a "API" statement, we select both the "definition" statement and "API" statement as snippet attention. We do not choose "API" without variables in "definition" as its argument.
2. Match between "definition" statement and "usage" statement: If there is a variable in a "definition" statement which matches a pointer/array variable in use in a "usage" statement, we select both the "definition" statement and "usage" statement as snippet attention.
3. If a statement is a "control" type, we extract it as snippet attention, which probably conduct proper bounds-checking and security screening.

With these principles, we can get snippet attention from code snippet, which is also depicted in Fig. 3. Since the code snippet is not able to be compiled, we implement an automate lexical-analysis-based program that analyzes token and context structures to obtain the statement types and match the principles from scratch. Non ASCII characters and comments are discarded. The algorithm to generate snippet attention is illustrated in Algorithm 1. The operation "∪" removes duplicated "definition" statement.

3.3   Vectorization

We use our lexical analysis to generate tokens (e.g., variables, operators, keywords) as corpus from code snippets and snippet attentions. Then we use the word2vec tool [20] with Continuous Bag-of-Words, to embed token sequences of code snippets/snippet attentions into vectors. To deal with different vector lengths, we set two different lengths τ_1 for code snippet and τ_2 for snippet attention. If vector lengths of code snippets (snippet attentions) are longer than τ_1 (τ_2), we cut the vectors at the ending part of code snippets (snippet attentions). If vector lengths of code snippets (snippet attentions) are shorter than τ_1 (τ_2), we pad zeros at the ending part to match τ_1 (τ_2).

Algorithm 1. Snippet Attention Generation

Input: "definition" statements D, "API" statements A, "usage" statements U and "control" statements C in a code snippet CS.
Output: Snippet attention SA in CS.
1: $SA \leftarrow \emptyset$
2: **for** each "definition" statement $d_i \in D$ **do**
3:    **for** each "API" statement $a_i \in A$ **do**
4:       **if** d_i and a_i match principle 1 **then**
5:          $SA \leftarrow SA \cup d_i \cup a_i$
6:       **end if**
7:    **end for**
8:    **for** each "usage" statement $u_i \in U$ **do**
9:       **if** d_i and u_i match principle 2 **then**
10:          $SA \leftarrow SA \cup d_i \cup u_i$
11:       **end if**
12:    **end for**
13: **end for**
14: **for** each "control" statement $c_i \in C$ **do**
15:    $SA \leftarrow SA \cup c_i$
16: **end for**
17: **return** SA

3.4   Deep Feature Representation

We propose a network that use BiGRU as a basic building block since it is able to catch long distance information from both forward and backward direction. The network, as highlighted in Fig. 4, is composed of global feature model, local feature mode and fusion model. The global feature model uses code snippets as input while the local feature model uses snippet attentions as input. Because these two kinds of features accommodate different kinds of information, we put a fusion model to get the comprehensive features. For global and local feature models, the global and local features vectors have lengths of τ_1 and τ_2 and we make the parameter *returnsequence* true to obtain a two-dimensional output for every code snippet and snippet attention. The *hidden_units* in global/feature models are 128 and the outputs of the BiGRU layers are concatenated to form a $(\tau_1 + \tau_2, 256)$ dimensional vector for each (code snippet, snippet attention) pair. The *hidden_units* in fusion model is set to 256. After max-pooling, we get a hidden representation of 512 dimension for each input pair. To obtain deep features, two steps are required:

- **Deep Model Pre-training**. In this step, we put a softmax in each model and train these three models separately. Code snippet and snippet attention pairs from all source projects are used to pre-train the parameters of the three models.
- **Deep Feature Extraction**. After pre-training, we feed data from all projects (source and target) into this pre-trained network to obtain the deep features for source and target domains.

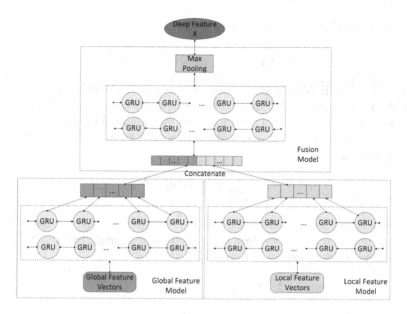

Fig. 4. Deep Feature Model.

3.5   Domain Adaptation and Classification

In this stage, we aim to wipe off the distribution discrepancy of deep features among different domains. If we combine multiple source domains, we extend the CORAL algorithm [21] to align these source domains, which is described in step 1. In the next step, the MTLF [22] is applied to draw close data distribution of the same class between source and target domains. In single source domain scenario, we skip step 1 and directly use step 2 to get close source and target domains.

- **Step1: Multi-Source Alignment**. Inspire by CORAL, we align the distribution of different source domains by exploring their second-order statistics—the covariance, which reflects the distribution of a dataset. CORAL whitens features in one domain and re-colors it with the covariance of the distribution in the other domain. The whitened domain is removed with its feature correlations and clustered around the normalized zero mean. Since we will draw close the distribution of the same class between source and target domains in the next step, we whiten all the source domains, discarding the re-coloring procedure. Our alignment method is illustrated in Algorithm 2.
- **Step2: Mahalanobis Distance Metric Learning**. This step aims to minimize the distribution gap between source and target domains, which means minimizing intra-class and maximizing inter-class distance. At the core of this algorithm is MTLF, which bridges the distributional divergences between source and target domains by learning an appropriate distance metric. This approach need a handful of labeled data for training in target domain. A

Mahalanobis distance metric, which is defined as follows:

$$dist_{ij} = \sqrt{(\mathbf{x}_i - \mathbf{x}_j)\mathbf{M}(\mathbf{x}_i - \mathbf{x}_j)^\mathsf{T}} \tag{1}$$

where the matrix $\mathbf{M}$ is positive semi-definite and can be decomposed as $\mathbf{M} = \mathbf{A}^\mathsf{T}\mathbf{A}$. Thus, learning the Mahalnobis distance metric can be substituted by learning $\mathbf{A}$. Moreover, if an instance is highly co-related to the other domain, the re-weighting function $\omega(x)$ gives high weight to it. The within-class loss to be minimize is defined as:

$$\ell_{in}(\mathbf{A}, \omega) = \sum_{y_i = y_j} \omega(x_i)\omega(x_j)\|\mathbf{A}(x_i - x_j)\|^2 \tag{2}$$

where x is the deep feature and y denotes the label. Samely, the loss function ℓ_{out} designed for inter-class data to be maximized is similarly defined. Finally, the overall training object is formulated as:

$$\min_{\mathbf{A}, \omega} tr(\mathbf{A}^\mathsf{T}\mathbf{A}) + \alpha \sum_{i \in D_s \bigcup D_t} \|\omega(x_i) - \omega_0(x_i)\|^2 + \\ \beta[\ell_{in}(\mathbf{A}, \omega) - \ell_{out}(\mathbf{A}, \omega)] \tag{3}$$

where $\omega_0(x_i) = \frac{P_T(x_i)}{P_S(x_i)}$ is the estimated density ratio. D_s and D_t are labeled data from source and target domains. By repeatedly updating ω and $\mathbf{A}$, we obtain the final Mahalanobis matrix $\mathbf{A}$.

- **Step3: Classification.** Since $\mathbf{A}(x_i - x_j)$ is same as $\mathbf{A}x_i - \mathbf{A}x_j$ and $\mathbf{A}x$ can be regarded as projecting x into a new space by matrix $\mathbf{A}$. In this new space, samples with same labels are get closed. Consequently, we put a KNN as the final classifier trained by all labeled data and pinpoint vulnerability types in target domain.

Algorithm 2. Multi-Source Alignment

Input: Multiple Source Data $D = \{S_1, ..., S_n\}$
Output: Adjusted Source Data $D^* = \{S_1^*, ..., S_n^*\}$
1: **for** $S_i \in D$ **do**
2:     $C_i = cov(S_i) + eye(size(S_i, 2))$
3:     $S_i^* = S_i * C_i^{\frac{-1}{2}}$
4: **end for**

---

4   Experiment Implementation

We raise four Research Questions (RQs) based on the detection scenarios mentioned in Sect. 1 and our experiments are designed to answer them.

- **RQ1**: Is our approach effective in cross-project multiclass vulnerability detection? How well is the detection performance compared with other state-of-the-art approaches?
- **RQ2**: When there's CWE inconsistency between source and target projects, is our system capable of achieving decent performance in this cross vulnerability type scenario?
- **RQ3**: When handling with multi-source scenario, is our system more efficient compared to other techniques and is the multi-source alignment able to improve detection capabilities?
- **RQ4**: Does our deep feature extracting strategy contribute to improving the detection effects?

We adopt Tensorflow-1.15.0 to implement the deep model. We use python 3.8.0 to implement our Multi-Source Alignment and the MTLF is implemented with Matlab R2021a. The genism package(version 4.0.1) is used to make code snippets/snippet attentions vectorized with Word2vec. We run our experiments in a server installed Ubuntu Linux 18.04, with NVIDIA GeForce RTX 2080Ti GPU and Intel(R) Xeon(R) Silver 4214 CPU @ 2.20 GHz.

4.1   Dataset and Experimental Setup

Dataset: We selected 5 real-world projects, which are Linux Kernel, Qemu, Wireshark, Firefox and FFmpeg. We reuse the original functions of these projects offered in [9] and collected more project functions to start the Code Snippet Generation phase. Finally we obtained code snippets of 6 different CWEs for each projects. Thus, including data with no vulnerability, there are 7 different classes in each project. The numbers of all the basic 7 vulnerability types and the corresponding labels in each projects are shown in Table 1.

Table 1. Basic Dataset and Label

CWE Type(Label)	Linux Kernel	Qemu	Wireshark	Firefox	FFmpeg
Non-vulnerable(0)	460	467	489	422	335
CWE-119(1)	174	80	294	183	202
CWE-399(2)	155	42	217	68	43
CWE-20(3)	107	33	91	29	42
CWE-189(4)	82	28	103	27	64
CWE-835(5)	106	98	75	10	20
CWE-416(6)	103	86	30	16	20

For RQ2 (i.e., scenario 2), we collect 3 more CWE types of Linux Kernel, which are CWE-200, CWE-787 and CWE-400. In this scenario, we substitute an basic CWE type in source project (i.e., Linux Kernel) with a new type. We make the new type has the same label with the substituted basic type. The numbers of CWE-200, CWE-787 and CWE-400 are 30, 40 and 93, respectively.

When experimenting to answer RQ3, there are two subcases in this multi-source scenario (i.e., scenario 3 mentioned in Sect. 1). Subcase 1 simulates label absence in source domains and combines different labels from different source domains. In this subcase, we use all data of the selected classes in a source project. For example, if we choose the label 4, 5 and 6 from Wireshark, we pick all samples of these labels. In subcase 2, there's no label absence in source domains but the samples of each class in each source domain is much less than target domain. To simulate this subcase, when combining Linux Kernel and Wireshark to detect Qemu, we discard 70% of the data in each source domain; when combining Linux Kernel and Qemu to detect Firefox, we also discard 70% of the data in each source domain; In addition, we further combine Firefox and FFmpeg to detect Wireshark, keeping all the data since samples of most classes in each source domain are already much smaller than target domain.

Experimental Setup: After the hyperparameter tuning, in the deep model, we set learning rate to 0.01. Epoch is 50 and batch size is 32 for all the three models. τ_1 is 500 and τ_2 is 300. The input shapes of global feature model and local feature model are (500,40) and (300,40). We make K equal 1 in KNN. Same as CD-VULD [15], we hold out 30% as labeled data in target domain to train Matrix **A** and the rest is for test.

4.2   Evaluation Metric and Baseline

Evaluation Metric: We use the multiclass classification evaluation metric [23], which includes Macro-Averaged False Positive Rate (M_FPR), Macro-Averaged False Negative Rate (M_FNR), Macro-Averaged F1-measure (M_F1), Weighted-Averaged False Positive Rate (W_FPR), Weighted-Averaged False Negative Rate(M_FNR) and Weighted-Averaged F1-measure(W_F1). The metric reflect the average or weighted-average indicators of each class such as FPRs, where the weight refers to the sample numbers in each class.

Baseline: We include state-of-the-art deep learning technique SySeVR [9] and multiclass vulnerability detection system AMVD [11] in our baseline. Further more, we also use CD-VULD, which is a state-of-the-art detection system using transfer learning and domain adaptation. For CD-VULD, we hold out exactly the identical labeled target data with our approach for training Matrix **A**. For SySeVR and CD-VULD, we apply softmax layer in the models to get the classification results. To answer RQ4, we use only code snippet and apply the deep feature extraction approach in CD-VULD, which is a RNN-based structure specially designed to extract deep feature for codes. Also, we apply the approach in [8] to get the deep feature at the same output layer corresponding with our system.

4.3   Experimental Results

1) Experiments for answering RQ1

We use one project as the source project that is of high quality labels to detect another target project. Results are shown in Table 2.

Table 2. Results for RQ1: Cross-Project Multiclass Vulnerability Detection

Source-Target	Methods	M_FPR	M_FNR	**M_F1**	W_FPR	W_FNR	W_F1
Linux Kernel-Firefox	SySeVR	0.1517	0.8178	0.1915	0.4541	0.5970	0.3840
	AMVD	0.1412	0.7660	0.2713	0.4282	0.5537	0.4406
	CD-VULD	0.0487	0.2749	0.6716	0.0697	0.2505	0.7903
	Ours	**0.0299**	**0.1814**	**0.7513**	**0.0415**	**0.1619**	**0.8583**
Linux Kernel-Qemu	SySeVR	0.1506	0.8161	0.1803	0.3350	0.6927	0.3205
	AMVD	0.1273	0.7266	0.2508	0.3084	0.6372	0.3727
	CD-VULD	0.0509	0.2877	0.6775	0.1034	0.2496	0.7619
	Ours	**0.0266**	**0.1562**	**0.8401**	**0.0604**	**0.1248**	**0.8769**
Wireshark-Firefox	SySeVR	0.1619	0.7664	0.2871	0.2410	0.6855	0.3846
	AMVD	0.1520	0.8087	0.2181	0.2461	0.6573	0.3952
	CD-VULD	0.0482	0.2743	0.7296	0.1000	0.2241	0.8067
	Ours	**0.0213**	**0.1729**	**0.8434**	**0.0563**	**0.0923**	**0.9012**
FFmpeg-Firefox	SySeVR	0.1477	0.7519	0.2949	0.3466	0.6290	0.4089
	AMVD	0.1459	0.7583	0.2881	0.4003	0.5951	0.4214
	CD-VULD	0.0664	0.4248	0.5856	0.0845	0.3313	0.7296
	Ours	**0.0439**	**0.2560**	**0.7208**	**0.0732**	**0.2182**	**0.8159**
Linux Kernel-FFmpeg	SySeVR	0.1937	0.7759	0.2353	0.0791	0.7629	0.3291
	AMVD	0.1278	0.6879	0.2799	0.1991	0.6235	0.4293
	CD-VULD	0.0397	0.2735	0.6466	0.0595	0.2191	0.7961
	Ours	**0.0200**	**0.1071**	**0.8559**	**0.0464**	**0.0936**	**0.9177**

The most intuitive conclusion is that pure deep learning without any transfer learning technique such as SySeVR completely fails because of the large divergency between projects. Although CD-VULD leverages domain adaptation and uses the same labeled target data with our system, its performance is obviously worse than ours. The average M_F1 and W_F1 of our system are 0.1401 and 0.0971 higher than CD-VULD. The average evaluation metric comparison with other approaches is shown in Fig. 5.

Insight1: Our system is effective in cross-project multiclass vulnerability detection and it outperforms other state-of-the-art approaches.

2) Experiments for answering RQ2

We use the prepared extra CWEs of Linux Kernel. In this scenario, we substitute one of the basic CWEs with a new one in each test case. The overall result is shown in Table 3.

The parentheses in the first column demonstrate the new vulnerability type and its corresponding label. In this scenario, source project and target project have one mismatched vulnerability type but with the same label. Our system

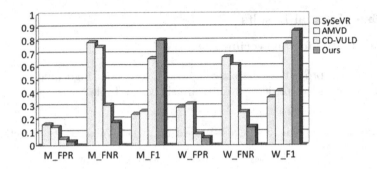

Fig. 5. Average metric for cross-project multiclass vulnerability detection.

Table 3. Results for RQ2: Cross CWE Type Detection

Source-Target	Methods	M_FPR	M_FNR	M_F1	W_FPR	W_FNR	W_F1
Linux Kernel(CWE-200 as label 6)-FFmpeg	SySeVR	0.1800	0.7505	0.2480	0.0765	0.7529	0.3335
	AMVD	0.1266	0.7278	0.2836	0.2053	0.6215	0.4361
	CD-VULD	0.0406	0.2993	0.6379	0.0589	0.2219	0.7966
	Ours	**0.0180**	**0.1208**	**0.8039**	**0.0121**	**0.1138**	**0.8933**
Linux Kernel(CWE-787 as label 2)-Qemu	SySeVR	0.1482	0.8018	0.1896	0.3207	0.6923	0.3232
	AMVD	0.1385	0.7488	0.2343	0.3157	0.6410	0.3658
	CD-VULD	0.0582	0.3443	0.6268	0.1096	0.2895	0.7204
	Ours	**0.0288**	**0.1498**	**0.8222**	**0.0559**	**0.1421**	**0.8631**
Linux Kernel(CWE-400 as label 3)-Firefox	SySeVR	0.1539	0.7894	0.2497	0.4197	0.6252	0.3896
	AMVD	0.1501	0.7796	0.2577	0.4528	0.5875	0.4099
	CD-VULD	0.0425	0.3085	0.6504	0.0625	0.2241	0.8066
	Ours	**0.0305**	**0.1892**	**0.7341**	**0.0349**	**0.1667**	**0.8710**

also achieves decent performance in this scenario compared with other methods. The average M_F1 and W_F1 are 0.1483 and 0.1013 higher than CD-VULD.

We also record the evaluation metric of the mismatched vulnerability type by our approach, as shown in Table 4.

Table 4. Performance of our system on mismatched types

Source-Target(Evaluated Label)	FPR	FNR	F1
Linux Kernel-FFmpeg(6)	0.0145	0.0714	0.7647
Linux Kernel-Qemu(2)	0.0126	0.1379	0.8197
Linux Kernel-Firefox(3)	0.0217	0.1000	0.7347

We can conclude from Tables 3 and 4 that in testcase 1, although the F1 of mismatched type is lower than the M_F1, we acquire better FPR and FNR for the mismatched type. In testcase 2 and testcase 3, the F1 of mismatched type is more or less compared with the M_F1.

Insight2: Our system also performs well in cross vulnerability type detection scenario and obtains the best results compared with other techniques.

3) Experiments for answering RQ3

Equipped with the dataset mentioned above, we are able to conduct detecting experiments for Multi-Source scenario. Experimental results for this scenario are displayed in Table 5.

Table 5. Results for RQ3: Multi-Source Scenario

Source-Target	Methods	M_FPR	M_FNR	M_F1	W_FPR	W_FNR	W_F1
Linux Kernel +Wireshark (0-3)+(4-6)-Firefox	SySeVR	0.1799	0.8534	0.1638	0.4523	0.7419	0.2712
	AMVD	0.1674	0.8074	0.2214	0.4648	0.6798	0.3278
	CD-VULD	0.0588	0.2935	0.6511	0.0759	0.2975	0.7560
	Ours	**0.0296**	**0.1728**	**0.7681**	**0.0481**	**0.1544**	**0.8631**
FFmpeg+ Linux Kernel (0-3)+(4-6) -Qemu	SySeVR	0.1546	0.8106	0.1812	0.3566	0.6997	0.3103
	AMVD	0.1494	0.7631	0.2380	0.3819	0.6461	0.3538
	CD-VULD	0.0571	0.3529	0.6675	0.0451	0.3042	0.7322
	Ours	**0.0299**	**0.1597**	**0.8181**	**0.0447**	**0.1555**	**0.8546**
Wireshark+ Linux Kernel (0-3)+(4-6) -FFmpeg	SySeVR	0.2612	0.7921	0.2214	0.0931	0.8028	0.2799
	AMVD	0.1569	0.7716	0.2333	0.0729	0.7410	0.3526
	CD-VULD	0.0479	0.3199	0.6076	0.0609	0.2629	0.7669
	Ours	**0.0197**	**0.1306**	**0.7816**	**0.0232**	**0.1135**	**0.8997**
FFmpeg +Firefox (0-6)+(0-6) -Wireshark	SySeVR	0.1511	0.7540	0.2384	0.1201	0.7486	0.3156
	AMVD	0.1441	0.7759	0.2144	0.0983	0.7304	0.3408
	CD-VULD	0.0718	0.3826	0.6178	0.0811	0.3777	0.6792
	S1-Wire	0.0358	0.1908	0.7661	0.0342	0.1991	0.8449
	S2-Wire	0.0400	0.2342	0.7428	0.0341	0.2231	0.8280
	Ours⁻	0.0337	0.1984	0.7681	0.0335	0.1898	0.8488
	Ours	**0.0298**	**0.1562**	**0.7959**	**0.0325**	**0.1650**	**0.8697**
Linux Kernel +Qemu (0-6)+(0-6) -Firefox	SySeVR	0.1668	0.7902	0.2228	0.4082	0.7348	0.2888
	AMVD	0.1649	0.7895	0.2206	0.4038	0.7273	0.2921
	CD-VULD	0.1026	0.3908	0.5884	0.1162	0.3700	0.6429
	S1-Firefox	0.0325	0.2201	0.7138	0.0547	0.1982	0.8211
	S2-Firefox	0.0433	0.2426	0.7003	0.0566	0.2019	0.8179
	Ours⁻	0.0338	0.2217	0.7159	0.0540	0.1996	0.8206
	Ours	**0.0301**	**0.1983**	**0.7402**	**0.0472**	**0.1736**	**0.8455**
Linux Kernel +Wireshark (0-6)+(0-6) -Qemu	SySeVR	0.1797	0.8013	0.2245	0.3268	0.7518	0.2922
	AMVD	0.1767	0.7883	0.2364	0.3337	0.7426	0.2986
	CD-VULD	0.0890	0.4428	0.5496	0.1791	0.4051	0.6223
	S1-Qemu	0.0337	0.2245	0.7702	0.0599	0.1709	0.8348
	S2-Qemu	0.0349	0.2088	0.7682	0.0603	0.1794	0.8266
	Ours⁻	0.0343	0.2236	0.7759	0.0589	0.1744	0.8338
	Ours	**0.0305**	**0.1776**	**0.8097**	**0.0576**	**0.1521**	**0.8531**

In this table, parentheses in the first column denote the corresponding label picked from each source domain. For example, "Linux Kernel + Wireshark (0-3) + (4-6)" means we pick label 0, 1, 2, 3 from Linux Kernel and label 4, 5, 6 from Wireshark. The upper part of this table demonstrated 3 testcases for subcase 1, and the lower part denotes subcase 2. "S1" in the "Methods" column represents using only the first source project with our system while "S2" represents using the second source project by our system, and "Wire" in the "Methods" column represents "Wireshark". "Ours⁻" refers to removing the multi-source alignment step from our system and is given to indicate the importance of the alignment step. In subcase 2, for detecting with single source, Ours⁻ and our whole system, we hold out the same part of target labeled samples. Obviously, the detecting performance improves when aligning multiple source domains than only single source. Also in subcase 2, there's a huge performance gap between CD-VULD and our system. Moreover, the performance of our system is apparently better than Ours⁻ due to the reason that the distribution divergency of the same class in each source domain interferes the intra-class clustering when drawing close with target domain. The average metric of the 6 testcases compared with other systems is shown in Fig. 6.

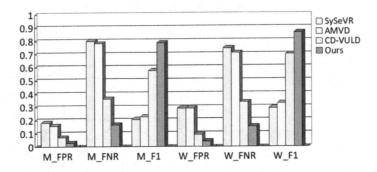

Fig. 6. Average evaluation metric of our system and other approaches in multi-source scenario.

Insight3: Our detecting system can hold the multi-source scenario and greatly improves the performance compared with other ones. Besides, the Multi-Source Alignment algorithm significantly raise the detection performance.

4) Experiments for answering RQ4

We conducted two contrast experiments using deep feature extracting approaches in [8] and CD-VULD as mentioned above. Experimental results are shown in Table 6.

In this table, "CD-VULD⁺" refers to using the deep feature extracting method of CD-VULD, which only takes code snippet as input. "μvuld⁺" represents adopting system of μvuldeepecker [8] to get deep features, which takes code snippet and the defined features in their paper as input. The rest part is the same with our system. We can infer that CD-VULD⁺ leads to the worst results and μvuld⁺ is also worse than ours since it fails to capture vulnerability patterns related to pointers and arrays. Besides, API calls with no variables as parameters are not related to vulnerabilities. The average M_F1 and W_F1 of our system are 0.0639 and 0.0278 higher than μvuld⁺ while the M_FNR and W_FNR of ours are 0.0454 and 0.0337 less than μvuld⁺. Compared with CD-VULD⁺, our system achieves 0.072 less M_FNR and 0.0656 less W_FNR. The average metric comparison with the other two approaches is shown in Fig. 7.

Insight4: Our deep feature extracting method helps to improve the detection effects and performs better than other deep feature extracting approaches.

Table 6. Results for RQ4: Evaluating our deep feature extraction strategy

Source-Target	Methods	M_FPR	M_FNR	M_F1	W_FPR	W_FNR	W_F1
Linux Kernel-Firefox	CD-VULD⁺	0.0409	0.2528	0.6706	0.0562	0.2178	0.8169
	μvuld⁺	0.0375	0.2272	0.6908	0.0557	0.1988	0.8291
	Ours	**0.0299**	**0.1814**	**0.7513**	**0.0415**	**0.1619**	**0.8583**
Linux Kernel-Qemu	CD-VULD⁺	0.0408	0.2547	0.7239	0.0813	0.1966	0.8231
	μvuld⁺	0.0364	0.2338	0.7473	0.0802	0.1709	0.8413
	Ours	**0.0266**	**0.1562**	**0.8401**	**0.0604**	**0.1248**	**0.8769**
Wireshark-Firefox	CD-VULD⁺	0.0345	0.2243	0.7045	0.0668	0.1723	0.8421
	μvuld⁺	0.0262	0.1944	0.7695	0.0626	0.1212	0.8812
	Ours	**0.0213**	**0.1729**	**0.8434**	**0.0563**	**0.0923**	**0.9012**
FFmpeg-Firefox	CD-VULD⁺	0.0558	0.3081	0.6470	0.0814	0.2765	0.7743
	μvuld⁺	0.0474	0.2979	0.6591	0.0802	0.2348	0.8010
	Ours	**0.0439**	**0.2560**	**0.7208**	**0.0732**	**0.2182**	**0.8159**
Linux Kernel-FFmpeg	CD-VULD⁺	0.0304	0.1935	0.7898	0.0537	0.1556	0.8581
	μvuld⁺	0.0260	0.1470	0.8251	**0.0452**	0.1337	0.8786
	Ours	**0.0200**	**0.1071**	**0.8559**	0.0464	**0.0936**	**0.9177**

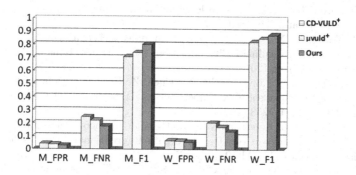

Fig. 7. Average evaluation metric of our and other deep feature extracting methods.

5   Conclusion and Future Work

We proposed a system to identify software vulnerability types across different projects at the granularity of code snippet. It's also the first system to figure out vulnerability types in cross-project settings. This work helps with the task when detecting a project with no high quality training data. Besides, we leverage snippet attention and adopt feature fusion model for high-level feature extracting. Moreover, different domain adaptation techniques are extended to eliminate distribution divergency between different projects. In the future, we will be engaged in vulnerability detection across different programming languages. Moreover, we will research and try more deep learning-based domain adaptation techniques to see if they work in our detecting scenarios.

Acknowledgment. This work is partially supported by the National Natural Science Foundation of China (No. 62172407), and the Youth Innovation Promotion Association CAS.

References

1. "CVE". https://cve.mitre.org/
2. "NVD". https://nvd.nist.gov/
3. "Checkmarx" (2019). https://www.checkmarx.com/
4. Xu, Z., Chen, B., Chandramohan, M., Liu, Y., Song, F.: Spain: security patch analysis for binaries towards understanding the pain and pills. In: IEEE/ACM 39th International Conference on Software, Engineering (ICSE), pp. 462–472, IEEE (2017)
5. Li, Y., et al.: Cerebro: context-aware adaptive fuzzing for effective vulnerability detection. In: Proceedings of the 2019 27th ACM Joint Meeting on European Software Engineering Conference and Symposium on the Foundations of Software Engineering, 2019, pp. 533–544 (2019)
6. Chen, H., et al.: Hawkeye: towards a desired directed grey-box fuzzer. In: Proceedings of the 2018 ACM SIGSAC Conference on Computer and Communications Security, pp. 2095–2108 (2018)
7. Li, Z., et al.: Vuldeepecker: a deep learning-based system for vulnerability detection. arXiv preprint arXiv:1801.01681 (2018)
8. Zou, D., Wang, S., Xu, S., Li, Z., Jin, H.: uvuldeepecker: a deep learning-based system for multiclass vulnerability detection. IEEE Transactions on Dependable and Secure Computing (2019)
9. Li, Z., Zou, D., Xu, S., Jin, H., Zhu, Y., Chen, Z.: Sysevr: a framework for using deep learning to detect software vulnerabilities. IEEE Trans. Dependable Secure Comput. 19(4), 2244–2258 (2021)
10. Duan, X., et al.: Vulsniper: focus your attention to shoot fine-grained vulnerabilities. In: Proceedings of the Twenty Eighth International Joint Conference on Artificial Intelligence, IJCAI-19, pp. 4665–4671 (2019)
11. Zhu, C., Du, G., Wu, T., Cui, N., Chen, L., Shi, G.: BERT-based vulnerability type identification with effective program representation. In: Wang, L., Segal, M., Chen, J., Qiu, T. (eds.) Wireless Algorithms, Systems, and Applications: 17th International Conference, WASA 2022, Dalian, China, November 24–26, 2022, Proceedings, Part I, pp. 271–282. Springer, Cham (2022). https://doi.org/10.1007/978-3-031-19208-1_23
12. Lin, G., Zhang, J., Luo, W., Pan, L., Xiang, Y.: Poster: vulnerability discovery with function representation learning from unlabeled projects. In: Proceedings of the 2017 ACM SIGSAC Conference on Computer and Communications Security (CCS), pp. 2539–2541. ACM (2017)
13. Lin, G., et al.: Software vulnerability discovery via learning multi-domain knowledge bases. IEEE Trans. Dependable Secure Comput. 18(5), 2469–2485 (2019)
14. Nguyen, V., Le, T., de Vel, O., Montague, P., Grundy, J., Phung, D.: Dual-component deep domain adaptation: a new approach for cross project software vulnerability detection. In: Lauw, H.W., Wong, R.C.-W., Ntoulas, A., Lim, E.-P., Ng, S.-K., Pan, S.J. (eds.) PAKDD 2020. LNCS (LNAI), vol. 12084, pp. 699–711. Springer, Cham (2020). https://doi.org/10.1007/978-3-030-47426-3_54
15. Liu, S., et al.: CD-VulD: cross-domain vulnerability discovery based on deep domain adaptation. IEEE Trans. Dependable Secure Comput. 19(1), 438–451 (2022)
16. Donahue, J., et al.: Decaf: a deep convolutional activation feature for generic visual recognition. In: International Conference On Machine Learning, pp. 647–655 (2014)

17. Behera, A., Wharton, Z., Hewage, P.R., Bera, A.: Context-aware Attentional Pooling (CAP) for Fine-grained Visual Classification. In: Proceedings of the AAAI Conference on Artificial Intelligence, pp. 929–937 (2021)
18. "Joern". https://joern.io/
19. "Common Weakness Enumeration". https://cwe.mitre.org/
20. "Word2vec". http://radimrehurek.com/gensim/models/word2vec.html
21. Sun, B., Feng, J., Saenko, K.: Return of frustratingly easy domain adaptation. In: AAAI, vol. 6, p. 8 (2016)
22. Xu, Y., et al.: A unified framework for metric transfer learning. IEEE Trans. Knowl. Data Eng. **29**(6), 1158–1171 (2017)
23. Model evaluation metrics (2019). https://scikit-learn.org/stable/modules/modelevaluation.html#model-evaluation

Author Index

B. Luo et al. (Eds.): ICONIP 2023, LNCS 14452, pp. 501–503, 2024.
https://doi.org/10.1007/978-981-99-8076-5

Printed in the United States
by Baker & Taylor Publisher Services